Chapter 4   Exponents, Polynomials and Additional Applications

Rules of Exponents

1. $x^m \cdot x^n = x^{m+n}$ **product rule**

2. $\dfrac{x^m}{x^n} = x^{m-n}$, $x \neq 0$ **quotient rule**

3. $(x^m)^n = x^{m \cdot n}$ **power rule**

4. $x^0 = 1$, $x \neq 0$ **zero exponent rule**

5. $x^{-m} = \dfrac{1}{x^m}$, $x \neq 0$ **negative exponent rule**

6. $\left(\dfrac{ax}{by}\right)^m = \dfrac{a^m x^m}{b^m y^m}$, $b \neq 0$, $y \neq 0$ **expanded power rule**

FOIL method (*First*, *Outer*, *Inner*, *Last*) of multiplying binomials: $(a+b)(c+d) = ac + ad + bc + bd$

Product of sum and difference of two quantities:
$(a+b)(a-b) = a^2 - b^2$

Squares of binomials: $(a+b)^2 = a^2 + 2ab + b^2$
$(a-b)^2 = a^2 - 2ab + b^2$

distance formula: $d = rt$

Chapter 5   Factoring

If $a \cdot b = c$, then a and b are **factors** of c.
Difference of two squares: $a^2 - b^2 = (a+b)(a-b)$
Sum of two cubes: $a^3 + b^3 = (a+b)(a^2 - ab + b^2)$
Difference of two cubes: $a^3 - b^3 = (a-b)(a^2 + ab + b^2)$

To Factor a Polynomial

1. Determine if the polynomial has a greatest common factor other than 1. If so, factor out the GCF from every term in the polynomial.
2. If the polynomial has two terms, determine if it is a difference of two squares or a sum or difference of two cubes. If so, factor using the appropriate formula.
3. If the polynomial has three terms, factor the trinomial using one of the procedures discussed.

4. If the polynomial has more than three terms, then try factoring by grouping.
5. As a final step, examine your factored polynomial to see if any factors listed have a common factor and can be factored further. If you find a common factor, factor it out at this point.

Quadratic equation: $ax^2 + bx + c = 0$, $a \neq 0$.
Zero-factor Property: If $ab = 0$, then $a = 0$ or $b = 0$.

To Solve A Quadratic Equation Using Factoring

1. Write the equation in standard form with the squared term positive. This will result in the one side of the equation being equal to 0.
2. Set each factor containing a variable equal to zero and find the solution.

Chapter 6   Rational Expressions and Equations

To Reduce Rational Expressions

1. Factor both the numerator and denominator as completely as possible.
2. Divide both the numerator and denominator by any common factors.

To Multiply Rational Expressions

1. Factor all numerators and denominators as completely as possible.
2. Divide out common factors.
3. Multiply the numerators together and multiply the denominators together.

To Add or Subtract Two Rational Expressions

1. Determine the least common denominator (LCD).

2. Rewrite each fraction as an equivalent fraction with the LCD.
3. Add or subtract the numerators while maintaining the LCD.
4. When possible, factor the remaining numerator and reduce the fraction.

To Solve Equations Containing Fractions

1. Determine the LCD of all fractions in the equation.
2. Multiply both sides of the equation by the LCD. This will result in every term in the equation being multiplied by the LCD.
3. Remove any parentheses and combine like terms on each side of the equation.
4. Solve the equation.
5. Check your solution in the original equation.

THIRD EDITION

Elementary Algebra for College Students

THIRD EDITION

Elementary Algebra for College Students

Allen R. Angel

Monroe Community College

PRENTICE HALL
Englewood Cliffs, New Jersey 07632

Library of Congress Cataloging-in-Publication Data

Angel, Allen R.
    Elementary algebra for college students / Allen R. Angel — 3rd
ed.
        p.   cm.
    Includes index.
    ISBN 0-13-259581-8
    1. Algebra.   I. Title.
QA152.2.A54   1992
512.9—dc20   91-22507
                                                              CIP

Editor in Chief: Tim Bozik
Acquisitions editor: Priscilla McGeehon
Marketing Manager: Paul R. Banks
Editorial/production supervision: Rachel J. Witty, Letter Perfect, Inc.
Development editor: Christine Peckaitis
Copy editor: Bill Thomas
Interior and cover design: Lee Goldstein
Photo editor: Lori Morris-Nantz
Photo researcher: Anita Dickhuth
Cover photo: © 1992 by Adam Peiperl
Prepress buyer: Paula Massenaro
Manufacturing buyer: Lori Bulwin

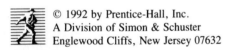

© 1992 by Prentice-Hall, Inc.
A Division of Simon & Schuster
Englewood Cliffs, New Jersey 07632

Printed in the United States of America
10 9 8 7 6 5 4

ISBN 0-13-259581-8

Prentice-Hall International (UK) Limited, *London*
Prentice-Hall of Australia Pty. Limited, *Sydney*
Prentice-Hall Canada Inc., *Toronto*
Prentice-Hall Hispanoamericana, S.A., *Mexico*
Prentice-Hall of India Private Limited, *New Delhi*
Prentice-Hall of Japan, Inc., *Tokyo*
Simon & Schuster Asia Pte. Ltd., *Singapore*
Editora Prentice-Hall do Brasil, Ltda., *Rio de Janeiro*

To my wife, Kathy,
and my sons, Robert and Steven

Contents

Preface xi

To the Student xv

1 Real Numbers 1

1.1 Study Skills Needed for Success in Mathematics, 2
1.2 Fractions, 6
1.3 The Real Number System, 15
1.4 Inequalities, 20
1.5 Addition of Real Numbers, 23
1.6 Subtraction of Real Numbers, 28
1.7 Multiplication and Division of Real Numbers, 33
1.8 An Introduction to Exponents, 40
1.9 Use of Parentheses and Order of Operations, 44
1.10 Properties of the Real Number System, 49
Summary, 53
Review Exercises, 55
Practice Test, 57

2 Solving Linear Equations 58

2.1 Combining Like Terms, 59
2.2 The Addition Property, 67
2.3 The Multiplication Property, 75
2.4 Solving Linear Equations with a Variable on Only One Side of the Equation, 81
2.5 Solving Linear Equations with the Variable on Both Sides of the Equation, 86
2.6 Ratios and Proportions, 92

2.7   Inequalities in One Variable, 102
Summary, 107
Review Exercises, 108
Practice Test, 109
Cumulative Review Test, 110

3 Formulas and Applications of Algebra   111

3.1   Formulas, 112
3.2   Changing Application Problems into Equations, 121
3.3   Solving Application Problems, 128
3.4   Geometric Problems, 135
Summary, 139
Review Exercises, 139
Practice Test, 140

4 Exponents, Polynomials, and Additional Applications   141

4.1   Exponents, 142
4.2   Negative Exponentials (Optional), 150
4.3   Scientific Notation (Optional), 157
4.4   Addition and Subtraction of Polynomials, 161
4.5   Multiplication of Polynomials, 167
4.6   Division of Polynomials, 175
4.7   Rate and Mixture Problems, 181
Summary, 192
Review Exercises, 193
Practice Test, 195
Cumulative Review Test, 196

5 Factoring   197

5.1   Factoring a Monomial from a Polynomial, 198
5.2   Factoring by Grouping, 204
5.3   Factoring Trinomials with $a = 1$, 209
5.4   Factoring Trinomials with $a \neq 1$, 217
5.5   Special Factoring Formulas and a General Review of Factoring, 229
5.6   Solving Quadratic Equations Using Factoring, 235
Summary, 241
Review Exercises, 242
Practice Test, 243

6 Rational Expressions and Equations   244

6.1   Reducing Rational Expressions, 245
6.2   Multiplication and Division of Rational Expressions, 250
6.3   Addition and Subtraction of Rational Expressions with a Common Denominator, 257
6.4   Finding the Least Common Denominator, 261
6.5   Addition and Subtraction of Rational Expressions, 264
6.6   Complex Fractions (Optional), 270
6.7   Solving Equations Containing Rational Expressions, 274
6.8   Applications of Rational Equations, 281
Summary, 290
Review Exercises, 290
Practice Test, 292
Cumulative Review Test, 293

7 Graphing Linear Equations   294

7.1   The Cartesian Coordinate System, 295
7.2   Graphing Linear Equations, 297
7.3   Slope of a Line, 310
7.4   Slope-Intercept Form of a Linear Equation, 317
7.5   Point-Slope Form of a Linear Equation (Optional), 326
7.6   Graphing Linear Inequalities, 329
Summary, 332
Review Exercises, 333
Practice Test, 336

8 Systems of Linear Equations   337

8.1   Introduction, 338
8.2   Solving Systems of Equations Graphically, 340
8.3   Solving Systems of Equations by Substitution, 346
8.4   Solving Systems of Equations by the Addition Method, 351
8.5   Applications of Systems of Equations, 358
8.6   Systems of Linear Inequalities, (Optional), 367
Summary, 371
Review Exercises, 371
Practice Test, 373
Cumulative Review Test, 374

9 Roots and Radicals   375

9.1   Introduction, 376
9.2   Multiplying and Simplifying Square Roots, 381

9.3   Dividing and Simplifying Square Roots, 386
9.4   Addition and Subtraction of Square Roots, 391
9.5   Solving Radical Equations, 397
9.6   Applications of Radicals, 402
9.7   Higher Roots and Fractional Exponents (Optional), 409
      Summary, 414
      Review Exercises, 414
      Practice Test, 416

10 Quadratic Equations   417

10.1   The Square Root Property, 418
10.2   Solving Quadratic Equations by Completing the Square, 422
10.3   Solving Quadratic Equations by the Quadratic Formula, 427
10.4   Graphing Quadratic Equations, 435
       Summary, 445
       Review Exercises, 446
       Practice Test, 447
       Cumulative Review Test, 447

Appendices   449

A   Review of Decimals and Percent, 449
B   Finding the Greatest Common Factor and Least
    Common Denominator, 452
C   Geometry, 455
D   Squares and Square Roots, 460

Answers   461

Index   495

Answers, 461
Index, 495

Preface

This book was written for college students and other adults who have never been exposed to algebra or those who have been exposed but need a refresher course. My primary goal was to write a book that students can read, understand, and enjoy. To achieve this goal I have used short sentences, clear explanations, and many detailed worked-out examples. I have tried to make the book relevant to college students by using practical applications of algebra throughout the text.

Features of the Text

Four-color Format: Color is used pedagogically in the following ways:

Important definitions and procedures are color screened.

Color screening or color type is used to make other important items stand out.

Errors that students commonly make are given in colored boxes as warnings for students.

Artwork is enhanced and clarified with use of multiple colors.

Other important items such as the Helpful Hints, Just for Fun Problems, Calculator Corners, and so on, are enhanced with color.

The four-color format allows for all these, and other features, to be presented in different forms and colors for easy identification by students.

The four-color format helps make the text more appealing and interesting to students.

Readability: One of the most important features of the text is its readability. The book is very readable, even for those with weak reading skills. Short, clear sentences are used and more easily recognized, and easy-to-understand language is used whenever possible. Because so many of our students now taking algebra are from different countries, this feature has become increasingly important.

Accuracy: Accuracy in a mathematics text is essential. To insure accuracy in this book, mathematicians from around the country have read the galleys carefully for typographical errors and have checked all the answers.

Spiral Approach to Learning: Many of our students do not thoroughly grasp new concepts the first time they are presented. In this text we use the spiral approach to learning. That is, we introduce a concept, then later in the text briefly reintroduce it and build upon it. Often an important concept is used in many sections of the text. Students are often reminded where the material was seen before, or where it will be used again. This also serves to emphasize the importance of the concept. Important concepts are also reinforced throughout the text in the Cumulative Review Exercises and Cumulative Review Test.

Keyed Section Objectives: Each section opens with a list of skills that the student should learn in that section. The objectives are then keyed to the appropriate portions of the sections with symbols such as ▶1.

Practical Applications: Practical applications of algebra are stressed throughout the text. Students need to learn how to translate application problems into algebraic symbols. The problem-solving approach used throughout this text gives students ample practice in setting up and solving application problems. The use of practical applications motivates students.

Detailed Worked-Out Examples: A wealth of examples have been worked out in a step-by-step, detailed manner. Important steps are highlighted in color, and no steps are omitted until after the student has seen a sufficient number of similar examples.

Study Skills Section: Many students taking this course have poor study skills in mathematics. Section 1.1, the first section of this text, discusses the study skills needed to be successful in mathematics. This section should be very beneficial for your students, and should help them to achieve success in mathematics.

Common Student Errors: Errors that students often make are illustrated. The reasons why certain procedures are wrong are explained, and the correct procedure for working the problem is illustrated. These common student error boxes will help prevent your students from making those errors we see so often.

Helpful Hints: The helpful hint boxes offer useful suggestions for problem solving and other varied topics. They are set off in a special manner so that students will be sure to read them.

Calculator Corners: The Calculator Corners, placed at appropriate intervals in the text, are written to reinforce the algebraic topics presented in the section and to give the student pertinent information on using the calculator to solve algebraic problems. No new algebraic information is given in the Calculator Corners.

Exercise Sets: Each exercise set is graded in difficulty. The early problems help develop the students' confidence, and then students are eased gradually into the more difficult problems. A sufficient number and variety of examples are given in the section for the student to successfully complete even the more difficult exercises. The number of exercises in each section is more than ample for student assignments and practice.

Writing Exercises: Many exercise sets now include exercises that require students to write out the answers in words. These exercises improve students' under-standing and comprehension of the material. Many of these exercises involve problem solving and help develop better reasoning and critical thinking skills. Writing exercises are indicated by the symbol✎.

Cumulative Review Exercises: All exercise sets contain questions from previous sections in the chapter and from previous chapters. These cumulative review exercises will reinforce topics that were previously covered and help students retain the earlier material, while they are learning the new material. For the students' benefit the Cumulative Review Exercises are keyed to the section where the material is covered.

Just for Fun Problems: At the end of many exercise sets are Just for Fun problems. These offer more challenging problems for the bright students in your class who want something extra. These problems present additional applications of algebra, material to be presented later in the text, or material to be covered in a later mathematics course. These exercises lend themselves nicely to group work in the classroom.

Chapter Summary: At the end of each chapter is a chapter summary which includes a glossary and important chapter facts. The terms in the glossary are keyed to the page where they are first introduced.

Review Exercises: At the end of each chapter are review exercises that cover all types of exercises presented in the chapter. The review exercises are keyed to the sections where the material was first introduced.

Practice Tests: The comprehensive end-of-chapter practice test will enable the students to see how well they are prepared for the actual class test. The Instructor's Resource Manual includes several forms of each chapter test that are similar to the student's practice test.

Cumulative Review Test: These tests, which appear at the end of each even-numbered chapter, test the students' knowledge of material from the beginning of the book to the end of that chapter. Students can use these tests for review, as well as for preparation for the final exam. These exams, like the cumulative review exercises, will serve to reinforce topics taught earlier.

Answers: Answers are provided to the following exercises: odd-numbered problems in the exercise sets, all cumulative review exercises, all Just for Fun problems, all review exercises, all practice tests, and all cumulative review tests.

Prerequisite

This text assumes no prior knowledge of algebra. However, a working knowledge of arithmetic skills is important. Fractions are reviewed early in the text, and decimals and percent are reviewed in Appendix A.

Modes of Instruction

The format of this book lends itself to many different modes of instruction. For students to be able to understand the material presented, the text must be readable. Short, clear sentences are used to make this text readable for students with weak reading skills. Wherever possible, common, easy-to-understand words are used.

The spiral approach, cumulative review exercises, and the cumulative review tests will continually reinforce important concepts and topics. The approach and the features of the text will result in greater understanding and retention of the material by your students.

The features of the text and the large variety of supplements available make this text suitable for many types of instructional modes including:

- lecture
- modified lecture
- learning laboratory
- self-paced instruction
- cooperative or group study

Changes in the Third Edition

When I wrote the third edition I considered the many letters and reviews I got from students and faculty alike. I would like to thank all of you who made suggestions for improving the third edition. I would also like to thank the many instructors and students who wrote to inform me of how much they enjoyed and appreciated the text.

Some of the changes made in the third edition of the text include:

- Applications of algebra are spread more evenly throughout the text. There is also less of a concentration of applications in chapter 3.
- A section on study skills necessary for success in mathematics has been added.
- More and more varied exercises in selected sections.
- Additional detailed worked-out examples have been added where needed.
- Addition of Cumulative Review Exercises after each exercise set.

- Addition of Cumulative Review Test after every even-numbered chapter.
- Greater emphasis on the spiral approach to learning.
- Addition of more Helpful Hints and Common Student Errors.
- More Calculator Corners.
- More exercises that require written student answers.
- Exponents are now introduced in two sections. The second section introduces the negative exponent rule (although listed as optional, if you do not intend to cover Section 4.3, scientific notation, this section may be omitted).
- More material on multiplying a monomial by a monomial.
- Factoring trinomials by trial and error is now introduced in Section 5.4 and covered in depth in Section 5.5. Factoring trinomials of the form $ax^2 + bx + c$, $a \neq 1$ is explained using both the grouping technique and the trial-and-error technique. The same examples are worked using both techniques. Students and instructors can select the method they wish to use.
- The graphing chapter has been broken down into smaller sections. The section titled Slope-Intercept and Point-Slope Forms of a Linear Equation in the second edition has been broken into three smaller sections covering slope, slope-intercept form of a line, and point-slope form of a line.
- The introductory section on radicals now includes changing from square root to exponential form.
- General fine-tuning of the text for greater clarity.

Supplements to the Third Edition

For Instructors

Annotated Instructor's Edition: Includes answers to every exercise on the same page.

Instructor's Resource Manual: Contains solutions to even-numbered exercises and eight tests per chapter (three are multiple choice)

PH Test Manager: Allows users to generate tests by chapter or section number, choosing from thousands of test questions and hundreds of algorithms, which generate different numbers for the same item. Editing and graphing capability are included.

Test Item File: Contains thousands of test items for use with PH Test-Manager.

Syllabus and Teaching Outlines (with Instructor's Disk): Contains suggested homework assignments keyed to objectives and teaching outlines integrating supplements into the course. All available on ASCII disk for individual customization in your course.

For Students

Math Master Tutor Software: Carefully keyed to the book, with page references, includes four modes of instruction: *Explorations* (including detailed, worked-out examples with explanation); *Summary; Exercises* (open-ended, algorithmically generated with step-by-step solutions); and *Quiz* (with a printout option). Available free with a qualified adoption for IBM and Macintosh.

Interactive Algebra Tutor: An alternative, generic software with multiple-choice questions, available on Apple, IBM, or Macintosh.

Videotapes: Closely tied to the book, these instructional tapes feature a lecture format with worked-out examples and exercises from each section of the book. A video on study skills is also included. One master set available with each adoption of 100 or more copies.

Study Guide: Includes additional worked-out examples, additional drill problems and practice tests and their answers. Important points are emphasized.

Student's Solutions Manual: Includes detailed step-by-step solutions to all odd-numbered problems in the Exercise Sets and Cumulative Review Exercises; also includes all solutions to the Just for Fun problems and Cumulative Review Tests.

Acknowledgments

Writing a textbook is a long and time-consuming project. Many people deserve thanks for encouraging and assisting me with this project. Most importantly I would like to thank my wife, Kathy, and sons, Robert and Steven. Without their constant encouragement and understanding, this project would not have become a reality.

I would like to thank my colleagues at Monroe Community College for helping with this project, especially Peter Collinge and Annette Leopard. I would like to thank Richard Semmler of Northern Virginia Community College for his many valuable suggestions. Judith Conturo Karas did an excellent job of typing the manuscript.

I would like to thank my students, and students and faculty from around the country, for using the second edition and offering valuable suggestions for the third edition.

I would like to thank my editors at Prentice Hall, Priscilla McGeehon and Christine Peckaitis, and production editor, Rachel J. Witty, Letter Perfect, Inc.

I would like to thank the following reviewers and proofreaders for their thoughtful comments and suggestions:

WAYNE BARBER, *Chemekata Community College;* JACK BARONE, *Baruch College;* RONALD BOHUSLOV, *Merritt College;* BETH BOREL, *University of S. W. Louisiana;* FRANCINE BORTZEL, *Seton Hall University;* HELEN BURRIER, *Kirkwood Community College;* FRANK CERRATO, *City College of San Francisco;* LAURA CLARKE, *Milwaukee Area Tech.;* BEN CORNELIUS, *Oregon Institute of Technology;* ARTHUR DULL, *Diablo Valley Community College;* DALE EWEN, *Parkland College;* PETER FREEDHAND, *New York University;* ROBERT GESELL, *Cleary College;* MARK GIDNEY, *Lees McRae College;* JAY GRAENING, *University of Arkansas—Main;* MARGARET GREENE, *Florida Community College at Jacksonville;* KEN HODGE, *Rose State College;* LARRY HOEHN, *Austin Peay State University;* JUDY KASABIAN, *El Camino College;* HERBERT KASUBE, *Bradley University;* MELVIN KIRKPATRICK, *Roane State Community College;* ADELE LEGERE, *Oakton Community College;* GLENN LIPELY, *Malone College;* CHARLES LUTTRELL, *Frederick Community College;* MERWIN LYNG, *Mayville State College;* P. WILLIAM MAGLIARO, *Bucks County Community College;* JACK MCCOWN, *Central Oregon Community College;* JOHN MICHAELS, *SUNY at Brockport;* LOIS MILLER, *Golden West College;* JULIE MONTE, *Daytona Beach Junior College;* CATHY PACE, *Louisiana Tech University;* C. V. PEELE, *Marshall University;* JAMES PERKINS, *Piedmont Community College;* MATTHEW PICKARD, *University of Puget Sound;* JON PLACHY, *Metropolitan State College;* RAYMOND PLUTA, *Castleton State College;* DOLORES SCHAFFNER, *University of South Dakota;* RICHARD SEMMLER, *Northern Virginia Community College;* KEN SEYDEL, *Skyline College;* EDITH SILVER, *Mercer County Community College;* FAY THAMES, *Lamar University;* TOMMY THOMPSON, *Brookhaven College;* LEE-ING TONG, *Southeastern Massachusetts University;* JOHN WENGER, *Loop College;* BRENDA WOOD, *Florida Community College at Jacksonville;* KARL ZILM, *Lewis and Clark Community College.*

To the Student

Algebra is a course that cannot be learned by observation. To learn algebra you must become an active participant. You must read the text, pay attention in class, and, most importantly, you must work the exercises. The more exercises you work, the better.

This text was written with you in mind. Short, clear sentences are used, and many examples are given to illustrate specific points. The text stresses useful applications of algebra. Hopefully, as you progress through the course, you will come to realize that algebra is not just another math course that you are required to take, but a course that offers a wealth of useful information and applications.

This text makes full use of color. The different colors are used to highlight important information. Important procedures, definitions, and formulas are placed within colored boxes.

The boxes marked **Common Student Errors** should be studied carefully. These boxes point out errors that students commonly make, and provide the correct procedures for doing these problems. The boxes marked **Helpful Hints** should also be studied carefully, for they also stress important information.

Ask your professor early in the course to explain the policy on when the calculator may be used. If your professor allows you to use a calculator, then pay particular attention to the **Calculator Corners.**

Other questions you should ask your professor early in the course include: What supplements are available for use? Where can help be obtained when the professor is not available? Supplements that may be available include: student's study guide, student's solutions manual, tutorial software, and video tapes, including a tape on the study skills needed for success in mathematics.

You may wish to form a study group with other students in your class. Many students find that working in small groups provides an excellent way to learn the material. By discussing and explaining the concepts and exercises to one another you reinforce your own understanding. Once guidelines and procedures are determined by your group, make sure to follow them.

One of the first things you should do is to read Section 1.1, Study Skills Needed for Success in Mathematics. Read this section slowly and carefully, and pay particular attention to the advice and information given. Occasionally, refer back to this section. This could be the most important section of the book. Carefully read the material on doing your homework and on attending class.

At the end of all exercise sets (after the first two) are **cumulative review exercises.** You should work these problems on a regular basis, even if they are not assigned. These problems are from earlier sections and chapters of the text, and they will refresh your memory and reinforce those topics. If you have a problem when working these exercises, read the appropriate section of the text or study your notes that correspond to that material. The section of the text where the Cumulative Review Exercises were introduced is indicated in brackets, [ ], to the left of the exercise. After reviewing the material, if you still have a problem, make an appointment to see your professor. Working the Cumulative Review Exercises throughout the semester will also help prepare you to take your final exam.

At the end of many exercise sets are **Just for Fun** problems. These exercises are not for everyone. They are for those students who are doing well in the course and are looking for more of a challenge. These exercises often present additional applications of algebra, material that will be presented in a later section, or material that will be presented in a later course.

At the end of each chapter are a **summary,** a set of **review exercises,** and a **practice test.** Before each examination you should review these sections carefully and take the practice test. If you do well on the practice test, you should do well on the class test. The questions in the review exercises are marked to indicate the section in which that material was first introduced. If you have a problem with a review exercise question, reread the section indicated. You may also wish to take the **Cumulative Review Test** that appears at the end of every even-numbered chapter.

In the back of the text there is an **answer section** which contains the answers to the odd-numbered exercises, all cumulative review exercises, Just for Fun problems, review exercises, practice tests, and cumulative review tests. The answers should be used only to check your work.

I have tried to make this text as clear and error free as possible. No text is perfect, however. If you find an error in the text, or an example or section that you believe can be improved, I would greatly appreciate hearing from you. If you enjoy the text, I would also appreciate hearing from you.

ALLEN R. ANGEL

CHAPTER 1

Real Numbers

1.1 Study Skills for Success in Mathematics

1.2 Fractions

1.3 The Real Number System

1.4 Inequalities

1.5 Addition of Real Numbers

1.6 Subtraction of Real Numbers

1.7 Multiplication and Division of Real Numbers

1.8 An Introduction to Exponents

1.9 Use of Parentheses and Order of Operations

1.10 Properties of the Real Number System

Summary

Review Exercises

Practice Test

See Section 1.6, Exercise 105.

1.1

Study Skills for Success in Mathematics

▸ **1** Recognize the goals of the text.

▸ **2** Prepare for class effectively.

▸ **3** Realize the importance of exams.

▸ **4** Determine how to find help.

You need to acquire certain study skills that will help you to complete this course. These study techniques will also help you succeed in any other mathematics course you take.

▸ **1** The goals of this text include:

1. teaching traditional algebra topics
2. preparing students to take more advanced mathematics courses
3. building confidence so students enjoy mathematics
4. improving reasoning and critical thinking skills
5. increasing understanding of how important mathematics is in solving real-life problems
6. encouraging students to think mathematically, so that they will feel comfortable translating real-life problems into mathematical equations, and then solving the problems

It is important to realize that this course is the foundation for more advanced mathematics courses. A thorough understanding of algebra will make it easier to be successful in later mathematics courses.

Have a Positive Attitude

You may be thinking to yourself, "I hate math," or "I wish I did not have to take this class." You may have picked up on the term "math anxiety" and feel you fit this category. The first thing to do to be successful in this course is to change your attitude to a more positive one. You must be willing to give this course, and yourself, a fair chance.

Based on past experiences in mathematics, you may feel this is difficult. However, mathematics is something you need to work at. Many of you reading this book are more mature now than when you took previous mathematics courses. This maturity factor, and the desire to learn, are extremely important, and can make a tremendous difference in your ability to succeed in mathematics. I believe you can be successful in this course, but you also need to believe it.

▸ **2** Prepare for Class Effectively

To be prepared for class, you need to do your homework. If you have difficulty with the homework, or some of the concepts, write down questions to ask your professor. Prior to class, you should spend a few minutes previewing any new material in the textbook. At this point, you don't have to understand everything you read. Just get a feeling for the definitions and concepts that will be discussed. This quick preview will help you understand what your instructor is explaining during class.

After the material is explained in class, read the corresponding sections of the text slowly and carefully, word by word.

Reading the Text

A mathematics text is not a novel. Mathematics textbooks should be read slowly and carefully. If you don't understand what you are reading, reread the material. When you come across a new concept or definition, you may wish to underline it, so that it stands out. This way, when looking for it later, it will be easier to find. When you come across a worked-out example, read and follow the example very carefully. Don't just skim it. Try working out the example yourself on another sheet of paper. Make notes of anything you don't understand to ask your instructor.

Doing Homework

Two very important commitments that you must make to be successful in this course are attending class and doing your homework regularly. Your assignments must be worked conscientiously and completely. Mathematics cannot be learned by observation. You need to practice what you have heard in class. It is through doing homework that you truly learn the material.

Don't forget to check the answers to your homework assignments. This book contains the answers to the odd-numbered exercises in the back of the book. In addition, the answers to all the cumulative review, Just for Fun, and end-of-chapter review exercises, practice tests, and cumulative review tests are in the back of the book.

Ask questions in class about homework problems you don't understand. You should not feel comfortable until you understand all the concepts needed to successfully work each and every assigned problem.

Make sure when you do your homework that you write it neatly and carefully. Pay particular attention to copying signs and exponents correctly. Do your homework in a step-by-step manner. This way you can refer to it later and still understand what is written.

Attending and Participating in Class

You should plan to attend every class. Most instructors will agree that there is an inverse relationship between absences and grades. That is, the more absences you have, the lower your grade will be. Every time you miss a class, you miss important information. If you need to miss a class, contact your instructor ahead of time, and get the reading assignment and homework.

While in class, pay attention to what your instructor is saying. If you don't understand something, ask your instructor to repeat the material. If you have read the assigned material before class and have questions that have not been answered, ask your instructor. If you don't ask questions, your instructor will not know that you have a problem understanding the material.

In class, take careful notes. Write numbers and letters clearly, so that you can read them later. It is not necessary to write down every word your instructor says. Copy the major points and the examples that do not appear in the text. You should not be taking notes so frantically that you lose track of what your instructor is saying. It is a mistake to believe that you can copy material in class without understanding it, and then figure it out when you get home.

Studying

Study in the proper atmosphere, in an area where you will not be constantly disturbed, so that your attention can be devoted to what you are reading. The area where you study should be well ventilated and well lit. You should have sufficient desk space to spread out all your materials. Your chair should be comfortable. There should be no loud music to distract you from studying.

When studying, you should not only understand how to work a problem, but also know why you follow the specific steps you do to work the problem. If you do not have an understanding of why you follow the specific process, you will not be able to transfer the process to solve similar problems.

Time Management

It is recommended that students study and do homework for at least two hours for each hour of class time. Some students require more time than others. Finding the necessary time to study is not always easy. Below are some suggestions that you may find helpful.

1. Plan ahead. Determine when you will have time to study and to do your homework. Do not schedule other activities for this time period. Try to space these times evenly over the week.
2. Be organized, so that you will not have to waste time looking for your books, your pen, your calculator, or your notes.
3. If you are allowed to use calculators, use one to perform tedious calculations.
4. When you stop studying, clearly mark where you stopped in the text.
5. Try not to take on added responsibilities. You must set your priorities. If your education is a top priority, as it should be, then you may have to cut the time spent on other activities.
6. If time is a problem, do not overburden yourself with too many courses. Consider taking fewer credits. If you do not have sufficient time to study, your understanding and your grade in all of your courses could be affected.

▶ **3**   Importance of Exams

Studying
for an Exam

If you study a little bit each day, you should not need to cram the night before an exam. If you wait until the last minute, you will not have time to seek the help you need. To review for an exam:

1. Read your class notes.
2. Review your homework assignments.
3. Study formulas, definitions, and procedures given in the text.
4. Read the Common Student Error boxes and Helpful Hint boxes carefully.
5. Read the summary at the end of each chapter.
6. Work the review exercises at the end of each chapter. If you have difficulties, restudy those sections. If you still have trouble, seek help.
7. Work the chapter practice test.

Taking an Exam

Make sure you get sufficient sleep the night before the test. If you studied properly, you should not have to stay up late preparing for a test. Arrive at the exam site early so that you have a few minutes to relax before the exam. If you need to rush to get to the exam, you will start out nervous and anxious. After you are given the exam, you should:

1. Carefully write down any formulas or ideas that you need to remember.
2. Look over the entire exam quickly to get an idea of its length. You will need to pace yourself to make sure you complete the entire exam. Be prepared to spend more time on problems worth more points.
3. Read the test directions carefully.
4. Read each question carefully. Answer each question completely, and make sure you have answered the specific question asked.
5. Work the questions you understand best first; then go back and work those you are not sure of. Do not spend too much time on any one problem.
6. Attempt each problem. You may be able to get at least partial credit.

7. Work carefully, and write clearly so that your instructor can read your work. When your writing is unclear, it is easy to make mistakes.

8. Check your work, and your answers, if you have time.

9. Do not be concerned if others finish the test before you. Don't be disturbed if you are the last to finish. Use all your extra time to check your work.

▶ **4** How to Find Help

Using Supplements

This text comes with a large variety of supplements. Find out from your instructor early in the semester which supplements are available and might be beneficial for you to use. Supplements should not replace reading the text, but should be used to enhance your understanding of the material.

Seeking Help

Be sure to get help as soon as you need it! Do not wait! In mathematics, one day's material is often based on the previous day's material. So, if you don't understand the material today, you will not be able to understand the material tomorrow.

Where should you seek help? There are often a number of resources on campus. Try to make a friend in the class with whom you can study. Often you can help one another. You may wish to form a study group with other students in your class. Discussing the concepts and homework with your peers will reinforce your own understanding of the material.

You should know your instructor's office hours, and you should not hesitate to seek help from your instructor when you need it. Make sure you have read the assigned material and attempted the homework prior to meeting with your instructor. Come prepared with specific questions to ask.

There are often other sources of help available. Many colleges have a mathematics lab or a mathematics learning center, where tutors are available. Ask your instructor early in the semester if tutoring is available. Find out how to arrange for tutorial help and work with a tutor as needed.

Exercise Set 1.1

Do you know:

1. Your professor's name and office hours?
2. How you can best reach your professor?
3. Where you can obtain help if your professor is not available?
4. When you can use your calculator in this class?
5. The name and phone number of a friend in your class?
6. What supplements are available to assist you in learning?

If you do not know the answer to any of the questions just asked, you should find out as soon as possible.

7. What are your reasons for taking this course?
8. What are your goals for this course?
9. Are you beginning this course with a positive attitude? It is important that you do!
10. List the things you need to do to prepare properly for class.
11. Explain how a mathematics text should be read.
12. For each hour of class time, how many hours outside of class are recommended for studying and doing homework?
13. Two very important commitments that you must make to be successful in this course are: **(a)** doing homework regularly and completely and **(b)** attending class regularly. Explain why these commitments are necessary.
14. Write a summary of the steps you should follow when taking an exam.
15. Have you given any thought to studying with a friend or a group of friends? Can you see any advantages in doing so? Can you see any disadvantages in doing so?

1.2

Fractions

▸**1** Learn multiplication symbols.

▸**2** Recognize factors.

▸**3** Reduce fractions to lowest terms.

▸**4** Multiply fractions.

▸**5** Divide fractions.

▸**6** Add and subtract fractions.

▸**7** Convert mixed numbers to fractions.

Students taking algebra for the first time often ask, "What is the difference between arithmetic and algebra?" When doing arithmetic, all the quantities used in the calculations are known. In algebra, however, one or more of the quantities are often unknown and must be found.

EXAMPLE 1   A recipe calls for 3 cups of flour. Mrs. Clark has 2 cups of flour. How many additional cups does she need?

Solution: The answer is 1 cup.   ■

Although very elementary, this is an example of an algebraic problem. The unknown quantity is the number of additional cups of flour needed.

An understanding of decimal numbers and fractions is essential to success in algebra. The procedures to add, subtract, multiply, and divide numbers containing decimal points are reviewed in Appendix A. Percents are also reviewed in Appendix A. You may wish to review this material now.

You will need to know how to reduce a fraction to its lowest terms and how to add, subtract, multiply, and divide fractions. We will review these topics in this section. We will also explain the meaning of factors.

▸**1** In algebra we often use letters called **variables** to represent numbers. A letter commonly used for a variable is the letter x. So that we do not confuse the variable x with the times sign, we use different notation to indicate multiplication.

Multiplication Symbols

If a and b stand for (or represent) any two mathematical quantities, then each of the following may be used to indicate the product of a and b ("a times b").

$$ab \qquad a \cdot b \qquad a(b) \qquad (a)b \qquad (a)(b)$$

Examples

3 times 4 may be written:	3 times x may be written:	x times y may be written:
	$3x$	xy
3(4)	3(x)	x(y)
(3)4	(3)x	(x)y
(3)(4)	(3)(x)	(x)(y)
3 · 4	3 · x	x · y

▶ **2** The numbers or variables multiplied in a multiplication problem are called **factors.**

If $a \cdot b = c$, then a and b are **factors** of c.

For example, in $3 \cdot 5 = 15$, the numbers 3 and 5 are factors of the product 15. As a second example, consider $2 \cdot 15 = 30$. The numbers 2 and 15 are factors of the product 30. Note that 30 has many other factors. Since $5 \cdot 6 = 30$, the numbers 5 and 6 are also factors of 30. Since $3x$ means 3 times x, both the 3 and the x are factors of $3x$.

▶ **3** Now we have the necessary information to discuss fractions. The top number of a fraction is called the **numerator,** and the bottom number is called the **denominator.** In the fraction $\frac{3}{5}$, the 3 is the numerator and the 5 is the denominator.

A fraction is **reduced to its lowest terms** when the numerator and denominator have no common factors other than 1. To reduce a fraction to its lowest terms, follow these steps.

To Reduce a Fraction to Its Lowest Terms

1. Find the largest number that will divide (without remainder) both the numerator and the denominator. This number is called the **greatest common factor.**
2. Then divide both the numerator and the denominator by the greatest common factor.

If you do not remember how to find the greatest common factor (GCF) of two or more numbers, read Appendix B.

EXAMPLE 2 Reduce $\dfrac{10}{25}$ to its lowest terms.

Solution: The largest number that divides both 10 and 25 is 5. Therefore, 5 is the greatest common factor. Divide both the numerator and the denominator by 5 to reduce the fraction to its lowest terms.

$$\frac{10}{25} = \frac{10 \div 5}{25 \div 5} = \frac{2}{5}$$

EXAMPLE 3   Reduce $\dfrac{6}{18}$ to its lowest terms.

Solution: Both 6 and 18 can be divided by 1, 2, 3, and 6. The largest of these numbers, 6, is the greatest common factor. Divide both the numerator and the denominator by 6.

$$\frac{6}{18} = \frac{6 \div 6}{18 \div 6} = \frac{1}{3}$$ ■

Note in Example 3 that the numerator and denominator could have both been written with a factor of 6. Then the common factor 6 is divided out.

$$\frac{6}{18} = \frac{1 \cdot 6}{3 \cdot 6} = \frac{1}{3}$$

When you work with fractions you should give your answers in lowest terms.

Multiplication of Fractions

▶ **4**   To multiply two or more fractions, multiply their numerators together and then multiply their denominators together.

Multiplication of Fractions

$$\frac{a}{b} \cdot \frac{c}{d} = \frac{ac}{bd}$$

EXAMPLE 4   Multiply $\dfrac{6}{13}$ by $\dfrac{5}{12}$.

Solution: $\dfrac{6}{13} \cdot \dfrac{5}{12} = \dfrac{6 \cdot 5}{13 \cdot 12} = \dfrac{30}{156} = \dfrac{5}{26}.$ ■

In Example 4, reducing $\frac{30}{156}$ to its lowest terms, $\frac{5}{26}$, is for many students more difficult than the multiplication itself. When multiplying fractions, to help avoid having to reduce an answer to its lowest terms, we often divide both a numerator and denominator by a common factor. **This process can be used only when multiplying fractions; it cannot be used when adding or subtracting fractions.**

EXAMPLE 5   Divide a numerator and a denominator by a common factor and then multiply.

$$\frac{6}{13} \cdot \frac{5}{12}$$

Solution: Since the numerator 6 and the denominator 12 can both be divided by the common factor 6, we divide out, as follows:

$$\frac{6}{13} \cdot \frac{5}{12} = \frac{\overset{1}{6}}{13} \cdot \frac{5}{\underset{2}{12}} = \frac{1 \cdot 5}{13 \cdot 2} = \frac{5}{26}$$ ■

Note that the answer obtained in Example 5 is identical to the answer obtained in Example 4.

EXAMPLE 6   Multiply $\dfrac{27}{40} \cdot \dfrac{16}{9}$.

Solution: $\dfrac{27}{40} \cdot \dfrac{16}{9} = \dfrac{\overset{3}{\cancel{27}}}{40} \cdot \dfrac{16}{\underset{1}{\cancel{9}}}$ Divide both 27 and 9 by 9.

$= \dfrac{\overset{3}{\cancel{27}}}{\underset{5}{\cancel{40}}} \cdot \dfrac{\overset{2}{\cancel{16}}}{\underset{1}{\cancel{9}}}$ Divide both 40 and 16 by 8.

$= \dfrac{3 \cdot 2}{5 \cdot 1} = \dfrac{6}{5}$.   ■

The numbers 0, 1, 2, 3, 4, . . . are called **whole numbers.** The three dots after the 4 indicate that the whole numbers continue indefinitely in the same manner. Thus the numbers 468 and 1,043 are also whole numbers. Whole numbers will be discussed further in Section 1.3. To multiply a whole number by a fraction, write the whole number with a denominator of 1 and then multiply.

EXAMPLE 7   Multiply $5 \cdot \dfrac{2}{15}$.

Solution: $\dfrac{5}{1} \cdot \dfrac{2}{15} = \dfrac{\overset{1}{\cancel{5}}}{1} \cdot \dfrac{2}{\underset{3}{\cancel{15}}} = \dfrac{2}{3}$.   ■

Division of Fractions   ▶ **5**   To divide one fraction by another, invert the divisor (second fraction if written with ÷) and proceed as in multiplication.

Division of Fractions
$$\dfrac{a}{b} \div \dfrac{c}{d} = \dfrac{a}{b} \cdot \dfrac{d}{c} = \dfrac{ad}{bc}$$

EXAMPLE 8   Divide $\dfrac{3}{5} \div \dfrac{5}{6}$.

Solution: $\dfrac{3}{5} \div \dfrac{5}{6} = \dfrac{3}{5} \cdot \dfrac{6}{5} = \dfrac{3 \cdot 6}{5 \cdot 5} = \dfrac{18}{25}$.   ■

Sometimes, rather than being asked to obtain the answer to a problem by adding, subtracting, multiplying, or dividing, you may be asked to evaluate an expression. To **evaluate** an expression means to obtain the answer to the problem using the operations given in the problem.

EXAMPLE 9   Evaluate $\dfrac{4}{7} \div \dfrac{5}{12}$.

Solution: $\dfrac{4}{7} \div \dfrac{5}{12} = \dfrac{4}{7} \cdot \dfrac{12}{5} = \dfrac{48}{35}$.

EXAMPLE 10   Evaluate $\dfrac{3}{8} \div 9$.

Solution: Write 9 as $\frac{9}{1}$.

$$\frac{3}{8} \div 9 = \frac{3}{8} \div \frac{9}{1} = \frac{\overset{1}{\cancel{3}}}{8} \cdot \frac{1}{\underset{3}{\cancel{9}}} = \frac{1}{24}$$

Addition and Subtraction of Fractions

▶ **6**   *Only fractions that have the same* (or common) *denominator can be added or subtracted.* To add (or subtract) fractions with the same denominator, add (or subtract) the numerators and keep the common denominator.

Addition and Subtraction of Fractions

$$\frac{a}{c} + \frac{b}{c} = \frac{a+b}{c} \quad \text{or} \quad \frac{a}{c} - \frac{b}{c} = \frac{a-b}{c}$$

EXAMPLE 11   Evaluate $\dfrac{9}{15} + \dfrac{2}{15}$.

Solution: $\dfrac{9}{15} + \dfrac{2}{15} = \dfrac{9+2}{15} = \dfrac{11}{15}$.

EXAMPLE 12   Evaluate $\dfrac{8}{13} - \dfrac{5}{13}$.

Solution: $\dfrac{8}{13} - \dfrac{5}{13} = \dfrac{8-5}{13} = \dfrac{3}{13}$.

To add (or subtract) fractions with unlike denominators, we must first rewrite each fraction with the same, or a common, denominator. The smallest number that is divisible by two or more denominators is called the **least common denominator.** If you have forgotten how to find the least common denominator, or LCD, review Appendix B now.

EXAMPLE 13   Add $\dfrac{1}{2} + \dfrac{1}{5}$.

Solution: We cannot add these fractions until we rewrite them with a common denominator. Since the lowest number that both 2 and 5 divide (without remainder) is 10, we will rewrite both fractions with the least common denominator of 10.

$$\frac{1}{2} = \frac{1}{2} \cdot \frac{5}{5} = \frac{5}{10} \quad \text{and} \quad \frac{1}{5} = \frac{1}{5} \cdot \frac{2}{2} = \frac{2}{10}$$

Now add.

$$\frac{1}{2} + \frac{1}{5} = \frac{5}{10} + \frac{2}{10} = \frac{7}{10}$$

Note that multiplying both the numerator and denominator by the same number is the same as multiplying by 1. Thus the value of the fraction does not change.

EXAMPLE 14 Subtract $\frac{3}{4} - \frac{2}{3}$.

Solution: The least common denominator is 12. Therefore, we rewrite both fractions with a denominator of 12.

$$\frac{3}{4} = \frac{3}{4} \cdot \frac{3}{3} = \frac{9}{12} \quad \text{and} \quad \frac{2}{3} = \frac{2}{3} \cdot \frac{4}{4} = \frac{8}{12}$$

Now subtract.

$$\frac{3}{4} - \frac{2}{3} = \frac{9}{12} - \frac{8}{12} = \frac{1}{12}$$

COMMON STUDENT ERROR

It is important that you realize that dividing out a common factor in the numerator of one fraction and the denominator of a different fraction can be performed only when multiplying fractions. **This process cannot be performed when adding or subtracting fractions.**

Correct

Multiplication problems

$$\frac{\overset{1}{\cancel{3}}}{5} \cdot \frac{1}{\cancel{3}}$$

$$\frac{\overset{2}{\cancel{8}} \cdot 3}{\underset{1}{\cancel{4}}}$$

Wrong

Addition problems

$$\frac{\overset{1}{\cancel{3}}}{5} + \frac{1}{\cancel{3}}$$

$$\frac{\overset{2}{\cancel{8}} + 3}{\underset{1}{\cancel{4}}}$$

▶ **7** Consider the number $5\frac{2}{3}$. This is an example of a **mixed number.** A mixed number consists of a whole number followed by a fraction. The mixed number $5\frac{2}{3}$ means $5 + \frac{2}{3}$. The mixed number $5\frac{2}{3}$ may be changed to a fraction as follows:

$$5\frac{2}{3} = 5 + \frac{2}{3} = \frac{15}{3} + \frac{2}{3} = \frac{17}{3}$$

Any fraction whose numerator is greater than its denominator may be changed to a mixed number. For example, $\frac{17}{3}$ may be changed to

$$\frac{17}{3} = \frac{15}{3} + \frac{2}{3} = 5 + \frac{2}{3} = 5\frac{2}{3}$$

The procedure used to change from a mixed number to a fraction can be simplified as follows.

> **To Change a Mixed Number to a Fraction**
>
> 1. Multiply the denominator of the fraction in the mixed number by the whole number preceding it.
> 2. Add the numerator of the fraction in the mixed number to the product obtained in step 1. This sum represents the numerator of the fraction we are seeking. The denominator of the fraction we are seeking is the same as the denominator of the fraction in the mixed number.

EXAMPLE 15 Change the mixed number $5\frac{2}{3}$ to a fraction.

Solution: Multiply the denominator, 3, by the whole number, 5, to get a product of 15. To this product add the numerator, 2. This sum, 17, represents the numerator of the fraction. The denominator of the fraction we are seeking is the same as the denominator of the fraction in the mixed number, 3. Thus $5\frac{2}{3} = \frac{17}{3}$.

$$5\frac{2}{3} = \frac{15 + 2}{3} = \frac{17}{3}$$ ∎

EXAMPLE 16 Change $6\frac{5}{9}$ to a fraction.

Solution: Multiply 9 by 6 to get 54; then add 5 to get 59. This is the numerator of the fraction we are seeking.

$$6\frac{5}{9} = \frac{54 + 5}{9} = \frac{59}{9}$$ ∎

> **To Change a Fraction Greater Than 1 to a Mixed Number**
>
> 1. Divide the numerator by the denominator. Note the quotient and remainder.
> 2. Write the mixed number. The quotient found in step 1 is the whole number part of the mixed number. The remainder is the numerator of the fraction in the mixed number. The denominator in the fraction of the mixed number will be the same as the denominator in the original fraction.

EXAMPLE 17 Convert $\frac{17}{3}$ to a mixed number.

Solution:

$$3\overline{)17} \begin{array}{c} 5 \longleftarrow \text{whole number} \\ \end{array}$$

Denominator $\longrightarrow$

$$\frac{15}{2} \longleftarrow \text{remainder}$$

$$\frac{17}{3} = 5\frac{2}{3} \begin{array}{l} \longleftarrow \text{remainder} \\ \longleftarrow \text{denominator (or divisor)} \end{array}$$
$$\text{whole number}$$

Thus $\dfrac{17}{3}$ converted to a mixed number is $5\dfrac{2}{3}$.

EXAMPLE 18 Convert $\dfrac{21}{5}$ to a mixed number.

Solution:

$$5\overline{)21} \begin{array}{c} 4 \\ \end{array} \qquad \frac{21}{5} = 4\frac{1}{5}$$
$$\frac{20}{1}$$

To add, subtract, multiply, or divide mixed numbers, we often change the mixed numbers to fractions.

EXAMPLE 19 Add $2\dfrac{1}{4} + \dfrac{1}{2}$.

Solution: Change $2\dfrac{1}{4}$ to $\dfrac{9}{4}$; then add.

$$2\frac{1}{4} + \frac{1}{2} = \frac{9}{4} + \frac{1}{2}$$

$$= \frac{9}{4} + \frac{2}{4}$$

$$= \frac{11}{4} \text{ or } 2\frac{3}{4}$$

EXAMPLE 20 Multiply $\left(3\dfrac{3}{4}\right)\left(4\dfrac{3}{5}\right)$.

Solution: Change both mixed numbers to fractions; then multiply.

$$\left(3\frac{3}{4}\right)\left(4\frac{3}{5}\right) = \frac{\overset{3}{\cancel{15}}}{4} \cdot \frac{23}{\underset{1}{\cancel{5}}} = \frac{69}{4} \quad \text{or} \quad 17\frac{1}{4}$$

EXAMPLE 21 Divide $\dfrac{4}{5} \div 2\dfrac{5}{8}$.

Solution: Change $2\dfrac{5}{8}$ to a fraction; then follow the procedure for dividing fractions.

$$\frac{4}{5} \div 2\frac{5}{8} = \frac{4}{5} \div \frac{21}{8}$$

$$= \frac{4}{5} \cdot \frac{8}{21} = \frac{32}{105}$$

Exercise Set 1.2

Reduce each fraction to its lowest terms. If a fraction is already in its lowest terms, state so.

1. $\dfrac{5}{15}$

2. $\dfrac{12}{20}$

3. $\dfrac{9}{12}$

4. $\dfrac{4}{5}$

5. $\dfrac{18}{36}$

6. $\dfrac{9}{30}$

7. $\dfrac{15}{35}$

8. $\dfrac{36}{72}$

9. $\dfrac{40}{64}$

10. $\dfrac{15}{120}$

11. $\dfrac{8}{15}$

12. $\dfrac{5}{35}$

13. $\dfrac{96}{72}$

14. $\dfrac{14}{28}$

Find the product or quotient. Write the answers in lowest terms.

15. $\dfrac{1}{2} \cdot \dfrac{3}{5}$

16. $\dfrac{3}{5} \cdot \dfrac{4}{7}$

17. $\dfrac{5}{8} \cdot \dfrac{2}{7}$

18. $\dfrac{5}{12} \cdot \dfrac{6}{5}$

19. $\dfrac{3}{8} \cdot \dfrac{2}{9}$

20. $\dfrac{15}{16} \cdot \dfrac{4}{3}$

21. $\dfrac{1}{3} \div \dfrac{1}{5}$

22. $\dfrac{2}{3} \cdot \dfrac{3}{5}$

23. $\dfrac{5}{12} \div \dfrac{4}{3}$

24. $\dfrac{4}{9} \div \dfrac{16}{5}$

25. $\dfrac{10}{3} \div \dfrac{5}{9}$

26. $\dfrac{12}{5} \div \dfrac{3}{7}$

27. $\dfrac{4}{9} \cdot \dfrac{15}{16}$

28. $\dfrac{3}{10} \cdot \dfrac{5}{12}$

29. $\dfrac{4}{15} \div \dfrac{12}{13}$

30. $\dfrac{15}{16} \div \dfrac{1}{2}$

31. $\dfrac{12}{7} \cdot \dfrac{19}{24}$

32. $\dfrac{28}{13} \cdot \dfrac{2}{7}$

33. $1\dfrac{4}{5} \cdot \dfrac{20}{3}$

34. $4\dfrac{4}{5} \div \dfrac{8}{15}$

35. $\left(\dfrac{3}{5}\right)\left(1\dfrac{2}{3}\right)$

36. $\left(\dfrac{5}{8}\right)\left(3\dfrac{1}{3}\right)$

37. $3\dfrac{2}{3} \div 1\dfrac{5}{6}$

38. $3\dfrac{1}{4} \div \dfrac{5}{6}$

Add or subtract. Write the answers in lowest terms.

39. $\dfrac{2}{5} + \dfrac{1}{5}$

40. $\dfrac{3}{10} + \dfrac{5}{10}$

41. $\dfrac{5}{12} - \dfrac{2}{12}$

42. $\dfrac{18}{36} - \dfrac{1}{36}$

43. $\dfrac{9}{13} + \dfrac{4}{13}$

44. $\dfrac{9}{10} - \dfrac{3}{10}$

45. $\dfrac{21}{29} - \dfrac{18}{29}$

46. $\dfrac{1}{3} + \dfrac{1}{5}$

47. $\dfrac{2}{5} + \dfrac{5}{6}$

48. $\dfrac{1}{9} - \dfrac{1}{18}$

49. $\dfrac{4}{12} - \dfrac{2}{15}$

50. $\dfrac{5}{6} - \dfrac{3}{7}$

51. $\dfrac{2}{10} + \dfrac{1}{15}$

52. $\dfrac{5}{8} - \dfrac{1}{6}$

53. $\dfrac{8}{9} - \dfrac{4}{6}$

54. $\dfrac{3}{8} + \dfrac{5}{12}$

55. $\dfrac{5}{6} + \dfrac{9}{24}$

56. $\dfrac{7}{15} - \dfrac{12}{30}$

57. $\dfrac{11}{12} - \dfrac{3}{4}$

58. $2\dfrac{1}{2} + \dfrac{1}{4}$

59. $3\dfrac{1}{4} + \dfrac{2}{3}$

60. $\dfrac{3}{10} + 2\dfrac{1}{3}$

61. $2\dfrac{1}{2} + 1\dfrac{1}{3}$

62. $\dfrac{4}{5} - \dfrac{2}{7}$

63. $4\dfrac{2}{3} - 1\dfrac{1}{5}$

64. $3\dfrac{1}{8} - \dfrac{3}{4} - \dfrac{1}{2}$

65. $1\dfrac{4}{5} - \dfrac{3}{4} + 3$

66. $2\dfrac{2}{3} + 1\dfrac{3}{5} - \dfrac{3}{12}$

Solve each problem.

67. John, a dressmaker, wishes to make 5 identical dresses. If each dress needs $2\dfrac{3}{4}$ yards of material, how much material will John need?

68. A board is $22\dfrac{1}{2}$ feet long. What is the length of each piece when cut in five equal lengths? (Ignore the thickness of the cuts.)

69. A length of $3\dfrac{1}{16}$ inches is cut from a piece of wood $16\dfrac{3}{4}$ inches long. What is the length of the remaining piece of wood?

70. At the beginning of the day a stock was selling for $11\frac{7}{8}$ dollars. At the close of the session it was selling for $13\frac{3}{4}$. How much did the stock gain that day?

71. A plumber connected two pieces of pipe measuring $3\frac{3}{8}$ feet and $5\frac{1}{16}$ feet, respectively. What is the total length of these two pieces of pipe?

72. A recipe calls for $2\frac{1}{2}$ cups of flour and another $1\frac{1}{3}$ cups of flour to be added later. How much flour does the recipe require?

73. At high tide the water level at a measuring stick is $20\frac{3}{4}$ feet. At low tide the water level dropped to $8\frac{7}{8}$ feet. How much did the water level fall?

74. Five lengths of wood each $4\frac{1}{2}$ feet long are placed end to end. What is the total length of the five pieces?

75. A recipe calls for $\frac{3}{4}$ teaspoon of teriyaki seasoning for each pound of beef. To cook $4\frac{1}{2}$ pounds of beef, how many teaspoons of teriyaki are needed?

76. Dawn cuts a piece of wood measuring $3\frac{1}{8}$ inches into two equal pieces. How long is each piece?

77. Tom wishes to subdivide a $4\frac{5}{8}$ acre lot into 3 equal size lots. What will be the acreage of each lot?

78. A nurse must give $\frac{1}{16}$ milligram of a drug for each kilogram of patient weight. If Mr. Duncan weighs (or has a mass of) 80 kg, find the amount of the drug Mr. Duncan should be given.

79. Find

(a) $\dfrac{5}{6} \cdot \dfrac{3}{8}$ **(b)** $\dfrac{5}{6} \div \dfrac{3}{8}$

(c) $\dfrac{5}{6} + \dfrac{3}{8}$ **(d)** $\dfrac{5}{6} - \dfrac{3}{8}$

80. Explain the procedure to reduce a fraction to its lowest terms.

81. Explain the procedure to multiply fractions.

82. Explain the procedure to divide fractions.

83. Explain the procedure to add or subtract fractions.

84. Explain the procedure to convert a mixed number to a fraction.

85. Explain the procedure for converting a fraction whose numerator is greater than its denominator into a mixed number.

JUST FOR FUN

Use the directions shown to find the amount of each ingredient needed to make three servings of Minute Rice.

**Amounts of RICE and WATER: Use equal amounts rice and water.
Minute Rice doubles in volume.**

TO MAKE	RICE & WATER (equal measures)	SALT	BUTTER OR MARGARINE (if desired)
2 servings	$\frac{2}{3}$ cup	$\frac{1}{4}$ tsp.	1 tsp.
4 servings	$1\frac{1}{3}$ cups	$\frac{1}{2}$ tsp.	2 tsp.

MINUTE® is a registered trademark of General Foods Corporation, White Plains, N.Y.

1.3

The Real Number System

▶ **1**   Identify some important sets of numbers.

▶ **2**   Know the structure of the real numbers.

We will be talking about and using various types of numbers throughout the text. This section introduces you to some of those numbers and to the structure of the real number system. This section is a quick overview. Some of the sets of numbers we mention in this section, such as rational and irrational numbers, are discussed in greater depth later in the text.

▶ **1** A **set** is a collection of **elements** listed within braces. The set {a, b, c, d, e} consists of five elements, namely a, b, c, d, and e. There are many different sets of numbers. Two important sets are the natural numbers and the whole numbers. The whole numbers were introduced earlier.

Natural numbers: {1, 2, 3, 4, 5, . . .}

Whole numbers: {0, 1, 2, 3, 4, 5, . . .}

An aid in understanding sets of numbers is the real number line (Fig. 1.1).

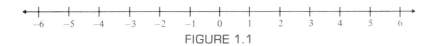

FIGURE 1.1

The real number line continues indefinitely in both directions. The numbers to the right of 0 are positive and those to the left of 0 are negative. Zero is neither positive nor negative (Fig. 1.2).

FIGURE 1.2

Figure 1.3 illustrates the natural numbers marked on the number line. The natural numbers are also called the **positive integers** or the **counting numbers.**

FIGURE 1.3

Another important set of numbers is the integers.

Integers: {. . . , −5, −4, −3, −2, −1, 0, 1, 2, 3, 4, 5, . . .}

negative integers   positive integers

The integers consist of the negative integers, 0, and the positive integers. The integers are marked on the number line in Figure 1.4.

FIGURE 1.4

Can you think of any numbers that are not integers? You probably said "fractions" or "decimal numbers." Fractions and decimal numbers belong to the set of rational numbers. The set of **rational numbers** consists of all the numbers that can be expressed as a quotient of two integers, with the denominator not 0.

Rational numbers: {quotient of two integers, denominator not 0}

The fraction $\frac{1}{2}$ is a quotient of two integers with the denominator not 0. Thus $\frac{1}{2}$ is a rational number. The decimal number 0.4 can be written $\frac{4}{10}$ and is therefore a rational number. All integers are also rational numbers since they can be written with a denominator of 1. For example $3 = \frac{3}{1}$, $-12 = \frac{-12}{1}$, and $0 = \frac{0}{1}$. Some rational numbers are illustrated on the number line in Figure 1.5.

FIGURE 1.5

Most of the numbers that we use are rational numbers; however, there are some numbers that are not rational. Numbers such as the square root of 2, written $\sqrt{2}$, are not rational numbers. Any number that can be represented on the number line that is not a rational number is called an **irrational number.** The $\sqrt{2}$ is *approximately* 1.41. Some irrational numbers are illustrated on the number line in Figure 1.6. Rational and irrational numbers will be discussed further in later chapters.

FIGURE 1.6

▶ **2** Notice that many different types of numbers can be illustrated on the number line. Any number that can be represented on the number line is a *real number*.

Real numbers: {all numbers that can be represented on the real number line}

All the numbers mentioned thus far are real numbers. The natural numbers, the whole numbers, the integers, the rational numbers, and the irrational numbers are all real numbers. There are some types of numbers that are not real numbers, but these numbers are beyond the scope of this book. Figure 1.7 illustrates the relationship between the various sets of numbers within the set of real numbers.

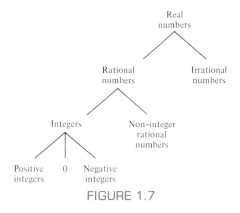

FIGURE 1.7

In Fig. 1.7 we can see that when we combine the rational numbers and the irrational numbers we get the real numbers. When we combine the integers with the non-integer rational numbers (such as $\frac{1}{2}$ and 0.42), we get the rational numbers. When we combine the positive integers (or natural numbers), 0, and the negative integers, we get the integers.

Consider the positive integer 5. If we follow the positive integer branch in Fig. 1.7 upward, we see that the number 5 is also an integer, a rational number, and a real number. Now consider the number $\frac{1}{2}$. It belongs to the non-integer rational numbers. If we follow this branch upward, we can see that $\frac{1}{2}$ is also a rational number and a real number.

EXAMPLE 1   Consider the following set of numbers:

$$\left\{-6, -0.5, 4\frac{1}{2}, -96, \sqrt{3}, 0, 9, -\frac{4}{7}, -2.9, \sqrt{7}, -\sqrt{5}\right\}$$

List the elements of the set that are

(a) Natural numbers.   (b) Whole numbers.
(c) Integers.   (d) Rational numbers.
(e) Irrational numbers.   (f) Real numbers.

Solution: (a) 9   (b) 0,9   (c) −6, −96, 0, 9

(d) $-6, -0.5, 4\frac{1}{2}, -96, 0, 9, -\frac{4}{7}, -2.9$

(e) $\sqrt{3}, \sqrt{7}, -\sqrt{5}$

(f) $-6, -0.5, 4\frac{1}{2}, -96, \sqrt{3}, 0, 9, -\frac{4}{7}, -2.9, \sqrt{7}, -\sqrt{5}$ ■

Exercise Set 1.3

List each set of numbers.

1. Integers

2. Counting numbers

3. Natural numbers

4. Positive numbers

5. Negative numbers

6. Whole numbers

In Exercises 7–38, state whether each statement is true or false.

7. −4 is a negative integer.

8. 0 is a whole number.

9. 0 is an integer.

10. −1 is an integer.

11. $\frac{1}{2}$ is an integer.

12. 0.5 is an integer.

13. $\sqrt{7}$ is a rational number.

14. $\sqrt{7}$ is a real number.

15. $-\frac{3}{5}$ is a rational number.

16. 0 is a rational number.

17. $-19\frac{1}{5}$ is an irrational number.

18. −7 is a real number.

19. −0.06 is a real number.

20. $2\frac{5}{8}$ is an irrational number.

21. 0 is a positive integer.

22. The natural numbers, counting numbers, and positive integers are different names for the same set of numbers.

23. When zero is added to the set of counting numbers, the set of whole numbers is formed.

24. When the negative integers, the positive integers, and 0 are combined, the integers are formed.

25. Any number to the left of zero on the number line is a negative number.

26. Every integer is a rational number.

27. Every integer is an irrational number.

28. Every rational number is a real number.

29. Every irrational number is a real number.

30. The number 0 is an irrational number.

31. Some real numbers are not rational numbers.

32. Some rational numbers are not real numbers.

33. Every natural number is positive.

34. Every integer is positive.

35. No rational numbers are integers.

36. All real numbers can be represented on the number line.

37. Irrational numbers cannot be represented on the number line.

38. Some rational numbers are negative integers.

39. Consider the set of numbers

$$\left\{-6, 7, 12.4, -\frac{9}{5}, -2\frac{1}{4}, \sqrt{3}, 0, 9, \sqrt{7}, 0.35\right\}$$

List those numbers that are
(a) Positive integers.
(b) Whole numbers.
(c) Integers.
(d) Rational numbers.
(e) Irrational numbers.
(f) Real numbers.

40. Consider the set of numbers

$$\left\{-\frac{5}{3}, 0, -2, 5, 5\frac{1}{2}, \sqrt{2}, -\sqrt{3}, 1.63, 207\right\}$$

List those numbers that are
(a) Positive integers.
(b) Whole numbers.
(c) Integers.
(d) Rational numbers.
(e) Irrational numbers.
(f) Real numbers.

41. Consider the set of numbers

$$\left\{\frac{1}{2}, \sqrt{2}, -\sqrt{2}, 4\frac{1}{2}, \frac{5}{12}, -1.67, 5, -300, -9\frac{1}{2}\right\}$$

List those numbers that are
(a) Positive integers.
(b) Whole numbers.
(c) Negative integers.
(d) Integers.
(e) Rational numbers.
(f) Irrational numbers.
(g) Real numbers.

In each of the following exercises, give three examples of numbers that satisfy the conditions.

42. A real number but not an integer.

43. A rational number but not an integer.

44. An integer but not a negative integer.

45. A real number but not a rational number.

46. An irrational number and a positive number.

47. An integer and a rational number.

48. A negative integer and a real number.

49. A negative integer and a rational number.

50. A real number but not a positive rational number.

51. A rational number but not a negative number.

52. An integer but not a positive integer.

53. A real number but not an irrational number.

54. Write a paragraph or two explaining the structure of the real number system. Include the whole numbers, counting numbers, integers, rational numbers, irrational numbers, and real numbers in your explanation.

Cumulative Review Exercises

[1.2] **55.** Convert $6\frac{2}{3}$ to a fraction.

56. Write $\frac{16}{3}$ as a mixed number.

57. Add $\frac{3}{5} + \frac{5}{8}$.

58. Multiply $\left(\frac{5}{9}\right)\left(4\frac{2}{3}\right)$.

1.4

Inequalities

▶ **1** Determine which is the greater of two numbers.

▶ **2** Find the absolute value of a number.

▶ **1**   The number line (Fig. 1.8) can be used to explain inequalities. When comparing two numbers, **the number to the right on the number line is the greater number, and the number to the left is the lesser number.** The symbol $>$ is used to represent the words "is greater than." The symbol $<$ is used to represent the words "is less than."

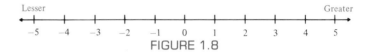

FIGURE 1.8

The statement that the number 3 is greater than the number 2 is written $3 > 2$. Notice that 3 is to the right of 2 on the number line. The statement that the number 0 is greater than the number -1 is written $0 > -1$. Notice that 0 is to the right of -1 on the number line.

Instead of stating that 3 is greater than 2, we could state that 2 is less than 3, written $2 < 3$. Notice that 2 is to the left of 3 on the number line. The statement that the number -1 is less than the number 0 is written $-1 < 0$. Notice that -1 is to the left of 0 on the number line.

EXAMPLE 1   Insert either $>$ or $<$ in the shaded area between the paired numbers to make a true statement.

(a) $-4 \ \ -2$ (b) $-\frac{3}{2} \ \ 2.5$ (c) $\frac{1}{2} \ \ \frac{1}{4}$ (d) $-2 \ \ 4$

Solution: The points given are shown on the number line (Fig. 1.9).

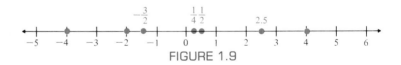

FIGURE 1.9

(a) $-4 < -2$; notice that -4 is to the left of -2.
(b) $-\frac{3}{2} < 2.5$; notice that $-\frac{3}{2}$ is to the left of 2.5.
(c) $\frac{1}{2} > \frac{1}{4}$; notice that $\frac{1}{2}$ is to the right of $\frac{1}{4}$.
(d) $-2 < 4$; notice that -2 is to the left of 4.   ■

EXAMPLE 2   Insert either $>$ or $<$ in the shaded area between the paired numbers to make a true statement.

(a) $-1 \ \ -2$ (b) $-1 \ \ 0$ (c) $-2 \ \ 2$ (d) $-4.09 \ \ -4.9$

Solution: The points given are shown on the number line (Fig. 1.10).

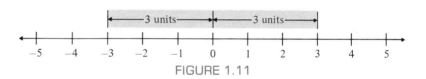

FIGURE 1.10

(a) $-1 > -2$; notice that -1 is to the right of -2.
(b) $-1 < 0$; notice that -1 is to the left of 0.
(c) $-2 < 2$; notice that -2 is to the left of 2.
(d) $-4.09 > -4.9$; notice that -4.09 is to the right of -4.9. ■

▶ **2** The concept of absolute value can be explained with the help of the number line shown in Figure 1.11. The **absolute value** of a number can be considered the distance between the number and 0 on the number line. Thus the absolute value of 3, symbolized by $|3|$, is 3 since it is 3 units from 0 on the number line. Similarly, the absolute value of the number -3, symbolized by $|-3|$, is also 3 since -3 is 3 units from 0.

$$|3| = 3 \quad \text{and} \quad |-3| = 3$$

FIGURE 1.11

Since the absolute value of a number measures the distance (without regard to direction) of a number from 0 on the number line, **the absolute value of every number will be either positive or zero.**

Number	Absolute Value of Number
6	$\lvert 6 \rvert = 6$
-6	$\lvert -6 \rvert = 6$
0	$\lvert 0 \rvert = 0$
$-\dfrac{1}{2}$	$\left\lvert -\dfrac{1}{2} \right\rvert = \dfrac{1}{2}$

EXAMPLE 3 Insert either $>$, $<$, or $=$ in the shaded area to make a true statement.
(a) $|3|$ ▢ 3 (b) $|-2|$ ▢ $|2|$ (c) -2 ▢ $|-4|$
(d) $|-5|$ ▢ 0 (e) $|12|$ ▢ $|-18|$

Solution: (a) $|3| = 3$.
(b) $|-2| = |2|$, since both $|-2|$ and $|2|$ equal 2.
(c) $-2 < |-4|$, since $|-4| = 4$.
(d) $|-5| > 0$, since $|-5| = 5$.
(e) $|12| < |-18|$, since $|12| = 12$ and $|-18| = 18$. ■

The concept of absolute value is very important in higher-level mathematics courses. If you take a course in intermediate algebra, you will be introduced to a more formal definition of absolute value. We will use absolute value in Section 1.5 in our explanation of adding and subtracting real numbers.

Exercise Set 1.4

Insert either $<$ or $>$ in the shaded area to make a true statement.

1. 2 ⬚ 3

2. 4 ⬚ -2

3. -3 ⬚ 0

4. -6 ⬚ -4

5. $\dfrac{1}{2}$ ⬚ $-\dfrac{2}{3}$

6. $\dfrac{3}{5}$ ⬚ $\dfrac{4}{5}$

7. 0.2 ⬚ 0.4

8. -0.2 ⬚ -0.4

9. $\dfrac{2}{5}$ ⬚ -1

10. 0 ⬚ -0.9

11. 4 ⬚ -4

12. $-\dfrac{3}{4}$ ⬚ -1

13. -2.1 ⬚ -2

14. -1.83 ⬚ -1.82

15. $\dfrac{5}{9}$ ⬚ $-\dfrac{5}{9}$

16. -9 ⬚ -12

17. $-\dfrac{3}{2}$ ⬚ $\dfrac{3}{2}$

18. -4.09 ⬚ -5.3

19. 0.49 ⬚ 0.43

20. -1.0 ⬚ -0.7

21. 5 ⬚ -7

22. 0.001 ⬚ 0.002

23. -0.006 ⬚ -0.007

24. $\dfrac{1}{2}$ ⬚ $-\dfrac{1}{2}$

25. $\dfrac{3}{5}$ ⬚ 1

26. $\dfrac{5}{3}$ ⬚ $\dfrac{3}{5}$

27. $-\dfrac{2}{3}$ ⬚ -3

28. -5 ⬚ -2

Insert either $<$, $>$, or $=$ in the shaded area to make a true statement.

29. 8 ⬚ $|-7|$

30. $|-8|$ ⬚ $|-7|$

31. $|0|$ ⬚ $\dfrac{2}{3}$

32. $|-4|$ ⬚ -3

33. $|-3|$ ⬚ $|-4|$

34. $|-1.9|$ ⬚ -1.8

35. 4 ⬚ $\left|-\dfrac{9}{2}\right|$

36. -5 ⬚ $|5|$

37. $\left|-\dfrac{6}{2}\right|$ ⬚ $\left|-\dfrac{2}{6}\right|$

38. $\left|\dfrac{2}{5}\right|$ ⬚ $|-0.40|$

39. What numbers are 4 units from 0 on the number line?

40. What numbers are 5 units from 0 on the number line?

41. What numbers are 2 units from 0 on the number line?

42. Are there any real numbers whose absolute value is not a positive number? Explain your answer.

43. What is the absolute value of a number?

Cumulative Review Exercises

[1.2] **44.** Subtract $1\frac{2}{3} - \frac{3}{8}$.

45. List the set of whole numbers.

46. List the set of counting numbers.

[1.3] **47.** Consider the set of numbers $\{5, -2, 0, \frac{1}{3}, \sqrt{3}, -\frac{5}{9}, 2.3\}$. List the numbers in the set that are:

(a) Natural numbers.

(b) Whole numbers.

(c) Integers.

(d) Rational numbers.

(e) Irrational numbers.

(f) Real numbers.

1.5

Addition of Real Numbers

▶ **1** Add real numbers using the number line.

▶ **2** Identify opposites or additive inverses.

▶ **3** Add using absolute values.

There are many practical uses for negative numbers. A submarine going below sea level, a bank account that has been overdrawn, a business spending more than it makes, and a temperature below zero are some examples.

The four basic **operations** of arithmetic are addition, subtraction, multiplication, and division. In the next few sections we will explain how to add, subtract, multiply, and divide numbers. We will consider both positive and negative numbers. In this section we discuss the operation of addition.

▶ **1** To add numbers, we make use of the number line. Represent the first number to be added (first *addend*) by an arrow starting at 0. The arrow is drawn to the right if the number is positive. If the number is negative, the arrow is drawn to the left. From the tip of the first arrow, draw a second arrow to represent the second addend. The second arrow is drawn to the right or left, as just explained. The sum of the two numbers is found at the tip of the second arrow. Note that *any number except 0 without a sign in front of it is positive*. For example, 3 means $+3$ and 5 means $+5$.

EXAMPLE 1 Evaluate $3 + (-4)$ using the number line.

Solution: *Always begin at 0.* Since the first addend, the 3, is positive, the first arrow starts at 0 and is drawn 3 units to the right (Fig. 1.12).

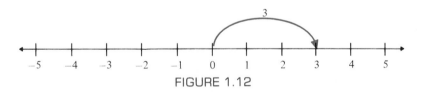

FIGURE 1.12

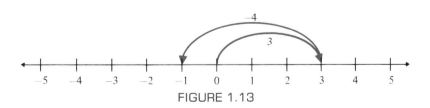

FIGURE 1.13

The second arrow starts at 3 and is drawn 4 units to the left, since the second addend is negative (Fig. 1.13). The tip of the second arrow is at -1. Thus

$$3 + (-4) = -1$$

■

EXAMPLE 2   Evaluate $-4 + 2$ using the number line.

Solution: Begin at 0. Since the first addend is negative, -4, the first arrow is drawn 4 units to the left. From there, since 2 is positive, the second arrow is drawn 2 units to the right. The second arrow ends at -2. See Figure 1.14.

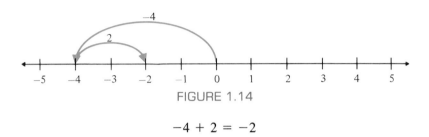

FIGURE 1.14

$$-4 + 2 = -2$$ ■

EXAMPLE 3   Evaluate $-3 + (-2)$ using the number line.

Solution: Start at 0. Since both numbers being added are negative, both arrows will be drawn to the left. See Figure 1.15.

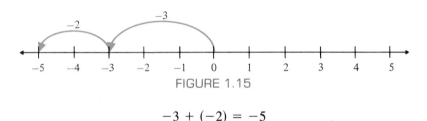

FIGURE 1.15

$$-3 + (-2) = -5$$ ■

In Example 3, we can think of the expression $-3 + (-2)$ as combining a *loss* of 3 and a *loss* of 2 to have a total *loss* of 5, or -5.

EXAMPLE 4   Add $5 + (-5)$.

Solution: The first arrow starts at 0 and is drawn 5 units to the right. The second arrow starts at 5 and is drawn 5 units to the left. The tip of the second arrow is at 0. Thus $5 + (-5) = 0$. See Figure 1.16.

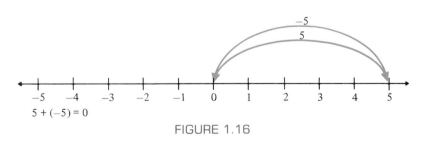

FIGURE 1.16

$$5 + (-5) = 0$$ ■

EXAMPLE 5   A submarine dives 250 feet. A short while later it dives an additional 190 feet. Find the depth of the submarine with respect to sea level (assume that depths below sea level are indicated by negative numbers).

Solution: It may be helpful to use a vertical number line (Fig. 1.17) to explain this problem.

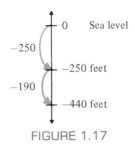

FIGURE 1.17

$$-250 + (-190) = -440 \text{ feet}$$ ■

▶ **2**

> Any two numbers whose sum is zero are said to be **opposites** (or **additive inverses**) of each other. In general, if we let a represent any real number, then its opposite is $-a$ and $a + (-a) = 0$.

   In Example 4 the sum of 5 and -5 is zero. Thus -5 is the opposite of 5 and 5 is the opposite of -5.

EXAMPLE 6   Find the opposite of each number.

   (a) 3   (b) -4

Solution: (a) The opposite of 3 is -3, since $3 + (-3) = 0$.
   (b) The opposite of -4 is 4, since $-4 + 4 = 0$.   ■

▶ **3**   Now that we have had some practice adding signed numbers on the number line, we will give a rule (in two parts) for using absolute value to add signed numbers. Remember that the absolute value of a nonzero number will always be positive.

> **To add real numbers with the same sign** (either both positive or both negative), add their absolute values. The sum has the same sign as the numbers being added.

EXAMPLE 7   Add $4 + 8$.

Solution: Since both numbers have the same sign, both positive, we add their absolute values: $|4| + |8| = 4 + 8 = 12$. Since both numbers being added are positive, the sum is positive. Thus $4 + 8 = 12$.   ■

EXAMPLE 8   Add $-6 + (-9)$.

Solution: Since both numbers have the same sign, both negative, we add their absolute values: $|-6| + |-9| = 6 + 9 = 15$. Since both numbers being added are negative, their sum is negative. Thus $-6 + (-9) = -15$.   ■

The sum of two positive numbers will always be positive and the sum of two negative numbers will always be negative.

> **To add two signed numbers with different signs** (one positive and the other negative), find the difference between the larger absolute value and the smaller absolute value. The answer has the sign of the number with the larger absolute value.

EXAMPLE 9   Add $10 + (-6)$.

Solution: The two numbers being added have different signs; thus we find the difference between the larger absolute value and the smaller: $|10| - |-6| = 10 - 6 = 4$. Since $|10|$ is greater than $|-6|$, and the sign of the 10 is positive, the sum is positive. Thus $10 + (-6) = 4$.   ■

EXAMPLE 10   Add $12 + (-18)$.

Solution: The numbers being added have different signs; thus we find the difference between the larger absolute value and the smaller: $|-18| - |12| = 18 - 12 = 6$. Since $|-18|$ is greater than $|12|$, and the sign of the -18 is negative, the sum is negative. Thus $12 + (-18) = -6$.   ■

EXAMPLE 11   Add $-24 + 19$.

Solution: The two numbers being added have different signs; thus we find the difference between the larger absolute value and the smaller: $|-24| - |19| = 24 - 19 = 5$. Since $|-24|$ is greater than $|19|$, the sum is negative. Therefore $-24 + 19 = -5$.   ■

The sum of two signed numbers with different signs may be either positive or negative. The sign of the sum will be the same as the sign of the number with the larger absolute value.

EXAMPLE 12   The ABC Company had a loss of $4000 for the first 6 months of the year and a profit of $15,500 for the second 6 months of the year. Find the net profit or loss for the year.

Solution: This problem can be represented as $-4000 + 15,500$. Since the two numbers being added have different signs, find the difference between the larger absolute value and the smaller.

$$|15,500| - |-4000| = 15,500 - 4000 = 11,500$$

Thus the net profit for the year was $11,500.   ■

Exercise Set 1.5

State the opposite of each number.

1. 18   **2.** -7 **3.** -32 **4.** 3

5. 0   **6.** 6   **7.** $\dfrac{5}{3}$ **8.** $-\dfrac{1}{2}$

9. $\dfrac{3}{5}$ **10.** -1 **11.** 0.63   **12.** -0.721

13. $2\dfrac{1}{2}$ **14.** $-3\dfrac{1}{4}$ **15.** -3.1 **16.** 5.26

Add as indicated.

17. $4 + 3$ **18.** $-4 + 3$ **19.** $4 + (-3)$ **20.** $4 + (-2)$
21. $-4 + (-2)$ **22.** $-3 + (-5)$ **23.** $6 + (-6)$ **24.** $-6 + 6$
25. $-4 + 4$ **26.** $-3 + 5$ **27.** $-8 + (-2)$ **28.** $6 + (-5)$
29. $-3 + 3$ **30.** $-8 + 2$ **31.** $-3 + (-7)$ **32.** $0 + (-3)$
33. $0 + 0$ **34.** $0 + (-0)$ **35.** $-6 + 0$ **36.** $-9 + 13$
37. $22 + (-19)$ **38.** $-13 + (-18)$ **39.** $-45 + 36$ **40.** $40 + (-25)$
41. $18 + (-9)$ **42.** $-7 + 7$ **43.** $-14 + (-13)$ **44.** $-27 + (-9)$
45. $-35 + (-9)$ **46.** $34 + (-12)$ **47.** $4 + (-30)$ **48.** $-16 + 9$
49. $-35 + 40$ **50.** $-12 + 17$ **51.** $180 + (-200)$ **52.** $-33 + (-92)$
53. $-105 + 74$ **54.** $183 + (-183)$ **55.** $184 + (-93)$ **56.** $-42 + 129$
57. $-452 + 312$ **58.** $-94 + (-98)$

Answer true or false.

59. The sum of two negative numbers is always a negative number.

60. The sum of two positive numbers is never a negative number.

61. The sum of a positive number and a negative number is always a positive number.

62. The sum of a negative number and a positive number is sometimes a negative number.

63. The sum of a positive number and a negative number is always a negative number.

Set up an expression that can be used to solve each problem and then solve.

64. Mr. Thorp owed $38 on his bank credit card. He charged another item costing $121. Find the amount that Mr. Thorp owed the bank.

65. Mr. Weber charged $193 worth of goods on his charge card. Find his balance after he made a payment of $112.

66. Mrs. Petrie paid $1424 in federal income tax. When she was audited, Mrs. Petrie had to pay an additional $503. What was her total tax?

67. Mr. Vela hiked down a canyon at Bryce Canyon National Park, a distance of 940 meters. He climbed back up 486 meters and then rested. Find his distance from the rim of the canyon.

68. A car accelerates to a speed of 60 miles per hour. It then decelerates by 20 miles per hour. Find the speed.

See Exercise 67.

69. An airplane at an altitude of 2400 feet above sea level drops a package into the ocean. The package settles at a point 200 feet below sea level. How far did the object fall?

70. A football team loses 18 yards on one play and then loses 3 yards on the following play. What is the total loss in yardage?

✎ **71.** Explain in your own words how to add two numbers with like signs.

✎ **72.** Explain in your own words how to add two numbers with unlike signs.

Cumulative Review Exercises

[1.1] **73.** Multiply $(\frac{3}{5})(1\frac{2}{3})$.

74. Subtract $3 - \frac{5}{16}$

[1.3] *Insert either* $<$, $>$, *or* $=$ *in the shaded area to make the statement true.*

75. $|-3| \quad 2$

76. $8 \quad |-7|$

1.6

Subtraction of Real Numbers

▶ **1** Subtract real numbers.

▶ **1** Any subtraction problem can be rewritten as an addition problem using the additive inverse.

Subtraction of Real Numbers

In general, if a and b represent any two real numbers, then

$$a - b = a + (-b)$$

This rule says that to subtract b from a, add the opposite or additive inverse of b to a.

EXAMPLE 1 Evaluate $9 - (+4)$.

Solution: In this example we are subtracting a positive 4 from 9. To accomplish this using the rule just given, we must add the opposite of $+4$, which is -4, to 9.

$$9 - (+4) = 9 + (-4) = 5$$

subtract   positive 4   add   negative 4

The $9 + (-4)$ was evaluated using the procedures for *adding* real numbers presented in the previous section. ∎

Often in a subtraction problem, when the number being subtracted is a positive number, the $+$ sign preceding the number being subtracted is not illustrated. For example, in the subtraction $9 - 4$,

$$9 - 4 \text{ means } 9 - (+4)$$

Thus, to evaluate $9 - 4$, we must add the opposite of 4 (or $+4$), which is -4, to 9.

$$9 - 4 = 9 + (-4) = 5$$

subtract   positive 4   add   negative 4

This procedure is illustrated in Example 2.

EXAMPLE 2   Evaluate $5 - 3$.

Solution: We must subtract a positive 3 from 5. To change this problem to an addition problem, add the opposite of 3, which is -3, to 5.

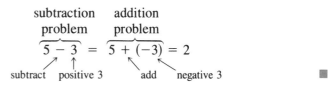

EXAMPLE 3   Evaluate $4 - 9$.

Solution: Add the opposite of 9, -9, to 4.

$$4 - 9 = 4 + (-9) = -5$$ ■

EXAMPLE 4   Evaluate $-4 - 2$.

Solution: Add the opposite of 2, -2, to -4.

$$-4 - 2 = -4 + (-2) = -6$$ ■

EXAMPLE 5   Evaluate $4 - (-2)$.

Solution: We are asked to subtract a negative 2 from 4. To do this, add the opposite of -2, 2, to 4.

$$4 - (-2) = 4 + 2 = 6$$

subtract negative 2   add   positive 2   ■

EXAMPLE 6   Evaluate $-6 - (-3)$.

Solution: Add the opposite of -3, 3, to -6.

$$-6 - (-3) = -6 + 3 = -3$$ ■

EXAMPLE 7   Subtract 12 from 3.

Solution: $3 - 12 = 3 + (-12) = -9$.   ■

EXAMPLE 8   Subtract 5 from 5.

Solution: $5 - 5 = 5 + (-5) = 0$.   ■

EXAMPLE 9   Subtract -6 from 4.

Solution: $4 - (-6) = 4 + 6 = 10$.   ■

HELPFUL HINT

By examining Example 9, we see that

$$4 - (-6) = 4 + 6$$

two negative   + sign
signs in a row

Whenever we *subtract a negative number*, we can replace the two negative signs in a row with a plus sign.

EXAMPLE 10   Evaluate: (a) $8 - (-5)$ (b) $-3 - (-9)$.

Solution: (a) $8 - (-5) = 8 + 5 = 13$.
(b) $-3 - (-9) = -3 + 9 = 6$.   ∎

EXAMPLE 11   Mary Jo Morin's checkbook indicated a balance of $125 before she wrote a check for $183. Find the balance in her checkbook.

Solution: $125 - 183 = 125 + (-183) = -58$. The negative indicates a deficit. Therefore, Mary Jo has a deficit of $58.   ∎

EXAMPLE 12   Janet made $4200 in the stock market, while Peter lost $3000. How much farther ahead is Janet than Peter financially?

Solution: Janet's gain is represented as a positive number. Peter's loss is represented as a negative number.

$$4200 - (-3000) = 4200 + 3000 = 7200$$

Janet is therefore $7200 ahead of Peter financially.   ∎

EXAMPLE 13   Evaluate each of the following:
(a) $12 + (-4)$ (b) $-16 - 3$ (c) $5 + (-4)$
(d) $6 - (-5)$ (e) $-12 - (-3)$ (f) $8 - 13$

Solution: Parts (a) and (c) are addition problems, while the other parts are subtraction problems. We can rewrite each subtraction problem as an addition problem to evaluate.
(a) $12 + (-4) = 8$ (b) $-16 - 3 = -16 + (-3) = -19$
(c) $5 + (-4) = 1$ (d) $6 - (-5) = 6 + 5 = 11$
(e) $-12 - (-3) = -12 + 3 = -9$ (f) $8 - 13 = 8 + (-13) = -5$ ∎

In the following Helpful Hint, the term "expression" is used. An **expression** is a general term for any collection of numbers, letters (called variables), grouping symbols (parentheses or brackets), and operations. Some examples of expressions are $5 - 3$, $2a + 3b$, $x + 2$, and $3(x + y)$. We will discuss expressions further in Chapter 2.

HELPFUL HINT

In earlier examples, we changed subtraction problems to addition problems. We did this because we know how to add real numbers. Later in the text, when we work out a subtraction problem, we will not show this part of the process in obtaining the answer. You need to practice and thoroughly understand how to add and subtract real numbers. You should understand this material so well that, when asked to evaluate an expression like $-4 - 6$, you will be able to compute the answer mentally. You should understand that $-4 - 6$ means the same as $-4 + (-6)$, but you should not need to write the addition to find the value of the expression, -10.

Let us evaluate a few subtraction problems without showing the process of changing the subtraction to addition.

EXAMPLE 14   Evaluate each of the following:
(a) $-7 - 5$ (b) $4 - 12$ (c) $18 - 25$ (d) $-20 - 12$.

Solution: (a) $-7 - 5 = -12$ (b) $4 - 12 = -8$ (c) $18 - 25 = -7$
(d) $-20 - 12 = -32$ ■

In Example 14 we may have reasoned that $-7 - 5$ meant $-7 + (-5)$, which is -12, but we did not need to show it.

In evaluating expressions involving more than one addition and subtraction, work from left to right unless grouping symbols are involved.

EXAMPLE 15   Evaluate each of the following:

(a) $-6 - 12 - 4$ (b) $-3 + 1 - 7$ (c) $8 - 10 + 2.$

Solution: We work from left to right.

(a) $\underbrace{-6 - 12} - 4$ (b) $\underbrace{-3 + 1} - 7$ (c) $\underbrace{8 - 10} + 2$

$= \quad -18 \quad - 4$ $= \quad -2 \quad - 7$ $= \quad -2 \quad + 2$

$= -22$ $= -9$ $= 0$ ■

After this section you will generally not see an expression like $3 + (-4)$. Instead the expression will be written as $3 - 4$. Recall that $3 - 4$ means $3 + (-4)$ by our definition of subtraction. **Whenever we see an expression of the form $a + (-b)$, we can write the expression as $a - b$.** For example, $12 + (-15)$ can be written $12 - 15$ and $-6 + (-9)$ can be written $-6 - 9$.

EXAMPLE 16   Evaluate each of the following:

(a) $-3 - (-4) + (-10) + (-5)$ (b) $-3 - (-4) - 10 - 5.$

Solution: (a) Again we work from left to right.

$$-3 - (-4) + (-10) + (-5) = -3 + 4 + (-10) + (-5)$$
$$= 1 + (-10) + (-5)$$
$$= -9 + (-5)$$
$$= -14$$

(b) This part is really the same problem as part (a), since $+(-10)$ can be written -10 and $+(-5)$ can be written -5.

$$-3 - (-4) - 10 - 5 = -3 + 4 - 10 - 5$$
$$= 1 - 10 - 5$$
$$= -9 - 5$$
$$= -14$$ ■

Exercise Set 1.6

Evaluate each expression.

1. $6 - 3$	**2.** $-6 - 4$	**3.** $4 - 5$	**4.** $5 - 3$
5. $3 - 3$	**6.** $-4 - 2$	**7.** $(-7) - (-4)$	**8.** $-4 - (-3)$
9. $-3 - 3$	**10.** $-4 - 4$	**11.** $3 - (-3)$	**12.** $4 - 4$
13. $0 - 6$	**14.** $6 - 6$	**15.** $0 - (-6)$	**16.** $9 - (-3)$
17. $-3 - 5$	**18.** $-5 - (-3)$	**19.** $-5 + 7$	**20.** $-7 - 9$
21. $5 - 3$	**22.** $5 - 12$	**23.** $6 - (-3)$	**24.** $6 - 10$
25. $8 - 8$	**26.** $-8 - 8$	**27.** $-8 - 10$	**28.** $4 - 12$
29. $-5 - (-3)$	**30.** $7 - 9$	**31.** $(-4) - (-4)$	**32.** $15 - 8$

33. $6 - 6$

34. $(-8) - (-12)$

35. $8 - 8$

36. $-6 - (-2)$

37. $4 - 5$

38. $-9 - 2$

39. $-2 - 3$

40. $9 - (-12)$

41. $-25 - 16$

42. $-20 - (-15)$

43. $37 - 40$

44. $40 - 37$

45. $-100 - 80$

46. $80 - 100$

47. $-20 - 90$

48. $-50 - (-40)$

49. $70 - (-70)$

50. $130 - (-90)$

51. $87 - 87$

52. $93 - (-93)$

53. $-45 - 37$

54. $-53 - (-7)$

55. Subtract 4 from 9.

56. Subtract 9 from 4.

57. Subtract 3 from -15.

58. Subtract -4 from -5.

59. Subtract 8 from -8.

60. Subtract 10 from -20.

61. Subtract 8 from 18.

62. Subtract 5 from -5.

63. Subtract -3 from -5.

64. Subtract 10 from -3.

65. Subtract -4 from 9.

66. Subtract 18 from -18.

67. Subtract 18 from 18.

68. Subtract 5 from 5.

69. Subtract 12 from 8.

70. Subtract -9 from 12.

71. Subtract -15 from -4.

72. Subtract -12 from 3.

73. Subtract -36 from 45.

74. Subtract 17 from -12.

Evaluate each expression.

75. $6 + 5 - (+4)$

76. $9 - (+6) - (+5)$

77. $-3 + (-4) + 5$

78. $9 - 7 + (-2)$

79. $-13 - (+5) + 3$

80. $7 - (+4) - (-3)$

81. $-9 - (-3) + 4$

82. $15 + (-7) - (-3)$

83. $5 - (+3) + (-2)$

84. $12 + (-5) - (-4)$

85. $25 + (+12) - (-6)$

86. $-7 + 6 - 3$

87. $-4 - 7 + 5$

88. $20 - 4 - 25$

89. $-4 + 7 - 12$

90. $-36 - 5 + 9$

91. $45 - 3 - 7$

92. $-2 + 7 - 9$

93. $-9 - 4 - 8$

94. $25 - 19 + 27$

95. $-4 - 13 + 5$

96. $(-4) + (-3) + 5 - 7$

97. $-9 - 3 - (-4) + 5$

98. $17 + (-3) - 9 - (-7)$

99. $32 + 5 - 7 - 12$

100. $-19 + (-3) - (-5) - (-2)$

101. $-7 - 4 - 3 + 5$

102. $19 + 4 - 20 - 25$

Solve the following problems.

103. An airplane is 2000 feet above sea level. A submarine is 1500 feet below sea level. How far above the submarine is the airplane?

104. A Jeep travels 162 miles due east. It then turns around and travels 83 miles due west. What is the Jeep's distance from its starting point?

105. The highest point on Earth, Mt. Everest, is 29,028 feet above sea level. The lowest point on Earth, the Marianas Trench, is 36,198 feet below sea level. How far above the Marianas Trench is the top of Mr. Everest?

106. The greatest change in temperature within a 24-hour period occurred at Browning, Montana, on January 23, 1916. The temperature fell from 44°F to $-56°F$. How much did the temperature drop?

107. **(a)** Will the statement $a + (-b) = a - b$ be true for all real numbers a and b?

(b) If $a = -3$ and $b = 5$, determine if $a + (-b) = a - b$.

See Exercise 104.

108. **(a)** Explain in your own words how to subtract -2 from 6.

(b) Subtract -2 from 6 following the procedure given in part (a).

109. **(a)** Explain in your own words how to subtract 6 from -9.

(b) Subtract 6 from -9 using the procedure given in part (a).

Cumulative Review Exercises _____

[1.3] **110.** List the set of integers.

✎ **111.** Explain the relationship between the set of rational numbers, the set of irrational numbers, and the set of real numbers.

[1.4] *Insert either* >, <, *or* = *in the shaded area to make each statement true.*

112. $|-3| \quad -5$

113. $|-6| \quad |-7|$

JUST FOR FUN _____

Find the sum.

1. $1 - 2 + 3 - 4 + 5 - 6 + 7 - 8 + 9 - 10$

2. $1 - 2 + 3 - 4 + 5 - 6 + \cdots + 99 - 100$

3. $-1 + 2 - 3 + 4 - 5 + 6 - \cdots - 99 + 100$

1.7

Multiplication and Division of Real Numbers

▸ **1** Multiply real numbers.

▸ **2** Divide real numbers.

▸ **3** Remove negative signs from denominators.

▸ **4** Learn the differences between $\frac{1}{0}$, $\frac{0}{1}$, and $\frac{0}{0}$.

Multiplication of Real Numbers

▸ **1** The following rules are used in determining the sign of the product when two numbers are multiplied.

> **Multiplication of Real Numbers**
> _____
> **1.** The product of two numbers with **like signs** is a **positive number.**
> **2.** The product of two numbers with **unlike** signs is a **negative number.**

By this rule, the product of two positive numbers or two negative numbers will be a positive number. The product of a positive number and a negative number will be a negative number.

EXAMPLE 1   Evaluate $3(-5)$.

Solution: Since the numbers have unlike signs, the product is negative.

$$3(-5) = -15$$

EXAMPLE 2   Evaluate $(-6)(7)$.

Solution: Since the numbers have unlike signs, the product is negative.

$$(-6)(7) = -42$$

EXAMPLE 3   Evaluate $(-7)(-5)$.

Solution: Since the numbers have like signs, both negative, the product is positive.

$$(-7)(-5) = 35$$ ∎

EXAMPLE 4   Evaluate each expression.
(a) $-6 \cdot 3$ (b) $(-4)(-8)$ (c) $4(-9)$
(d) $0 \cdot 4$ (e) $0(-2)$ (f) $-3(-6)$

Solution: (a) $-6 \cdot 3 = -18$ (b) $(-4)(-8) = 32$ (c) $4(-9) = -36$
(d) $0 \cdot 4 = 0$ (e) $0(-2) = 0$ (f) $-3(-6) = 18$

Note that zero times any real number equals zero.   ∎

EXAMPLE 5   Multiply $\left(\dfrac{-1}{8}\right)\left(\dfrac{-3}{5}\right)$.

Solution: $\left(\dfrac{-1}{8}\right)\left(\dfrac{-3}{5}\right) = \dfrac{(-1)\cdot(-3)}{8\cdot 5} = \dfrac{3}{40}$.   ∎

EXAMPLE 6   Evaluate $\left(\dfrac{3}{20}\right)\left(\dfrac{-3}{10}\right)$.

Solution: $\left(\dfrac{3}{20}\right)\left(\dfrac{-3}{10}\right) = \dfrac{3(-3)}{20\cdot 10} = \dfrac{-9}{200}$.   ∎

Sometimes you may be asked to perform more than one multiplication in a given problem. When this happens, the sign of the final product can be determined by counting the number of *negative* numbers being multiplied. **The product of an even number of negative numbers will always be positive. The product of an odd number of negative numbers will always be negative.** Can you explain why?

EXAMPLE 7   Evaluate $(-2)(3)(-2)(-1)$.

Solution: Since there are three negative numbers (an odd number of negatives), the product will be negative, as illustrated.

$$
\begin{aligned}
(-2)(3)(-2)(-1) &= (-6)(-2)(-1) \\
&= (12)(-1) \\
&= -12
\end{aligned}
$$ ∎

EXAMPLE 8   Evaluate $(-3)(2)(-1)(-2)(-4)$.

Solution: Since there are four negative numbers (an even number), the product will be positive.

$$
\begin{aligned}
(-3)(2)(-1)(-2)(-4) &= (-6)(-1)(-2)(-4) \\
&= (6)(-2)(-4) \\
&= (-12)(-4) \\
&= 48
\end{aligned}
$$ ∎

Division of Real
Numbers

▶ **2** The rules for dividing numbers are very similar to those used in multiplying numbers.

Division of Real Numbers

---

1. The quotient of two numbers with **like** signs is a **positive** number.
2. The quotient of two numbers with **unlike** signs is a **negative** number.

Therefore, the quotient of two positive numbers or two negative numbers will be a positive number. The quotient of a positive and a negative number will be a negative number.

EXAMPLE 9   Evaluate $\dfrac{20}{-5}$.

Solution: Since the numbers have unlike signs, the quotient is negative.

$$\frac{20}{-5} = -4$$

EXAMPLE 10   Evaluate $\dfrac{-36}{4}$.

Solution: Since the numbers have unlike signs, the quotient is negative.

$$\frac{-36}{4} = -9$$

EXAMPLE 11   Evaluate $\dfrac{-30}{-5}$.

Solution: Since the numbers have like signs, both negative, the quotient is positive.

$$\frac{-30}{-5} = 6$$

EXAMPLE 12   Evaluate $-16 \div (-2)$.

Solution: $\dfrac{-16}{-2} = 8.$

EXAMPLE 13   Evaluate $\dfrac{-2}{3} \div \dfrac{-5}{7}$.

Solution: Invert the *divisor*, $\dfrac{-5}{7}$, and then multiply.

$$\frac{-2}{3} \div \frac{-5}{7} = \left(\frac{-2}{3}\right)\left(\frac{7}{-5}\right)$$

$$= \frac{-14}{-15}$$

$$= \frac{14}{15}$$

▶ **3**　We now know that the quotient of a positive and a negative number is a negative number. The fractions $-\frac{3}{4}$, $\frac{-3}{4}$, and $\frac{3}{-4}$ all represent the same negative number, negative three-fourths.

If a and b represent any real numbers, $b \neq 0$, then

$$\frac{a}{-b} = \frac{-a}{b} = -\frac{a}{b}$$

In mathematics we generally do not write a fraction with a negative sign in the denominator. When a negative sign appears in a denominator, we can move it to the numerator or place it in front of the fraction. For example, the fraction $\frac{5}{-7}$ should be written as either $-\frac{5}{7}$ or $\frac{-5}{7}$.

EXAMPLE 14　Evaluate $\dfrac{2}{5} \div \dfrac{-8}{15}$.

Solution:　$\dfrac{2}{5} \div \dfrac{-8}{15} = \dfrac{\overset{1}{\cancel{2}}}{\underset{1}{\cancel{5}}} \cdot \dfrac{\overset{3}{\cancel{15}}}{\underset{4}{-\cancel{8}}}$

$$= \frac{1(3)}{1(-4)} = \frac{3}{-4} = -\frac{3}{4} \qquad ■$$

HELPFUL HINT

For multiplication and division of real numbers:

$\left.\begin{array}{l} (+)(+) = + \\[6pt] (-)(-) = + \end{array}\right\}$　$\left.\begin{array}{l} \dfrac{(+)}{(+)} = + \\[6pt] \dfrac{(-)}{(-)} = + \end{array}\right\}$　Like signs give positive products and quotients.

$\left.\begin{array}{l} (+)(-) = - \\[6pt] (-)(+) = - \end{array}\right\}$　$\left.\begin{array}{l} \dfrac{(+)}{(-)} = - \\[6pt] \dfrac{(-)}{(+)} = - \end{array}\right\}$　Unlike signs give negative products and quotients

---

COMMON STUDENT ERROR

At this point some students begin confusing problems like $-2 - 3$ with $(-2)(-3)$ and problems like $2 - 3$ with problems like $2(-3)$. If you do not understand the difference between problems like $-2 - 3$ and $(-2)(-3)$, make an appointment to see your instructor as soon as possible.

Subtraction Problems	*Multiplication Problems*
$-2 - 3 = -5$	$(-2)(-3) = 6$
$2 - 3 = -1$	$(2)(-3) = -6$

The operations on real numbers are summarized in Table 1.1.

TABLE 1.1. Summary of Operations on Real Numbers

Signs of Numbers	Addition	Subtraction	Multiplication	Division
Both Numbers Are Positive	Sum Is Always Positive	Difference May Be Either Positive or Negative	Product Is Always Positive	Quotient Is Always Positive
Examples				
6 and 2	$6 + 2 = 8$	$6 - 2 = 4$	$6 \cdot 2 = 12$	$6 \div 2 = 3$
2 and 6	$2 + 6 = 8$	$2 - 6 = -4$	$2 \cdot 6 = 12$	$2 \div 6 = \frac{1}{3}$
One Number Is Positive and the Other Number Is Negative	Sum May Be Either Positive or Negative	Difference May Be Either Positive or Negative	Product Is Always Negative	Quotient Is Always Negative
Examples				
6 and -2	$6 + (-2) = 4$	$6 - (-2) = 8$	$6(-2) = -12$	$6 \div (-2) = -3$
-6 and 2	$-6 + 2 = -4$	$-6 - (2) = -8$	$-6(2) = -12$	$-6 \div 2 = -3$
Both Numbers Are Negative	Sum Is Always Negative	Difference May Be Either Positive or Negative	Product Is Always Positive	Quotient Is Always Positive
Examples				
-6 and -2	$-6 + (-2) = -8$	$-6 - (-2) = -4$	$-6(-2) = 12$	$-6 \div (-2) = 3$
-2 and -6	$-2 + (-6) = -8$	$-2 - (-6) = 4$	$-2(-6) = 12$	$-2 \div (-6) = \frac{1}{3}$

▶ **4** Now let us look at division involving the number 0. What is $\frac{0}{1}$ equal to? Note that $\frac{6}{3} = 2$ because $3 \cdot 2 = 6$. We can follow the same procedure to determine the value of $\frac{0}{1}$. Suppose that $\frac{0}{1}$ is equal to some number, which we will designate by a question mark.

$$\text{If } \frac{0}{1} = ? \quad \text{then} \quad 1 \cdot ? = 0$$

Since only $1 \cdot 0 = 0$, the ? must be 0. Thus $\frac{0}{1} = 0$. Using the same technique, we can show that zero divided by any nonzero number is zero.

$$\frac{0}{a} = 0, \qquad a \neq 0$$

What is $\frac{1}{0}$ equal to?

$$\text{If } \frac{1}{0} = ? \quad \text{then} \quad 0 \cdot ? = 1$$

But since 0 multiplied by any number will be 0, there is no value that can replace ?. We say that $\frac{1}{0}$ is *undefined*. Using the same technique, we can show that any real number, except 0, divided by 0 is undefined.

$$\frac{a}{0} \text{ is } \textbf{undefined,} \qquad a \neq 0$$

What is $\frac{0}{0}$ equal to?

$$\text{If } \frac{0}{0} = ? \quad \text{then} \quad 0 \cdot ? = 0$$

But since the product of any number and 0 is 0, the ? can be replaced by any real number. For this reason we say that $\frac{0}{0}$ is *indeterminate*.

$$\frac{0}{0} \text{ is \textbf{indeterminate}}$$

Summary of Division Involving Zero

$$\frac{0}{a} = 0, a \neq 0 \qquad \frac{a}{0} \text{ is undefined, } a \neq 0 \qquad \frac{0}{0} \text{ is indeterminate}$$

Exercise Set 1.7

Find the product.

1. $(-4)(-3)$ **2.** $-4 \cdot 2$ **3.** $3(-3)$ **4.** $6(-2)$

5. $(-4)(8)$ **6.** $(-3)(2)$ **7.** $9(-1)$ **8.** $-1(8)$

9. $-4(-3)$ **10.** $0(4)$ **11.** $-9(-4)$ **12.** $(-12)(-3)$

13. $-6 \cdot 5$ **14.** $-9(-3)$ **15.** $5(-12)$ **16.** $(-9)(-8)$

17. $-4(0)$ **18.** $0(8)$ **19.** $(-4)(-4)$ **20.** $(-6)(-6)$

21. $8 \cdot 3$ **22.** $-4(-6)$ **23.** $5(-3)$ **24.** $-4 \cdot 7$

25. $8(12)$ **26.** $(-5)(-6)$ **27.** $-9(-9)$ **28.** $(15)(-4)$

29. $-2(5)$ **30.** $6(-12)$ **31.** $(-6)(2)(-3)$ **32.** $5(-2)(-8)$

33. $0(3)(8)$ **34.** $2(-3)(7)$ **35.** $(-1)(-1)(-1)$ **36.** $2(4)(-2)(-5)$

37. $-5(-3)(8)(-1)$ **38.** $(-3)(-4)(-5)(-1)$ **39.** $4(3)(1)(-1)$

40. $(-3)(2)(5)(3)$ **41.** $(-4)(3)(-7)(1)$ **42.** $(-1)(3)(0)(-7)$

Find the product.

43. $\left(\frac{-1}{2}\right)\left(\frac{3}{5}\right)$ **44.** $\left(\frac{2}{3}\right)\left(\frac{-3}{5}\right)$ **45.** $\left(\frac{-8}{9}\right)\left(\frac{-7}{12}\right)$ **46.** $\left(\frac{-5}{12}\right)\left(\frac{-6}{11}\right)$

47. $\left(\frac{6}{-3}\right)\left(\frac{4}{-2}\right)$ **48.** $\left(\frac{8}{-11}\right)\left(\frac{6}{-5}\right)$ **49.** $\left(\frac{5}{-7}\right)\left(\frac{6}{8}\right)$ **50.** $\left(\frac{9}{10}\right)\left(\frac{7}{-8}\right)$

Find the quotient.

51. $\frac{6}{2}$ **52.** $9 \div (-3)$ **53.** $-16 \div (-4)$ **54.** $\frac{-24}{8}$

55. $\frac{-36}{-9}$ **56.** $-45 \div 5$ **57.** $\frac{-16}{4}$ **58.** $\frac{36}{-2}$

59. $\frac{18}{-1}$ **60.** $\frac{-12}{-1}$ **61.** $-15 \div (-3)$ **62.** $12 \div (-6)$

63. $\frac{-6}{-1}$ **64.** $\frac{60}{-12}$ **65.** $\frac{-25}{-5}$ **66.** $\frac{36}{-4}$

67. $\dfrac{1}{-1}$ **68.** $\dfrac{-1}{1}$ **69.** $\dfrac{-48}{12}$ **70.** $\dfrac{50}{-5}$

71. $\dfrac{-18}{-2}$ **72.** $\dfrac{100}{-5}$ **73.** $\dfrac{0}{-1}$ **74.** $-200 \div (-20)$

75. $(-30) \div (-30)$ **76.** $(-180) \div 20$ **77.** Divide 0 by 3.

78. Divide -16 by -2.   **79.** Divide 20 by -5.   **80.** Divide 30 by -10.

81. Divide -30 by -10.   **82.** Divide -180 by 30.   **83.** Divide -60 by 5.

84. Divide -25 by -5.   **85.** Divide 80 by -20.   **86.** Divide -60 by 12.

87. Divide -90 by -2.   **88.** Divide 125 by -25.

Find the quotient.

89. $\dfrac{5}{12} \div \left(\dfrac{-5}{9}\right)$ **90.** $(-3) \div \dfrac{5}{19}$ **91.** $\dfrac{3}{-10} \div (-8)$ **92.** $\dfrac{-4}{9} \div \left(\dfrac{-6}{7}\right)$

93. $\dfrac{-15}{21} \div \left(\dfrac{-15}{21}\right)$ **94.** $\dfrac{8}{-15} \div \left(\dfrac{-9}{10}\right)$ **95.** $(-12) \div \dfrac{5}{12}$ **96.** $\dfrac{-16}{3} \div \left(\dfrac{5}{-9}\right)$

97. $6 \div \left(\dfrac{-5}{6}\right)$ **98.** $-12 \div \left(\dfrac{-2}{3}\right)$

Indicate whether each of the following is 0, undefined, or indeterminate.

99. $0 \div 6$ **100.** $-4 \div 0$ **101.** $\dfrac{0}{0}$ **102.** $\dfrac{-2}{0}$

103. $\dfrac{0}{1}$ **104.** $0 \div (-2)$ **105.** $8 \div 0$ **106.** $\dfrac{0}{4}$

107. $\dfrac{0}{-6}$ **108.** $\dfrac{0}{-1}$

Answer true or false.

109. The product of two negative numbers is a negative number.

110. The product of a positive number and a negative number is a negative number.

111. The quotient of two negative numbers is a positive number.

112. The quotient of two numbers with unlike signs is a positive number.

113. The product of an even number of negative numbers is a positive number.

114. The product of an odd number of negative numbers is a negative number.

115. Zero divided by 0 is 1.

116. Six divided by 0 is 0.

117. Zero divided by 0 is 0.

118. Zero divided by 1 is undefined.

119. One divided by 0 is undefined.

120. Zero divided by 0 is indeterminate.

121. Write out the rules for determining the sign of the product or quotient of two numbers.

122. Explain why the product of an even number of negative numbers is a positive number.

123. Will the product of $(1)(-2)(3)(-4)(5)(-6) \cdots (33)(-34)$ be a positive number or a negative number? Explain how you determined your answer.

Cumulative Review Exercises

[1.1] **124.** Find the quotient $\dfrac{5}{7} \div \dfrac{1}{5}$.

[1.5] **125.** Subtract -18 from -20.

Evaluate each of the following.

126. $6 - 3 - 4 - 2$

127. $5 - (-2) + 3 - 7$

JUST FOR FUN

Find the quotient.

1. $\dfrac{1 - 2 + 3 - 4 + 5 - \cdots + 99 - 100}{1 - 2 + 3 - 4 + 5 - \cdots + 99 - 100}.$

2. $\dfrac{-1 + 2 - 3 + 4 - 5 + \cdots - 99 + 100}{1 - 2 + 3 + 4 + 5 - \cdots + 99 - 100}.$

1.8

An Introduction to Exponents

▶ **1** Identify exponents.

▶ **2** Evaluate expressions containing exponents.

▶ **3** Learn the difference between $-x^2$ and $(-x)^2$.

▶ **1** To understand certain topics in algebra, you must understand exponents. Exponents will be introduced in this section and will be discussed in more detail in Chapter 4.

In the expression 4^2, the 4 is called the **base,** and the 2 is called the **exponent.** The number 4^2 is read "4 squared" or "4 to the second power" and means

$$\underbrace{4 \cdot 4}_{\text{2 factors of 4}} = 4^2$$

The number 4^3 is read "4 cubed" or "4 to the third power" and means

$$\underbrace{4 \cdot 4 \cdot 4}_{\text{3 factors of 4}} = 4^3$$

In general, the number b to the nth power, written b^n, means

$$\underbrace{b \cdot b \cdot b \cdot \cdots \cdot b}_{n \text{ factors of } b} = b^n$$

Thus $b^4 = b \cdot b \cdot b \cdot b$ or $bbbb$ and $x^3 = x \cdot x \cdot x$ or xxx.

EXAMPLE 1 ▶ **2** Evaluate each expression.

(a) 3^2 (b) 2^5 (c) 1^5 (d) 4^3 (e) $(-3)^2$ (f) $(-2)^3$ (g) $\left(\frac{2}{3}\right)^2$

Solution   (a) $3^2 = 3 \cdot 3 = 9$
(b) $2^5 = 2 \cdot 2 \cdot 2 \cdot 2 \cdot 2 = 32$
(c) $1^5 = 1 \cdot 1 \cdot 1 \cdot 1 \cdot 1 = 1$ (1 raised to any real power equals 1; why?)
(d) $4^3 = 4 \cdot 4 \cdot 4 = 64$
(e) $(-3)^2 = (-3)(-3) = 9$
(f) $(-2)^3 = (-2)(-2)(-2) = -8$
(g) $\left(\frac{2}{3}\right)^2 = \left(\frac{2}{3}\right)\left(\frac{2}{3}\right) = \frac{4}{9}$ ■

COMMON STUDENT ERROR

Students should realize that $a^b \neq ba$.
2^5 means $2 \cdot 2 \cdot 2 \cdot 2 \cdot 2$, not $5 \cdot 2$. Thus $2^5 = 32$, and not 10.

Other examples of exponential notation are:

(a) $x \cdot x \cdot x \cdot x = x^4$ (b) $aabbb = a^2b^3$

(c) $x \cdot x \cdot y = x^2y$ (d) $aaabb = a^3b^2$

(e) $xyxx = x^3y$ (f) $xyzzy = xy^2z^2$

(g) $3 \cdot x \cdot x \cdot y = 3x^2y$ (h) $5xyyyy = 5xy^4$

(i) $3 \cdot 3 \cdot x \cdot x = 3^2x^2$ (j) $5 \cdot 5 \cdot 5 \cdot xxy = 5^3x^2y$

Notice in parts (e) and (f) that the order of the factors does not matter.

It is not necessary to write exponents of 1. Thus, when writing xxy, we write x^2y and not x^2y^1. **Whenever we see a letter or number without an exponent, we always assume that letter or number has an exponent of 1.**

EXAMPLE 2   Write each expression as a product of factors.

(a) x^2y (b) xy^3z (c) $3x^2yz^3$ (d) 2^3xy (e) $3^2x^3y^2$

Solution: (a) $x^2y = xxy$ (b) $xy^3z = xyyyz$ (c) $3x^2yz^3 = 3xxyzzz$

(d) $2^3xy = 2 \cdot 2 \cdot 2xy$ (e) $3^2x^3y^2 = 3 \cdot 3xxxyy$ ∎

▶ **3**   Note that **an exponent refers to only the number or letter that directly precedes it unless parentheses are used to indicate otherwise.** For example, in the expression $3x^2$, only the x is squared. In the expression $-x^2$ only the x is squared. To help explain this concept, we will write $-x^2$ as $-1x^2$. This can be done since any real number may be multiplied by 1 without affecting its value.

$$-x^2 = -1x^2$$

By looking at $-1x^2$ we can see that only the x is squared, not the -1. If the entire expression $-x$ was to be squared, we would need to use parentheses and write $(-x)^2$. Note the following difference:

$$-x^2 = -(x)(x)$$
$$(-x)^2 = (-x)(-x)$$

Consider the expressions -3^2 and $(-3)^2$. How do they differ?

$$-3^2 = -(3)(3) = -9$$
$$(-3)^2 = (-3)(-3) = 9$$

EXAMPLE 3   Evaluate each expression.

(a) -5^2 (b) $(-5)^2$ (c) -2^3 (d) $(-2)^3$

Solution: (a) $-5^2 = -(5)(5) = -25$ (b) $(-5)^2 = (-5)(-5) = 25$

(c) $-2^3 = -(2)(2)(2) = -8$ (d) $(-2)^3 = (-2)(-2)(-2) = -8$ ∎

EXAMPLE 4   Evaluate (a) -2^4 and (b) $(-2)^4$.

Solution: (a) $-2^4 = -(2)(2)(2)(2) = -16$ (b) $(-2)^4 = (-2)(-2)(-2)(-2) = 16$ ∎

EXAMPLE 5   Evaluate (a) x^2 and (b) $-x^2$ for $x = 3$.

Solution: Substitute 3 for x.

(a) $x^2 = 3^2 = 3 \cdot 3 = 9$ (b) $-x^2 = -3^2 = -(3)(3) = -9$ ∎

EXAMPLE 6   Evaluate (a) y^2 and (b) $-y^2$ for $y = -4$.

Solution: Substitute -4 for y.

(a) $y^2 = (-4)^2 = (-4)(-4) = 16$
(b) $-y^2 = -(-4)^2 = -(-4)(-4) = -16$ ∎

Note that $-x^2$ will always be a negative number for any nonzero value of x, and $(-x)^2$ will always be a positive number for any nonzero value of x. Can you explain why? See Exercises 86 and 87.

COMMON STUDENT ERROR

When asked to evaluate $-x^2$ for any real number x, many students will incorrectly treat $-x^2$ as $(-x)^2$.

Evaluate $-x^2$ when $x = 5$.

Correct   *Wrong*

$-5^2 = -(5)(5)$ $-5^2 = (-5)(-5)$

$= -25$ $= 25$

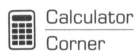 **Calculator Corner**

To evaluate 2^3 on the calculator, press

c 2 ✕ 2 ✕ 2 = 8

If your calculator contains a y^x key, use this key to evaluate 2^3 in the following manner:

c 2 y^x 3 = 8

To evaluate an exponential expression, key in the base, press the y^x key, then key in the exponent. After the = is pressed, the answer is displayed. To evaluate 4^8, use c 4 y^x 8 = 65536.

Exercise Set 1.8

Evaluate each expression.

1. 3^2	**2.** 4^2	**3.** 2^3	**4.** 1^5
5. 3^3	**6.** -5^2	**7.** 6^3	**8.** $(-2)^2$
9. $(-2)^3$	**10.** -3^4	**11.** $(-1)^3$	**12.** 6^2
13. 3^3	**14.** 2^5	**15.** -6^2	**16.** 5^3
17. $(-6)^2$	**18.** $(-3)^3$	**19.** 2^4	**20.** $(-3)^4$
21. 5^1	**22.** -3^2	**23.** $(-2)^4$	**24.** -1^4
25. -2^4	**26.** $(-1)^4$	**27.** $(-4)^3$	**28.** $3^2(4)^2$
29. $5^2 \cdot 3^2$	**30.** $(-1)^4(3)^3$	**31.** $5(4^2)$	**32.** $2^3 \cdot 5^1$
33. $2^1 \cdot 4^2$	**34.** $(-2)^4(-1)^3$	**35.** $3(-5^2)$	**36.** $9(-2)^2$
37. $3^2 \cdot 2^4$			

Express in exponential form.

38. $x \cdot x \cdot y \cdot y$ **39.** $x \cdot y \cdot z \cdot z$ **40.** $xyyyz$

41. $xxxxz$ **42.** $yyzzz$ **43.** $aabbab$

44. $xyxyz$ **45.** $x \cdot x \cdot y \cdot z \cdot z$ **46.** $a \cdot x \cdot a \cdot x \cdot y$

47. $x \cdot x \cdot x \cdot y \cdot y$ **48.** $x \cdot y \cdot y \cdot z \cdot z \cdot z$ **49.** $xyyyy$

50. $3xyy$ **51.** $5 \cdot 5yyz$ **52.** $2 \cdot 2 \cdot 2 \cdot xxyyy$

Express as a product of factors.

53. $x^2 y$ **54.** $y^2 z$ **55.** xy^3

56. $x^2 yz$ **57.** $xy^2 z^3$ **58.** $2x^2 y^2$

59. $3^2 yz$ **60.** $2^3 y^3$ **61.** $2^3 x^3 y$

62. $3^3 xy^3$ **63.** $(-2)^2 y^3 z$ **64.** $(-1)^2 x^3 y^2$

Evaluate (a) x^2 and (b) $-x^2$ for each of the following values of x.

65. 3   **66.** 2   **67.** 4

68. 1   **69.** -2 **70.** 5

71. 7   **72.** 8   **73.** -1

74. -5 **75.** $-\dfrac{1}{2}$ **76.** $\dfrac{3}{4}$

Answer true or false.

77. $(-4)^{20}$ is a negative number.

78. $(-4)^{19}$ is a negative number.

79. $-(-3)^{15}$ is a negative number.

80. $-(-2)^{14}$ is a negative number.

81. $x^2 y$ means $x^2 y^1$.

82. $3xy^4$ means $3^1 x^1 y^4$.

83. $2x^5 y$ means $2^1 x^5 y^1$.

84. When a number is written without an exponent, the exponent on the number is 0.

85. When a variable is written without an exponent, the exponent on the variable is 1.

86. Explain why $-x^2$ will always be a negative number for any nonzero value of x.

87. Explain why $(-x)^2$ will always be a positive number for any nonzero value of x.

88. Will the expression $(-6)^{15}$ be a positive or a negative number? Explain how you determine your answer.

89. Will the expression $(-1)^{100}$ be a positive or a negative number? Explain how you determined your answer.

Cumulative Review Exercises

[1.5] **90.** Subtract -6 from 12.

91. Evauate $-4 - 3 + 9 - 7$.

[1.6] *Evaluate each of the following.*

92. $\left(\dfrac{-5}{7}\right) \div \left(\dfrac{-3}{14}\right)$ **93.** $\dfrac{0}{4}$

JUST FOR FUN

Simplify each expression. Leave the answers in exponential form.

1. $2^2 \cdot 2^3$ **2.** $3^2 \cdot 3^3$ **3.** $x^m \cdot x^n$

4. $\dfrac{2^3}{2^2}$ **5.** $\dfrac{3^4}{3^2}$ **6.** $\dfrac{x^m}{x^n}$

7. $(2^3)^2$ **8.** $(3^3)^2$ **9.** $(x^m)^n$

10. $(2x)^2$ **11.** $(3x)^2$ **12.** $(ax)^2$

General rules that may be used to solve problems of this type will be discussed in Chapter 4.

1.9

Use of Parentheses and Order of Operations

▶ **1**   Learn the order of operations.

▶ **2**   Evaluate expressions for given values of the variable.

▶ **1**   Evaluate $2 + 3 \cdot 4$. Is it 20? Is it 14? To be able to answer questions of this type, we must know the order of operations to follow when evaluating a mathematical expression. You will often have to evaluate expressions containing multiple operations.

To Evaluate Mathematical Expressions, Use the Following Order

1. First, evaluate the information within **parentheses,** ( ), or brackets, [ ]. If the expression contains nested parentheses (one pair of parentheses within another pair), evaluate the information in the innermost parentheses first.
2. Next, evaluate all **exponents.**
3. Next, evaluate all **multiplications** or **divisions** in the order in which they occur, working from left to right.
4. Finally, evaluate all **additions** or **subtractions** in the order in which they occur, working from left to right.

We can now answer the question posed previously. Since multiplications are performed before additions,

$$2 + 3 \cdot 4 \quad \text{means} \quad 2 + (3 \cdot 4) = 2 + 12 = 14$$

Parentheses may be used (1) to change the order of operations to be followed in evaluating an algebraic expression or (2) to help clarify the understanding of an expression.

In the example, $2 + 3 \cdot 4$, if we wished to have the addition performed before the multiplication, we could indicate this by placing parentheses about the $2 + 3$:

$$(2 + 3) \cdot 4 = 5 \cdot 4 = 20$$

Consider the expression $1 \cdot 3 + 2 \cdot 4$. According to the order, multiplications are to be performed before additions. We can rewrite this expression as $(1 \cdot 3) + (2 \cdot 4)$. Note that the order of operations was not changed. The parentheses were used only to help clarify the order to be followed.

HELPFUL HINT	If parentheses are not used to change the order of operations, multiplications and divisions are always performed before additions and subtractions. When a problem has only multiplications and divisions, work from left to right. Similarly, when a problem has only additions and subtractions, work from left to right.

EXAMPLE 1   Evaluate $2 + 3 \cdot 5^2 - 7$.

Solution: Color shading is used to indicate the order in which the expression is to be evaluated.

$$2 + 3 \cdot 5^2 - 7$$
$$= 2 + 3 \cdot 25 - 7$$
$$= 2 + 75 - 7$$
$$= 77 - 7$$
$$= 70$$

EXAMPLE 2   Evaluate $6 + 3[(12 \div 4) + 5]$.

Solution:
$$6 + 3[(12 \div 4) + 5]$$
$$= 6 + 3[3 + 5]$$
$$= 6 + 3(8)$$
$$= 6 + 24$$
$$= 30$$

EXAMPLE 3   Evaluate $(4 \div 2) + 4(5 - 2)^2$.

Solution:
$$(4 \div 2) + 4(5 - 2)^2$$
$$= 2 + 4(3)^2$$
$$= 2 + 4 \cdot 9$$
$$= 2 + 36$$
$$= 38$$

EXAMPLE 4   Evaluate $5 + 2^2 \cdot 3 - 3^2$.

Solution:
$$5 + 2^2 \cdot 3 - 3^2$$
$$= 5 + 4 \cdot 3 - 9$$
$$= 5 + 12 - 9$$
$$= 17 - 9$$
$$= 8$$

EXAMPLE 5   Evaluate $-8 - 81 \div 9 \cdot 2^2 + 7$.

Solution:
$$-8 - 81 \div 9 \cdot 2^2 + 7$$
$$= -8 - 81 \div 9 \cdot 4 + 7$$
$$= -8 - 9 \cdot 4 + 7$$
$$= -8 - 36 + 7$$
$$= -44 + 7$$
$$= -37$$

EXAMPLE 6   Evaluate each expression.

(a) $-4^2 + 6 \div 3$ (b) $(-4)^2 + 6 \div 3$

Solution: (a) $\quad -4^2 + 6 \div 3$ (b) $\quad (-4)^2 + 6 \div 3$

$\qquad\qquad = -16 + 6 \div 3 \qquad\qquad = 16 + 6 \div 3$

$\qquad\qquad = -16 + 2 \qquad\qquad\quad\; = 16 + 2$

$\qquad\qquad = -14 \qquad\qquad\qquad\; = 18$ ■

EXAMPLE 7   Evaluate $\dfrac{3}{8} - \dfrac{2}{5} \cdot \dfrac{1}{12}$

Solution: First perform the multiplication.

$$\frac{3}{8} - \left(\frac{\overset{1}{2}}{5} \cdot \frac{1}{\underset{6}{12}}\right)$$

$$= \frac{3}{8} - \frac{1}{30}$$

$$= \frac{45}{120} - \frac{4}{120}$$

$$= \frac{41}{120}$$ ■

EXAMPLE 8   Write the following statements as mathematical expressions using parentheses and brackets and then evaluate: Multiply 5 by 3. To this product add 6. Multiply this sum by 7.

Solution:

$\qquad\qquad 5 \cdot 3 \qquad\qquad$ Multiply 5 by 3.

$\qquad\quad (5 \cdot 3) + 6 \qquad$ Add 6.

$\qquad 7[(5 \cdot 3) + 6] \qquad$ Multiply the sum by 7.

Now evaluate to determine the answer.

$$7[(5 \cdot 3) + 6]$$
$$= 7[15 + 6]$$
$$= 7(21)$$
$$= 147$$ ■

Sometimes brackets are used in place of parentheses to help avoid confusion. The preceding expression could have been written using only parentheses as $7((5 \cdot 3) + 6)$.

EXAMPLE 9   Write the following statements as mathematical expressions using parentheses and brackets and then evaluate: Subtract 3 from 15. Divide this difference by 2. Multiply this quotient by 4.

Solution:

$\qquad\qquad 15 - 3 \qquad\qquad$ Subtract 3 from 15.

$\qquad\quad (15 - 3) \div 2 \qquad$ Divide by 2.

$\qquad 4[(15 - 3) \div 2] \qquad$ Multiply the quotient by 4.

Now evaluate.
$$4[(15 - 3) \div 2]$$
$$= 4[12 \div 2]$$
$$= 4(6)$$
$$= 24$$ ■

▶ **2**   Now we will evaluate some expressions for given values of the variables.

EXAMPLE 10   Evaluate $7x - 2$ when $x = 2$.

Solution: Substitute 2 for each x in the expression.

$$7x - 2 = 7(2) - 2 = 14 - 2 = 12$$ ∎

EXAMPLE 11   Evaluate $(3x + 1) + 2x^2$ when $x = 4$.

Solution:
$$
\begin{aligned}
(3x + 1) + 2x^2 &= [3(4) + 1] + 2(4)^2 \\
&= [12 + 1] + 2(4)^2 \\
&= 13 + 2(16) \\
&= 13 + 32 \\
&= 45
\end{aligned}
$$ ∎

EXAMPLE 12   Evaluate $-y^2 + 3(x + 2) - 5$ when $x = -3$ and $y = -2$.

Solution:
$$
\begin{aligned}
-y^2 + 3(x + 2) - 5 &= -(-2)^2 + 3(-3 + 2) - 5 \\
&= -(-2)^2 + 3(-1) - 5 \\
&= -(4) + 3(-1) - 5 \\
&= -4 - 3 - 5 \\
&= -7 - 5 \\
&= -12
\end{aligned}
$$ ∎

Calculator Corner

We now know that $2 + 3 \times 4$ means $2 + (3 \times 4)$ and has a value of 14. What will a calculator display if you key in the following?

$$\boxed{c}\ 2\ \boxed{+}\ 3\ \boxed{\times}\ 4\ \boxed{=}$$

The answer depends on your calculator. *Scientific calculators* will evaluate an expression following the rules stated earlier in this section.

Scientific calculator: $\boxed{c}\ 2\ \boxed{+}\ 3\ \boxed{\times}\ 4\ \boxed{=}\ 14$

Nonscientific calculators will perform operations in the order they are entered.

Nonscientific calculator: $\boxed{c}\ 2\ \boxed{+}\ 3\ \boxed{\times}\ 4\ \boxed{=}\ 20$

Remember that in algebra, unless otherwise instructed by parentheses, we always perform multiplications and divisions before additions and subtractions.

Is your calculator a scientific calculator?

To calculate $2 + (3 \times 4)$ on a nonscientific calculator, we first enter the multiplication and then the addition, as follows:

$$\boxed{c}\ 3\ \boxed{\times}\ 4\ \boxed{+}\ 2\ \boxed{=}\ 14$$

Scientific calculators are not much more expensive than nonscientific calculators. You should give some thought to purchasing a scientific calculator, especially if you plan to take more mathematics or science courses.

Exercise Set 1.9

Evaluate each expression.

1. $3 + 4 \cdot 5$
2. $2 - 3^2 + 4$
3. $2 - 2 + 5$
4. $(6^2 \div 3) - (6 - 4)$
5. $1 + 3 \cdot 2^2$
6. $4 \cdot 3^2 - 2 \cdot 5$
7. $-3^2 + 5$
8. $(-2)^3 + 8 \div 4$
9. $(4 - 3) \cdot (5 - 1)^2$
10. $20 - 6 - 3 - 2$
11. $3 \cdot 7 + 4 \cdot 2$
12. $6 + 9(3 + 4)$
13. $[1 - (4 \cdot 5)] + 6$
14. $[12 - (4 \div 2)] - 5$
15. $4^2 - 3 \cdot 4 - 6$
16. $5 - 3 + 4^2 - 6$
17. $-3[-4 + (6 - 8)]$
18. $(-3)^2 + (3 - 4)^3 - 5$
19. $(6 \div 3)^3 + 4^2 \div 8$
20. $5^2 - 2^2(4 - 2)^2$
21. $-4^2 + 8 \div 2 \cdot 5 + 3$
22. $-4 - (-12 + 4) \div 2 + 1$
23. $3 + (4^2 - 10)^2 - 3$
24. $[-(1 - 4)^2]^2 + 9$
25. $-[12 - (-4 - 5)]^2$
26. $(-2)^2 + 4^2 \div 2^2 + 3$
27. $(3^2 - 1) \div (3 + 1)^2$
28. $-4(5 - 2)^2 + 5$
29. $2[(36 \div 9) + 1]$
30. $3[(4 + 6) \div 2]$
31. $2[3(8 - 2^2) - 6]$
32. $(13 + 5) - (4 - 2)^2$
33. $10 - [8 - (3 + 4)]^2$
34. $6 - 8 \cdot 2 \div 4 \div 2 + 5$
35. $[4 + ((5 - 2)^2 \div 3)^2]^2$
36. $2[((6 \div 3)^2 + 4)^2 - 3]$
37. $[-2(4 - 6)^2]^2 - 4[-3(6 \div 3)^2]$
38. $[7 - [3(8 \div 4)]^2 + 9 \cdot 4]^2$
39. $(14 \div 7 \cdot 7 \div 7 - 7)^2$
40. $2.5 + 7.56 \div 2.1 + (9.2)^2$
41. $(8.4 + 3.1)^2 - (3.64 - 1.2)$
42. $2[1.63 + 5(4.7)] - 3.15$
43. $(4.3)^2 + 2(5.3) - 3.05$
44. $\frac{2}{3} + \frac{3}{8} \cdot \frac{4}{5}$
45. $\left(\frac{2}{7} + \frac{3}{8}\right) - \frac{3}{112}$
46. $\left(\frac{5}{6} \cdot \frac{4}{5}\right) + \left(\frac{2}{3} \cdot \frac{5}{8}\right)$
47. $\frac{3}{4} - 4 \cdot \frac{5}{40}$
48. $\frac{2}{3} + 4 \div 3^2$
49. $2\left(3 + \frac{2}{5}\right) \div \left(\frac{3}{5}\right)^2$
50. $64 \cdot \frac{1}{2} \div 8 + \frac{3}{4}$

Write the following statements as mathematical expressions using parentheses and brackets and then evaluate.

51. Multiply 6 by 3. From this product, subtract 4. From this difference, subtract 2.
52. Add 4 to 9. Divide this sum by 2. Add 10 to this quotient.
53. Divide 20 by 5. Add 12 to this quotient. Subtract 8 from this sum. Multiply this difference by 9.
54. Multiply 6 by 3. To this product, add 27. Divide this sum by 8. Multiply this quotient by 10.
55. Add $\frac{4}{5}$ to $\frac{3}{7}$. Multiply this sum by $\frac{2}{3}$.
56. Multiply $\frac{3}{8}$ by $\frac{4}{5}$. To this product, add $\frac{7}{120}$. From this sum, subtract $\frac{1}{60}$.

Evaluate each expression for the values given.

57. $x + 4$, when $x = -2$.
58. $2x - 4x + 5$, when $x = 1$.
59. $3x - 2$, when $x = 4$.
60. $3(x - 2)$, when $x = 5$.
61. $x^2 - 6$, when $x = -3$.
62. $x^2 + 4$, when $x = 5$.
63. $-3x^2 - 4$, when $x = 1$.
64. $2x^2 + x$, when $x = 3$.
65. $-4x^2 - 2x + 5$, when $x = -3$.
66. $-3x^2 + 6x + 5$, when $x = 5$.
67. $3(x - 2)^2$, when $x = 7$.
68. $4(x + 1)^2 - 6x$, when $x = 5$.
69. $2(x - 3)(x + 4)$, when $x = 1$.
70. $3x^2(x - 1) + 5$, when $x = -4$.
71. $-6x + 3y$, when $x = 2$ and $y = 4$.
72. $6x + 3y^2 - 5$, when $x = 1$ and $y = -3$.
73. $x^2 - y^2$, when $x = -2$, and $y = -3$.
74. $x^2 - y^2$, when $x = 2$ and $y = -4$.

75. $4(x + y)^2 + 4x - 3y$, when $x = 2$ and $y = -3$.

76. $(4x - 3y)^2 - 5$, when $x = 4$ and $y = -2$.

77. $3(a + b)^2 + 4(a + b) - 6$, when $a = 4$ and $b = -1$.

78. $4xy - 6x + 3$, when $x = 5$ and $y = 2$.

79. $x^2y - 6xy + 3x$, when $x = 2$ and $y = 3$.

80. $\dfrac{6x^2}{3} + \dfrac{2x^2}{2}$, when $x = 2$.

81. $6x^2 + 3xy - y^2$, when $x = 2$ and $y = -3$.

82. $3(x - 4)^2 - (3x - 4)^2$, when $x = -1$.

83. $5(2x - 3)^2 - 4(6 - y)^2$, when $x = -2$ and $y = -1$.

84. $[2(x - 3) + (y + 2)]^2 - 6x^2$, when $x = 3$ and $y = -2$.

85. In your own words, write the order of operations to follow to evaluate a mathematical expression.

86. (a) Write in your own words the procedure you would use to evaluate $[9 - (8 \div 2)]^2 - 6^3$.

   (b) Evaluate the expression in part (a).

87. (a) Write in your own words the procedure you would use to evaluate the expression $-4x^2 + 3x - 6$ when x has a value of 5.

   (b) Evaluate the expression in part (a) when $x = 5$.

Cumulative Review Exercises

[1.7] **88.** Evaluate $(-2)(-4)(6)(-1)(-3)$.

[1.8] **89.** Evaluate **(a)** x^2 and **(b)** $-x^2$ when $x = -5$.

Evaluate each of the following.

90. $(-2)^4$

91. -2^4

JUST FOR FUN

Evaluate each expression for the given values.

1. $4([3(x - 2)]^2 + 4)$, when $x = 4$.

2. $[(3 - 6)^2 + 4]^2 + 3 \cdot 4 - 12 \div 3$.

3. $-2[(3x^2 + 4)^2 - (3x^2 - 2)^2]$, when $x = -2$.

1.10

Properties of the Real Number System

▶ **1** Identify the commutative property.

▶ **2** Identify the associative property.

▶ **3** Identify the distributive property.

Here, we introduce various properties of the real number system. We will use these properties throughout the text.

▶ **1**   The *commutative property of addition* states that the order in which any two real numbers are added does not matter.

Commutative Property of Addition

If a and b represent any two real numbers, then

$$a + b = b + a$$

Notice the commutative property involves a change in *order*. For example,

$$4 + 3 = 3 + 4$$
$$7 = 7$$

The *commutative property of multiplication* states that the order in which any two real numbers are multiplied does not matter.

Commutative Property of Multiplication

If a and b represent any two real numbers, then
$$a \cdot b = b \cdot a$$

For example,
$$6 \cdot 3 = 3 \cdot 6$$
$$18 = 18$$

The commutative property **does not hold** *for subtraction or division.* For example, $4 - 6 \neq 6 - 4$ and $6 \div 3 \neq 3 \div 6$.

▶ **2**   The *associative property of addition* states that, in the addition of three or more numbers, parentheses may be placed around any two adjacent numbers without changing the results.

Associative Property of Addition

If a, b, and c represent any three real numbers, then
$$(a + b) + c = a + (b + c)$$

Notice that the associative property involves a change of *grouping*. For example,
$$(3 + 4) + 5 = 3 + (4 + 5)$$
$$7 + 5 = 3 + 9$$
$$12 = 12$$

In this example the 3 and 4 are grouped together on the left, and the 4 and 5 are grouped together on the right.

The *associative property of multiplication* states that, in the multiplication of three or more numbers, parentheses may be placed around any two adjacent numbers without changing the results.

Associative Property of Multiplication

If a, b, and c represent any three real numbers, then
$$(a \cdot b) \cdot c = a \cdot (b \cdot c)$$

For example,
$$(6 \cdot 2) \cdot 4 = 6 \cdot (2 \cdot 4)$$
$$12 \cdot 4 = 6 \cdot 8$$
$$48 = 48$$

Notice that the associative property involves a change of grouping. When the associative property is used, the content within the parentheses changes.

The associative property **does not hold** *for subtraction or division.* For example, $(4 - 1) - 3 \neq 4 - (1 - 3)$ and $(8 \div 4) \div 2 \neq 8 \div (4 \div 2)$.

▶ **3**   A very important property of the real numbers is the *distributive property of multiplication over addition.*

> **Distributive Property**
>
> If a, b, and c represent any three real numbers, then
>
> $$a(b + c) = ab + ac$$

For example, if we let $a = 2$, $b = 3$, and $c = 4$, then

$$2(3 + 4) = (2 \cdot 3) + (2 \cdot 4)$$
$$2 \cdot 7 = 6 + 8$$
$$14 = 14$$

Therefore, we may either add first and then multiply, or multiply first and then add. The distributive property will be discussed in more detail in Chapter 2.

HELPFUL HINT

Commutative property: change in order
Associative property: change in grouping
Distributive property: two operations, multiplication and addition

---

The following are additional illustrations of the commutative, associative, and distributive properties. If we assume that x represents any real number, then:

$x + 4 = 4 + x$ by the commutative property of addition.
$x \cdot 4 = 4 \cdot x$ by the commutative property of multiplication.
$(x + 4) + 7 = x + (4 + 7)$ by the associative property of addition.
$(x \cdot 4) \cdot 6 = x \cdot (4 \cdot 6)$ by the associative property of multiplication.
$3(x + 4) = (3 \cdot x) + (3 \cdot 4)$ or $3x + 12$ by the distributive property.

EXAMPLE 1   Name the following properties.

(a) $4 + (-2) = -2 + 4$ (c) $x \cdot y = y \cdot x$
(b) $x + y = y + x$ (d) $(-12 + 3) + 4 = -12 + (3 + 4)$

Solution: (a) Commutative property of addition. (c) Commutative property of multiplication.
(b) Commutative property of addition. (d) Associative property of addition.   ■

EXAMPLE 2   Name the following properties.

(a) $2(x + 2) = (2 \cdot x) + (2 \cdot 2) = 2x + 4$
(b) $4(x + y) = (4 \cdot x) + (4 \cdot y) = 4x + 4y$
(c) $3x + 3y = (3 \cdot x) + (3 \cdot y) = 3(x + y)$
(d) $(3 \cdot 6) \cdot 5 = 3 \cdot (6 \cdot 5)$

Solution: (a) Distributive property.   (c) Distributive property (in reverse order).
(b) Distributive property.   (d) Associative property of multiplication.   ■

EXAMPLE 3   Name the following properties.
(a) $(3 + 4) + 5 = (4 + 3) + 5$ (b) $(2 + 3) + (4 + 5) = (4 + 5) + (2 + 3)$
(c) $3(x + 4) = 3(4 + x)$ (d) $3(x + 4) = (x + 4)3$

Solution: (a) Commutative property of addition. $3 + 4$ was changed to $4 + 3$; the same numbers remain within parentheses.
(b) Commutative property of addition. The order of parentheses was changed; however, the same numbers remain within the parentheses.
(c) Commutative property of addition. $x + 4$ was changed to $4 + x$.
(d) Commutative property of multiplication. The information within parentheses is not changed.   ■

HELPFUL HINT   Do not confuse the distributive property with the associative property of multiplication. Make sure you understand the difference.

Distributive Property	*Associative Property of Multiplication*
$3(4 + x) = 3 \cdot 4 + 3 \cdot x$	$3(4 \cdot x) = (3 \cdot 4)x$
$= 12 + 3x$	$= 12x$

For the distributive property to be used, there must be two *terms*, separated by a plus or minus sign, within the parentheses, as in $3(4 + x)$.

EXAMPLE 4   Name the property used to go from one step to the next.
(a) $9 + 4(x + 5)$
(b) $= 9 + 4x + 20$
(c) $= 9 + 20 + 4x$
(d) $= 29 + 4x$ addition facts
(e) $= 4x + 29$

Solution: (a to b)   Distributive property.
(b to c)   Commutative property of addition; $4x + 20 = 20 + 4x$.
(d to e)   Commutative property of addition; $29 + 4x = 4x + 29$.   ■

The distributive property can be expanded in the following manner:
$$a(b + c + d + \cdots + n) = ab + ac + ad + \cdots + an$$
For example, $3(x + y + 5) = 3x + 3y + 15$.

Exercise Set 1.10

Name the property illustrated.

1. $3(4 + 2) = 3(4) + 3(2)$

2. $3 + y = y + 3$

3. $5 \cdot y = y \cdot 5$

4. $1(x + 3) = (1)(x) + (1)(3) = x + 3$

5. $2(x + 4) = 2x + 8$

6. $3(4 + x) = 12 + 3x$

7. $x \cdot (y \cdot z) = (x \cdot y) \cdot z$

8. $1(x + 4) = x + 4$

9. $1(x + 3) = x + 3$

10. $3 + (4 + x) = (3 + 4) + x$

Complete each exercise using the given property.

11. $3 + 4 =$
commutative property of addition

12. $-3 + 4 =$
commutative property of addition

13. $-6 \cdot (4 \cdot 2) =$
associative property of multiplication
15. $(6)(y) =$
commutative property of multiplication
17. $1(x + y) =$
distributive property
19. $4x + 3y =$
commutative property of addition
21. $5x + 5y =$
distributive property (in reverse order)
23. $(x + 2)3 =$
commutative property of multiplication
25. $(3x + 4) + 6 =$
associative property of addition
27. $3(x + y) =$
commutative property of multiplication
29. $4(x + y + 3) =$
distributive property

14. $-4 + (5 + 3) =$
associative property of addition
16. $4(x + 3) =$
distributive property
18. $6(x + y) =$
distributive property
20. $3(x + y) =$
distributive property
22. $(3 + x) + y =$
associative property of addition
24. $2x + 2z =$
distributive property (in reverse order)
26. $3(x + y) =$
commutative property of addition
28. $(3x)y =$
associative property of multiplication
30. $3(x + y + 2) =$
distributive property

Name the property illustrated to go from one step to the next. See Example 4.

31. $(3 + x) + 4 = (x + 3) + 4$
32. $= x + (3 + 4)$
       $= x + 7$ addition facts

33. $6 + 5(x + 3) = 6 + 5x + 15$
34. $= 6 + 15 + 5x$
       $= 21 + 5x$ addition facts
35. $= 5x + 21$

36. $(x + 4)5 = 5(x + 4)$
37. $= 5x + 20$
38. $= 20 + 5x$

In Exercises 39–43 indicate if the given situation illustrates the commutative property. That is, does changing the order in which the items are done result in the same final outcome? Explain your answer.

39. Putting sugar and then cream in coffee; putting cream and then sugar in coffee.
40. Putting on suntan lotion and then sunning yourself; sunning yourself and then putting on suntan lotion.
41. Putting on your socks and then your shoes; putting on your shoes and then your socks.
42. Brushing your teeth and then washing your face; washing your face and then brushing your teeth.

Cumulative Review Exercises

[1.2] **43.** Add $2\frac{3}{5} + \frac{2}{3}$.
44. Subtract $3\frac{5}{8} - 2\frac{3}{16}$.

[1.9] *Evaluate each of the following.*
45. $12 - 24 \div 8 + 4 \cdot 3^2$
46. $-4x^2 + 6xy + 3y^2$, when $x = 2$ and $y = -3$.

SUMMARY

GLOSSARY

Absolute value *(21):* The distance between a number and 0 on the number line. The absolute value of any nonzero number will be positive.
Additive inverses or opposites *(25):* Two numbers whose sum is zero.

Denominator *(7):* The bottom number of a fraction.
Evaluate *(9):* To evaluate a problem means to find the numerical answer to the problem.
Expression *(30):* An expression is any collection of numbers, letters, grouping symbols, and operations.

Factor *(7):* If $a \cdot b = c$ then a and b are factors of c.
Greatest common factor *(7):* The largest number that divides two or more numbers.
Least common denominator *(10):* The smallest number divisible by two or more denominators.
Numerator *(7):* The top number of a fraction.
Operation *(23):* The basic operations of arithmetic are addition, subtraction, multiplication, and division.
Reduced to its lowest terms *(7):* A fraction is reduced to its lowest terms when its numerator and denominator have no common factor other than 1.
Set *(16):* A collection of elements listed within braces.
Variable *(6):* A letter used to represent a number.

IMPORTANT FACTS

Fractions: $\dfrac{a}{c} + \dfrac{b}{c} = \dfrac{a+b}{c}$ $\qquad$ $\dfrac{a}{c} - \dfrac{b}{c} = \dfrac{a-b}{c}$

$\dfrac{a}{b} \cdot \dfrac{c}{d} = \dfrac{ac}{bd}$ $\qquad$ $\dfrac{a}{b} \div \dfrac{c}{d} = \dfrac{a}{b} \cdot \dfrac{d}{c} = \dfrac{ad}{bc}$

Sets of Numbers

Natural numbers: $\{1, 2, 3, 4, \ldots\}$

Whole numbers: $\{0, 1, 2, 3, 4, \ldots\}$

Integers: $\{\ldots, -3, -2, -1, 0, 1, 2, 3, \ldots\}$

Rational numbers: {quotient of two integers, denominator not 0}

Real numbers: {all numbers that can be represented on the number line}

Irrational numbers: {real numbers that are not rational numbers}

To *add real numbers with the same sign*, add their absolute values. The sum has the same sign as the numbers being added.

To *add real numbers with different signs*, find the difference between the larger absolute value and the smaller absolute value. The answer has the sign of the number with the larger absolute value.

To subtract b from a, add the opposite of b to a.

$$a - b = a + (-b)$$

The *products* and *quotients* of numbers with *like signs* will be *positive*. The *products* and *quotients* of numbers with *unlike signs* will be *negative*.

Division Involving 0

$$\dfrac{0}{a} = 0, \ a \neq 0$$

$\dfrac{a}{0}$ is undefined, $\qquad a \neq 0$

$\dfrac{0}{0}$ is undeterminate

Exponents

$$b^n = \underbrace{b \cdot b \cdot b \cdot \cdots \cdot b}_{n \text{ factors of } b}$$

Order of Operations

1. Evaluate within parentheses.
2. Evaluate all exponents.
3. Perform multiplications or divisions working left to right.
4. Perform additions or subtractions working left to right.

Properties of the Real Number Systems

Property	Addition	Multiplication
Commutative	$a + b = b + a$	$ab = ba$
Associative	$(a + b) + c = a + (b + c)$	$(ab)c = a(bc)$
Distributive	$a(b + c) = ab + ac$	

Review Exercises

[1.2] *Perform the indicated operations. Reduce answers to lowest terms.*

1. $\dfrac{3}{5} \cdot \dfrac{5}{6}$

2. $\dfrac{2}{5} \div \dfrac{10}{9}$

3. $\dfrac{5}{12} \div \dfrac{3}{5}$

4. $\dfrac{5}{6} + \dfrac{1}{3}$

5. $\dfrac{3}{8} - \dfrac{1}{9}$

6. $2\dfrac{1}{3} - 1\dfrac{1}{5}$

[1.3]

7. List the set of natural numbers.
8. List the set of whole numbers.
9. List the set of integers.
10. Describe the set of rational numbers.
11. Describe the set of real numbers.
12. Consider the set of numbers

$$\left\{3,\ -5,\ -12,\ 0,\ \dfrac{1}{2},\ -0.62,\ \sqrt{7},\ 426,\ -3\dfrac{1}{4}\right\}$$

List those that are
(a) Positive integers.
(b) Whole numbers.
(c) Integers.
(d) Rational numbers.

(e) Irrational numbers.
(f) Real numbers.
13. Consider the set of numbers

$$\left\{-2.3,\ -8,\ -9,\ 1\dfrac{1}{2},\ \sqrt{2},\ -\sqrt{2},\ 1,\ -\dfrac{3}{17}\right\}$$

List those that are
(a) Natural numbers.
(b) Whole numbers.
(c) Negative integers.
(d) Integers.
(e) Rational numbers.
(f) Real numbers.

[1.4] *Insert either $<$, $>$, or $=$ in the shaded area to make a true statement.*

14. $-3 \ \blacksquare\ -5$

15. $-2 \ \blacksquare\ 1$

16. $0.6 \ \blacksquare\ -1.3$

17. $-2.6 \ \blacksquare\ -3.6$

18. $0.50 \ \blacksquare\ 0.509$

19. $4.6 \ \blacksquare\ 4.06$

20. $-3.2 \ \blacksquare\ -3.02$

21. $5 \ \blacksquare\ |-3|$

22. $-3 \ \blacksquare\ |-7|$

23. $|-2.5| \ \blacksquare\ \left|\dfrac{5}{2}\right|$

[1.5–1.6] *Evaluate as indicated.*

24. $-3 + 6$

25. $-4 + (-5)$

26. $-6 + 6$

27. $4 + (-9)$

28. $0 + (-3)$

29. $-10 + 4$

30. $-8 - (-2)$

31. $-9 - (-4)$

32. $4 - (-4)$

33. $0 - 2$

34. $-8 - 1$

35. $2 - 12$

36. $7 - 2$

37. $2 - 7$

38. $0 - (-4)$

39. $-7 - 5$

Evaluate as indicated.

40. $6 - 4 + 3$

41. $-5 + 7 - 6$

42. $-5 - 4 - 3$

43. $-2 + (-3) - 2$

44. $-(-4) + 5 - (+3)$

45. $7 - (+4) - (-3)$

46. $5 - 2 - 7 + 3$

47. $4 - (-2) + 3$

[1.7] *Evaluate as indicated.*

48. $-4(7)$

49. $(-9)(-3)$

50. $4(-9)$

51. $-2(3)$

52. $\left(\dfrac{3}{5}\right)\left(\dfrac{-2}{7}\right)$

53. $\left(\dfrac{10}{11}\right)\left(\dfrac{3}{-5}\right)$

54. $\left(\dfrac{-5}{8}\right)\left(\dfrac{-3}{7}\right)$

55. $0 \cdot \dfrac{4}{9}$

56. $4(-2)(-6)$

57. $(-1)(-3)(4)$

58. $-5(2)(7)$

59. $(-3)(-4)(-5)$

60. $-1(-2)(3)(-4)$

61. $(-4)(-6)(-2)(-3)$

Evaluate as indicated.

62. $15 \div (-3)$

63. $6 \div (-2)$

64. $-20 \div 5$

65. $-36 \div (-2)$

66. $0 \div 4$

67. $0 \div (-4)$

68. $72 \div (-9)$

69. $-40 \div (-8)$

70. $-4 \div \left(\dfrac{-4}{9}\right)$

71. $\dfrac{15}{32} \div (-5)$

72. $\dfrac{3}{8} \div \left(\dfrac{-1}{2}\right)$

73. $\dfrac{28}{-3} \div \dfrac{9}{-2}$

74. $\dfrac{14}{3} \div \left(\dfrac{-6}{5}\right)$

75. $\left(\dfrac{-5}{12}\right) \div \left(\dfrac{-5}{12}\right)$

Indicate whether each of the following is 0, undefined, or indeterminate.

76. $0 \div 4$

77. $0 \div (-6)$

78. $8 \div 0$

79. $-4 \div 0$

80. $0 \div 0$

81. $0 \div 1$

[1.5–1.7] *Evaluate as indicated.*

82. $-4(2 - 8)$

83. $2(4 - 8)$

84. $(3 - 6) + 4$

85. $(-4 + 3) - (2 - 6)$

86. $[4 + 3(-2)] - 6$

87. $(-4 - 2)(-3)$

88. $[4 + (-4)] + (6 - 8)$

89. $9[3 + (-4)] + 5$

90. $-4(-3) + [4 \div (-2)]$

91. $(-3 \cdot 4) \div (-2 \cdot 6)$

92. $(-3)(-4) + 6 - 3$

93. $[-2(3) + 6] - 4$

[1.8] *Evaluate each expression.*

94. 4^2

95. 6^2

96. 9^3

97. 1^5

98. 3^4

99. 2^4

100. $(-3)^3$

101. $(-1)^9$

102. $(-2)^5$

103. $\left(\dfrac{2}{7}\right)^2$

104. $\left(\dfrac{-3}{5}\right)^2$

105. $\left(\dfrac{2}{5}\right)^3$

Express in exponential form.

106. xxy

107. xyy

108. $xxyyx$

109. $yyzz$

110. $2 \cdot 2 \cdot 3 \cdot 3 \cdot 3xyy$

111. $5 \cdot 7 \cdot 7 \cdot xxy$

112. $xyxyz$

Express as a product of factors.

113. x^2y

114. xz^3

115. y^3z

116. $2x^3y^2$

Evaluate each expression for the values given.

117. $-x^2$, when $x = 3$

118. $-x^2$, when $x = -4$

119. $-x^3$, when $x = 3$

120. $-x^4$, when $x = -2$

[1.9] *Evaluate each expression.*

121. $3 + 5 \cdot 4$

122. $7 - 3^2$

123. $3 \cdot 5 + 4 \cdot 2$

124. $(3 - 7)^2 + 6$

125. $6 + 4 \cdot 5$

126. $8 - 36 \div 4 \cdot 3$

127. $6 - 3^2 \cdot 5$

128. $2 - (8 - 3)$

129. $[6 - (3 \cdot 5)] + 5$

130. $3[9 - (4^2 + 3)] \cdot 2$

131. $(-3^2 + 4^2) + (3^2 \div 3)$

132. $2^3 \div 4 + 6 \cdot 3$

133. $(4 \div 2)^4 + 4^2 \div 2^2$

134. $(15 - 2^2)^2 - 4 \cdot 3 + 10 \div 2$

135. $4^3 \div 4^2 - 5(2 - 7) \div 5$

136. $10 - 15 \div 5 + 10(-3) - 12$

Evaluate each expression for the values given.

137. $4x - 6$, when $x = 5$

138. $8 - 3x$, when $x = 2$

139. $6 - 4x$, when $x = -5$

140. $x^2 - 5x + 3$, when $x = 6$

141. $5y^2 + 3y - 2$, when $y = -1$

142. $-x^2 + 2x - 3$, when $x = 2$

143. $-x^2 + 2x - 3$, when $x = -2$

144. $-3x^2 - 5x + 5$, when $x = 1$

145. $3xy - 5x$, when $x = 3$ and $y = 4$

[1.10] *Name the property illustrated.*

146. $(4 + 3) + 9 = 4 + (3 + 9)$

147. $6 \cdot x = x \cdot 6$

148. $4(x + 3) = 4x + 12$

149. $(x + 4)3 = 3(x + 4)$

150. $6x + 3x = 3x + 6x$

151. $(x + 7) + 4 = x + (7 + 4)$

152. $-6x + 3 = 3 + (-6x)$

Practice Test

1. Consider the set of numbers

$$\left\{-6, 42, -3\frac{1}{2}, 0, 6.52, \sqrt{5}, \frac{5}{9}, -7, -1\right\}$$

List those that are

(a) Natural numbers.

(b) Whole numbers.

(c) Integers.

(d) Rational numbers.

(e) Irrational numbers.

(f) Real numbers.

Insert either $<$, $>$, or $=$ in the shaded area to make a true statement.

2. $-6 \quad -3$

3. $|-3| \quad |-2|$

Evaluate each expression.

4. $-4 + (-8)$

5. $-6 - 5$

6. $4 - (-12)$

7. $5 - 12 - 7$

8. $(-4 + 6) - 3(-2)$

9. $(-4)(-3)(2)(-1)$

10. $\left(\dfrac{-2}{9}\right) \div \left(\dfrac{-7}{8}\right)$

11. $\left(-12 \cdot \dfrac{1}{2}\right) \div 3$

12. $3 \cdot 5^2 - 4 \cdot 6^2$

13. $(4 - 6^2) \div [4(2 + 3) - 4]$

14. $-6[(-2 - 3)] \div 5 \cdot 2$

15. $(-3)^4$

16. $\left(\dfrac{3}{5}\right)^3$

17. Write $2 \cdot 2 \cdot 5 \cdot 5 \cdot yyzzz$ in exponential form.

18. Write $2^2 3^3 x^4 y^2$ as a product of factors.

Evaluate each expression for the values given.

19. $2x^2 - 6$, when $x = -4$

20. $6x - 3y^2 + 4$, when $x = 3$ and $y = -2$

21. $-x^2 - 6x + 3$, when $x = -2$

Name the property illustrated.

22. $x + 3 = 3 + x$

23. $4(x + 9) = 4x + 36$

24. $(2 + x) + 4 = 2 + (x + 4)$

25. $5(x + y) = (x + y)5$

2.1 Combining Like Terms

2.2 The Addition Property

2.3 The Multiplication Property

2.4 Solving Linear Equations with a Variable on Only One Side of the Equation

2.5 Solving Linear Equations with the Variable on Both Sides of the Equation

2.6 Ratios and Proportions

2.7 Inequalities in One Variable

Summary

Review Exercises

Practice Test

Cumulative Review Test

See Section 2.6, Exercise 32.

2.1

Combining Like Terms

▶**1** Identify terms.

▶**2** Identify like terms.

▶**3** Combine like terms.

▶**4** Use the distributive property to remove parentheses.

▶**5** Remove parentheses when they are preceded by a plus or minus sign.

▶**6** Simplify an expression.

▶**1** In Section 1.2 and other sections of the text, we indicated that letters called **variables** (or **literal numbers**) are used to represent numbers.

As was indicated in Chapter 1, an **expression** (sometimes referred to as an **algebraic expression**) is a collection of numbers, variables, grouping symbols, and operation symbols. Examples of expressions are:

$$5, \quad x^2 - 6, \quad 4x - 3, \quad 2(x + 5) + 6, \quad \frac{x + 3}{4}$$

When an algebraic expression consists of several parts, the parts that are added or subtracted are called the **terms** of the expression. The expression $2x - 3y - 5$ has three terms: $2x$, $-3y$, and -5. The expression

$$3x + 2xy + 5(x + y)$$

also has three terms: $3x$, $2xy$, and $5(x + y)$.

The $+$ and $-$ signs that break the expression into terms are a part of the term. However, when listing the terms of an expression, it is not necessary to list the $+$ sign at the beginning of a term.

Expression	*Terms*
$-2x + 3y - 8$	$-2x, \quad 3y, \quad -8$
$3y - 2x + \dfrac{1}{2}$	$3y, \quad -2x, \quad \dfrac{1}{2}$
$7 + x + 4 - 5x$	$7, \quad x, \quad 4, \quad -5x$
$3(x - 1) - 4x + 2$	$3(x - 1), \quad -4x, \quad 2$
$\dfrac{x + 4}{3} - 5x + 3$	$\dfrac{x + 4}{3}, \quad -5x, \quad 3$

The numerical part of a term is called its **numerical coefficient** or simply its **coefficient.** In the term $6x$, the 6 is the numerical coefficient. Note that $6x$ means the variable x is multiplied by 6.

	Term	*Numerical coefficient*
	$3x$	3
	$-\dfrac{1}{2}x$	$-\dfrac{1}{2}$
	$4(x - 3)$	4
	$\dfrac{2x}{3}$	$\dfrac{2}{3}$, since $\dfrac{2x}{3}$ means $\dfrac{2}{3}x$
	$\dfrac{x + 4}{3}$	$\dfrac{1}{3}$, since $\dfrac{x + 4}{3}$ means $\dfrac{1}{3}(x + 4)$

Whenever a term appears without a numerical coefficient, we assume that the numerical coefficient is 1.

Examples

x means $1x$	$-x$ means $-1x$
x^2 means $1x^2$	$-x^2$ means $-1x^2$
xy means $1xy$	$-xy$ means $-1xy$
$(x + 2)$ means $1(x + 2)$	$-(x + 2)$ means $-1(x + 2)$

If an expression has a term that is a number (without a variable), we refer to that number as a **constant term,** or simply a **constant.** In the expression $x^2 + 3x - 4$, the -4 is a constant term, or a constant.

▸ **2 Like terms** are terms that have the same variables with the same exponents. The following are examples of like terms and unlike terms. Note that if two terms are like terms then only their numerical coefficients may differ.

Like Terms	*Unlike Terms*	
$3x,\quad -4x$	$3x,\quad 2$	(variables differ)
$4y,\quad 6y$	$3x,\quad 4y$	(variables differ)
$5,\quad -6$	$x,\quad 3$	(variables differ)
$3(x + 1),\quad -2(x + 1)$	$2x,\quad 3xy$	(variables differ)
$3x^2,\quad 4x^2$	$3x,\quad 4x^2$	(exponents differ)

EXAMPLE 1 Determine if there are any like terms in each algebraic expression.
(a) $2x + 3x + 4$ (b) $2x + 3y + 2$ (c) $x + 3 + y - \frac{1}{2}$

Solution (a) $2x$ and $3x$ are like terms.
(b) No like terms.
(c) 3 and $-\frac{1}{2}$ are like terms. ■

EXAMPLE 2 Determine if there are any like terms in each algebraic expression.
(a) $5x - x + 6$ (b) $3 - 2x + 4x - 6$ (c) $12 + x + 7$

Solution: (a) $5x$ and $-x$ (or $-1x$) are like terms.
(b) 3 and -6 are like terms; and $-2x$ and $4x$ are like terms.
(c) 12 and 7 are like terms. ■

▶ **3** Often, we would like to simplify expressions by combining like terms. **To combine like terms** means to add or subtract the like terms in an expression. To combine like terms, we can use the procedure that follows.

To Combine Like Terms

---

1. Determine which terms are like terms.
2. Add or subtract the coefficients of the like terms.
3. Multiply the number found in step 2 by the common variables.

Examples 3 through 9 illustrate this procedure.

EXAMPLE 3   Combine like terms: $4x + 3x$.

Solution: $4x$ and $3x$ are like terms with the common variable x.
Since $4 + 3 = 7$, then $4x + 3x = 7x$. ■

EXAMPLE 4   Combine like terms: $\dfrac{3}{5}x - \dfrac{2}{3}x$.

Solution: Since $\dfrac{3}{5} - \dfrac{2}{3} = \dfrac{9}{15} - \dfrac{10}{15} = -\dfrac{1}{15}$, then $\dfrac{3}{5}x - \dfrac{2}{3}x = -\dfrac{1}{15}x$. ■

EXAMPLE 5   Combine like terms: $5.23a - 7.45a$.

Solution: Since $5.23 - 7.45 = -2.22$, then $5.23a - 7.45a = -2.22a$. ■

EXAMPLE 6   Combine like terms: $3x + x + 5$.

Solution: The $3x$ and x are like terms.

$$3x + x + 5$$
$$\text{means} \quad 3x + 1x + 5$$
$$\text{which equals} \quad 4x + 5$$ ■

EXAMPLE 7   Combine like terms: $12 + x + 7$.

Solution: The 12 and 7 are like terms. We can rearrange the terms to get

$$x + 12 + 7 \quad \text{or} \quad x + 19$$ ■

EXAMPLE 8   Combine like terms: $3y + 4x - 3 - 2x$.

Solution: $4x$ and $-2x$ are the only like terms.

$$\text{Rearranging terms:} \quad 4x - 2x + 3y - 3$$
$$\text{Combining like terms:} \quad 2x + 3y - 3$$ ■

EXAMPLE 9   Combine like terms: $-2x + 3y - 4x + 3 - y + 5$.

Solution: $-2x$ and $-4x$ are like terms.
$3y$ and $-y$ are like terms.
3 and 5 are like terms.

Grouping the like terms together gives

$$-2x - 4x + 3y - y + 3 + 5$$
$$-6x \quad + \quad 2y \quad + \quad 8$$

∎

The commutative and associative properties were used to rearrange the terms in Examples 7, 8, and 9. The order of the terms in the answer is not critical. Thus $2y - 6x + 8$ is also an acceptable answer to Example 9.

COMMON STUDENT ERROR

Students often misinterpret the meaning of a term like $3x$. What does $3x$ mean?

Correct   *Wrong*

$$3x = x + x + x \qquad 3x = x \cdot x \cdot x$$

Just as $2 + 2 + 2$ can be expressed as $3 \cdot 2$, $x + x + x$ can be expressed as $3 \cdot x$ or $3x$. Note that when we combine like terms in $x + x + x$ we get $3x$. Also note that $x \cdot x \cdot x = x^3$, not $3x$.

▶ **4** We introduced the distributive property in Section 1.10. Because this property is so important, we will study it again. But before we do, let us go back briefly to the subtraction of real numbers. Recall from Section 1.6 that

$$6 - 3 = 6 + (-3)$$

For any real numbers a and b,

$$a - b = a + (-b)$$

We will use the fact that $a + (-b)$ means $a - b$ in discussing the distributive property.

Distributive Property

For any real numbers a, b, and c,

$$a(b + c) = ab + ac$$

EXAMPLE 10   Use the distributive property to remove parentheses.

(a) $2(x + 4)$ (b) $-2(x + 4)$

Solution: (a) $2(x + 4) = 2x + 2(4) = 2x + 8$
(b) $-2(x + 4) = -2x + (-2)(4) = -2x + (-8) = -2x - 8$

Note in part (b) that, instead of leaving the answer $-2x + (-8)$, we wrote it as $-2x - 8$.

∎

EXAMPLE 11   Use the distributive property to remove parentheses.

(a) $3(x - 2)$ (b) $-2(4x - 3)$

Solution: (a) By the definition of subtraction, we may write $x - 2$ as $x + (-2)$.

$$3(x - 2) = 3[x + (-2)] = 3x + 3(-2)$$
$$= 3x + (-6)$$
$$= 3x - 6$$

(b) $-2(4x - 3) = -2[4x + (-3)] = -2(4x) + (-2)(-3) = -8x + 6$ ■

The distributive property is used often in algebra, and so you need to understand it well. You should understand it so well that you will be able to simplify an expression using the distributive property without having to write down all the steps that we listed in working Examples 10 and 11. Study closely the Helpful Hint that follows.

HELPFUL HINT

With a little practice, you will be able to eliminate some of the intermediate steps when you use the distributive property to remove parentheses. When using the distributive property, there are eight possibilities with regard to signs. Study and learn the eight possibilities that follow.

Positive Coefficient

$2(x) = 2x$
(a) $2(x + 3) = 2x + 6$
$2(+3) = +6$

$2(x) = 2x$
(b) $2(x - 3) = 2x - 6$
$2(-3) = -6$

$2(-x) = -2x$
(c) $2(-x + 3) = -2x + 6$
$2(+3) = +6$

$2(-x) = -2x$
(d) $2(-x - 3) = -2x - 6$
$2(-3) = -6$

Negative Coefficient

$(-2)(x) = -2x$
(e) $-2(x + 3) = -2x - 6$
$(-2)(+3) = -6$

$(-2)(x) = -2x$
(f) $-2(x - 3) = -2x + 6$
$(-2)(-3) = +6$

$(-2)(-x) = 2x$
(g) $-2(-x + 3) = 2x - 6$
$(-2)(+3) = -6$

$(-2)(-x) = 2x$
(h) $-2(-x - 3) = 2x + 6$
$(-2)(-3) = +6$

The distributive property can be expanded as follows:

$$a(b + c + d + \cdots + n) = ab + ac + ad + \cdots + an$$

Examples of the expanded distributive property are

$$3(x + y + z) = 3x + 3y + 3z$$
$$2(x + y - 3) = 2x + 2y - 6$$

EXAMPLE 12   Use the distributive property to remove parentheses.

(a) $4(x - 3)$ (b) $-2(2x - 4)$ (c) $-\frac{1}{2}(4x + 5)$ (d) $-2(3x - 2y + 4z)$

Solution: (a) $4(x - 3) = 4x - 12$ (b) $-2(2x - 4) = -4x + 8$

(c) $-\frac{1}{2}(4x + 5) = -2x - \frac{5}{2}$ (d) $-2(3x - 2y + 4z) = -6x + 4y - 8z$ ∎

The distributive property can also be used from the right, as illustrated in Example 13.

EXAMPLE 13   Use the distributive property to remove parentheses from the expression $(2x - 8y)4$.

Solution: Basically we will follow the same procedure as in Example 12. However, we will distribute the 4 on the right side of the parentheses over the terms within the parentheses.

$$(2x - 8y)4 = 2x(4) - 8y(4)$$
$$= 8x - 32y$$ ∎

Example 13 could have been rewritten as $4(2x - 8y)$ by the commutative property of multiplication, and then the 4 could have been distributed from the left to obtain the same answer, $8x - 32y$.

▶ **5**   Consider the expression $(4x + 3)$. How do we remove parentheses? Recall that the coefficient of a term is assumed to be 1 if none is present. Therefore, we may write

$$(4x + 3) = 1(4x + 3)$$
$$= 1(4x) + (1)(3)$$
$$= 4x + 3$$

Note that $(4x + 3) = 4x + 3$. **When no sign or a plus sign precedes parentheses, the parentheses may be removed without having to change the expression inside the parentheses.**

Examples

$$(x + 3) = x + 3$$
$$(2x - 3) = 2x - 3$$
$$+(2x - 5) = 2x - 5$$
$$+(x + 2y - 6) = x + 2y - 6$$

Now consider the expression $-(4x + 3)$. How do we remove parentheses? Here, the number in front of the parentheses is -1, and we write

$$-(4x + 3) = -1(4x + 3)$$
$$= -1(4x) + (-1)(3)$$
$$= -4x + (-3)$$
$$= -4x - 3$$

Note that $-(4x + 3) = -4x - 3$. **When a minus sign precedes parentheses, the signs of all the terms within the parentheses are changed when the parentheses are removed.**

Examples

$$-(x + 4) = -x - 4$$
$$-(-2x + 3) = 2x - 3$$
$$-(5x - y + 3) = -5x + y - 3$$
$$-(-2x - 3y - 5) = 2x + 3y + 5$$

▶ **6**

> **To Simplify an Expression Means to:**
>
> **1.** Remove any parentheses using the distributive property.
> **2.** Combine like terms.

EXAMPLE 14 Simplify $6 - (2x + 3)$.

Solution: $6 - (2x + 3) = 6 - 2x - 3$ Use the distributive property.

$ = -2x + 3$ Combine like terms.   ▨

Note: $3 - 2x$ is the same as $-2x + 3$; however, we generally write the term containing the variable first.

EXAMPLE 15 Simplify $6x + 4(2x + 3)$.

Solution: $6x + 4(2x + 3) = 6x + 8x + 12$ Use the distributive property.

$ = 14x + 12$ Combine like terms.   ▨

EXAMPLE 16 Simplify $2(x - 1) + 9$.

Solution: $2(x - 1) + 9 = 2x - 2 + 9$ Use the distributive property.

$ = 2x + 7$ Combine like terms.   ▨

EXAMPLE 17 Simplify $2(x + 3) - 3(x - 2) - 4$.

Solution: $2(x + 3) - 3(x - 2) - 4 = 2x + 6 - 3x + 6 - 4$ Use the distributive property.

$ = 2x - 3x + 6 + 6 - 4$ Rearrange terms.

$ = -x + 8$ Combine like terms.   ▨

HELPFUL HINT

It is important for you to have a clear understanding of the concepts of term and factor. When two or more expressions are **multiplied**, each expression is a **factor** of the product. For example, since $4 \cdot 3 = 12$, the 4 and the 3 are factors of 12. Since $3 \cdot x = 3x$, the 3 and the x are factors of $3x$. Similarly, in the expression $5xyz$, the 5, x, y, and z are all factors of the expression.

In an expression, the parts that are **added or subtracted** are the **terms** of the expression. For example, in the expression $2x^2 + 3x - 4$, there are three terms, $2x^2$, $3x$, and -4. Note that the terms of an expression may have factors. For example, in the term $2x^2$, the 2 and the x^2 are factors of that term because they are multiplied together.

Exercise Set 2.1

Combine like terms when possible.

1. $3x + 5x$	**2.** $4x + 3$	**3.** $2x - 3x$
4. $4x + 3y$	**5.** $12 + x - 3$	**6.** $-2x - 3x$
7. $-4x + 7x$	**8.** $4x - 7x + 4$	**9.** $x + 3x - 7$
10. $3 + 2x - 5$	**11.** $6 - 3 + 2x$	**12.** $2 + 2x + 3x$
13. $-7 + 5x + 12$	**14.** $-2x - 3x - 2 - 3$	**15.** $5x + 2y + 3 + y$
16. $-x + 2 - x - 2$	**17.** $4x - 2x + 3 - 7$	**18.** $x - 4x + 3$
19. $4 + x + 3$	**20.** $x + 2x + y + 2$	**21.** $-3x + 2 - 5x$
22. $x + 4 - 6$	**23.** $5 + 2x - 4x + 6$	**24.** $3x + 4x - 2 + 5$
25. $x - 2 - 4 + 2x$	**26.** $2x + 4 - 3 + x$	**27.** $2 - 3x - 2x + 1$
28. $3x - x + 4 - 6$	**29.** $2y + 4y + 6$	**30.** $6 - x - x$
31. $x - 6 + 3x - 4$	**32.** $-2x + 4x - 3$	**33.** $4 - x + 4x - 8$
34. $x + 4 + \frac{3}{5}$	**35.** $x + \frac{3}{4} - \frac{1}{3}$	**36.** $5.23x + 1.42 - 4.61x$
37. $68.2x - 19.7x + 8.3$	**38.** $\frac{1}{2}x + 3y + 1$	**39.** $x + \frac{1}{2}y - \frac{3}{8}y$
40. $2x + 3 + 4x + 5$	**41.** $-4x - 3.1 - 5.2$	**42.** $-x + 2x + y$
43. $1 + x + 6 - 3x$	**44.** $2x - 7 - 5x + 2$	**45.** $3x - 7 - 9 + 4x$
46. $x - y - 2y + 3$	**47.** $4x + 6 + 3x - 7$	**48.** $-y - 6 - 3y - y$
49. $-4 + x - 6 + 2$	**50.** $x - 3y + 2x + 4$	**51.** $-19.36 + 40.02x + 12.25 - 18.3x$
52. $52x - 52x - 63.5 - 63.5$	**53.** $\frac{3}{5}x - 3 - \frac{7}{4}x - 2$	**54.** $\frac{1}{5}y + 3x - 2x - \frac{2}{3}y$

Use the distributive property to remove parentheses.

55. $2(x + 4)$	**56.** $3(x - 2)$	**57.** $4(x + 5)$
58. $-2(x + 3)$	**59.** $-2(x - 4)$	**60.** $3(-x + 5)$
61. $-\frac{1}{2}(2x - 4)$	**62.** $-5(x + 6)$	**63.** $1(-4 + x)$
64. $3(y + 3)$	**65.** $\frac{1}{4}(x - 12)$	**66.** $5(x + y + 4)$
67. $-0.6(3x - 5)$	**68.** $-(x - 3)$	**69.** $\frac{1}{2}(-2x + 6)$
70. $-2(x + y - z)$	**71.** $0.4(2x - 0.5)$	**72.** $-(x + 4y)$
73. $-(-x + y)$	**74.** $(3x + 4y - 6)$	**75.** $-(2x - 6y + 8)$
76. $-(-2x + 6 - y)$	**77.** $3(4 - 2x + y)$	**78.** $-2(-x + 3y + 5)$
79. $2(\frac{1}{2}x - 4y + \frac{1}{4})$	**80.** $2(3 - \frac{1}{2}x + 4y)$	**81.** $(x + 3y - 9)$
82. $(-x + 5 - 2y)$	**83.** $-(-x + 4 + 2y)$	

Simplify when possible.

84. $2(x + 3) + 4$	**85.** $4(x - 2) - x$	**86.** $6 - (x + 3)$
87. $-2(3 - x) + 1$	**88.** $-(2x + 3) + 5$	**89.** $6x + 2(4x + 9)$
90. $3(x + y) + 2y$	**91.** $2(x - y) + 2x + 3$	**92.** $6 + (x - 8) + 2x$
93. $(x + y) - 2x + 3$	**94.** $4 - (2x + 3) + 5$	**95.** $8x - (x - 3)$
96. $-(x - 5) - 3x + 4$	**97.** $2(x - 3) - (x + 3)$	**98.** $3y - (2x + 2y) - 6x$
99. $4(x - 3) + 2(x - 2) + 4$	**100.** $4(x + 3) - 2x$	**101.** $2(x - 4) - 3x + 6$
102. $6 - 2(x + 3) + 5x$	**103.** $-3(x - 4) + 2x - 6$	**104.** $-(x + 2) + 3x - 6$
105. $4(x - 3) + 4x - 7$	**106.** $-3(x + 2y) + 3y + 4$	**107.** $0.4 + (x + 5) - 0.6 + 2$
108. $4 - (2 - x) + 3x$	**109.** $9 - (-3x + 4) - 5$	**110.** $2y - 6(y - 2) + 3$
111. $4(x + 2) - 3(x - 4) - 5$	**112.** $4 - (y - 5) + 2x + 3$	**113.** $-0.2(2 - x) + 4(y + 0.2)$

114. $-5(-y + 2) + 3(2 - x) - 4$

115. $-6x + 3y - (6 + x) + (x + 3)$

116. $(x + 3) + (x - 4) - 6x$

117. $-(x + 3) + (2x + 4) - 6$

118. $\frac{1}{2}(x + 3) + \frac{1}{3}(3x + 6)$

119. $\frac{2}{3}(x - 2) - \frac{1}{2}(x + 4)$

120. When no sign or a plus sign precedes an expression within parentheses, explain how to remove the parentheses.

121. When a minus sign precedes an expression within parentheses, explain how to remove the parentheses.

122. Explain the differences between a factor and a term.

123. Consider the expression $2x^2 + 3x - 5$.

  (a) List the terms of this expression. Explain why each is a term.

  (b) List the factors of the term $2x^2$. Explain why each is a factor of the term.

Cumulative Review Exercises

[1.4] *Evaluate the following.*

124. $|-7|$

125. $-|-16|$

[1.9] **126.** Write a paragraph explaining the order of operations.

127. Evaluate $-x^2 + 5x - 6$ when $x = -1$.

JUST FOR FUN

Simplify each of the following expressions.

1. $4x + 5y + 6(3x - 5y) - 4x + 3$

2. $2x^2 - 4x + 8x^2 - 3(x + 2) - x^2 - 2$

3. $x^2 + 2y - y^2 + 3x + 5x^2 + 6y^2 + 5y$

4. $2[3 + 4(x - 5)] - [2 - (x - 3)]$

2.2

The Addition Property

▶ **1** Identify equations.

▶ **2** Check solutions.

▶ **3** Identify and define equivalent equations.

▶ **4** Use the addition property to solve equations.

▶ **1** A statement that shows two algebraic expressions are equal is called an **equation.** For example, $4x + 3 = 2x - 4$ is an equation. In this chapter we learn procedures used to solve *linear equations in one variable*.

A **linear equation** in one variable is an equation of the form
$$ax + b = c$$
for real numbers a, b, and c, $a \neq 0$.

Examples of of linear equations in one variable are

$$x + 4 = 7$$
$$2x - 4 = 6$$

▶ **2** The **solution of an equation** is the number or numbers that make the equation a true statement. For example, the solution to $x + 4 = 7$ is 3. We will shortly learn how to find the solution to an equation, or to **solve an equation.** But before we do this we will learn how to *check* the solution of an equation.

The solution to an equation may be **checked** by substituting the value that is believed to be the solution back into the original equation. If the substitution results in a true statement, your solution is probably correct. If the substitution results in a false statement, then either your solution or your check is incorrect, and you need to go back and find your error. Try to check all your solutions.

To check to see if 3 is the solution to $x + 4 = 7$, we substitute 3 for each x in the equation.

Check: $x = 3$

$$x + 4 = 7$$
$$3 + 4 = 7$$
$$7 = 7 \qquad \text{true}$$

Since the check results in a true statement, 3 is a solution to the equation $x + 4 = 7$.

EXAMPLE 1 Consider the equation $2x - 4 = 6$. Determine whether

(a) 3 is a solution.
(b) 5 is a solution.

Solution: (a) To determine whether 3 is a solution to the equation, substitute 3 for x.

Check: $x = 3$

$$2x - 4 = 6$$
$$2(3) - 4 = 6$$
$$6 - 4 = 6$$
$$2 = 6 \qquad \text{not a true statement}$$

Since the value 3 does not check, 3 is not a solution.

(b) *Check:* $x = 5$

$$2x - 4 = 6$$
$$2(5) - 4 = 6$$
$$10 - 4 = 6$$
$$6 = 6 \qquad \text{a true statement}$$

Since the value 5 checks, 5 is a solution to the equation. ■

We can use the same procedures to check more complex equations as shown in Examples 2 and 3.

EXAMPLE 2   Determine whether 18 is a solution to the equation $3x - 2(x + 3) = 12$.

Solution: To determine whether 18 is a solution, substitute 18 for each x in the equation. If the substitution results in a true statement, then 18 is a solution.

$$3x - 2(x + 3) = 12$$
$$3(18) - 2(18 + 3) = 12$$
$$54 - 2(21) = 12$$
$$54 - 42 = 12$$
$$12 = 12 \quad \text{true}$$

Since we obtain a true statement, 18 is a solution to the equation $3x - 2(x + 3) = 12$. ∎

 Calculator Corner

Calculators can be used to check solutions to equations. For example, to check to see if $\frac{-10}{3}$ is a solution to the equation $2x + 3 = 5(x + 3) - 2$, we perform the following steps:

1. Substitute $\frac{-10}{3}$ for each x.

$$2x + 3 = 5(x + 3) - 2$$
$$2\left(\frac{-10}{3}\right) + 3 = 5\left(\frac{-10}{3} + 3\right) - 2$$

2. Evaluate each side of the equation separately using your calculator. If you obtain the same value on both sides, then your solution checks. The procedures for evaluating the left and right sides of the equation depend on whether or not your calculator is a scientific calculator (see page 47). If you do not have a scientific calculator, remember that you need to work within parentheses first, and then do your multiplications and divisions from left to right before your additions and subtractions. In the following steps we assume you do not have a scientific calculator.

To evaluate the left side of the equation, $2(\frac{-10}{3}) + 3$, press the following keys:

$$\boxed{C} \; \underbrace{10 \; \boxed{+/-} \; \boxed{\div} \; 3}_{\text{gives value of } \frac{-10}{3}} \; \boxed{\times} \; 2 \; \boxed{+} \; 3 \; \boxed{=} \; -3.6666666$$

To evaluate the right side of the equation, $5(\frac{-10}{3} + 3) - 2$, press the following keys:

$$\boxed{C} \; \underbrace{10 \; \boxed{+/-} \; \boxed{\div} \; 3 \; \boxed{+} \; 3 \; \boxed{=}}_{\text{gives value within parentheses}} \; \boxed{\times} \; 5 \; \boxed{-} \; 2 \; \boxed{=} \; -3.6666665$$

Since both sides give the same value (there may be a slight roundoff error), the solution checks. Note that calculators may differ, so you should follow the instructions that come with your calculator.

EXAMPLE 3   Determine whether $-\dfrac{3}{2}$ is a solution to the equation $3(x + 3) = 6 + x$.

Solution: Substitute $-\dfrac{3}{2}$ for each x in the equation.

$$3(x + 3) = 6 + x$$

$$3\left(-\frac{3}{2} + 3\right) = 6 + \left(-\frac{3}{2}\right)$$

$$3\left(-\frac{3}{2} + \frac{6}{2}\right) = \frac{12}{2} - \frac{3}{2}$$

$$3\left(\frac{3}{2}\right) = \frac{9}{2}$$

$$\frac{9}{2} = \frac{9}{2} \qquad \text{true}$$

Thus $-\dfrac{3}{2}$ is a solution to the equation.   ■

▶ **3**   Now that we know how to check a solution to an equation we will discuss the procedure for solving equations. Complete procedures for solving equations will be given shortly. For now, you need to understand that **to solve an equation, it is necessary to get the variable all by itself on one side of the equal sign. We say that we isolate the variable.** To isolate the variable, we make use of two properties: addition and multiplication. Look first at Figure 2.1.

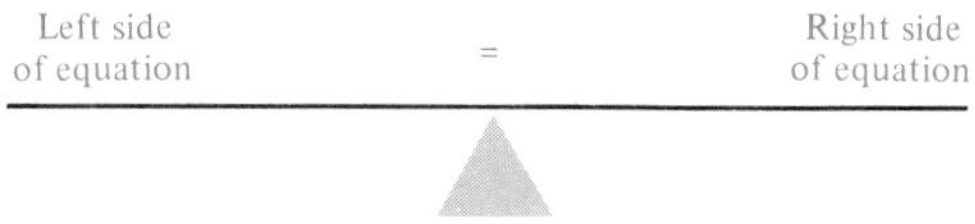

| Left side of equation | = | Right side of equation |

FIGURE 2.1

Think of an equation as a balanced statement whose left side is balanced by its right side. When solving an equation, we must make sure that the equation remains balanced at all times. That is, both sides must always remain equal. **We ensure that an equation always remains equal by doing the same thing to both sides of the equation.** For example, if we decide to add something to the left side of the equation, we must add exactly the same amount to the right side. If we decide to multiply the right side of the equation by some number, we must multiply the left side by the same number.

When we add the same number to both sides of an equation or multiply both sides of an equation by the same nonzero number, we do not change the solution to the equation, just the form. Two or more equations with the same solution are called **equivalent equations.** The equations $2x - 4 = 2$, $2x = 6$, and $x = 3$ are equivalent, since the solution to each is 3.

Check: $x = 3$

$2x - 4 = 2$	$2x = 6$	$x = 3$
$2(3) - 4 = 2$	$2(3) = 6$	$3 = 3$ true
$6 - 4 = 2$	$6 = 6$ true	
$2 = 2$ true		

When solving an equation, we use the addition and multiplication properties to express a given equation as simpler equivalent equations until we obtain the solution.

▶**4** In this section we use the addition property to solve equations. In Section 2.3 we use the multiplication property to solve equations.

Addition Property

If $a = b$, then $a + c = b + c$ for any real numbers a, b, and c.

This property implies that the same number can be added to both sides of an equation without changing the solution. **The addition property is used to solve equations of the form $x + a = b$.** To isolate the variable x in equations of this form, add the opposite or additive inverse of a, $-a$, to both sides of the equation.

To isolate the variable when solving equations of the form $x + a = b$, we use the addition property to eliminate the number **on the same side of the equal sign as the variable.** Study the following examples carefully.

Equation	*To solve, use the addition property to eliminate the number*
$x + 8 = 10$	8
$x - 7 = 12$	-7
$5 = x - 12$	-12
$-4 = x + 9$	9

Now let us work some problems.

EXAMPLE 4 Solve the equation $x - 4 = 3$.

Solution: To isolate the variable, x, we must eliminate the -4 from the left side of the equation. To do this we add 4, the opposite of -4, to *both sides* of the equation.

$$x - 4 = 3$$
$$x - 4 + 4 = 3 + 4 \qquad \text{Add 4 to both sides of the equation.}$$
$$x + 0 = 7$$
$$x = 7$$

Note how the process helps to isolate x.

Check:
$$x - 4 = 3$$
$$7 - 4 = 3$$
$$3 = 3 \qquad \text{true}$$

EXAMPLE 5 Solve the equation $y - 3 = -5$.

Solution: To solve this equation, we must isolate the variable, y. To eliminate the -3 from the left side of the equation, we add its opposite, 3, to *both sides* of the equation.

$$y - 3 = -5$$
$$y - 3 + 3 = -5 + 3 \qquad \text{Add 3 to both sides of the equation.}$$
$$y + 0 = -2$$
$$y = -2$$

Note that we did not check the solution to Example 5. Space limitations prevent us from showing all checks. However, you should check all of your answers.

EXAMPLE 6   Solve the equation $x + 5 = 9$.

Solution: To solve this equation, we must isolate the x. Therefore, we must eliminate the 5 from the left side of the equation. To do this, we add the opposite of 5, -5, to both sides of the equation.

$$x + 5 = 9$$
$$x + 5 + (-5) = 9 + (-5) \qquad \text{Add } -5 \text{ to both sides of the equation.}$$
$$x + 0 = 4$$
$$x = 4 \qquad \blacksquare$$

In Example 6 we added -5 to both sides of the equation. From Section 1.6 we know that $5 + (-5) = 5 - 5$. Thus we can see that adding a negative 5 to both sides of the equation is equivalent to subtracting a 5 from both sides of the equation. The addition property says the same number may be *added* to both sides of an equation. **Since subtraction is defined in terms of addition, the addition property also allows us to *subtract* the same number from both sides of the equation.** Thus Example 6 could have also been worked as follows:

$$x + 5 = 9$$
$$x + 5 - 5 = 9 - 5 \qquad \text{Subtract 5 from both sides of the equation.}$$
$$x + 0 = 4$$
$$x = 4$$

In this text, unless there is a specific reason to do otherwise, rather than adding a negative number to both sides of the equation, we will subtract a number from both sides of the equation.

EXAMPLE 7   Solve the equation $x + 7 = -3$.

Solution:
$$x + 7 = -3$$
$$x + 7 - 7 = -3 - 7 \qquad \text{Subtract 7 from both sides of the equation.}$$
$$x + 0 = -10$$
$$x = -10$$

Check:
$$x + 7 = -3$$
$$-10 + 7 = -3$$
$$-3 = -3 \qquad \text{true} \qquad \blacksquare$$

EXAMPLE 8   Solve the equation $4 = x - 5$.

Solution: In this example the variable x is on the right side of the equation. To isolate the x, we must eliminate the -5 from the right side of the equation. This can be accomplished by adding 5 to both sides of the equation.

$$4 = x - 5$$
$$4 + 5 = x - 5 + 5 \qquad \text{Add 5 to both sides of the equation.}$$
$$9 = x + 0$$
$$9 = x$$

Thus the solution is 9. $\qquad \blacksquare$

EXAMPLE 9   Solve the equation $-6.25 = x + 12.78$.

Solution: The variable is on the right side of the equation. Subtract 12.78 from both sides of the equation to isolate the variable.

$$-6.25 = x + 12.78$$
$$-6.25 - 12.78 = x + 12.78 - 12.78 \qquad \text{Subtract 12.78 from both}$$
$$\qquad\qquad\qquad\qquad\qquad\qquad\qquad \text{sides of the equation.}$$
$$-19.03 = x + 0$$
$$-19.03 = x$$

The solution is -19.03.   ∎

COMMON STUDENT ERROR

When solving equations, our goal is to get the variable all by itself on one side of the equal sign. Consider the equation $x + 3 = -4$. How do we solve it?

Correct	*Wrong*
Remove the 3 from the left side of the equation.	Remove the -4 from the right side of the equation.

$$x + 3 = -4 \qquad\qquad\qquad x + 3 = -4$$
$$x + 3 - 3 = -4 - 3 \qquad x + 3 + 4 = -4 + 4$$
$$x = -7 \qquad\qquad\qquad\qquad x + 7 = 0$$

Variable is now isolated.   Variable is **not** isolated.

Use the addition property to **remove the number that is on the same side of the equal sign as the variable.**

Consider the following two problems:

(a) $\qquad\qquad x - 5 = 12$ $\qquad$ (b) $\qquad 15 = x + 3$

$$x - 5 + 5 = 12 + 5 \qquad\qquad 15 - 3 = x + 3 - 3$$
$$x + 0 = 12 + 5 \qquad\qquad\quad 15 - 3 = x + 0$$
$$x = 17 \qquad\qquad\qquad\qquad 12 = x$$

Note in these problems how the number on the same side of the equal sign as the variable is transferred to the opposite side of the equal sign when the addition property is used. Also note that the sign of the number changes when transferred from one side of the equal sign to the other.

When you feel comfortable using the addition property, you may wish to do some of the steps mentally to reduce some of the written work. For example, the preceding two problems may be shortened as follows.

(a) $\qquad\qquad\qquad\qquad\qquad\qquad\qquad\qquad$ *Shortened Form*

$$x - 5 = 12 \qquad\qquad\qquad\qquad x - 5 = 12$$
$$x - 5 + 5 = 12 + 5 \xleftarrow{\text{Do this}} \qquad x = 12 + 5$$
$$\qquad\qquad\qquad\qquad \text{step mentally.}$$
$$x = 12 + 5 \qquad\qquad\qquad\qquad x = 17$$
$$x = 17$$

(b)   *Shortened Form*

$$15 = x + 3$$ $$15 = x + 3$$

$$15 - 3 = x + 3 - 3 \longleftarrow \begin{array}{l}\text{Do this}\\ \text{step mentally.}\end{array}$$ $$15 - 3 = x$$

$$15 - 3 = x$$ $$12 = x$$

$$12 = x$$

Exercise Set 2.2

By checking, determine if the number following the equation is a solution to the equation.

1. $3x - 1 = 11, 4$

2. $2x + 1 = x - 5, -6$

3. $2x - 5 = 5(x + 2), -3$

4. $2(x - 3) = 3(x + 1), 1$

5. $3x - 5 = 2(x + 3) - 11, 0$

6. $-2(x - 3) = -5x + 3 - x, -2$

7. $5(x + 2) - 3(x - 1) = 4, 2.3$

8. $x + 3 = 3x + 2, \frac{1}{2}$

9. $4x - 4 = 2x - 3, \frac{1}{2}$

10. $3x + 4 = -2x + 9, \frac{1}{2}$

11. $3(x + 2) = 5(x - 1), \frac{11}{2}$

12. $-(x + 3) - (x - 6) = 3x - 4, 5$

Solve each equation, and then check your solution.

13. $x + 3 = 8$

14. $x - 4 = 9$

15. $x + 7 = -3$

16. $x - 4 = -8$

17. $x + 3 = -4$

18. $x - 16 = 36$

19. $x + 43 = -18$

20. $6 + x = 9$

21. $-6 + x = 12$

22. $13 = 5 + x$

23. $27 = x - 16$

24. $-9 = x - 25$

25. $-13 = x - 1$

26. $4 = 11 + x$

27. $29 = -43 + x$

28. $-18 = -14 + x$

29. $7 + x = -19$

30. $9 + x = 9$

31. $x + 29 = -29$

32. $4 + x = -9$

33. $6 = x - 4$

34. $5 + x = 12$

35. $x + 7 = -5$

36. $6 = 4 + x$

37. $9 + x = 12$

38. $-4 = x - 3$

39. $-5 = 4 + x$

40. $12 = 16 + x$

41. $30 = x - 19$

42. $15 + x = -5$

43. $x - 12 = -9$

44. $x + 6 = -12$

45. $4 + x = 9$

46. $-6 = 9 + x$

47. $-8 = -9 + x$

48. $-12 = 8 + x$

49. $2 = x - 9$

50. $2 = x + 9$

51. $-50 = x - 24$

52. $-29 + x = -15$

53. $16 + x = -20$

54. $-25 = 18 + x$

55. $40.2 + x = -7.3$

56. $-27.23 + x = 9.77$

57. $-37 + x = 9.5$

58. $7.2 + x = 7.2$

59. $x - 8.42 = -30$

60. $6.2 + x = 5.7$

61. $9.75 = x + 9.75$

62. $139 = x - 117$

63. $600 = x - 120$

64. $427 = x - 963$

65. Explain what is meant by equivalent equations.

66. When solving the equation $x - 4 = 6$, would you add 4 to both sides of the equation or subtract 6 from both sides of the equation? Explain your answer.

67. When solving the equation $5 = x + 3$, would you subtract 5 from both sides of the equation or subtract 3 from both sides of the equation? Explain your answer.

Cumulative Review Exercises

[1.9] *Evaluate*

68. $3x + 4(x - 3) + 2$ when $x = 4$.

69. $6x - 2(2x + 1)$ when $x = -3$.

[2.1] *Simplify*

70. $4x + 3(x - 2) - 5x - 7$.

71. $-(x - 3) + 7(2x - 5) - 3x$.

JUST FOR FUN

1. By checking, determine which of the following are solutions to the equation $2(x + 3) = 2x + 6$.

(a) -1 (b) 5 (c) $\dfrac{1}{2}$

(d) Select any other number not given in parts (a), (b), or (c) and determine if that number is a solution to the equation.

2. By checking, determine which of the following are solutions to $2x^2 - 7x + 3 = 0$.

(a) 3 (b) 2 (c) $\dfrac{1}{2}$

2.3

The Multiplication Property

▶ **1** Identify reciprocals.

▶ **2** Use the multiplication property to solve equations.

▶ **3** Solve equations of the form $-x = a$.

▶ **1** Before we discuss the multiplication property, let us discuss what is meant by the **reciprocal** of a number. Two numbers are reciprocals of each other when their product is 1. Some examples of numbers and their reciprocals follow.

Number	Reciprocal	Product
3	$\dfrac{1}{3}$	$(3)\left(\dfrac{1}{3}\right) = 1$
-2	$-\dfrac{1}{2}$	$(-2)\left(-\dfrac{1}{2}\right) = 1$
$\dfrac{1}{4}$	4	$\left(\dfrac{1}{4}\right)(4) = 1$
$\dfrac{-3}{5}$	$\dfrac{-5}{3}$	$\left(\dfrac{-3}{5}\right)\left(\dfrac{-5}{3}\right) = 1$
-1	-1	$(-1)(-1) = 1$

The reciprocal of a positive number is a positive number and the reciprocal of a negative number is a negative number. Note that 0 has no reciprocal.

In general, if a represents any number, its reciprocal is $1/a$. For example, the reciprocal of 3 is $\frac{1}{3}$, the reciprocal of -2 is $\frac{1}{-2}$ or $-\frac{1}{2}$, and the reciprocal of $-\frac{3}{5}$ is $1 \div \left(\frac{-3}{5}\right)$ or $-\frac{5}{3}$.

▶ **2** In Section 2.2 we used the addition property to solve equations of the form $x + a = b$. In this section we use the multiplication property to solve equations of the form $ax = b$. Notice the difference in the two forms of equations.

Multiplication Property

If $a = b$, then $a \cdot c = b \cdot c$ for any numbers a, b, and c.

The multiplication property implies that both sides of an equation can be multiplied by the same number without changing the solution. **The multiplication property can be used to solve equations of the form $ax = b$.** We can isolate the variable in equations of this form by multiplying both sides of the equation by the reciprocal of a, $\frac{1}{a}$. By doing so the numerical coefficient of the variable, x, becomes 1, which can be omitted when we write the variable. By following this process we say that we *eliminate* the coefficient from the variable.

Equation	*To solve, use the multiplication property to eliminate the coefficient*
$4x = 9$	4
$-5x = 20$	-5
$15 = \frac{1}{2}x$	$\frac{1}{2}$
$7 = -9x$	-9

Now let us work some problems.

EXAMPLE 1   Solve the equation $3x = 6$.

Solution: To isolate the variable, x, we must eliminate the 3 from the left side of the equation. To do this, we multiply both sides of the equation by the reciprocal of 3, which is $\frac{1}{3}$.

$$3x = 6$$

$$\frac{1}{3} \cdot 3x = \frac{1}{3} \cdot 6 \qquad \text{Multiply both sides of the equation by } \frac{1}{3}.$$

$$\frac{1}{\cancel{3}} \cdot \overset{1}{\cancel{3}}x = \frac{1}{\cancel{3}} \cdot \overset{2}{\cancel{6}} \qquad \text{Divide out the common factors.}$$

$$1x = 2$$

$$x = 2 \qquad \blacksquare$$

Notice in Example 1 that the $1x$ is replaced by an x in the next step. The step where we list the $1x$ can be omitted to save time and space.

EXAMPLE 2   Solve the equation $\dfrac{x}{2} = 4$.

Solution: Since dividing by 2 is the same as multiplying by $\frac{1}{2}$, the equation $\frac{x}{2} = 4$ is the same as $\frac{1}{2}x = 4$. We will therefore multiply both sides of the equation by the reciprocal of $\frac{1}{2}$, which is 2.

$$\frac{x}{2} = 4$$

$$\overset{1}{\cancel{2}} \left(\frac{x}{\cancel{2}} \right) = 2 \cdot 4 \qquad \text{Multiply both sides of the equation by 2.}$$

$$x = 2 \cdot 4$$

$$x = 8$$

Check: $\dfrac{x}{2} = 4$

$\dfrac{8}{2} = 4$

$4 = 4$ true

EXAMPLE 3   Solve the equation $\dfrac{2}{3}x = 6$.

Solution: The reciprocal of $\dfrac{2}{3}$ is $\dfrac{3}{2}$. Multiply both sides of the equation by $\dfrac{3}{2}$.

$$\frac{2}{3}x = 6$$

$$\frac{3}{2} \cdot \frac{2}{3}x = \frac{3}{2} \cdot 6$$

$$1x = 9$$

$$x = 9$$

In Example 1, $3x = 6$, we multiplied both sides of the equation by $\frac{1}{3}$ to isolate the variable. We could have also isolated the variable by dividing both sides of the equation by 3, as follows:

$$3x = 6$$

$$\frac{\overset{1}{\cancel{3}}x}{\underset{1}{\cancel{3}}} = \frac{\overset{2}{\cancel{6}}}{\underset{1}{\cancel{3}}} \qquad \text{Divide both sides of the equation by 3.}$$

$$x = 2$$

We can do this because dividing by 3 is equivalent to multiplying by $\frac{1}{3}$. **Since division can be defined in terms of multiplication ($\frac{a}{b}$ means $a \cdot \frac{1}{b}$), the multiplication property also allows us to divide both sides of an equation by the same nonzero number.**

EXAMPLE 4   Solve the equation $8p = 5$.

Solution: $8p = 5$

$\dfrac{8p}{8} = \dfrac{5}{8}$ Divide both sides of the equation by 8.

$p = \dfrac{5}{8}$

EXAMPLE 5   Solve the equation $-12 = -3x$.

Solution: In this equation the variable, x, is on the right side of the equal sign. To isolate x, we divide both sides of the equation by -3.

$$-12 = -3x$$

$$\frac{-12}{-3} = \frac{-3x}{-3} \qquad \text{Divide both sides of the equation by } -3.$$

$$4 = x \qquad \blacksquare$$

EXAMPLE 6   Solve the equation $0.32x = 1.28$

Solution: We begin by dividing both sides of the equation by 0.32 to isolate the variable x.

$$0.32x = 1.28$$

$$\frac{0.32x}{0.32} = \frac{1.28}{0.32} \qquad \text{Divide both sides of the equation by } 0.32$$

$$x = 4 \qquad \blacksquare$$

If your instructor permits the use of a calculator, working problems involving decimal numbers on a calculator will probably save you time.

HELPFUL HINT   When solving an equation of the form $ax = b$, we can isolate the variable by

1. multiplying both sides of the equation by the reciprocal of a, $\frac{1}{a}$, as was done in Examples 1, 2, and 3, or
2. dividing both sides of the equation by a, as was done in Examples 4, 5, and 6.

Either method may be used to isolate the variable. However, if the equation contains a fraction, or fractions, you will arrive at a solution more quickly by multiplying by the reciprocal of a. This is illustrated in Example 7.

EXAMPLE 7   Solve the equation $-2x = \dfrac{3}{5}$.

Solution: Since this equation contains a fraction, we will isolate the variable by multiplying both sides of the equation by $-\frac{1}{2}$, the reciprocal of -2.

$$-2x = \frac{3}{5}$$

$$\left(-\frac{1}{2}\right)(-2x) = \left(-\frac{1}{2}\right)\left(\frac{3}{5}\right) \qquad \text{Multiply both sides of the equation by } -\frac{1}{2}.$$

$$1x = \left(-\frac{1}{2}\right)\left(\frac{3}{5}\right)$$

$$x = -\frac{3}{10} \qquad \blacksquare$$

In Example 7, if you wished to solve the equation by dividing both sides of the equation by -2, you would have to divide the fraction $\frac{3}{5}$ by -2.

▶ **3** When solving an equation in the following sections, we may obtain an equation like $-x = 7$. This is *not* a solution. The solution to an equation will be of the form $x =$ some number. When given an equation of the form $-x = 7$, we can solve for x by multiplying both sides of the equation by -1, as illustrated in the following example.

EXAMPLE 8   Solve the equation $-x = 7$.

Solution: $-x = 7$ means that $-1x = 7$. To obtain a positive x, we can multiply both sides of the equation by -1.

$$-x = 7$$
$$-1x = 7$$
$$(-1)\,(-1x) = (-1)\,(7) \qquad \text{Multiply both sides of the equation by } -1.$$
$$1x = -7$$
$$x = -7$$

Check: $-x = 7$
$$-(-7) = 7$$
$$7 = 7 \qquad \text{true}$$

Thus, the solution is -7.

Whenever we have the negative of a variable equal to a quantity, as in Example 7, we can solve for the variable by multiplying both sides of the equation by -1.

EXAMPLE 9   Solve the equation $-x = -5$.

Solution:
$$-x = -5$$
$$-1x = -5$$
$$(-1)\,(-1x) = (-1)\,(-5)$$
$$1x = 5$$
$$x = 5$$

HELPFUL HINT

For any real number a, $a \neq 0$,

If $-x = a$ then $x = -a$

Examples: $-x = 7$ $-x = -2$
$$x = -7 \qquad\qquad x = -(-2)$$
$$x = 2$$

When you feel comfortable using the multiplication property, you may wish to do some of the steps mentally to reduce some of the written work. Now we illustrate two examples worked out in detail, along with their shortened form.

EXAMPLE 10   Solve the equation $-3x = -21$.

Solution: $-3x = -21$

$\dfrac{-3x}{-3} = \dfrac{-21}{-3}$ ← Do this step mentally.

$x = \dfrac{-21}{-3}$

$x = 7$

Shortened Form

$-3x = -21$

$x = \dfrac{-21}{-3}$

$x = 7$ ■

EXAMPLE 11   Solve the equation $\dfrac{1}{3}x = 9$.

Solution: $\dfrac{1}{3}x = 9$

$3\left(\dfrac{1}{3}x\right) = 3(9)$ ← Do this step mentally.

$x = 3(9)$

$x = 27$

Shortened Form

$\dfrac{1}{3}x = 9$

$x = 3(9)$

$x = 27$ ■

In the previous section we discussed the addition property and in this section we discussed the multiplication property. It is important that you understand the difference between the two. The following helpful hint should be studied carefully.

HELPFUL HINT

The **addition property** is used to solve equations of the form $x + a = b$.

$x + 3 = -6$ $x - 5 = -2$

$x + 3 - 3 = -6 - 3$ $x - 5 + 5 = -2 + 5$

$x = -9$ $x = 3$

The **multiplication property** is used to solve equations of the form $ax = b$.

$3x = 6$ $\dfrac{x}{2} = 4$ $\dfrac{2}{5}x = 12$

$\dfrac{3x}{3} = \dfrac{6}{3}$ $2\left(\dfrac{x}{2}\right) = 2(4)$ $\left(\dfrac{5}{2}\right)\left(\dfrac{2}{5}x\right) = \left(\dfrac{5}{2}\right)(12)$

$x = 2$ $x = 8$ $x = 30$

Note: The *addition property* is used when a number is *added to or subtracted from* a variable. The *multiplication property* is used when a variable is *multiplied or divided by a number*.

Exercise Set 2.3

Solve each equation, and then check your solution.

1. $3x = 9$ **2.** $4x = 16$ **3.** $\dfrac{x}{2} = 4$ **4.** $\dfrac{x}{3} = 12$

5. $-4x = 8$ **6.** $8 = 16y$ **7.** $\dfrac{x}{6} = -2$ **8.** $\dfrac{x}{3} = -2$

9. $\dfrac{x}{5} = 1$ **10.** $-2x = 12$ **11.** $-32x = -96$ **12.** $16 = -4y$

13. $-6 = 4z$

14. $\dfrac{x}{8} = -3$

15. $-x = -4$

16. $-x = 9$

17. $-2 = -y$

18. $-3 = \dfrac{x}{5}$

19. $-\dfrac{x}{7} = -7$

20. $4 = \dfrac{x}{9}$

21. $9 = -18x$

22. $12y = -15$

23. $-\dfrac{x}{3} = -2$

24. $-\dfrac{a}{8} = -7$

25. $19x = 35$

26. $-24x = -18$

27. $-4.2x = -8.4$

28. $-3.72 = 1.24y$

29. $7x = -7$

30. $3x = \dfrac{3}{5}$

31. $5x = -\dfrac{3}{8}$

32. $-2b = -\dfrac{4}{5}$

33. $15 = -\dfrac{x}{5}$

34. $\dfrac{x}{16} = -4$

35. $-\dfrac{x}{5} = -25$

36. $-x = -\dfrac{5}{9}$

37. $\dfrac{x}{5} = -7$

38. $-3x = -18$

39. $6 = \dfrac{x}{4}$

40. $-3 = \dfrac{x}{-5}$

41. $6c = -30$

42. $\dfrac{2}{7}x = 7$

43. $\dfrac{y}{-2} = -6$

44. $-2x = \dfrac{3}{5}$

45. $\dfrac{-3}{8}x = 6$

46. $-x = \dfrac{4}{7}$

47. $\dfrac{1}{3}x = -12$

48. $6 = \dfrac{3}{5}x$

49. $-4 = -\dfrac{2}{3}z$

50. $-8 = \dfrac{-4}{5}x$

51. $-1.4x = 28.28$

52. $-0.42x = -2.142$

53. $2x = -\dfrac{5}{2}$

54. $6x = \dfrac{8}{3}$

55. $\dfrac{2}{3}x = 6$

56. $-\dfrac{1}{2}x = \dfrac{2}{3}$

57. When solving the equation $3x = 5$, would you divide both sides of the equation by 3 or by 5? Explain why.

58. When solving the equation $-2x = 5$, would you add 2 to both sides of the equation or divide both sides of the equation by -2. Explain why.

59. Consider the equation $\dfrac{2}{3}x = 4$. This equation could be solved by multiplying both sides of the equation by $\dfrac{3}{2}$, the reciprocal of $\dfrac{2}{3}$, or by dividing both sides of the equation by $\dfrac{2}{3}$. Which method do you feel would be easier? Explain your answer. Find the solution to the equation.

60. Consider the equation $4x = \dfrac{3}{5}$. Would it be easier to solve this equation by dividing both sides of the equation by 4 or by multiplying both sides of the equation by $\dfrac{1}{4}$, the reciprocal of 4? Explain your answer. Find the solution to the problem.

61. Consider the equation $\dfrac{3}{7}x = \dfrac{4}{5}$. Would it be easier to solve this equation by dividing both sides of the equation by $\dfrac{3}{7}$ or by multiplying both sides of the equation by $\dfrac{7}{3}$, the reciprocal of $\dfrac{3}{7}$? Explain your answer. Find the solution to the equation.

Cumulative Review Exercises

[1.6] **62.** Subtract -6 from -15.

63. Evaluate $6 - (-3) - 5 - 4$.

[2.1] **64.** Simplify $-(x + 3) - 5(2x - 7) + 6$.

[2.2] **65.** Solve the equation $-48 = x + 9$.

2.4

Solving Linear Equations with a Variable on Only One Side of the Equation

▶1 Solve linear equations that contain a variable on only one side of the equal sign.

▶1 In this section we discuss how to solve linear equations when the variable appears on only one side of the equal sign. In Section 2.5 we will discuss how to solve linear equations when the variable appears on both sides of the equal sign.

No one method is the "best" to solve all linear equations. Following is a general procedure that can be used to solve linear equations when the variable appears on only one side of the equation and the equation does not contain fractions.

> **To Solve Linear Equations
> with a Variable on Only One Side of the Equal Sign**
>
> 1. Use the distributive property to remove parentheses.
> 2. Combine like terms on the same side of the equal sign.
> 3. Use the addition property to obtain an equation with the term containing the variable on one side of the equal sign and a constant on the other side. This will result in an equation of the form $ax = b$.
> 4. Use the multiplication property to isolate the variable. This will give an answer of the form $x = \dfrac{b}{a}\left(\text{or } 1x = \dfrac{b}{a}\right)$.
> 5. Check the solution in the *original* equation.

Equations containing fractions will be solved using a different procedure. We will discuss how to solve equations containing fractions in Section 6.7.

When solving an equation remember that our goal is to obtain the variable all by itself on one side of the equation.

EXAMPLE 1   Solve the equation $2x - 5 = 9$.

Solution: We will follow the procedure just outlined for solving equations. Since the equation contains no parentheses and since there are no like terms to be combined, we start with step 3.

Step 3 $2x - 5 = 9$

$2x - 5 \boxed{+\ 5} = 9 \boxed{+\ 5}$ Add 5 to both sides of the equation.

$2x = 14$

Step 4 $\dfrac{2x}{2} = \dfrac{14}{2}$ Divide both sides of the equation by 2.

$x = 7$

Step 5   *Check:* $2x - 5 = 9$

$2(7) - 5 = 9$

$14 - 5 = 9$

$9 = 9$ true

Since the check is true, the solution is 7. Note that after completing step 3 we obtain $2x = 14$ which is an equation of the form $ax = b$. And after completing step 4 we obtain the answer in the form $x =$ some number. ■

HELPFUL HINT
_____

When solving an equation that does not contain fractions, **the addition property (step 3) is to be used before the multiplication property (step 4).** If you use the multiplication property before the addition property, it is still possible to obtain the correct answer. However, you will usually have to do more work, and you may end up working with fractions. What would happen if you tried to solve Example 1 using the multiplication property before the addition property? In Chapter 6 we will discuss at length how to solve equations that contain fractions.

EXAMPLE 2   Solve the equation $-2x - 6 = -3$.

Solution:

$$-2x - 6 = -3$$

Step 3 $-2x - 6 + 6 = -3 + 6$ Add 6 to both sides of the equation.

$$-2x = 3$$

Step 4 $$\frac{-2x}{-2} = \frac{3}{-2}$$ Divide both sides of the equation by -2.

$$x = -\frac{3}{2}$$

Step 5   *Check:* $-2x - 6 = -3$

$$-2\left(-\frac{3}{2}\right) - 6 = -3$$

$$3 - 6 = -3$$

$$-3 = -3 \qquad \text{true}$$

The solution is $-\dfrac{3}{2}$.

Note that checks are always made with the original equation. In some of the following examples the check will be omitted to save space.

EXAMPLE 3   Solve the equation $16 = 4x + 6 - 2x$.

Solution: Again we must isolate the variable x. Since the right side of the equation has two like terms containing the variable x, we will first combine these like terms.

Step 2 $16 = 4x + 6 - 2x$

$16 = 2x + 6$ Like terms were combined.

Step 3 $16 - 6 = 2x + 6 - 6$ Subtract 6 from both sides of equation.

$10 = 2x$

Step 4 $$\frac{10}{2} = \frac{2x}{2}$$ Divide both sides of equation by 2.

$5 = x$

The preceding solution can be condensed as follows.

$16 = 4x + 6 - 2x$

$16 = 2x + 6$ Like terms were combined.

$10 = 2x$ 6 was subtracted from both sides of equation.

$5 = x$ Both sides of equation were divided by 2.

In Chapter 3, Applications of Algebra, we will be solving many equations that contain decimal numbers. To solve such equations we follow the same procedure as outlined earlier. Example 4 illustrates the solution to an equation that contains decimal numbers.

HELPFUL HINT

In the first two chapters you have been introduced to a variety of mathematics terms. Some of the most commonly used terms are "evaluate," "simplify," "solve," and "check." Make sure you understand what each term means and when each term is used.

Evaluate: To *evaluate an expression* means to give a numerical answer for the problem. For example,

Evaluate: $16 \div 2^2 + 36 \div 4$ Evaluate: $-x^2 + 3x - 2$ when $x = 4$
$$= 16 \div 4 + 36 \div 4 \qquad\qquad = -4^2 + 3(4) - 2$$
$$= 4 + 36 \div 4 \qquad\qquad\quad = -16 + 12 - 2$$
$$= 4 + 9 = 13 \qquad\qquad\quad = -4 - 2 = -6$$

Simplify: To *simplify an expression* usually means to combine like terms. For example,

Simplify: $3(x - 2) - 4(2x + 3) = 3x - 6 - 8x - 12$
$$= -5x - 18$$

Note that when you simplify an expression containing variables you do not generally end up with just a numerical value unless all the variable terms happen to add to zero.

Solve: To *solve an equation* means to find the value or the values that make the equation a true statement. For example,

Solve: $2x + 3(x + 1) = 18$
$$2x + 3x + 3 = 18$$
$$5x + 3 = 18$$
$$5x = 15$$
$$x = 3$$

Check: To *check an equation,* we substitute the value believed to be the solution back into the original equation. If this substitution results in a true statement, then we say the answer checks. For example, to check the solution of the equation just solved, we do the following:

Check: $x = 3$ $2x + 3(x + 1) = 18$
$$2(3) + 3(3 + 1) = 18$$
$$6 + 3(4) = 18$$
$$6 + 12 = 18$$
$$18 = 18 \qquad \text{true}$$

Since we obtained a true statement, the 3 checks.

It is important to realize that *expressions can be evaluated or simplified* (depending on the type of problem) and *equations are solved and then checked.*

EXAMPLE 4   Solve the equation $x + 1.24 - 0.07x = 4.96$.

Solution:
$$x + 1.24 - 0.07x = 4.96$$
$$0.93x + 1.24 = 4.96 \qquad \text{Like terms were combined,} \\ 1x - 0.07x = 0.93x.$$
$$0.93x + 1.24 - 1.24 = 4.96 - 1.24 \qquad \text{Subtract 1.24 from both sides of equation.}$$
$$0.93x = 3.72$$
$$\frac{0.93x}{0.93} = \frac{3.72}{0.93} \qquad \text{Divide both sides of equation by 0.93.}$$
$$x = 4$$

EXAMPLE 5   Solve the equation $2(x + 4) - 5x = -3$.

Solution:
$$2(x + 4) - 5x = -3$$
$$2x + 8 - 5x = -3 \qquad \text{The distributive property was used.}$$
$$-3x + 8 = -3 \qquad \text{Like terms were combined.}$$
$$-3x + 8 - 8 = -3 - 8 \qquad \text{Subtract 8 from both sides of equation.}$$
$$-3x = -11$$
$$\frac{-3x}{-3} = \frac{-11}{-3} \qquad \text{Divide both sides of equation by } -3.$$
$$x = \frac{11}{3}$$

The preceding solution can be condensed as follows:

$$2(x + 4) - 5x = -3$$
$$2x + 8 - 5x = -3 \qquad \text{The distributive property was used.}$$
$$-3x + 8 = -3 \qquad \text{Like terms were combined.}$$
$$-3x = -11 \qquad \text{8 was subtracted from both sides of equation.}$$
$$x = \frac{11}{3} \qquad \text{Both sides of equations were divided by } -3.$$

EXAMPLE 6   Solve the equation $2x - (x + 2) = 6$.

Solution:
$$2x - (x + 2) = 6$$
$$2x - x - 2 = 6 \qquad \text{The distributive property was used.}$$
$$x - 2 = 6 \qquad \text{Like terms were combined.}$$
$$x = 8 \qquad \text{2 was added to both sides of equation.}$$

Exercise Set 2.4

Solve each equation.

1. $2x + 3 = 7$

2. $2x - 4 = 8$

3. $-2x - 5 = 7$

4. $-4x + 5 = -3$

5. $5x - 6 = 19$

6. $6 - 3x = 18$

7. $5x - 2 = 10$

8. $-9x + 3 = 15$

9. $-x - 4 = 8$

10. $6 = 2x - 3$

11. $12 - x = 9$

12. $-3x - 3 = -12$

13. $9 + 2x = 24$

14. $-7x + 3 = -12$

15. $32x + 9 = -12$

16. $14 = 18 + 7x$

17. $-44 = 9x + 12$

18. $-18 + 18x = -18$

19. $6x - 9 = 21$

20. $-x + 4 = -8$

21. $12 = -6x + 5$

22. $15 = 7x + 1$

23. $-2x - 7 = -13$

24. $-2 - x = -12$

25. $x + 0.05x = 21$

26. $x + 0.07x = 16.05$

27. $2.3x - 9.34 = 6.3$

28. $-2.3 = -1.4 + 0.6x$

29. $28.8 = x - 0.10x$

30. $32.76 = 2.45x - 8.75x$

31. $2(x + 1) = 6$

32. $3(x - 2) = 12$

33. $4(3 - x) = 12$

34. $-2(x + 3) = -9$

35. $-4 = -(x + 5)$

36. $-3(2 - 3x) = 9$

37. $12 = 4(x + 3)$

38. $-2(x + 4) + 5 = 1$

39. $5 = 2(3x + 6)$

40. $-2 = 5(3x + 1) - 12x$

41. $2x + 3(x + 2) = 11$

42. $4 = -2(x + 3)$

43. $x - 3(2x + 3) = 11$

44. $3(4 - x) + 5x = 9$

45. $5x + 3x - 4x - 7 = 9$

46. $-(x + 2) = 4$

47. $0.7(x + 3) = 4.2$

48. $12 + (x + 9) = 7$

49. $1.4(5x - 4) = -1.4$

50. $0.1(2.4x + 5) = 1.7$

51. $3 - 2(x + 3) + 2 = 1$

52. $2(3x - 4) - 4x = 12$

53. $1 - (x + 3) + 2x = 4$

54. $5x - 2x - 7x = -20$

55. $4 - 6x + 9 - 3 = -8$

56. $-4(x + 2) - 3x = 20$

57. When solving equations that do not contain fractions, do we normally use the addition or multiplication property first in the process of isolating the variable? Explain your answer.

58. **(a)** Explain, in a step-by-step manner, how to solve the equation $2(3x + 4) = -4$.

(b) Solve the equation by following the steps listed in part (a).

59. **(a)** Explain, in a step-by-step manner, how to solve the equation $4x - 2(x + 3) = 4$.

(b) Solve the equation by following the steps listed in part (a).

Cumulative Review Exercises

[1.2] **60.** Add $\dfrac{5}{8} + \dfrac{3}{5}$.

[1.9] **61.** Evaluate $[5(2 - 6) + 3(8 \div 4)^2]^2$.

[2.2] **62.** To solve an equation, what do you need to do to the variable?

[2.3] **63.** To solve the equation $7 = -4x$, would you add 4 to both sides of the equation or divide both sides of the equation by -4? Explain your answer.

JUST FOR FUN

Solve each equation.

1. $3(x - 2) - (x + 5) - 2(3 - 2x) = 18$.

2. $-6 = -(x - 5) - 3(5 + 2x) - 4(2x - 4)$.

3. $4[3 - 2(x + 4)] - (x + 3) = 13$.

2.5

Solving Linear Equations with the Variable on Both Sides of the Equation

▶ **1** Solve equations when the variable appears on both sides of the equal sign.

▶ **2** Identify and define identities.

▶ **1** The equation $4x + 6 = 2x + 4$ contains the variable x on both sides of the equal sign. To solve equations of this type, we must use the appropriate properties to rewrite the equation with all terms containing the variable on only one side of the equal sign and all terms not containing the variable on the other side of the equal sign. This will allow us to isolate the variable, which is our goal. Following is a general procedure, similar to the one outlined in Section 2.4, that can be used to solve linear equations with the variable on both sides of the equal sign.

To Solve Linear Equations with the Variable on Both Sides of the Equal Sign

1. Use the distributive property to remove parentheses.
2. Combine like terms on the same side of the equal sign.
3. Use the addition property to rewrite the equation with all terms containing the variable on one side of the equal sign and all terms not containing the variable on the other side of the equal sign. It may be necessary to use the addition property a number of times to accomplish this. Repeated use of the addition property will eventually result in an equation of the form $ax = b$.
4. Use the multiplication property to isolate the variable. This will give an answer of the form $x =$ some number.
5. Check the solution in the original equation.

Whenever possible, you should check your work. We will not show all the checks to save space.

EXAMPLE 1 Solve the equation $4x + 6 = 2x + 4$.

Solution: Remember that our goal is always to get all terms with the variable on one side of the equal sign and all terms without the variable on the other side of the equal sign. Many methods can be used to isolate the variable. We will illustrate two. In method 1, we will isolate the variable on the left side of the equation. In method 2, we will isolate the variable on the right side of the equation. In both methods, we will follow the steps given in the preceding box. Since this equation does not contain parentheses, and since there are no like terms on the same side of the equal sign, we begin with step 3.

Method 1:

$$4x + 6 = 2x + 4$$

Step 3 $\quad 4x - 2x + 6 = 2x - 2x + 4 \qquad$ Subtract $2x$ from both sides of the equation.

$$2x + 6 = 4$$

Step 3 $\quad 2x + 6 - 6 = 4 - 6 \qquad$ Subtract 6 from both sides of the equation.

$$2x = -2$$

Step 4 $\qquad \dfrac{2x}{2} = \dfrac{-2}{2} \qquad$ Divide both sides of the equation by 2.

$$x = -1$$

Method 2:

$$4x + 6 = 2x + 4$$

Step 3 $\quad 4x - 4x + 6 = 2x - 4x + 4 \qquad$ Subtract $4x$ from both sides of the equation.

$$6 = -2x + 4$$

Step 3 $6 \boxed{- 4} = -2x + 4 \boxed{- 4}$ Subtract 4 from both sides of the
   $2 = -2x$ equation.

Step 4 $\dfrac{2}{-2} = \dfrac{-2x}{-2}$ Divide both sides of the equation
   by -2.
   $-1 = x$

The same answer is obtained using both methods.

Step 5   *Check:* $4x + 6 = 2x + 4$
   $4(-1) + 6 = 2(-1) + 4$
   $-4 + 6 = -2 + 4$
   $2 = 2$ true ∎

EXAMPLE 2   Solve the equation $2x - 3 - 5x = 13 + 4x - 2$.

Solution: Since there are like terms *on the same side of the equal sign,* we will begin by combining these like terms.

Step 2 $2x - 3 - 5x = 13 + 4x - 2$
   $-3x - 3 = 4x + 11$ Like terms were combined.

Step 3 $-3x \boxed{+ 3x} - 3 = 4x \boxed{+ 3x} + 11$ Add $3x$ to both sides of the equation.
   $-3 = 7x + 11$

Step 3 $-3 \boxed{- 11} = 7x + 11 \boxed{- 11}$ Subtract 11 from both sides of
   $-14 = 7x$ the equation.

Step 4 $\dfrac{-14}{7} = \dfrac{7x}{7}$ Divide both sides of the
   equation by 7.
   $-2 = x$

Step 5   *Check:* $2x - 3 - 5x = 13 + 4x - 2$
   $2(-2) - 3 - 5(-2) = 13 + 4(-2) - 2$
   $-4 - 3 + 10 = 13 - 8 - 2$
   $-7 + 10 = 5 - 2$
   $3 = 3$ true

Since the check is true, the solution is -2. ∎

The solution to Example 2 could be condensed as follows:

$2x - 3 - 5x = 13 + 4x - 2$
$-3x - 3 = 4x + 11$ Like terms were combined.
$-3 = 7x + 11$ $3x$ was added to both sides of equation.
$-14 = 7x$ 11 was subtracted from both sides of equation.
$-2 = x$ Both sides of equation were divided by 7.

EXAMPLE 3   Solve the equation $5.74x + 5.42 = 2.24x - 9.28$.

Solution: We first notice that there are no like terms on the same side of the equal sign that can be combined. We will elect to collect the terms with the variables on the left side of the equation.

$$5.74x + 5.42 = 2.24x - 9.28$$

$$5.74x - 2.24x + 5.42 = 2.24x - 2.24x - 9.28 \qquad \text{Subtract } 2.24x \text{ from both sides of equation.}$$

$$3.5x + 5.42 = -9.28$$

$$3.5x + 5.42 - 5.42 = -9.28 - 5.42 \qquad \text{Subtract } 5.42 \text{ from both sides of equation.}$$

$$3.5x = -14.7$$

$$\frac{3.5x}{3.5} = \frac{-14.7}{3.5} \qquad \text{Divide both sides of equation by } 3.5.$$

$$x = -4.2 \qquad \blacksquare$$

EXAMPLE 4 Solve the equation $2(p + 3) = -3p + 10$.

Solution:

$$2(p + 3) = -3p + 10$$

Step 1
$$2p + 6 = -3p + 10 \qquad \text{Distributive property was used.}$$

Step 3
$$2p + 3p + 6 = -3p + 3p + 10 \qquad \text{Add } 3p \text{ to both sides of equation.}$$

$$5p + 6 = 10$$

Step 3
$$5p + 6 - 6 = 10 - 6 \qquad \text{Subtract 6 from both sides of equation.}$$

$$5p = 4$$

Step 4
$$\frac{5p}{5} = \frac{4}{5} \qquad \text{Divide both sides of equation by 5.}$$

$$p = \frac{4}{5} \qquad \blacksquare$$

The solution to Example 4 could be condensed as follows:

$$2(p + 3) = -3p + 10$$

$$2p + 6 = -3p + 10 \qquad \text{Distributive property was used.}$$

$$5p + 6 = 10 \qquad 3p \text{ was added to both sides of equation.}$$

$$5p = 4 \qquad 6 \text{ was subtracted from both sides of equation.}$$

$$p = \frac{4}{5} \qquad \text{Both sides of equation were divided by 5.}$$

HELPFUL HINT

After the distributive property was used in Step 1, Example 4, we obtained the equation $2p + 6 = -3p + 10$. At this point we had to decide whether to collect terms with the variable on the left or the right side of the equal sign. If we wish the sum of the variable terms to be positive, we use the addition property to eliminate the variable, with the *smaller* numerical coefficient from one side of the equation. Since -3 is smaller than 2, we added $3p$ to both sides of the equation. This eliminated the $-3p$ from the right side of the equation and resulted in the sum of the variable terms on the left side of the equation, $5p$, being positive.

EXAMPLE 5 Solve the equation $2(x - 5) + 3 = 3x + 9$.

Solution:
$$2(x - 5) + 3 = 3x + 9$$

Step 1	$2x - 10 + 3 = 3x + 9$	Distributive property was used.
Step 2	$2x - 7 = 3x + 9$	Like terms were combined.
Step 3	$-7 = x + 9$	$2x$ was subtracted from both sides of equation.
Step 3	$-16 = x$	9 was subtracted from both sides of equation. ∎

EXAMPLE 6 Solve the equation $7 - 2x + 5x = -2(-3x + 4)$.

Solution:
$$7 - 2x + 5x = -2(-3x + 4)$$

Step 1	$7 - 2x + 5x = 6x - 8$	Distributive property was used.
Step 2	$7 + 3x = 6x - 8$	Like terms were combined.
Step 3	$7 = 3x - 8$	$3x$ was subtracted from both sides of equation.
Step 3	$15 = 3x$	8 was added to both sides of equation.
Step 4	$5 = x$	Both sides of equation were divided by 3.

The solution is 5. ∎

▸ **2** Thus far all the equations we have solved have had a single value for a solution. Equations of this type are called **conditional equations,** for they are only true under specific conditions. Some equations, as in Example 7, are true for all values of x. Equations that are true for all values of x are called **identities.** A third type of equation, as in Example 8, has no solution.

EXAMPLE 7 Solve the equation $2x + 6 = 2(x + 3)$.

Solution:
$$2x + 6 = 2(x + 3)$$
$$2x + 6 = 2x + 6$$

Since the same expression appears on both sides of the equal sign, the statement is true for all values of x. If we proceeded to solve this equation further, we might obtain

| $2x = 2x$ | 6 was subtracted from both sides of equation. |
| $0 = 0$ | $2x$ was subtracted from both sides of equation. |

Note: The solution process could have been terminated at $2x + 6 = 2x + 6$. Since one side is identical to the other side, the equation is true for all values of x. **Therefore, the solution to this equation is all real numbers.** ∎

EXAMPLE 8 Solve the equation $-3x + 4 + 5x = 4x - 2x + 5$.

Solution:
$$-3x + 4 + 5x = 4x - 2x + 5$$

$2x + 4 = 2x + 5$	Like terms were combined.
$2x - 2x + 4 = 2x - 2x + 5$	Subtract $2x$ from both sides of equation.
$4 = 5$ false	

When solving an equation, if you obtain an obviously false statement, as in this example, the equation has no solution. No value of x will make the equation a true statement. **Therefore, when giving the answer to this problem, you should use the words *no solution*.** An answer left blank may be marked wrong. ∎

COMMON STUDENT ERROR

At this point some students will begin to confuse combining like terms with using the addition property. Remember that *when combining terms you work on only one side of the equal sign at a time*, as in

$$3x + 4 - x = 4x - 8$$
$$2x + 4 = 4x - 8 \qquad \text{The } 3x \text{ and } -x \text{ were combined.}$$

When using the addition property, you add (or subtract) the same quantity to (from) **both sides of the equation.**

Correct

$$2x + 4 = 4x - 8$$
$$2x - 2x + 4 = 4x - 2x - 8 \qquad 2x \text{ was subtracted from both sides of equation.}$$
$$4 = 2x - 8$$
$$4 + 8 = 2x - 8 + 8 \qquad 8 \text{ was added to both sides of equation.}$$
$$12 = 2x$$
$$x = 6$$

Wrong

$$3x + 4 - x = 4x - 8$$
$$3x + x + 4 - x + x = 4x - 8 \qquad \begin{array}{l}\text{Wrong use of the addition property;}\\ \text{note } x \text{ was not added to } \textit{both} \text{ sides}\\ \text{of the equation.}\end{array}$$

Ordinarily, when solving an equation, combining like terms is done before the use of the addition property.

Exercise Set 2.5

Solve each equation.

1. $2x + 4 = 3x$

2. $x + 5 = 2x - 1$

3. $-4x + 10 = 6x$

4. $6x = 4x + 8$

5. $5x + 3 = 6$

6. $-6x = 2x + 16$

7. $15 - 3x = 4x - 2x$

8. $9 - 12x = 4x + 15$

9. $x - 3 = 2x + 18$

10. $-5x = -4x + 9$

11. $3 - 2x = 9 - 8x$

12. $124.8 - 9.4x = 4.8x + 32.5$

13. $4 - 0.6x = 2.4x - 8.48$

14. $8 + y = 2y - 6 + y$

15. $5x = 2(x + 6)$

16. $8x - 4 = 3(x - 2)$

17. $x - 25 = 12x + 9 + 3x$

18. $5y + 6 = 2y + 3 - y$

19. $2(x + 2) = 4x + 1 - 2x$

20. $4x = 10 - 2(x - 4)$

21. $-(x + 2) = -6x + 32$

22. $15(4 - x) = 5(10 + 2x)$

23. $4 - (2x + 5) = 6x + 31$

24. $4(2x - 3) = -2(3x + 16)$

25. $0.1(x + 10) = 0.3x - 4$

26. $3y - 6y + 2 = 8y + 6 - 5y$

27. $2(x + 4) = 4x + 3 - 2x + 5$

28. $5(2.9x - 3) = 2(x + 4)$

29. $9(-y + 3) = -6y + 15 - 3y + 12$

30. $-4(-y + 3) = 12y + 8 - 2y$

31. $-(3 - x) = -(2x + 3)$

32. $12 - 2x - 3(x + 2) = 4x + 6 - x$

33. $-(x + 4) + 5 = 4x + 1 - 5x$

34. $19x + 3(4x + 9) = -6x - 38$

35. $35(2x + 12) = 7(x - 4) + 3x$

36. $10(x - 10) + 5 = 5(2x - 20)$

37. $0.4(x + 0.7) = 0.6(x - 4.2)$

38. $3(x - 4) = 2(x - 8) + 5x$

39. $-(x - 5) + 2 = 3(4 - x) + 5x$

40. $1.2(6x - 8) = 2.4(x - 5)$

41. $2(x - 6) + 3(x + 1) = 4x + 3$

42. $-2(-3x + 5) + 6 = 4(x - 2)$

43. $5 + 2x = 6(x + 1) - 5(x - 3)$

44. $4 - (6x + 3) = -(-2x + 3)$

45. $5 - (x - 5) = 2(x + 3) - 6(x + 1)$

46. $12 - 6x + 3(2x + 3) = 2x + 5$

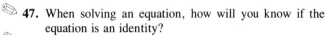 **47.** When solving an equation, how will you know if the equation is an identity?

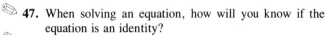 **48.** When solving an equation, how will you know if the equation has no real solution?

49. (a) Explain, in a step by step manner, how to solve the equation $4(x + 3) = 6(x - 5)$.

　　　 (b) Solve the equation by following the steps listed in part (a).

50. (a) Explain, in a step by step manner, how to solve the equation $4x + 3(x + 2) = 5x - 10$.

　　　 (b) Solve the equation by following the steps listed in part (a).

Cumulative Review

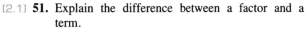 [2.1] **51.** Explain the difference between a factor and a term.

52. Simplify $2(x - 3) + 4x - (4 - x)$.

[2.4] **53.** Solve $2(x - 3) + 4x - (4 - x) = 0$.

54. Solve the equation $(x + 4) - (4x - 3) = 16$.

JUST FOR FUN

1. Solve the equation $-2(x + 3) + 5x = -3(5 - 2x) + 3(x + 2) + 6x$.

2. Solve the equation $4(2x - 3) - (x + 7) - 4x + 6 = 5(x - 2) - 3x + 7(2x + 2)$

3. Solve the equation $4 - [5 - 3(x + 2)] = x - 3$.

2.6

Ratios and Proportions

▶ **1**　Understand ratios.

▶ **2**　Solve proportions using cross-multiplication.

▶ **3**　Solve practical application problems.

▶ **4**　Use proportions to change units.

▶ **5**　Use proportions in geometric problems.

▶ **1**　A **ratio** is a quotient of two quantities with the same units. Ratios provide a way to compare two numbers or quantities. The ratio of the number a to the number b may be written

$$a \text{ to } b, \qquad a:b, \qquad \text{or} \qquad \frac{a}{b}$$

where a and b are called the **terms** of the ratio.

EXAMPLE 1   An algebra class consists of 11 males and 15 females.

(a) Find the ratio of males to females.

(b) Find the ratio of females to the entire class.

Solution: (a) $11:15$ (b) $15:26$ ■

In Example 1, part (a) could also have been written $\frac{11}{15}$ or "11 to 15." Part (b) could also have been written $\frac{15}{26}$ or "15 to 26."

EXAMPLE 2   There are two types of cholesterols, low density lipoprotein, (LDL—considered the bad type of cholesterol) and high density lipoprotein (HDL—considered the good type of cholesterol). Some doctors recommend that the ratio of low density to high density cholesterol be less than or equal to $4:1$. A cholesterol test given to Mr. Duncan showed that his low density cholesterol measured 167 milligrams per deciliter, and his high density cholesterol measured 40 milligrams per deciliter. Is Mr. Duncan's low-to-high density-level ratio less than or equal to the recommended $4:1$ ratio?

Solution: The ratio of low density to high density cholesterol is $\frac{167}{40}$. If we divide 167 by 40 we obtain 4.175. Thus, Mr. Duncan's ratio is equivalent to $4.175:1$. Therefore, his ratio is not less than or equal to the desired 4:1 ratio.   ■

EXAMPLE 3   Find the ratio of 8 feet to 20 yards.

Solution: To express this as a ratio, both quantities must be in the same units. Since 1 yard equals 3 feet, 20 yards equals 60 feet. Thus the ratio is $\frac{8}{60}$. The ratio in lowest terms is $\frac{2}{15}$ (or $2:15$).   ■

▶ **2**   A **proportion** is a special type of equation. It is a statement of equality between two ratios. One way of denoting a proportion is $a:b = c:d$, which is read "a is to b as c is to d." In this text we write proportions as

$$\frac{a}{b} = \frac{c}{d}$$

The a and d are referred to as the **extremes,** and the b and c are referred to as the **means** of the proportion. One method that can be used in evaluating proportions is **cross-multiplication:**

Cross-multiplication

If $\dfrac{a}{b} = \dfrac{c}{d}$ then $ad = bc$.

Note that the product of the means is equal to the product of the extremes.

If any three of the four quantities of a proportion are known, the fourth quantity can easily be found.

EXAMPLE 4   Solve for x by cross-multiplying $\dfrac{x}{3} = \dfrac{25}{15}$.

Solution:

$$\dfrac{x}{3} = \dfrac{25}{15}$$

$$x \cdot 15 = 3 \cdot 25$$

$$15x = 75$$

$$x = \dfrac{75}{15} = 5$$

Check:

$$\dfrac{x}{3} = \dfrac{25}{15}$$

$$\dfrac{5}{3} = \dfrac{25}{15}$$

$$\dfrac{5}{3} = \dfrac{5}{3} \quad \text{true}$$

EXAMPLE 5   Solve for x by cross-multiplying $\dfrac{-8}{3} = \dfrac{64}{x}$.

Solution:

$$\dfrac{-8}{3} = \dfrac{64}{x}$$

$$-8x = 3 \cdot 64$$

$$-8x = 192$$

$$\dfrac{-8x}{-8} = \dfrac{192}{-8}$$

$$x = -24$$

Check:

$$\dfrac{-8}{3} = \dfrac{64}{x}$$

$$\dfrac{-8}{3} = \dfrac{\overset{8}{\cancel{64}}}{\underset{3}{\cancel{-24}}}$$

$$\dfrac{-8}{3} = \dfrac{8}{-3}$$

$$\dfrac{-8}{3} = \dfrac{-8}{3} \quad \text{true}$$

▶ **3**   Often, practical applications can be solved using proportions. To solve such problems using proportions, use the following procedure.

To Solve Problems Using Proportions

1. Represent the unknown quantity by a letter.
2. Set up the proportion by listing the given ratio on the left side of the equal sign, and the unknown and other given quantity on the right side of the equal sign. When setting up the right side of the proportion, the same respective quantities should occupy the same respective positions on the left and right. For example, an acceptable proportion might be

$$\text{Given ratio} \left\{ \dfrac{\text{miles}}{\text{hour}} = \dfrac{\text{miles}}{\text{hour}} \right.$$

3. Once the proportion is correctly written, drop the units and cross-multiply.
4. Solve the resulting equation.
5. Answer the questions asked.

Note that the ratios must have the same units. For example, if one ratio is given in miles/hour and the second ratio is given in feet/hour, one of the ratios must be changed before setting up the proportion.

EXAMPLE 6   A 30-pound bag of fertilizer will cover an area of 2500 square feet.

(a) How many pounds are needed to cover an area of 16,000 square feet?
(b) How many bags of fertilizer are needed?

Solution: (a) The given ratio is 30 pounds per 2500 square feet. The unknown quantity is the number of pounds necessary to cover 16,000 square feet.

Step 1   Let x = number of pounds.

Step 2   Given ratio $\left\{\dfrac{30 \text{ pounds}}{2500 \text{ square feet}} = \dfrac{x \text{ pounds} \longleftarrow \text{unknown}}{16,000 \text{ square feet} \leftarrow \text{given quantity}}\right.$

Note how the pounds and the area are given in the same relative positions.

Step 3
$$\frac{30}{2500} = \frac{x}{16,000}$$
$$30(16,000) = 2500x$$

Step 4
$$480,000 = 2500x$$
$$\frac{480,000}{2500} = x$$
$$192 = x$$

Step 5   One hundred ninety-two pounds of fertilizer are needed.

(b) Since each bag weighs 30 pounds, the number of bags is found by division.

$$192 \div 30 = 6.4 \text{ bags}$$

The number of bags needed is therefore 7, since you can purchase only whole bags.   ■

EXAMPLE 7   In Washington County the property tax rate is $8.065 per $1000 of assessed value. If a house and its property have been assessed at $124,000, find the tax the owner will have to pay.

Solution: The unknown quantity is the tax the property owner must pay. Let us call this unknown x.

$$\frac{\text{tax}}{\text{assessed value}} = \frac{\text{tax}}{\text{assessed value}}$$

Given tax rate $\left\{\dfrac{8.065}{1000} = \dfrac{x}{124,000}\right.$

$$(8.065)(124,000) = 1000x$$
$$1,000,060 = 1000x$$
$$\$1000.06 = x$$

The owner will have to pay $1000.06 tax.   ■

COMMON STUDENT ERROR

When you plan the proportion the same units should not be multiplied by themselves during cross-multiplication. *The following is wrong.*

Wrong

▶ **4**   Proportions can also be used to convert from a quantity given in one unit to a quantity in a different unit. For example, you can use a proportion to convert a measurement in feet to a measurement in meters, or to convert from American dollars to Mexican pesos. The following examples illustrate how this may be done.

EXAMPLE 8   Convert 18.36 inches to feet.

Solution: We know that 1 foot is 12 inches. We use this known fact in one ratio of our proportion. In the second ratio we set the quantities with the same units in the same respective positions.

$$\text{Known} \begin{cases} \\ \end{cases} \text{ratio} \quad \frac{1 \text{ foot}}{12 \text{ inches}} = \frac{x \text{ feet}}{18.36 \text{ inches}}$$

Since we are given 18.36 inches, we place this in the denominator of the second ratio. The unknown quantity is the number of feet, which we will call x. Note that both numerators contain the same units and both denominators contain the same units. Now drop the units and solve for x by cross-multiplying.

$$\frac{1}{12} = \frac{x}{18.36}$$

$$1(18.36) = 12x$$

$$18.36 = 12x$$

$$\frac{18.36}{12} = \frac{12x}{12}$$

$$1.53 = x$$

Thus 18.36 inches equals 1.53 feet.   ■

EXAMPLE 9   Given that 1 kilogram = 2.2 pounds. (a) Find the weight in pounds of a poodle that weighs 7.48 kilograms. (b) Mary Jo weighs 121 pounds. How many kilograms does Mary Jo weigh?

Solution: (a) We use the fact that 1 kilogram = 2.2 pounds when setting up our proportion. The unknown quantity is the number of pounds. We will call this quantity x.

$$\text{Known} \begin{cases} \\ \end{cases} \text{ratio} \quad \frac{1 \text{ kilogram}}{2.2 \text{ pounds}} = \frac{7.48 \text{ kilograms}}{x \text{ pounds}}$$

$$\frac{1}{2.2} = \frac{7.48}{x}$$

$$1x = (2.2)(7.48)$$

$$x = 16.456$$

Thus the poodle weighs 16.456 pounds.

(b) The unknown quantity is the number of kilograms. We will call the unknown quantity x.

$$\frac{1 \text{ kilogram}}{2.2 \text{ pounds}} = \frac{x \text{ kilograms}}{121 \text{ pounds}}$$

$$\frac{1}{2.2} = \frac{x}{121}$$

$$1(121) = 2.2x$$

$$121 = 2.2x$$

$$\frac{121}{2.2} = x$$

$$55 = x$$

Thus Mary Jo weighs 55 kilograms.   ■

EXAMPLE 10   Marisa exchanged 15 U.S. dollars for 41,250 Mexican pesos at a bank in Cancun, Mexico.

(a) What is the conversion rate per U.S. dollar (therefore, what is 1 U.S. dollar worth in pesos)?

(b) At the straw market in downtown Cancun, Marisa purchased a handmade Mayan calendar for 220,000 pesos. What is the cost of the calendar in U.S. dollars?

Solution: (a) We know that 15 U.S. dollars equals 41,250 Mexican pesos. We use this fact in our proportion. The unknown quantity is the number of pesos equal to 1 U.S. dollar.

$$\text{Known} \left\{ \frac{15 \text{ dollars}}{41{,}250 \text{ pesos}} = \frac{1 \text{ dollar}}{x \text{ pesos}} \right.$$

$$15x = 41{,}250$$

$$x = \frac{41{,}250}{15} = 2750$$

Thus $1 can be converted to 2750 pesos.

(b) We need to convert 220,000 pesos to U.S. dollars. We will call the amount of U.S. dollars x.

$$\frac{15 \text{ dollars}}{41{,}250 \text{ pesos}} = \frac{x \text{ dollars}}{220{,}000 \text{ pesos}}$$

$$3{,}300{,}000 = 41{,}250x$$

$$\frac{3{,}300{,}000}{41{,}250} = x$$

$$80 = x$$

Thus the cost of the Mayan calendar is 80 U.S. dollars.

HELPFUL HINT

Some of the problems we have just worked using proportions could have been done without using proportions. However, when working problems of this type, students often have difficulty in deciding whether to multiply or divide to obtain the correct answer. By setting up a proportion, you may be better able to understand the problem and have more success at obtaining the correct answer.

▶ **5**   Proportions can also be used to solve problems in geometry and trigonometry. The following examples illustrate how proportions may be used to solve problems involving **similar figures.** Two figures are said to be *similar* when their respective angles are equal and their respective sides are in proportion.

EXAMPLE 11   If the following are two similar figures, find the length of side x.

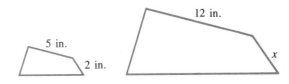

Solution: We set up a proportion of corresponding sides to find the length of side x.

lengths from
smaller figure

lengths from
larger figure

5 inches and 12 inches are corresponding
sides of similar figures. $\longrightarrow$

2 inches and x are corresponding
sides of similar figures. $\longrightarrow$

$$\frac{5}{2} = \frac{12}{x}$$

$$5x = 24$$

$$x = \frac{24}{5} = 4.8$$

Thus side x is 4.8 inches in length.

Note in Example 11 that the proportion could have also been set up as

$$\frac{5}{12} = \frac{2}{x}$$

EXAMPLE 12   Triangles ABC and $AB'C'$ are similar triangles. Use a proportion to find the length of side AB'.

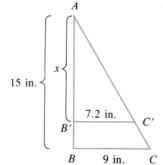

Solution: We set up a proportion of corresponding sides to find the length of side AB'. We will let x represent the length of side AB'. One proportion we can use is

$$\frac{\text{Length of } AB}{\text{Length of } BC} = \frac{\text{length of } AB'}{\text{length of } B'C'}$$

Now insert the proper values and solve for the variable x.

$$\frac{15}{9} = \frac{x}{7.2}$$

$$(15)(7.2) = 9x$$

$$108 = 9x$$

$$12 = x$$

Thus the length of side AB' is 12 inches.

Exercise Set 2.6

The results of a mathematics examination are 5 A's, 6 B's, 8 C's, 4 D's, and 2 F's. Write the ratios of the following:

1. A's to C's.
2. A's to total grades.
3. D's to F's.

4. Grades better than C to total grades.
5. Total grades to D's.
6. Grades better than C to grades less than C.

Determine the following ratios. Write each ratio in lowest terms.

7. 5 feet to 3 feet

8. 60 dollars to 80 dollars

9. 20 hours to 60 hours

10. 100 people to 80 people

11. 4 hours to 40 minutes

12. 6 feet to 4 yards

13. 26 ounces to 4 pounds

14. 7 dimes to 12 nickels

Solve for the variable by cross-multiplying.

15. $\dfrac{4}{x} = \dfrac{5}{20}$

16. $\dfrac{x}{4} = \dfrac{12}{48}$

17. $\dfrac{5}{3} = \dfrac{75}{x}$

18. $\dfrac{x}{32} = \dfrac{-5}{4}$

19. $\dfrac{90}{x} = \dfrac{-9}{10}$

20. $\dfrac{-3}{8} = \dfrac{x}{40}$

21. $\dfrac{1}{9} = \dfrac{x}{45}$

22. $\dfrac{y}{6} = \dfrac{7}{42}$

23. $\dfrac{3}{z} = \dfrac{2}{-20}$

24. $\dfrac{3}{12} = \dfrac{-1.4}{z}$

25. $\dfrac{15}{20} = \dfrac{x}{8}$

26. $\dfrac{12}{3} = \dfrac{x}{-100}$

Write a proportion that can be used to solve the problem. Solve the problem and find the desired value.

27. A car can travel 32 miles on 1 gallon of gasoline. How far can it travel on 12 gallons of gasoline?

28. A car can travel 23 miles on 1 gallon of gasoline. How far can it travel on 297 gallons?

29. A quality control worker can check 12 units in 2.5 minutes. How long will it take her to check 60 units?

30. If 100 feet of wire has an electrical resistance of 7.3 ohms, find the electrical resistance of 40 feet of wire.

31. The property tax in the town of Plainview, Texas, is $8.235 per $1000 of assessed value. If the Litton's house is assessed at $122,000, how much property tax will they pay?

32. A blueprint of a shopping mall is in the scale of 1 : 150. Thus 1 foot on a blueprint represents 150 feet of actual length. One part of the mall is to be 190 feet long. How long will this part be on the blueprint?

33. A model railroad is made in the scale 1 : 87. A box-car measures 12.2 meters. How large will the box-car be in the model railroad set?

34. A photograph shows a boy standing next to a tall cactus. If the boy, who is actually 48 inches tall, measures 0.6 inches in the photograph, how tall is the cactus that measures 3.25 inches in the photo?

35. If a 40-pound bag of fertilizer covers 5000 square feet of area, how many pounds of fertilizer are needed to cover an area of 26,000 square feet?

36. The instructions on a bottle of liquid insecticide say "use 3 teaspoons of insecticide per gallon of water." If your sprayer has an 8-gallon capacity, how much insecticide should be used to fill the sprayer?

37. A recipe for McGillicutty stew calls for $4\frac{1}{2}$ pounds of beef. If the recipe is for 20 servings, how much beef is needed to make 12 servings?

38. A recipe for a pancake mix states that two eggs should be used for each 6 cups of pancake mix. The Green

See Exercise 34.

County Fire Department is planning a Sunday brunch for the community. How many eggs will they use if they plan to use 120 cups of the pancake mix?

39. The counter on a video cassette recorder goes from 0 to 5.5 in 1 minute. What number will be on the VCR counter at the end of an hour and a half movie if the counter is set to 0 when the movie begins?

40. You are taping a 60-minute television show with your video cassette recorder. If the counter of the VCR starts at 0 and reads 570 at the end of the 60-minute program, find the number that would be on the counter after 4 minutes of play.

Use a proportion to answer each of the following questions. Round your answers to two decimal places.

41. Convert 57 inches to feet.

42. Convert 17.2 yards to feet.

43. Convert 17,952 feet to miles (5280 feet = 1 mile).

44. Convert 26.1 square feet to square yards (9 square feet = 1 square yard).

45. Convert 146.4 ounces to pounds.

46. One inch equals 2.54 centimeters. Find the length of a book in inches if it measures 26.67 centimeters.

47. One liter equals approximately 1.06 quarts. Find the volume in quarts of a 5-liter container.

48. One cubic foot equals approximately 0.03 cubic meter. Find the volume in cubic feet of 6 cubic meters of cement.

49. One mile equals approximately 1.6 kilometers. Find the distance in miles of a 25-kilometer kangaroo crossing.

50. One mile equals approximately 1.6 kilometers. Find the distance in kilometers from San Diego, California, to San Francisco, California, a distance of 520 miles.

51. In chemistry, we learn that 1 troy ounce equals 480 grains. If gold is selling for $500 per troy ounce, what is the cost per grain?

52. In chemistry, we learn that 100 torr equals 0.13 atmosphere. Find the number of torr in 0.39 atmospheres.

53. In a statistics course, we find that for one particular set of scores 16 points equals 3.2 standard deviations. How many points equals 1 standard deviation?

54. When Fong visited the United States from Canada, he converted 10 Canadian dollars for 8.60 U.S. dollars. If

See Exercise 49.

he converts his remaining 2000 Canadian dollars to U.S. dollars, how much more will he receive?

55. Antonio, who is visiting the United States from Italy, wishes to obtain 1200 U.S. dollars. If one Italian lira can be converted for 0.00081 U.S. dollars, how many lire will he need to convert to U.S. dollars?

56. Ms. Johnson spent an evening in a hotel in London, England. When she checked out, she was charged 90 pounds. What was the U.S. dollar equivalence of her hotel bill if 1 English pound could be converted to 1.64 U.S. dollars?

Pairs of similar figures are given in Exercises 57–62. For each pair, find the length of side x.

57.

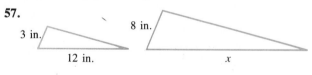

3 in.   8 in.   12 in.   x

58.

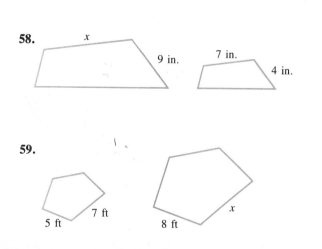

x   9 in.   7 in.   4 in.

59.

7 ft   5 ft   8 ft   x

60.

2 ft   1.8 ft   0.8 ft   x

61.

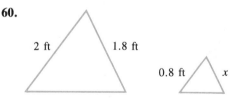

20 in.   14 in.   x   8 in.

62.

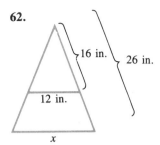

63. Mrs. Sanchez's low density cholesterol level is 127 mg/dl. Her high density cholesterol level is 60 mg/dl. Is Mrs. Sanchez's ratio of low density to high density cholesterol level less than or equal to the 4 : 1 recommended level? See Example 2.

64. (a) Another ratio used by some doctors when measuring cholesterol level is the ratio of total cholesterol to high density cholesterol.[1] Is this ratio increased or decreased if the total cholesterol remains the same but the high density level is increased? Explain how you determined your answer.

(b) Doctors recommend that the ratio of total cholesterol to high density cholesterol be less than or equal to 4.5 : 1. If Mike's total cholesterol is 220 mg/dl and his high density cholesterol is 50 mg/dl, is his ratio less than or equal to the 4.5 : 1? Explain your answer.

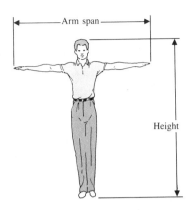

65. (a) Find the ratio of your height to your arm span when your arms are extended horizontally outward. You will need help in getting these measurements.

(b) If a box were to be drawn about your body with your arms extended, would the box be a square or a rectangle? If a rectangle, would the larger length be your arm span or your height measurement? Explain your answer.

66. A special ratio in mathematics is called the Golden Ratio. Do research in a history of mathematics book, or another book recommended by your professor, and explain what the golden ratio is and why it is important.

Cumulative Review Exercises

[1.10] *Name the properties illustrated.*

67. $x + 3 = 3 + x$

68. $3(xy) = (3x)y$

69. $2(x - 3) = 2x - 6$

[2.5] **70.** Solve the following equation:
$-(2x + 6) = 2(3x - 6)$.

JUST FOR FUN

1. The recipe to make a deep-dish apple pie includes:

12 cups sliced apples	$\frac{1}{4}$ teapoon salt
$\frac{1}{2}$ cup flour	2 tablespoons butter or margarine
1 teaspoon nutmeg	$1\frac{1}{2}$ cups sugar
1 teaspoon cinnamon	

Determine the amount of each of the other ingredients that should be used if only 8 cups of apples are available.

2. Insulin comes in 10-cubic-centimeter (cc) vials labeled in the number of units of insulin per cubic centimeter. Thus a vial labeled U40 means there are 40 units of insulin per cubic centimeter of fluid. If a patient needs 25 units of insulin, how many cubic centimeters of fluid should be drawn up into a syringe from the U40 vial?

---

[1] Total cholesterol includes both low and high density cholesterol, plus other types of cholesterol.

2.7

Inequalities in One Variable

▶ **1**   Solve inequalities.

▶ **1**   The greater-than symbol, $>$, and less-than symbol, $<$, were introduced in Section 1.4. The symbol $\geq$ means greater than or equal to and $\leq$ means less than or equal to. A mathematical statement containing one or more of these symbols is called an **inequality.** The direction of the symbol is sometimes called the **sense** or order of the inequality.

Examples of Inequalities in One Variable

$$x + 3 < 5, \qquad x + 4 \geq 2x - 6, \qquad 4 > -x + 3$$

To solve an inequality, we must get the variable by itself on one side of the inequality symbol. To do this, we make use of properties very similar to those used to solve equations. Here are four properties used to solve inequalities. Later in this section we will introduce two additional properties.

Properties Used to Solve Inequalities

For real numbers a, b, and c:

1. If $a > b$, then $a + c > b + c$.
2. If $a > b$, then $a - c > b - c$.
3. If $a > b$ **and** $c > 0$, then $ac > bc$.
4. If $a > b$ **and** $c > 0$, then $\dfrac{a}{c} > \dfrac{b}{c}$.

Property 1 says the same number may be added to both sides of an inequality. Property 2 says the same number may be subtracted from both sides of an inequality. Property 3 says the same *positive* number may be used to multiply both sides of an inequality. Property 4 says the same *positive* number may be used to divide both sides of an inequality. When any of these four properties is used, the direction of the inequality symbol does not change.

EXAMPLE 1   Solve the inequality $x - 4 > 7$, and graph the solution on the real number line.

Solution: To solve this inequality, we need to isolate the variable, x. Therefore, we must eliminate the -4 from the left side of the inequality. To do this, we add 4 to both sides of the inequality.

$$x - 4 > 7$$
$$x - 4 + 4 > 7 + 4 \qquad \text{Add 4 to both sides of the inequality.}$$
$$x > 11$$

FIGURE 2.2

The solution is all real numbers greater than 11. We can illustrate the solution on the number line by placing an open circle at 11 on the number line and drawing an arrow to the right; see Fig. 2.2.

The open circle at the 11 indicates that the 11 is *not* part of the solution. The arrow going to the right indicates that all the values greater than 11 are solutions to the inequality. ∎

EXAMPLE 2   Solve the inequality $2x + 6 \leq -2$, and graph the solution on the real number line.

Solution: To isolate the variable, we must eliminate the $+6$ from the left side of the inequality. We do this by subtracting 6 from both sides of the inequality.

$$2x + 6 \leq -2$$
$$2x + 6 - 6 \leq -2 - 6 \qquad \text{Subtract 6 from both sides of the inequality.}$$
$$2x \leq -8$$
$$\frac{2x}{2} \leq -\frac{8}{2} \qquad \text{Divide both sides of the equation by 2.}$$
$$x \leq -4$$

−4   0

FIGURE 2.3

The solution is all real numbers less than or equal to -4. We can illustrate the solution on the number line by placing a closed, or darkened, circle at -4 and drawing an arrow to the left; see Fig. 2.3.

The darkened circle at -4 indicates that -4 *is* a part of the solution. The arrow going to the left indicates that all the values less than -4 are also solutions to the inequality. ∎

Notice in properties 3 and 4 that we specified that $c > 0$. What happens when an inequality is multiplied or divided by a negative number? Examples 3 and 4 will illustrate this.

EXAMPLE 3   Multiply both sides of the inequality $8 > -4$ by -2.

Solution:
$$8 > -4$$
$$-2(8) < -2(-4)$$
$$-16 < 8$$
∎

EXAMPLE 4   Divide both sides of the inequality $8 > -4$ by -2.

Solution:
$$8 > -4$$
$$\frac{8}{-2} < \frac{-4}{-2}$$
$$-4 < 2$$
∎

Examples 3 and 4 illustrate that **when an inequality is multiplied or divided by a negative number, the sense (or direction) of the inequality changes.**

Additional Properties Used to Solve Inequalities

5. If $a > b$ and $c < 0$, then $ac < bc$.

6. If $a > b$ and $c < 0$, then $\dfrac{a}{c} < \dfrac{b}{c}$.

EXAMPLE 5   Solve the inequality $-2x > 6$, and graph the solution on the real number line.

Solution: To isolate the variable, we must eliminate the -2 on the left side of the inequality. To do this, we can divide both sides of the inequality by -2. When we do this, however, we must remember to change the sense of the inequality.

$$-2x > 6$$

$$\frac{-2x}{-2} < \frac{6}{-2} \qquad \text{Divide both sides of the inequality by } -2 \text{ and change the sense of the inequality.}$$

$$x < -3$$

FIGURE 2.4

The solution is all real numbers less than -3. The solution is graphed on the number line in Fig. 2.4. ■

EXAMPLE 6   Solve the inequality $4 \geq -5 - x$, and graph the solution on the real number line.

Solution: *Method 1:*

$$4 \geq -5 - x$$

$$4 + 5 \geq -5 + 5 - x \qquad \text{Add 5 to both sides of the inequality.}$$

$$9 \geq -x$$

$$-1(9) \leq -1(-x) \qquad \text{Multiply both sides of the inequality by } -1 \text{ and change the sense of the inequality.}$$

$$-9 \leq x$$

The inequality $-9 \leq x$ can also be written $x \geq -9$.

Method 2:

$$4 \geq -5 - x$$

$$4 + x \geq -5 - x + x \qquad \text{Add } x \text{ to both sides of the inequality.}$$

$$4 + x \geq -5$$

$$4 - 4 + x \geq -5 - 4 \qquad \text{Subtract 4 from both sides of the inequality.}$$

$$x \geq -9$$

FIGURE 2.5

The solution is graphed on the number line in Fig. 2.5. Other methods could also be used to solve this problem. ■

Notice in Example 6, method 1, we wrote $-9 \leq x$ as $x \geq -9$. Although the solution $-9 \leq x$ is correct, it is customary to write the solution to an inequality with the variable on the left. One reason we write the variable on the left is that it often makes it easier to graph the solution on the number line. How would you graph $-3 > x$? How would you graph $-5 \leq x$? If you rewrite these inequalities with the variable on the left side, the answer becomes clearer.

$$-3 > x \quad \text{means} \quad x < -3$$

and

$$-5 \leq x \quad \text{means} \quad x \geq -5$$

Notice that you can change an answer from a greater-than statement to a less-than statement or from a less-than statement to a greater-than statement. When you change the answer from one form to the other, remember that the inequality symbol must point to the symbol to which it was pointing originally.

HELPFUL HINT

$a > x$ means $x < a$ (Note that both inequality symbols point to x.)

$a < x$ means $x > a$ (Note that both inequality symbols point to a.)

Examples: $-3 > x$ means $x < -3$

$-5 \le x$ means $x \ge -5$

EXAMPLE 7   Solve the inequality $2x + 4 < -x + 12$, and graph the solution on the real number line.

Solution:

$$2x + 4 < -x + 12$$

$2x + x + 4 < -x + x + 12$ Add x to both sides of the inequality.

$$3x + 4 < 12$$

$3x + 4 - 4 < 12 - 4$ Subtract 4 from both sides of the inequality.

$$3x < 8$$

$\dfrac{3x}{3} < \dfrac{8}{3}$ Divide both sides of the inequality by 3.

$$x < \dfrac{8}{3}$$

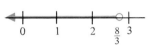

FIGURE 2.6

The number line is shown in Fig. 2.6.

EXAMPLE 8   Solve the inequality $-5x + 9 < -2x + 6$, and graph the solution on the real number line.

Solution:

$-5x + 9 < -2x + 6$

$-5x < -2x - 3$ 9 was subtracted from both sides of the inequality.

$-3x < -3$ $2x$ was added to both sides of the inequality.

$x > 1$ Both sides of the inequality were divided by -3 and the sense of the inequality was changed.

FIGURE 2.7

The number line is shown in Fig. 2.7.

EXAMPLE 9   Solve the inequality $2(x + 3) \le 5x - 3x + 8$, and graph the solution on the real number line.

Solution:

$2(x + 3) \le 5x - 3x + 8$

$2x + 6 \le 5x - 3x + 8$ Distributive property was used.

$2x + 6 \le 2x + 8$ Like terms were combined.

$2x - 2x + 6 \le 2x - 2x + 8$ Subtract $2x$ from both sides of the inequality.

$6 \le 8$

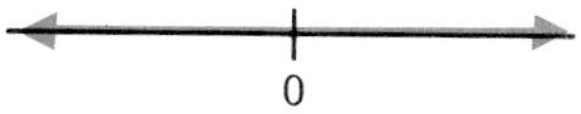

FIGURE 2.8

Since 6 is always less than or equal to 8, the solution is **all real numbers** (Fig. 2.8).

EXAMPLE 10   Solve the inequality $4(x + 1) > x + 5 + 3x$, and graph the solution on the real number line.

$$\textit{Solution:} \qquad 4(x + 1) > x + 5 + 3x$$

$$4x + 4 > x + 5 + 3x \qquad \text{Distributive property was used.}$$

$$4x + 4 > 4x + 5 \qquad \text{Like terms were combined.}$$

$$4x - 4x + 4 > 4x - 4x + 5 \qquad \text{Subtract } 4x \text{ from both sides of the inequality.}$$

$$4 > 5$$

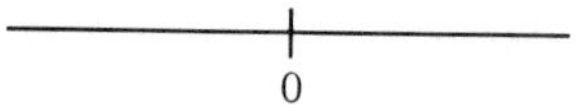

FIGURE 2.9   Since 4 is never greater than 5, the answer is **no solution** (Fig. 2.9).   ■

Exercise Set 2.7

Solve each inequality, and graph the solution on the real number line.

1. $x + 3 > 7$

2. $x - 4 > -3$

3. $x + 5 \geq 3$

4. $4 - x \geq 3$

5. $-x + 3 < 8$

6. $4 < 3 + x$

7. $6 > x - 4$

8. $-4 \leq -x - 3$

9. $8 \leq 4 - x$

10. $2x < 4$

11. $-2x < 3$

12. $6 \geq -3x$

13. $2x + 3 \leq 5$

14. $-4x - 3 > 5$

15. $12x + 24 < -12$

16. $3x - 4 \leq 9$

17. $4 - 6x > -5$

18. $8 < 4 - 2x$

19. $15 > -9x + 50$

20. $3x - 4 < 5$

21. $4 < 3x + 12$

22. $-4x > 2x + 12$

23. $6x + 2 \leq 3x - 9$

24. $-2x - 4 \leq -5x + 12$

25. $x - 4 \leq 3x + 8$

26. $-3x - 5 \geq 4x - 29$

27. $-x + 4 < -3x + 6$

28. $2(x - 3) < 4x + 10$

29. $-3(2x - 4) > 2(6x - 12)$

30. $-(x + 3) \leq 4x + 5$

31. $x + 3 < x + 4$

32. $x + 5 \geq x - 2$

33. $6(3 - x) < 2x + 12$

34. $2(3 - x) + 4x < -6$

35. $-21(2 - x) + 3x > 4x + 4$

36. $-(x + 3) \geq 2x + 6$

37. $4x - 4 < 4(x - 5)$

38. $-2(-5 - x) > 3(x + 2) + 4 - x$

39. $5(2x + 3) \geq 6 + (x + 2) - 2x$

40. $-3(-2x + 12) < -4(x + 2) - 6$

41. When solving an inequality, if you obtain the result $3 < 5$, what is the solution?

42. When solving an inequality, if you obtain the result $4 \geq 2$, what is the solution?

43. When solving an inequality, if you obtain the result $5 < 2$, what is the solution?

44. When solving an inequality, if you obtain the result $-4 \geq -2$, what is the solution?

45. When solving an inequality, under what conditions will it be necessary to change the sense of the inequality?

46. List the 6 rules used to solve inequalities.

Cumulative Review Exercises

[1.8] **47.** Evaluate $-x^2$ for $x = 3$.

   48. Evaluate $-x^2$ for $x = -5$.

[2.5] **49.** Solve $4 - 3(2x - 4) = 5 - (x + 3)$.

[2.6] **50.** The Milford electric company charges $0.174 per kilowatt hour of electricity. The Cisneros's monthly electric bill was $87 for the month of July. How many kilowatt hours of electricity did the Cisneros use in July?

JUST FOR FUN

1. Solve the inequality
$3(2 - x) - 4(2x - 3) \le 6 + 2x - 6(x - 5) + 2x$.

2. Solve the inequality
$-(x + 4) + 6x - 5 > -4(x + 3) + 2(x + 6) - 5x$.

3. The inequality symbols discussed so far are $<$, $\le$, $>$, and $\ge$. Can you name an inequality symbol that we have not mentioned in this section?

SUMMARY

GLOSSARY

Algebraic expression *(59):* A collection of numbers, variables, grouping symbols, and operation symbols.

Check *(68):* A procedure where the value believed to be the solution to an equation is substituted back into the equation.

Coefficient or numerical coefficient *(59):* The numerical part of a term.

Constant or constant term *(60):* A term in an expression that does not contain a variable.

Equivalent equations *(70):* Two or more equations that have the same solution.

Equation *(67):* A statement that two algebraic expressions are equal.

Identity *(90):* An equation that is true for all values of the variable.

Inequality *(102):* A mathematical statement containing one or more inequality symbols ($>$, $\ge$, $<$, $\le$).

Like terms *(60):* Terms that have the same variables with the same exponents.

Linear equation in one variable *(67):* Equation of the form $ax + b = c$, $a \ne 0$.

Proportion *(93):* A statement of equality between two ratios.

Ratio *(92):* A quotient of two quantities with the same units.

Reciprocal of a *(75):* $\frac{1}{a}$, $a \ne 0$.

Similar figures *(97):* Two figures are similar when their respective angles are equal and their respective sides are in proportion.

Simplify *(65):* To simplify an expression means to combine like terms in the expression.

Solution *(68):* The value or values that make an equation a true statement.

Solve *(68):* To find the solution to an equation.

Term *(59):* The parts that are added or subtracted in an algebraic expression.

IMPORTANT FACTS

Distributive property:
$$a(b + c) = ab + ac.$$

Addition property:
If $a = b$, then $a + c = b + c$.

Multiplication property:
If $a = b$, then $a \cdot c = b \cdot c$.

Cross-multiplication:
If $\frac{a}{b} = \frac{c}{d}$, then $ad = bc$.

Properties used to solve inequalities

1. If $a > b$, then $a + c > b + c$.

2. If $a > b$, then $a - c > b - c$.

3. If $a > b$ and $c > 0$, then $ac > bc$.

4. If $a > b$ and $c > 0$, then $\frac{a}{c} > \frac{b}{c}$.

5. If $a > b$ and $c < 0$, then $ac < bc$.

6. If $a > b$ and $c < 0$, then $\frac{a}{c} < \frac{b}{c}$.

Review Exercises

[2.1] *Use the distributive property to remove parentheses.*

1. $2(x + 4)$ **2.** $3(x - 2)$ **3.** $2(4x - 3)$

4. $-2(x + 4)$ **5.** $-(x + 2)$ **6.** $-(x - 2)$

7. $-4(4 - x)$ **8.** $3(6 - 2x)$ **9.** $4(5x - 6)$

10. $-3(2x - 5)$ **11.** $6(6x - 6)$ **12.** $4(-x + 3)$

13. $-3(x + y)$ **14.** $-2(3x - 2)$ **15.** $-(3 + 2y)$

16. $-(x + 2y - z)$ **17.** $3(x + 3y - 2z)$ **18.** $-2(2x - 3y + 7)$

Simplify where possible.

19. $2x + 3x$ **20.** $4y + 3y + 2$ **21.** $4 - 2y + 3$

22. $1 + 3x + 2x$ **23.** $6x + 2y + y$ **24.** $-2x - x + 3y$

25. $2x + 3y + 4x + 5y$ **26.** $6x + 3y + 2$ **27.** $2x - 3x - 1$

28. $5x - 2x + 3y + 6$ **29.** $x + 8x - 9x + 3$ **30.** $-4x - 8x + 3$

31. $3(x + 2) + 2x$ **32.** $-2(x + 3) + 6$ **33.** $2x + 3(x + 4) - 5$

34. $4(3 - 2x) - 2x$ **35.** $6 - (-x + 3) + 4x$ **36.** $2(2x + 5) - 10 - 4$

37. $-6(4 - 3x) - 18 + 4x$ **38.** $6 - 3(x + y) + 6x$ **39.** $3(x + y) - 2(2x - y)$

40. $3x - 6y + 2(4y + 8)$ **41.** $3 - (x - y) + (x - y)$ **42.** $(x + y) - (2x + 3y) + 4$

[2.2–2.5] *Solve each equation.*

43. $2x = 4$ **44.** $x + 3 = -5$ **45.** $x - 4 = 7$

46. $\dfrac{x}{3} = -9$ **47.** $2x + 4 = 8$ **48.** $14 = 3 + 2x$

49. $8x - 3 = -19$ **50.** $6 - x = 9$ **51.** $-x = -12$

52. $2(x + 2) = 6$ **53.** $-3(2x - 8) = -12$ **54.** $4(6 + 2x) = 0$

55. $3x + 2x + 6 = -15$ **56.** $4 = -2(x + 3)$ **57.** $27 = 46 + 2x - x$

58. $4x + 6 - 7x + 9 = 18$ **59.** $4 + 3(x + 2) = 12$ **60.** $-3 + 3x = -2(x + 1)$

61. $3x - 6 = -5x + 30$ **62.** $-(x + 2) = 2(3x - 6)$ **63.** $2x + 6 = 3x + 9$

64. $-5x + 3 = 2x + 10$ **65.** $3x - 12x = 24 - 6x$ **66.** $2(x + 4) = -3(x + 5)$

67. $4(2x - 3) + 4 = 9x + 2$ **68.** $6x + 11 = -(6x + 5)$ **69.** $2(x + 7) = 6x + 9 - 4x$

70. $-6(4 - 2x) = -11 + 12x - 13$ **71.** $4(x - 3) - (x + 5) = 0$ **72.** $-2(4 - x) = 6(x + 2) + 3x$

73. $-3(2x - 5) + 5x = 4x - 7$

[2.6] *Determine the following ratios. Write each ratio in lowest terms.*

74. 15 feet to 20 feet   **75.** 80 ounces to 12 pounds

76. 5 quarters to 12 nickels   **77.** 32 ounces to 2 pounds

Solve each proportion.

78. $\dfrac{x}{9} = \dfrac{6}{18}$ **79.** $\dfrac{15}{100} = \dfrac{x}{20}$ **80.** $\dfrac{3}{x} = \dfrac{15}{45}$ **81.** $\dfrac{20}{45} = \dfrac{15}{x}$

82. $\dfrac{6}{2} = \dfrac{-12}{x}$ **83.** $\dfrac{x}{9} = \dfrac{8}{-3}$ **84.** $\dfrac{-4}{9} = \dfrac{-16}{x}$ **85.** $\dfrac{x}{-15} = \dfrac{30}{-5}$

The following figures are pairs of similar figures. Find the length of side x.

86.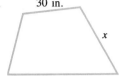
30 in.

x

6 in.

8 in.

87.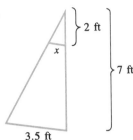

2 ft

x

7 ft

3.5 ft

[2.7] *Solve each inequality, and graph the solution on the real number line.*

88. $2x + 4 \geq 8$

89. $6 - 2x > 4x - 12$

90. $6 - 3x \leq 2x + 18$

91. $2(x + 4) \leq 2x - 5$

92. $2(x + 3) > 6x - 4x + 4$

93. $x + 6 > 9x + 30$

94. $x - 2 \leq -4x + 7$

95. $-(x + 2) < -2(-2x + 5)$

96. $2(x + 3) < -(x + 3) + 4$

97. $-6x - 3 \geq 2(x - 4) + 3x$

98. $-2(x - 4) \leq 3x + 6 - 5x$

99. $2(2x + 4) > 4(x + 2) - 6$

100. $4(x - 2) + 3 \leq 2(x + 3)$

Set up a proportion that can be used to solve each problem. Solve the proportion and find the value desired.

101. If a car traveling at a specified speed can travel 45 miles in 60 minutes, how many miles can it travel in 90 minutes?

102. If the scale of a map is 1 inch to 60 miles, what distance on the map represents 380 miles?

103. Bryce builds a model car to a scale of 1 inch to 0.9 feet. If the completed model is 10.5 inches, what is the size of the actual car?

104. If 20 U.S. dollars can be exchanged for 22.98 Canadian dollars, how many Canadian dollars can Carlos get in exchange for his 985 U.S. dollars?

105. If one U.S. dollar can be exchanged for 2788 Mexican pesos, find the value of 1 peso in terms of U.S. dollars.

106. If 3 radians equal 171.9 degrees, find the number of degrees in 1 radian.

107. In a physics course, we learn that 3 slugs equals approximately 96.6 pounds of force. To what is 1 pound of force equal, to the nearest hundredth, in terms of slugs? 0.03 slugs

Practice Test

Use the distributive property to remove parentheses.

1. $-2(4 - 2x)$

2. $-(x + 3y - 4)$

Simplify where possible.

3. $3x - x + 4$

4. $4 + 2x - 3x + 6$

5. $y - 2x - 4x - 6$

6. $x - 4y + 6x - y + 3$

7. $2x + 3 + 2(3x - 2)$

Solve each equation.

8. $2x + 4 = 12$

9. $-x - 3x + 4 = 12$

10. $4x - 2 = x + 4$

11. $3(x - 2) = -(5 - 4x)$

12. $2x - 3(-2x + 4) = -13 + x$

13. $3x - 4 - x = 2(x + 5)$

14. $-3(2x + 3) = -2(3x + 1) - 7$

15. $\dfrac{9}{x} = \dfrac{3}{-15}$.

Solve each inequality, and graph the solution on the real number line.

16. $2x - 4 < 4x + 10$

17. $3(x + 4) \geq 5x - 12$

18. $4(x + 3) + 2x < 6x - 3$

20. If 6 gallons of insecticide can treat 3 acres of land, how many gallons of insecticide are needed to treat 75 acres?

19. The following figures are similar figures. Find the length of side x.

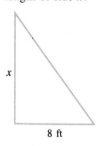

x

8 ft

4 ft

3 ft

Cumulative Review Test

1. Multiply $\dfrac{16}{20} \cdot \dfrac{4}{5}$.

2. Divide $\dfrac{8}{24} \div \dfrac{2}{3}$.

3. Insert $<$, $>$, or $=$ in the shaded area to make a true statement: $|-2| \quad 1$.

4. Evaluate $-6 - (-3) + 5 - 8$.

5. Subtract -4 from -12.

6. Evaluate $16 - 6 \div 2 \cdot 3$.

7. Evaluate $3[6 - (4 - 3^2)] - 30$.

8. Evaluate $-3x^2 - 4x + 5$ when $x = -2$.

9. Name the property illustrated:
$(x + 4) + 6 = x + (4 + 6)$.

Simplify each expression.

10. $6x + 2y + 4x - y$

11. $3x - 2x + 16 + 2x$

Solve each equation for x.

12. $4x - 2 = 10$

13. $\frac{1}{4}x = -10$

14. $6x + 5x + 6 = 28$

15. $3(x - 2) = 5(x - 1) + 3x + 4$

16. $\dfrac{15}{30} = \dfrac{3}{x}$

Solve for x, and graph the solution on the number line.

17. $x - 4 > 6$

18. $2x - 7 \leq 3x + 5$

19. A 36 pound bag of fertilizer can fertilize an area of 5000 square feet. How many pounds of fertilizer will Marisa need to fertilize her 22,000 square foot lawn?

20. If Samuel earns $10.50 after working for 2 hours scrubbing boats at the marina, how much does he earn after 8 hours?

3

Formulas and Applications of Algebra

3.1 Formulas

3.2 Changing Application Problems into Equations

3.3 Solving Application Problems

3.4 Geometric Problems

Summary

Review Exercises

Practice Test

See Section 3.3, Exercise 32.

3.1

Formulas

▶ **1** Use the simple interest formula.

▶ **2** Use geometric formulas.

▶ **3** Solve for a variable in a formula.

A **formula** is an equation commonly used to express a specific physical concept mathematically. For example, the formula for the area of a rectangle is

$$\text{area} = \text{length} \cdot \text{width} \quad \text{or} \quad A = lw$$

In this section we will discuss how to evaluate a formula. We will introduce some special formulas, including the simple interest formula and geometric formulas. After this we will discuss how to solve for a variable in a formula.

To **evaluate a formula,** substitute the appropriate numerical values for the variables and perform the indicated operations.

Simple Interest Formula

▶ **1** Examples 1 and 2 involve the simple interest formula.

Simple Interest Formula

$$\text{interest} = \text{principal} \cdot \text{rate} \cdot \text{time} \quad \text{or} \quad i = prt$$

This formula is used to determine the simple interest, i, obtained in some savings accounts or the simple interest an individual must pay on certain loans. In the simple interest formula $i = prt$, p is the principal (the amount invested or borrowed), r is the interest rate in decimal form, and t is the amount of time of the investment or loan.

EXAMPLE 1   Avery borrows $2000 from a bank for a 3-year period. The bank is charging 12% simple interest per year for the loan. Determine the interest Avery will owe the bank.

Solution: The principal, p, is $2000, the rate, r, is 12% or 0.12 in decimal form, and the time, t, is 3 years. Substituting these values in the simple interest formula gives

$$i = prt$$
$$i = 2000(0.12)(3) = 720$$

The simple interest is $720. When Avery repays his loan he will pay the principal, $2000, plus the interest, $720, for a total of $2720.   ∎

EXAMPLE 2   Amber invests $5000 in a savings account paying simple interest for 2 years. If the interest from the account is $800, find the rate.

Solution: We use the simple interest formula, $i = prt$. We are given the principal, p, the time,

t, and the interest, i. We are asked to find the rate, r. We substitute the given values in the simple interest formula and solve the resulting equation for r.

$$i = prt$$
$$800 = 5000(r)(2)$$
$$800 = 10,000r$$
$$\frac{800}{10,000} = \frac{10,000r}{10,000}$$
$$0.08 = r$$

Thus the simple interest rate is 0.08, or 8%.

Geometric Formulas ▸**2** The **perimeter**, P, is the sum of the lengths of the sides of a figure. Perimeters are measured in the same common unit as the sides. For example, perimeter may be measured in centimeters, inches, or feet. The **area**, A, is the total surface within the figure's boundaries. Areas are measured in square units. For example, area may be measured in square centimeters, square inches, or square feet. Table 3.1 gives the formulas for finding the areas and perimeters of triangles and quadrilaterals. **Quadrilateral** is a general name for a four-sided figure.

TABLE 3.1 Formulas for Areas and Perimeters of Quadrilaterals and Triangles

Figure	Sketch	Area	Perimeter
Square		$A = s^2$	$P = 4s$
Rectangle		$A = lw$	$P = 2l + 2w$
Parallelogram		$A = lh$	$P = 2l + 2w$
Trapezoid		$A = \frac{1}{2}h(b + d)$	$P = a + b + c + d$
Triangle		$A = \frac{1}{2}bh$	$P = a + b + c$

EXAMPLE 3 Find the perimeter of a rectangle when $l = 6$ feet and $w = 2$ feet.

Solution: Substitute 6 for l and 2 for w in the formula for the perimeter of a rectangle.

$$P = 2l + 2w$$
$$= 2(6) + 2(2) = 12 + 4 = 16 \text{ feet}$$

EXAMPLE 4   Find the length of a rectangle when the perimeter is 22 inches and the width is 3 inches.

Solution: Substitute 22 for P and 3 for w in the formula $P = 2l + 2w$; then solve for the length, l.

$$P = 2l + 2w$$
$$22 = 2l + 2(3)$$
$$22 = 2l + 6$$
$$22 - 6 = 2l + 6 - 6$$
$$16 = 2l$$
$$\frac{16}{2} = \frac{2l}{2}$$
$$8 = l$$

The length is 8 inches.

EXAMPLE 5   Find the height of a triangle if it has an area of 30 square feet and a base of 12 feet.

Solution: $A = \dfrac{1}{2}bh$

$$30 = \frac{1}{2}(12)h$$
$$30 = 6h$$
$$\frac{30}{6} = \frac{6h}{6}$$
$$5 = h$$

The height of the triangle is 5 feet.

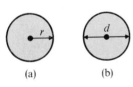

(a)   (b)

FIGURE 3.1

Another figure that we see and use daily is the circle. The **circumference, C,** is the length (or perimeter) of the curve that forms a circle. The **radius, r,** is the line segment from the center of the circle to any point on the circle (see Fig. 3.1a). A **diameter** of a circle is a line segment through the center with both end points on the circle (see Fig. 3.1b). *Note that the diameter is twice the radius.*

The formulas for both the area and the circumference of a circle are given in Table 3.2.

TABLE 3.2 Formulas for Circles

Circle	Area	Circumference
⭕	$A = \pi r^2$	$C = 2\pi r$

Pi, symbolized by the Greek lowercase letter π, has a value of *approximately* 3.14.

Calculator Corner

Most scientific calculators have a Pi key, $\boxed{\pi}$. If you press $\boxed{\pi}$, your calculator will display 3.1415927. This is still only a close approximation of π since π is an irrational number. If you own a scientific calculator you should use the π key instead of using 3.14 when working problems that involve π. In this book we will use 3.14 for π since not every student owns a scientific calculator. If you use the π key, your answers will be slightly more accurate than ours, but still approximate.

EXAMPLE 6   Determine the area and circumference of a circle with a diameter of 16 inches.

Solution: The radius is half the diameter; $r = \dfrac{16}{2} = 8$ inches.

$$A = \pi r^2 \qquad\qquad\qquad C = 2\pi r$$
$$A = 3.14(8)^2 \qquad\qquad C = 2(3.14)(8)$$
$$A = 3.14(64) \qquad\qquad C = 50.24 \text{ inches}$$
$$A = 200.96 \text{ square inches}$$

If you used a scientific calculator and used the $\boxed{\pi}$ key, your answer to part (a) would be 201.06193 and your answer to part (b) would be 50.265482. ∎

Table 3.3 gives formulas for finding the volume of certain three-dimensional figures. Volume is measured in cubic units, such as cubic centimeters or cubic feet.

TABLE 3.3 Formulas for Volumes of Three-Dimensional Figures

Figure	Sketch	Volume
Rectangular solid		$V = lwh$
Right circular cylinder		$V = \pi r^2 h$
Right circular cone		$V = \dfrac{1}{3}\pi r^2 h$
Sphere		$V = \dfrac{4}{3}\pi r^3$

EXAMPLE 7   Find the volume of a basketball if its diameter is 18 inches.

Solution: Since its diameter is 18 inches, its radius is 9 inches.

$$V = \frac{4}{3}\pi r^3$$

$$V = \frac{4}{3}(3.14)(9)^3 = \frac{4}{3}(3.14)(729) = 3052.08$$

Therefore a basketball has a volume of 3,052.08 cubic inches. If you used the $\boxed{\pi}$ key on your calculator your answer would be 3,053.6281. ∎

EXAMPLE 8   Find the height of a right circular cylinder if its volume is 904.32 cubic inches and its radius is 6 inches.

Solution:

$$V = \pi r^2 h$$

$$904.32 = (3.14)(6)^2 h$$

$$904.32 = (3.14)(36)h$$

$$904.32 = 113.04h$$

$$\frac{904.32}{113.04} = \frac{\cancel{113.04}h}{\cancel{113.04}}$$

$$8 = h$$

Thus the height is 8 inches.   ∎

Let us do one more problem that involves evaluating a formula.

EXAMPLE 9   The number of diagonals, d, in a polygon of n sides is given by the formula $d = \frac{1}{2}n^2 - \frac{3}{2}n$.

(a) How many diagonals has a quadrilateral (4 sides)?
(b) How many diagonals has an octagon (8 sides)?

Solution: (a) $n = 4$ (b) $n = 8$

Check: for (a)

$$d = \frac{1}{2}(4)^2 - \frac{3}{2}(4) \qquad d = \frac{1}{2}(8)^2 - \frac{3}{2}(8)$$

$$= \frac{1}{2}(16) - 6 \qquad\qquad = \frac{1}{2}(64) - 12$$

$$= 8 - 6 = 2 \qquad\qquad = 32 - 12 = 20$$ ∎

Solving for a Variable in a Formula or Equation

▶ 3   Often in this course and in other mathematics and science courses, you will be given an equation or formula solved for one variable and have to solve it for a different variable. We will now learn how to do this. This material will reinforce the concepts and procedures you used when solving equations in Chapter 2. We will use the procedures learned here to solve problems in many other sections of the text.

To solve for a variable in a formula, treat each of the quantities, except the one you are solving for, as if they were constants. Then solve for the desired variable by isolating it on one side of the equation, using the properties discussed previously.

EXAMPLE 10   Solve the formula $A = lw$ for w.

Solution: We must get w by itself on one side of the equation. We begin by removing the l from the right side of the equation to isolate the w.

$$A = lw$$

$$\frac{A}{l} = \frac{\cancel{l}w}{\cancel{l}} \qquad \text{Divide both sides of equation by } l.$$

$$\frac{A}{l} = w$$ ∎

EXAMPLE 11   Solve the formula $P = 2l + 2w$ for l.

Solution: We must get l all by itself on one side of the equation. We begin by removing the $2w$ from the right side of the equation to isolate the term containing the l.

$$P = 2l + 2w$$
$$P - 2w = 2l + 2w - 2w \qquad \text{Subtract } 2w \text{ from both sides of equation.}$$
$$P - 2w = 2l$$
$$\frac{P - 2w}{2} = \frac{2l}{2} \qquad \text{Divide both sides of equation by 2.}$$
$$\frac{P - 2w}{2} = l \qquad \left(\text{or } l = \frac{P}{2} - w \right)$$

EXAMPLE 12   An equation we will use in Chapter 7 is $y = mx + b$. Solve for m.

Solution: We must get the m all by itself on one side of the equal sign.

$$y = mx + b$$
$$y - b = mx + b - b \qquad \text{Subtract } b \text{ from both sides of equation.}$$
$$y - b = mx$$
$$\frac{y - b}{x} = \frac{mx}{x} \qquad \text{Divide both sides of equation by } x.$$
$$\frac{y - b}{x} = m \qquad \left(\text{or } m = \frac{y}{x} - \frac{b}{x} \right)$$

EXAMPLE 13   Solve the equation $2x + 3y = 12$ for y; then find y when $x = 6$.

Solution:
$$2x + 3y = 12$$
$$2x - 2x + 3y = 12 - 2x \qquad \text{Subtract } 2x \text{ from both sides of equation.}$$
$$3y = 12 - 2x$$
$$\frac{3y}{3} = \frac{12 - 2x}{3} \qquad \text{Divide both sides of equation by 3.}$$
$$y = \frac{12 - 2x}{3}$$

Now we substitute $x = 6$ and determine the value of y.

$$y = \frac{12 - 2x}{3}$$
$$y = \frac{12 - 2(6)}{3} = \frac{12 - 12}{3} = \frac{0}{3} = 0$$

We see that when $x = 6$, $y = 0$.

EXAMPLE 14   Solve the simple interest formula $i = prt$ for p.

Solution: We must isolate the p. Since p is multiplied by both r and t, we divide both sides of the equation by rt.

$$i = prt$$
$$\frac{i}{rt} = \frac{prt}{rt}$$
$$\frac{i}{rt} = p$$

Some formulas contain fractions. When a formula contains a fraction, we can eliminate the fraction by multiplying both sides of the equation by the denominator, as illustrated in Example 15.

EXAMPLE 15   Solve the formula $A = \dfrac{m + n}{2}$ for m.

Solution: We begin by multiplying both sides of the equation by 2 to eliminate the fraction. Then we isolate the variable m.

$$A = \frac{m + n}{2}$$

$$2\,A = 2\left(\frac{m + n}{2}\right) \qquad \text{Multiply both sides of equation by 2.}$$

$$2A = m + n$$

$$2A - n = m + n - n \qquad \text{Subtract } n \text{ from both sides of equation.}$$

$$2A - n = m \qquad\qquad\qquad ■$$

Exercise Set 3.1

Use the formula to find the value of the indicated variable for the values given. Use a calculator if its use is permitted.

1. $A = s^2$; find A when $s = 4$.

2. $P = a + b + c$; find P when $a = 4$, $b = 3$, and $c = 5$.

3. $P = 2l + 2w$; find P when $l = 8$ and $w = 5$.

4. $A = \dfrac{1}{2}bh$; find A when $b = 12$ and $h = 8$.

5. $A = \dfrac{1}{2}h(b + d)$; find A when $h = 6$, $b = 18$, and $d = 24$.

6. $A = \pi r^2$; find A when $r = 6$ and $\pi = 3.14$.

7. $C = 2\pi r$; find C when $\pi = 3.14$ and $r = 2$.

8. $p = i^2 r$; find r when $p = 4000$ and $i = 2$.

9. $A = \dfrac{1}{2}bh$; find h when $A = 20$ and $b = 4$.

10. $V = \dfrac{1}{3}Bh$; find h when $V = 40$ and $B = 12$.

11. $V = lwh$; find l when $V = 18$, $w = 1$ and $h = 3$.

12. $T = \dfrac{RS}{R + S}$; find T when $R = 50$ and $S = 50$.

13. $A = P(1 + rt)$; find A when $P = 1000$, $r = 0.08$, and $t = 1$.

14. $P = 2l + 2w$; find l when $P = 28$ and $w = 6$.

15. $M = \dfrac{a + b}{2}$; find b when $M = 36$ and $a = 16$.

16. $F = \dfrac{9}{5}C + 32$; find F when C = 10.

17. $C = \dfrac{5}{9}(F - 32)$; find C when F = 41.

18. $z = \dfrac{x - m}{s}$; find z when $x = 115$, $m = 100$, and $s = 15$.

19. $z = \dfrac{x - m}{s}$; find x when $z = 2$, $m = 50$, and $s = 5$.

20. $z = \dfrac{x - m}{s}$; find s when $z = 3$, $x = 80$, and $m = 59$.

21. $K = \dfrac{1}{2}mv^2$; find m when $K = 288$ and $v = 6$.

22. $A = P(1 + rt)$; find r when $A = 1500$, $t = 1$, and $P = 1000$.

23. $v = \pi r^2 h$; find h when $v = 678.24$, $\pi = 3.14$ and $r = 6$.

24. $v = \dfrac{4}{3}\pi r^3$; find v when $\pi = 3.14$ and $r = 6$.

Solve each equation for y; then find the value of y for the given value of x. See Example 13.

25. $2x + y = 8$, when $x = 2$.

26. $6x + 2y = -12$, when $x = -3$.

27. $2x = 6y - 4$, when $x = 10$.

28. $-3x - 5y = -10$, when $x = 0$.

29. $2y = 6 - 3x$, when $x = 2$.

30. $15 = 3y - x$, when $x = 3$.

31. $-4x + 5y = -20$, when $x = 4$.
32. $3x - 2y = -18$, when $x = -1$.
33. $-3x = 18 - 6y$, when $x = 0$.

34. $-12 = -2x - 3y$, when $x = -2$.
35. $-8 = -x - 2y$, when $x = -4$.
36. $2x + 5y = 20$, when $x = -5$.

Solve for the variable indicated.

37. $d = rt$, for t
38. $d = rt$, for r
39. $i = prt$, for p
40. $i = prt$, for r
41. $C = \pi d$, for d
42. $v = lwh$, for w
43. $A = \dfrac{1}{2}bh$, for b
44. $E = IR$, for I
45. $P = 2l + 2w$, for w
46. $PV = KT$, for T
47. $4n + 3 = m$, for n
48. $3t - 4r = 25$, for t
49. $y = mx + b$, for b
50. $y = mx + b$, for x

51. $I = P + Prt$, for r
52. $A = \dfrac{m + d}{2}$, for m
53. $A = \dfrac{m + 2d}{3}$, for d
54. $R = \dfrac{l + 3w}{2}$, for w
55. $d = a + b + c$, for b
56. $A = \dfrac{a + b + c}{3}$, for b
57. $ax + by = c$, for y
58. $ax + by + c = 0$, for y
59. $V = \pi r^2 h$, for h
60. $V = \dfrac{1}{3}\pi r^2 h$, for h

In exercises 61 and 62, use the formula in Example 9, $d = \frac{1}{2}n^2 - \frac{3}{2}n$, to find the number of diagonals in a figure with the given number of sides.

61. 10 sides   **62.** 6 sides

In exercises 63 and 64, use the formula $C = \frac{5}{9}(F - 32)$ to find the Celsius temperature (C) for the given Fahrenheit temperature (F).

63. $F = 50$ **64.** $F = 86$

In exercises 65 and 66, use the formula $F = \frac{9}{5}C + 32$ to find the Fahrenheit temperature (F) for the given Celsius temperature (C).

65. $C = 35$ **66.** $C = 10$

A formula in the study of chemistry is $P = \dfrac{KT}{V}$, where P is pressure, T is temperature, V is volume, and K is a constant. In exercises 67 through 70, find the missing quantity.

67. $T = 10$, $K = 1$, $V = 1$
69. $P = 80$, $T = 100$, $V = 5$

68. $T = 30$, $P = 3$, $K = 0.5$
70. $P = 100$, $K = 2$, $V = 6$

The sum of the first n *even numbers can be found by the formula $S = n^2 + n$. In exercises 71 and 72, find the sum of the numbers indicated.*

71. First 5 even numbers.

72. First 10 even numbers.

In Exercises 73 through 76, use the simple interest formula. See Examples 1 and 2.

73. Mr. Thongsophaporn borrowed $4000 for 3 years at 12% simple interest per year. How much interest did he pay?

74. Ms. Rodriguez lent her brother $4000 for a period of 2 years. At the end of the 2 years, her brother repaid the $4000 plus $640 interest. What simple interest rate did her brother pay?

75. Ms. Levy invested a certain amount of money in a sav-

ings account paying 7% simple interest per year. When she withdrew her money at the end of 3 years, she received $1050 in interest. How much money did Ms. Levy place in the savings account?

76. Mr. O'Connor borrowed $6000 at $7\frac{1}{2}$% simple interest per year. When he withdrew his money, he received $1800 in interest. How long had his money been left in the account?

Use the formulas given in Tables 3.1, 3.2 and 3.3 to work Exercises 77 through 90. See Examples 3 through 8.

77. Find the perimeter of a triangle whose sides are 5 inches, 12 inches, and 13 inches.

78. Find the area of a rectangle whose length is 9 inches and whose width is 4 inches.

79. Find the area of a triangle whose base is 6 centimeters and whose height is 8 centimeters.

80. Find the perimeter of a rectangle whose length is 5 meters and whose width is 3 meters.

81. Find the area of a circle whose radius is 4 inches. Use 3.14 for π.

82. Find the area of a circle whose diameter is 6 centimeters.

83. Find the circumference of a circle whose diameter is 8 inches.

84. Find the area of a trapezoid whose height is 2 feet and whose bases are 6 feet and 4 feet.

85. The area of the smallest post office in America (in Ochopee, Florida) is 48 square feet. If the length of the post office is 6 feet, find the width of the post office.

86. A sail on a sailboat is in the shape of a triangle. If the area of the sail is 36 square feet and the height of the sail is 12 feet, find the base of the sail.

87. The largest banyon tree in the continental United States is at the Edison House in Fort Meyers, Florida. The cir-

See Exercise 87.

cumference of the aerial roots of the tree is 390 feet. Find (a) the radius, to the nearest tenth of a foot, of the aerial roots, and (b) the diameter of the aerial roots to the nearest tenth of a foot.

88. Donovan's garden is in the shape of a trapezoid. If the height of the trapezoid is 12 meters, one base is 15 meters, and the area is 126 square meters, find the length of the other base.

89. An oil drum has a height of 4 feet and a diameter of 22 inches. Find the volume of the drum in cubic inches.

90. Find the volume of an ice cream cone (cone only) if its diameter is 3 inches and its height is 5 inches.

91. By using any formula for area, explain why area is measured in square units.

92. By using any formula for volume, explain why volume is measured in cubic units.

93. (a) Consider the formula for the circumference of a circle, $C = 2\pi r$. If you solve this formula for π, what will you obtain?

(b) If you take the ratio of the circumference of a circle to its diameter, about what numerical value will you obtain? Explain how you determined your answer.

(c) Carefully draw a circle, make it at least 4 inches in diameter. Use a piece of string and a ruler to determine the circumference and diameter of the circle. Find the ratio of the circumference to the diameter. When you divide the circumference by the diameter, what value do you obtain?

Cumulative Review Exercises

[1.9] **94.** Evaluate $\left[4(12 \div 2^2 - 3)^2\right]^2$.

[2.6] **95.** A stable has 4 Morgan and 6 Arabian horses. Find the ratio of Arabians to Morgans.

96. It takes 3 minutes to siphon 25 gallons of water out of a swimming pool. How long will it take to empty a 13,500 gallon swimming pool by siphoning? Write a proportion that can be used to solve the problem, and then find the desired value.

[2.7] **97.** Solve $2(x - 4) \geq 3x + 9$

JUST FOR FUN

1. (a) Using the formulas presented in this section, write an equation in d that can be used to find the shaded area in the figure shown.

(b) Find the shaded area when $d = 4$ feet.
(c) Find the shaded area when $d = 6$ feet.

2. A cereal box is to be made by folding the cardboard along the dashed lines as shown in the figure.
(a) Using the formula

$$\text{volume} = \text{length} \cdot \text{width} \cdot \text{height}$$

write an equation for the volume of the box.
(b) Find the volume of the box when $x = 7$ cm.

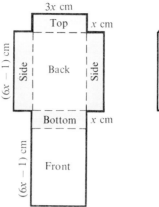
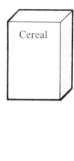

(c) Write an equation for the surface area of the box.
(d) Find the surface area when $x = 7$ cm.

3.2

Changing Application Problems into Equations

▸**1**   Translate phrases into mathematical expressions.

▸**2**   Translate application problems into equations.

▸**1**   One practical advantage of knowing algebra is that you can use it to solve every-day problems involving mathematics. For algebra to be useful to you in solving everyday problems, you must first be able to transform application problems into mathematical language. The purpose of this section is to help you take a verbal or word problem and write it as a mathematical expression.

Often the most difficult part of solving an application problem is translating it into an equation. Here are examples of statements represented as algebraic expressions.

Verbal	*Algebraic*
5 more than a number	$x + 5$
a number increased by 3	$x + 3$
7 less than a number	$x - 7$
a number decreased by 12	$x - 12$
twice a number	$2x$
the product of 6 and a number	$6x$
one-eighth of a number	$\frac{1}{8}x$ or $\frac{x}{8}$
a number divided by 3	$\frac{1}{3}x$ or $\frac{x}{3}$
4 more than twice a number	$2x + 4$
5 less than three times a number	$3x - 5$
3 times the sum of a number and 8	$3(x + 8)$
twice the difference of a number and 4	$2(x - 4)$

To give you more practice with the mathematical terms, we will also convert some algebraic expressions into verbal expressions. Often an algebraic expression can be written in several different ways. Following is a list of some of the possible verbal expressions that can be used to represent the given algebraic expression.

Algebraic	*Verbal*
$2x + 3$	Three more than twice a number The sum of twice a number and three Twice a number, increased by three Three added to twice a number
$3x - 4$	Four less than three times a number Three times a number, decreased by four The difference of three times a number and four Four subtracted from three times a number

EXAMPLE 1   Express each phrase as an algebraic expression.

(a) The distance, d, increased by 10 miles.
(b) 6 less than twice the area.
(c) 3 pounds more than four times the weight.
(d) Twice the sum of the height plus 3 feet.
(e) The cost increased by 6%.

Solution: (a) $d + 10$ (b) $2a - 6$
(c) $4w + 3$ (d) $2(h + 3)$
(e) $c + 0.06c$; note that 6% is written as a decimal. If the cost is c, 6% of the cost is $0.06c$.   ∎

In Example 1, the letter x (or any other letter) could have been used in place of those selected.

EXAMPLE 2   Write three different verbal statements to represent the following expressions: (a) $5x - 2$; (b) $2x + 7$.

Solution: (a) 1. Two less than five times a number.
2. Five times a number, decreased by two.
3. The difference of five times a number and two.
(b) 1. Seven more than twice a number.
2. Two times a number, increased by seven.
3. The sum of two times a number and seven.   ∎

EXAMPLE 3   Write a verbal statement to represent each expression: (a) $3x - 4$; (b) $3(x - 4)$.

Solution: (a) One possible statement is: four less than three times a number.
(b) Three times the difference of a number and four.   ∎

Sometimes in a problem, two numbers are related to each other in a certain way. We often represent the simplest, or most basic number that needs to be expressed, as a variable, and the other as an expression containing that variable. Some examples follow.

Verbal	One Number	Second Number
two numbers differ by 3	x	$x + 3$
John's age now and John's age in 6 years	x	$x + 6$
one number is six times the other number	x	$6x$
the sum of two numbers is 10	x	$10 - x$
a 25-foot tree cut in two pieces	x	$25 - x$
one number is 12% less than the other	x	$x - 0.12x$

Note that often more than one pair of expressions can be used to represent the two numbers. For example, two numbers differ by 3 can also be expressed as x and $x - 3$.

Consider the phrase "a 25-foot tree cut in two pieces." If we call one length x, then the other length must be $25 - x$. For example, if one length is 6 feet, then the other length must be $25 - 6$ or 19 feet.

EXAMPLE 4   For each relationship, select a variable to represent one quantity and express the other quantity in terms of the first.

(a) A boy is 15 years older than his brother.
(b) The speed of the second car is 1.4 times the speed of the first.
(c) $75 is divided between two people.
(d) John has $5 more than three times the amount of money that Dee has.
(e) The length of a rectangle is 3 units less than four times its width.
(f) A number, and the number increased by 6%.

Solution: (a) $x, x + 15$ (b) $x, 1.4x$ (c) $x, 75 - x$ (d) $x, 3x + 5$
(e) $x, 4x - 3$ (f) $x, x + 0.06x$ ■

Consider the statement "the cost of 3 items at $5 each." How would you represent this quantity using mathematical symbols? You would probably reason that the cost would be 3 times $5 and write $3 \cdot 5$ or $3(5)$.

Now consider the statement "the cost of x items at $5 each." How would you represent this statement using mathematical symbols? If you use the same reasoning, you might write $x \cdot 5$ or $x(5)$. Another way to write this product is $5x$. Thus, the cost of x items at $5 each could be represented as $5x$.

Finally, consider the statement "the cost of x items at y dollars each." How would you represent this statement using mathematical symbols? Following the reasoning used in the previous two illustrations, you might write $x \cdot y$ or $x(y)$. Since these products can be written as xy, the cost of x items at y dollars each can be represented as xy.

EXAMPLE 5   Write each of the following as an algebraic expression.

(a) The cost of purchasing x items at $2 each.
(b) Five percent commission on x dollars in sales.
(c) The number of calories in x potato chips, where each potato chip has 8 calories.
(d) The increase in population in n years for a city growing at a rate of 300 per year.
(e) The distance traveled in t hours when 55 miles are traveled each hour.

Solution: (a) We can reason like this: one item would cost $1(2)$ dollars, two items would cost $2(2)$ dollars, three items $3(2)$, four items $4(2)$, and so on. Continuing this reasoning process, we can see that x items would cost $x(2)$ or $2x$ dollars.

(b) A 5% commission on $1 sales would be 0.05(1), on $2 sales 0.05(2), on $3 sales 0.05(3), on $4 sales 0.05(4), and so on. Therefore, the commission on x dollar sales would be 0.05(x) or 0.05x.

(c) $8x$

(d) $300n$

(e) $55t$

EXAMPLE 6   A slice of white bread contains 65 calories and a slice of whole-wheat bread contains 55 calories. Write an algebraic expression to represent the total number of calories in x slices of white and y slices of whole-wheat bread.

Solution: x slices of white bread contain 65x calories.
y slices of whole-wheat bread contain 55y calories.
Together they contain $65x + 55y$ calories.

EXAMPLE 7   Write an algebraic expression for each phrase.

(a) The number of ounces in x pounds.
(b) The number of cents in a dimes and b nickels.
(c) The number of seconds in x hours, y minutes, and z seconds.

Solution: (a) Since each pound contains 16 ounces, x pounds is $16 \cdot x$ or $16x$ ounces.
(b) $10a + 5b$
(c) $3600x + 60y + z$ (3600 seconds = 1 hour)

COMMON STUDENT ERROR

In Example 4(f) we asked you to represent a number increased by 6%. Note the answer is $x + 0.06x$. Often students write the answer to this question as $x + 0.06$. It is important to realize that a percent of a quantity must always be a percent multiplied by some number or letter. Some phrases involving the word percent and the correct and incorrect interpretations follow.

Phrase	Correct	Wrong
The cost, c, increased by 7%	$c + 0.07c$	$c + 0.07$
The cost, c, reduced by 25%	$c - 0.25c$	$c - 0.25$
A $7\frac{1}{2}$% sales tax on c dollars	$0.075c$	0.075
The cost, c, plus a $7\frac{1}{2}$% sales tax	$c + 0.075c$	$c + 0.075$
A number, x, increased by 20%	$x + 0.20x$	$x + 0.20$

Some terms that we will be using are consecutive integers, consecutive even integers, and consecutive odd integers. **Consecutive integers** are integers that differ by 1 unit. For example, the integers 6 and 7 are two consecutive integers. Two consecutive integers may be represented as x and $x + 1$. **Consecutive even integers** are even integers that differ by 2 units. For example, 6 and 8 are two consecutive even integers. **Consecutive odd integers** are odd integers that differ by 2 units. For example, 7 and 9 are two consecutive odd integers. Two consecutive even integers, or two consecutive odd integers, may be represented as x and $x + 2$.

▶ **2** The word *is* in a verbal problem means *is equal to* and is represented by an equal sign. Some examples of verbal problems written as equations are as follows:

Verbal	*Equation*
6 more than twice a number *is* 4	$2x + 6 = 4$
a number decreased by 4 *is* 3 more than twice the number	$x - 4 = 2x + 3$
the product of two consecutive integers *is* 56	$x(x + 1) = 56$
one number is 4 more than three times the other number; their sum *is* 60	$x + (3x + 4) = 60$
a number increased by 15% *is* 120	$x + 0.15x = 120$
the sum of two consecutive odd integers *is* 24	$x + (x + 2) = 24$

Before we begin writing equations, let us translate some equations into verbal statements. Some examples of equations written as verbal statements follow. We will write only two verbal statements for each equation, but remember there are other ways these equations can be written.

Equation	*Verbal*
$3x - 4 = 4x + 3$	Four less than three times a number *is* three more than four times the number.
	Three times a number, decreased by four *is* four times the number, increased by three.
$3(x - 2) = 6x - 4$	Three times the difference of a number and two *is* four less than six times the number.
	The product of three and the difference of a number and two *is* six times a number, decreased by four.

EXAMPLE 8 Write two verbal statements to represent the equation $x - 2 = 3x - 5$.

Solution: **1.** A number decreased by two *is* five less than three times a number.
2. The difference of a number and two *is* the difference of three times the number and five.

EXAMPLE 9 Write a verbal statement to represent the equation $x + 2(x - 4) = 6$.

Solution: The sum of a number and twice the difference of a number and four *is* six.

EXAMPLE 10 Write each problem as an equation.
(a) One number is four less than twice the other. Their sum is 14.
(b) For two consecutive integers, the sum of the smaller and three times the larger is 23.

Solution: (a) Let x = one number.

Then $2x - 4$ = second number.

$$\text{First number} + \text{second number} = 14$$
$$x + (2x - 4) = 14$$

(b) Let x = smaller consecutive integer.
   Then $x + 1$ = larger consecutive integer.

$$\text{Smaller} + \text{three times the larger} = 23$$
$$x + 3(x + 1) = 23 \qquad \blacksquare$$

EXAMPLE 11   Write the following problem as an equation. One train travels 3 miles more than twice the distance another train travels. The total distance traveled by both trains is 800 miles.

Solution: Let x = distance traveled by one train.
   Then $2x + 3$ = distance traveled by second train.

$$\text{Distance of train 1} + \text{distance of train 2} = \text{total distance}$$
$$x + (2x + 3) = 800 \qquad \blacksquare$$

EXAMPLE 12   Express each of the following as an equation.
(a) The cost of renting a snow blower for x days at \$12 per day is \$60.
(b) The population of the town of Newton is increasing at a rate of 500 people per year. The increase in population in t years is 2500.
(c) The distance Dawn and Jack traveled for x days at 600 miles per day is 1500 miles.
(d) The number of cents in d dimes is 120.

Solution: (a) $12x = 60$ (b) $500t = 2500$
   (c) $600x = 1500$ (d) $10d = 120$ $\blacksquare$

Exercise Set 3.2

Write as an algebraic expression.

1. Five more than a number.
2. Seven less than a number.
3. Four times a number.
4. The product of a number and eight.
5. 70% of a number x.
6. 8% of a number y.
7. A 10% sales tax on a piano costing c dollars.
8. A $7\frac{1}{2}$% sales tax on a car costing p dollars.
9. Three less than six times a number.

10. Six times the difference of a number and 3.
11. Seven plus three-fourths of a number.
12. Four times a number, decreased by two.
13. Twice the sum of a number and 8.
14. Seventeen decreased by x.
15. The number of cents in x quarters.
16. The number of cents in x quarters and y dimes.
17. The number of inches in x feet.
18. The number of inches in x feet and y inches.
19. The number of ounces in a pounds and b ounces.

Express as a verbal statement. (There are many acceptable answers.)

20. $x - 6$
21. $x + 3$
22. $4x + 1$
23. $3x - 4$
24. $5x - 7$
25. $2x - 3$
26. $4x - 2$
27. $5 - x$
28. $2 - 3x$
29. $4 + 6x$
30. $2(x - 1)$
31. $3(x + 2)$

Select a variable to represent one quantity, and express the second quantity in terms of the first.

32. Eileen's salary is $45 more than Martin's salary.

33. A boy is 12 years older than his brother.

34. A number is one-third of another.

35. Two consecutive integers.

36. Two consecutive even integers.

37. One hundred dollars divided between two people.

38. Two numbers differ by 12.

39. A number is 5 less than four times another number.

40. A number is 3 more than one-half another number.

41. A Cadillac costs 1.7 times as much as a Ford.

42. A number is 4 less than three times another number.

43. An 80-foot tree cut into two pieces.

44. Two consecutive odd integers.

45. A number and the number increased by 12%.

46. A number and the number decreased by 15%.

47. The cost of an item and the cost increased by a 7% sales tax.

48. The cost of an item and the cost reduced by 25%.

Write as an algebraic expression.

49. The cost of purchasing x items at $4 each.

50. The rental fee for subscribing to Home Box Office for x months at a fee of $12 per month.

51. The cost in dollars of traveling x miles at 23 cents per mile.

52. The cost of paying a consultant $75 per day for t days.

53. The cost of paying a $15 per hour tennis court fee for x hours.

54. The distance traveled in t hours when traveling 30 miles per hour.

55. The number of employees hired when 10 new employees are hired per day for x days.

56. The cost of renting a telephone for b months at a cost of $3.12 per month.

57. The population growth of a city in n years if the city is growing at a rate of 300 per year.

58. The population decline of a city in n years if the city is losing 400 residents a year.

59. The sales tax on x dollars if the sales tax rate is 7.5%.

60. The number of ounces in y pounds.

61. The number of cents in a dimes.

62. The number of seconds in m minutes.

63. The number of dollars in p 5 dollar bills.

64. The number of cents in a dimes and b quarters.

Express as an equation.

65. One number is five times another. The sum of the two numbers is 18.

66. Marie is 6 years older than Denise. The sum of their ages is 48.

67. The sum of two consecutive integers is 47.

68. The product of two consecutive even integers is 48.

69. Twice a number, decreased by 8 is 12.

70. For two consecutive integers, the sum of the smaller and twice the larger is 29.

71. One-fifth of the sum of a number and 10 is 150.

72. One train travels six times as far as another. The total distance traveled by both trains is 700 miles.

73. One train travels 8 miles less than twice the other. The total distance traveled by both trains is 1000 miles.

74. One number is 3 greater than six times the other. Their product is 408.

75. A number increased by 8% is 92.

76. The cost of a car plus a 7% tax is $13,600.

77. The cost of a jacket at a 25% off sale is $65.

78. The cost of a meal plus a 15% tip is $18.

79. The cost of a video cassette recorder reduced by 20% is $215.

80. The product of a number and the number plus 5% is 120.

81. One number is 3 less than twice another number. Their sum is 21.

82. The cost of renting a phone at a cost of $2.37 per month for x months is $27.

83. The distance traveled by a car going 40 miles per hour for t hours is 180 miles.

84. The cost of traveling x miles at 23 cents per mile is $12.80.

85. The number of calories in y French fried potatoes at 15 calories per French fry is 215.

86. Milltown is increasing at a rate of 200 per year. The increase in population in t years is 2400.

87. The number of cents in q quarters is 150.

88. The number of ounces in p pounds is 64.

In Exercises 89–100, express each equation as a verbal statement. (There are many acceptable answers.)

89. $x + 3 = 6$

90. $x - 5 = 2x$

91. $3x - 1 = 2x + 4$

92. $x - 3 = 2x + 3$

93. $4(x - 1) = 6$

94. $3x + 2 = 2(x - 3)$

95. $5x + 6 = 6x - 1$

96. $x - 3 = 2(x + 1)$

97. $x + (x + 4) = 8$

98. $x + (2x + 1) = 5$

99. $2x + (x + 3) = 5$

100. $2x - (x + 3) = 6$

101. Explain why the cost of purchasing x items at 6 dollars each is represented as $6x$.

102. Explain why the cost of purchasing x items at y dollars each is represented as xy.

Cumulative Review Exercises

[2.6] *Write a proportion that can be used to solve each problem. Solve each problem, and find the desired values.*

103. A recipe for chicken stew calls for $\frac{1}{2}$ teaspoon of thyme for each pound of meat. If the meat for the stew weighs 6.7 pounds, how much thyme should be used?

104. Melinda mixes water with dry cat chow for her cat Max. If the directions say to mix 1 cup of water with every 3 cups of dry cat chow, how much water will Melinda add to $\frac{1}{2}$ cup of dry cat chow?

[3.1] **105.** $P = 2l + 2w$; find l when $P = 40$ and $w = 5$.

106. Solve $3x - 2y = 6$ for y. Then find the value of y when x has a value of 6.

JUST FOR FUN

1. *Learn to measure fat.* The percentage of calories from fat can be determined by following these instructions. First multiply the grams of fat, f, in one serving by 9. (This will give you the number of calories contributed by fat in one serving.) Then divide that number by the total number of calories per serving, c, and multiply that result by 100. This is the percentage of calories from fat, P.

(a) Write a formula for determining P, the percentage of calories from fat.

(b) Use this formula to determine the percentage of calories from fat in whole milk. One serving (1 cup) of whole milk contains 8 grams of fat and 150 calories.

(c) A king-size Snickers bar contains 24 grams of fat and 510 calories. Determine the percentage of calories from fat in a Snickers bar.

2. (a) Write an algebraic expression for the number of seconds in d days, h hours, m minutes, and s seconds.

(b) Use the expression found in part (a) to determine the number of seconds in 4 days, 6 hours, 15 minutes, and 25 seconds.

3.3

Solving Application Problems

▶**1** Set up and solve verbal problems.

There are many types of application problems that can be solved using algebra. In this section we introduce several types of application problems. In Section 3.4, we introduce additional types of application problems. Application problems are also presented in many other sections and exercise sets throughout the book. Because of time limitations, it is possible that your instructor may not be able to cover all the applications given in this book. If not, you may wish to spend a little time on your own reading those problems just to get a feel for the types of applications presented.

To be prepared for this section, it is necessary that you understand the material presented in Section 3.2. The best way to learn to set up a verbal or word problem is to practice. The more verbal problems you study and attempt, the easier it will become to solve them.

▶ **1** Transforming verbal problems into mathematical terms is something we do all the time without realizing it. For example, if you need 3 cups of milk for a recipe and the measuring cup holds only 2 cups, you reason that you need 1 additional cup of milk after the initial 2 cups. You may not realize it, but when you do this simple operation, you are using algebra.

Let x = number of additional cups of total milk needed

Thought process: initial 2 cups + number of additional cups = total milk needed

Equation to represent problem: $2 + x = 3$

When we solve for x, we get 1 cup of milk.

You probably said to yourself: Why do I have to go through all this when I know that the answer is $3 - 2$ or 1 cup? When you perform this subtraction, you have mentally solved the equation $2 + x = 3$.

$$2 + x = 3$$
$$2 - 2 + x = 3 - 2$$
$$x = 3 - 2$$
$$x = 1$$

Let's look at another example.

EXAMPLE 1   Suppose that you are at a supermarket, and your purchases so far total $13.20. In addition to groceries, you wish to purchase as many packages of gum as possible, but you have a total of only $18. If a package of gum costs $1.15, how many can you purchase?

Solution: How can we represent this problem as an equation? We might reason as follows. We need to find the number of packages of gum. Let us call this unknown quantity x.

Let x = number of packages of gum

Thought process: cost of groceries + cost of gum = total cost

Now substitute $13.20 for the cost of groceries and $18 for the total cost to get

$$13.20 + \text{cost of gum} = 18$$

At this point you might be tempted to replace the cost of gum with the letter x. But look at what x represents. The variable x represents the *number* of packages of gum, *not the cost of the gum*. In the preceding section we learned that the cost of x packages of gum at $1.15 per package is $1.15x$. Now substitute the cost of the x packages of gum, $1.15x$, into the equation to obtain

Equation to represent problem: $13.20 + 1.15x = 18$

When we solve this equation, we obtain $x = 4.2$ packages (to the nearest tenth). Since you cannot purchase a part of a pack of gum, we reason that only 4 packages of gum can be purchased.

Now let us look at the procedure for setting up and solving a word problem.

To Solve a Word Problem

1. Read the question carefully.
2. If possible, draw a sketch to help visualize the problem.
3. Determine which quantity you are being asked to find. Choose a letter to represent this unknown quantity. Write down exactly what this letter represents. If there is more than one unknown quantity, represent all unknown quantities in terms of this variable.
4. Write the word problem as an equation.
5. Solve the equation for the unknown quantity.
6. Answer the question or questions asked.
7. Check the solution in the original stated problem.

Let us now set up and solve some word problems using this procedure.

EXAMPLE 2   Two subtracted from four times a number is 10. Find the number.

Solution: We are asked to find the number. We designate the unknown number by the letter x. Let $x =$ unknown number.

$$2 \text{ subtracted from } 4 \text{ times a number is } 10$$

Write the equation: $4x - 2 = 10$
Solve the equation: $4x = 12$
Answer the question: $x = 3$

Check: Substitute 3 for the number in the original problem.
Two subtracted from four times a number is 10.
$$4(3) - 2 = 10$$
$$10 = 10 \qquad \text{true}$$

Since the solution checks, the unknown number is 3.   ■

EXAMPLE 3   The sum of two numbers is 17. Find the two numbers if the larger is five more than twice the smaller number.

Solution: We are asked to find *two* numbers. We will call the smaller number x. Then we will represent the larger number in terms of x.

$$\text{Let } x = \text{smaller number}$$
$$\text{then } 2x + 5 = \text{larger number}$$

The sum of the two numbers is 17. Therefore, we write the equation

$$\text{first number} + \text{second number} = 17$$
$$x + (2x + 5) = 17$$

Now solve the equation.

$$3x + 5 = 17$$
$$3x = 12$$
$$x = 4$$

Answer the questions: smaller number $= 4$

larger number $= 2x + 5$

$$= 2(4) + 5 = 13$$

Check: Sum of two numbers $= 17$

$$4 + 13 = 17$$
$$17 = 17 \quad \text{true}$$

■

EXAMPLE 4 The population of a growing city is 40,000. If the population is increasing by 300 a year, in how many years will the population reach 44,500 people?

Solution: We are asked to find the number of years.

Let $n =$ number of years

then $300n =$ increase in population over n years

Present population $+ \left(\begin{array}{c} \text{increase in population} \\ \text{over } n \text{ years} \end{array} \right) =$ future population

$$40,000 + 300n = 44,500$$
$$300n = 4500$$
$$n = \frac{4500}{300}$$
$$n = 15 \text{ years}$$

A check will show that 15 years is the correct answer. ■

EXAMPLE 5 The monthly rental fee for a telephone from the Southern Bell Telephone Company is $2.63. Radio Shack is selling new telephones for $34.99. In how many months will the rental fee equal the cost of the new telephone?

Solution: We are asked to find the number of months.

Let $x =$ number of months

then $2.63x =$ cost of renting phone for x months

Cost of renting phone for x months $=$ cost of new phone

$$2.63x = 34.99$$
$$x = \frac{34.99}{2.63} = 13.3 \text{ months}$$

The rental fee will equal the cost of a new telephone in 13.3 months. ■

EXAMPLE 6 The cost of renting an automobile is $25 a day plus 20 cents a mile. Find the maximum mileage Janet can drive in 1 day if she has only $56.

Solution: We are asked to find the number of miles Janet can drive.

$$\text{Let } x = \text{number of miles Janet can drive}$$
$$\text{then } 0.20x = \text{cost of driving } x \text{ miles}$$

$$\text{Total cost} = \text{daily fee} + \text{mileage cost}$$
$$56 = 25 + 0.20x$$
$$31 = 0.20x$$
$$\frac{31}{0.20} = x$$
$$155 = x$$

Janet can drive a maximum of 155 miles in 1 day.   ■

EXAMPLE 7   United Airlines wishes to keep its airfare, including a 7% tax, between Dallas, Texas, and Los Angeles, California, at exactly $160. Find the cost of the ticket before tax.

Solution: We are asked to find the cost of the ticket before tax.

$$\text{Let } x = \text{cost of the ticket before tax}$$
$$\text{then } 0.07x = \text{tax on the ticket}$$

$$\left(\begin{array}{c}\text{Cost of ticket} \\ \text{before tax}\end{array}\right) + \left(\begin{array}{c}\text{tax on} \\ \text{the ticket}\end{array}\right) = 160$$
$$x + 0.07x = 160$$
$$1.07x = 160$$
$$x = \frac{160}{1.07}$$
$$x = 149.53$$

Thus if United prices the ticket at $149.53, the total cost including a 7% tax will be $160.   ■

EXAMPLE 8   A major university is doing research to determine the most cost-efficient method to set up a number of computer terminals throughout the university. The university is considering two options. Option 1 is an $80,000 minicomputer whose terminals cost $1000 each. Option 2 is a $20,000 network system whose terminals cost $2500 each. How many terminals would the university have to install to make the total cost of the network system equal to the total cost of the minicomputer?

Solution: The network system has a much smaller initial cost ($20,000 versus $80,000); however, their terminals cost more ($2500 versus $1000). We are asked to find the number of terminals where both systems would have the same total cost.

$$\text{Let } n = \text{number of terminals}$$
$$\text{Then } 1000n = \text{cost of } n \text{ terminals for the minicomputer, and}$$
$$2500n = \text{cost of } n \text{ terminals for the network}$$

$$\text{total cost of Option 1} = \text{total cost of Option 2}$$
$$\text{cost of minicomputer} + \text{cost of terminals} = \text{cost of network} + \text{cost of terminals}$$
$$80,000 + 1000n = 20,000 + 2500n$$
$$80,000 = 20,000 + 1500n$$
$$60,000 = 1500n$$
$$40 = n$$

The total cost of the network system equals the cost of the minicomputer when 40 terminals are used. If the university plans to have fewer than 40 terminals, the network system would be less expensive. ▨

EXAMPLE 9 Four brothers must divide a total of $6000. How much will each receive if the two older brothers are each to receive twice as much as the two younger brothers?

Solution: We are asked to find how much each of the four brothers receives.

$$\text{Let } x = \text{amount each of the two younger brothers receives}$$
$$\text{then } 2x = \text{amount each of the two older brothers receives}$$

The total received by the four brothers is $6000. Thus the equation we use is

2 younger brothers   2 older brothers
each receive x each receive $2x$

$$\overbrace{x + x} \quad + \quad \overbrace{2x + 2x} \quad = 6000$$
$$6x = 6000$$
$$x = 1000$$

The two younger brothers each receive $1000 and the two older brothers each receive $2(1000) = \$2000$. ▨

Exercise Set 3.3

Set up an algebraic equation that can be used to solve each problem. Solve the equation, and find the values desired.

1. The sum of two consecutive integers is 71. Find the numbers.

2. The sum of two consecutive even integers is 134. Find the numbers.

3. The sum of two consecutive odd numbers is 76. Find the numbers.

4. One number is 3 more than twice a second number. Their sum is 27. Find the numbers.

5. One number is 5 less than three times a second number. Their sum is 43. Find the numbers.

6. The sum of three consecutive integers is 39. Find the three integers.

7. The sum of three consecutive odd integers is 87. Find the three integers.

8. The sum of three integers is 29. Find the three numbers if one number is twice the smallest and the third number is 4 more than twice the smallest.

9. The larger of two integers is 4 less than three times the smaller. When the smaller number is subtracted from the larger, the difference is 12. Find the two numbers.

10. The larger of two integers is 8 less than twice the smaller. When the smaller number is subtracted from the larger, the difference is 17. Find the two numbers.

11. The village of Naples, which has a population of 5200, is growing by 300 each year. In how many years will the population reach 9400?

12. A city with a population of 50,000 is decreasing by 300 per year. In how many years will the population drop to 44,000?

13. Southern Bell's monthly rental fee for a telephone is $3.50. The telephone store is selling telephones for $14. How long will it take for the monthly rental fee to equal the cost of a new phone?

14. It cost Teshanna $4.50 a week to wash and dry her clothing at the corner laundry. If a washer and dryer cost a total of $612, how many weeks would it take for Teshanna's laundry cost to equal the cost of purchasing a washer and dryer?

15. The cost of renting a car is $15 a day plus 20 cents per mile. How far can Milt drive in one day if he has only $55?

16. State College reimburses its employees $40 a day plus 21 cents a mile when they use their own vehicles on college business. If Prof. Kohn takes a 1-day trip and is reimbursed $103 how far did she travel?

17. B.J.'s Warehouse has a plan whereby for a yearly fee of $60 you save 8% of the price of all items purchased in the store. What would be the total Mary would need to spend during the year so that her savings equals the yearly fee?

18. At a one day 20% off sale, Tan purchased a hat for $15.99. What is the regular price of the hat?

19. During the 1992 contract negotiations the City School Board approved a 8% pay increase for its teachers effective in 1993. If Paul, a first grade teacher, projects his 1993 annual salary to be $31,320 what is his present salary?

20. Mr. Murphy receives a weekly salary of $210. He also receives a 6% commission on the total dollar volume of all sales he makes. What must his dollar volume be in a week if he is to make a total of $450?

21. Eunice, a hot dog vendor, wishes to price her hot dogs such that the total cost of the hot dog, including a 5% tax, is $1.25. What will be the price of a hot dog before tax?

22. Orange County has a 7% sales tax. How much does Robert's car cost before tax, if the total cost of the car plus its sales tax is $9200?

23. Maxwell is on a diet and can have only 500 calories for lunch. He orders a hamburger without a roll and French fries. How many French fries can he eat if the hamburger has 300 calories and each French fry has 20 calories?

24. Marilyn's investment of $7200 on solar heating equipment results in a yearly savings of $600 in her heating bills. How many years will it take for the savings to equal the investment?

25. Ninety-one hours of overtime must be split among four workers. The two younger workers are to be assigned the same number of hours. The third worker is to be assigned twice as much as each of the younger workers. The fourth worker is to be assigned three times as much as each of the younger workers. How much overtime should be assigned to each worker?

26. Gary Gutchell worked a 55-hour week last week. He is not sure of his hourly rate, but knows that he is paid $1\frac{1}{2}$ times his regular hourly rate for all hours over a 40-hour week. His pay last week was $400. What was his hourly rate?

27. After Mrs. Wisniewski is seated in a restaurant, she realizes that she has only $20.00. If from this $20 she must pay a 7% tax and she wishes to leave a 15% tip, what is the maximum price for a meal that she can afford to pay?

28. The fine for speeding in a certain community is $5 per mile per hour over the speed limit plus a $15 administrative charge. Michelle received a speeding ticket and had to pay a fine of $65. How many miles per hour in excess of the speed limit was she traveling?

29. The Main Street Racquetball Club has two payment plans for its members. One plan is a flat $35 per month. The second is to pay $7.25 per hour for court time. If court time can be rented only in 1-hour intervals, how many hours would you have to play per month to make it advantageous to select the flat fee?

30. The Midtown Tennis Club has two payment plans for its members. Plan 1 is a monthly fee of $20 plus $8 per hour court rental time. Plan 2 is no monthly fee, but court time is $16.25 per hour. If court time can be rented only in 1-hour intervals, how many hours would you have to play per month so that plan 1 becomes advantageous?

31. During the first week of a going out of business sale, the ski shop reduced the price of all items by 20%. During the second week of the sale, they reduced the price of all items over $100 by an additional $25. If Helga purchases a pair of Head skis during the second week for $231, what is the regular price of the skis?

32. The Holiday Health Club has reduced its yearly membership fee by 10%. In addition, if you sign up on a Monday, they will take an additional $20 off the already reduced price. If Jorge purchases a year's membership on a Monday and pays an annual fee of $250, what is the regular membership fee?

33. If the minicomputer system in Example 8 costs $65,000 plus $1500 per terminal, and the network system costs $20,000 plus $3000 per terminal, how many terminals would have to be ordered for the cost of the minicomputer system to equal the cost of the network system?

34. To run a chain saw you need to use a mixture of gasoline and oil. For each part oil, you need 15 parts gasoline. If a total of 4 gallons of the oil-gas mixture is to be made, how much oil and how much gas will need to be mixed?

35. Make up your own realistic word problem that can be solved using algebra. Express the problem as an equation and solve the equation. Make sure you answer the question that was asked in the problem.

Cumulative Review Exercises

[1.9] **36.** Evaluate $\dfrac{1}{4} + \dfrac{3}{4} \div \dfrac{1}{2} - \dfrac{1}{3}$.

Name each of the properties represented.

[1.10] **37.** $(x + y) + 5 = x + (y + 5)$

38. $xy = yx$

39. $x(x + y) = x^2 + xy$

[2.6] **40.** At a firemen's chicken barbecue, the chef estimates that he will need $\frac{1}{2}$ pound of coleslaw for each 5 people. If he expects 560 residents to attend, how many pounds of coleslaw will he need?

[3.1] **41.** Solve the formula $M = \dfrac{a + b}{2}$ for b.

JUST FOR FUN

1. To find the **average** of a set of values, you find the sum of the values and divide the sum by the number of values. **(a)** If Paul's first three test grades are 74, 88, and 76, write an equation that can be used to find the grade that Paul must get on his fourth exam to have an 80 average. **(b)** Solve the equation from part (a) and determine the grade Paul must receive.

2. In their opening round of the NCAA 1990 basketball tournament, Syracuse University scored 78 points against Coppin State College. Syracuse made 12 free throws (1 point each). Syracuse also made 4 times as many 2-point field goals as 3-point field goals (field goals made from more than 18 feet from the basket). How many 2-point field goals and how many 3-point field goals did Syracuse make?

3. Pick any number: say, 9
   Multiply the number by 4: $9 \cdot 4 = 36$
   Add 6 to the product: $36 + 6 = 42$
   Divide the sum by 2: $42 \div 2 = 21$
   Subtract 3 from the $21 - 3 = 18$
     quotient:

The solution is twice the number you started with. Show that when you select n to represent the given number the solution will always be $2n$.

3.4

Geometric Problems   ▶1   Solve geometric problems

This section serves two purposes. One is to reinforce the geometric formulas introduced in Section 3.1. The second is to reinforce procedures for setting up and solving word problems discussed in Sections 3.2 and 3.3. The more practice you have at setting up and solving the word problems, the better you will become at solving them.

EXAMPLE 1   Mrs. O'Connor is planning to build a sandbox for her daughter. She has 26 feet of wood to build the perimeter. What should be the dimensions of the rectangular sandbox if the length is to be 3 feet longer than the width?

Solution: We are asked to find the dimensions of the sandbox.

$$\text{Let } x = \text{width of sandbox}$$
$$\text{then } x + 3 = \text{length of sandbox (Fig. 3.2)}$$

From Section 3.1 we know that $P = 2l + 2w$. We have called the width of the sandbox x, and the length $x + 3$. We substitute these expressions into the equation as follows.

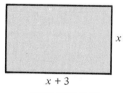

$x + 3$

FIGURE 3.2

$$P = 2l + 2w$$
$$26 = 2(x + 3) + 2x$$
$$26 = 2x + 6 + 2x$$
$$26 = 4x + 6$$
$$20 = 4x$$
$$5 = x$$

Thus the width is 5 feet, and the length $= x + 3 = 5 + 3 = 8$ feet.

Check: $P = 2l + 2w$
$$26 = 2(8) + 2(5)$$
$$26 = 16 + 10$$
$$26 = 26 \qquad \text{true}$$

EXAMPLE 2   The sum of the angles of a triangle measure 180 degrees (180°). If two angles are the same and the third is 30° greater than the other two, find the three angles of the triangle.

Solution: We are asked to find the three angles.

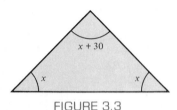

FIGURE 3.3

$$\text{Let } x = \text{each smaller angle}$$
$$\text{then } x + 30 = \text{larger angle (Fig. 3.3)}$$
$$\text{Sum of the 3 angles} = 180$$
$$x + x + (x + 30) = 180$$
$$3x + 30 = 180$$
$$3x = 150$$
$$x = \frac{150}{3} = 50°$$

Therefore, the three angles are 50°, 50°, and 50° + 30° or 80°.

Check: $50° + 50° + 80° = 180°$
$$180° = 180° \qquad \text{true} \qquad \blacksquare$$

Recall from Section 3.1 that a quadrilateral is a four-sided figure. Quadrilaterals include squares, rectangles, parallelograms, and trapezoids. The sum of the measures of the angles of any quadrilateral is 360°. We use this information in Example 3.

EXAMPLE 3   In a parallelogram the opposite angles have the same measures. If the two larger angles in a parallelogram are each 20° less than three times the smaller angles, find the measure of each angle.

Solution:
$$\text{Let } x = \text{the measure of each of the two smaller angles}$$
$$\text{then } 3x - 20 = \text{the measure of each of the two larger angles}$$

A diagram of the parallelogram is given in Figure 3.4.

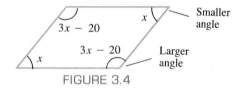

FIGURE 3.4

$$\left(\begin{array}{c}\text{measure of the}\\\text{two smaller angles}\end{array}\right) + \left(\begin{array}{c}\text{measure of the}\\\text{two larger angles}\end{array}\right) = 360°$$
$$x + x + (3x - 20) + (3x - 20) = 360$$
$$x + x + 3x - 20 + 3x - 20 = 360$$
$$8x - 40 = 360$$
$$8x = 400$$
$$x = 50$$

Thus each of the two smaller angles is 50° and each of the two larger angles is $3x - 20 = 3(50) - 20 = 130°$. As a check, $50° + 50° + 130° + 130° = 360°$. $\blacksquare$

EXAMPLE 4   A bookcase is to have four shelves, including the top, as shown in Figure 3.5. The height of the bookcase is to be 3 feet more than the width. Find the dimensions of the bookcase if only 30 feet of wood is available.

Solution: We are asked to find the dimensions of the bookcase.

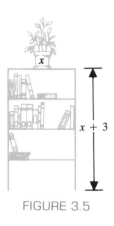

Let x = length of a shelf

then $x + 3$ = height of bookcase

4 shelves + 2 sides = total wood available

$$4x + 2(x + 3) = 30$$
$$4x + 2x + 6 = 30$$
$$6x + 6 = 30$$
$$6x = 24$$
$$x = 4$$

FIGURE 3.5

The length of a shelf is 4 feet and the height of the bookcase is $4 + 3$ or 7 feet.

Check: $4 + 4 + 4 + 4 + 7 + 7 = 30$

$$30 = 30 \qquad \text{true}$$

Exercise Set 3.4

Solve the following geometric problems.

1. An **equilateral triangle** is a triangle that has three sides of the same length. If the perimeter of an equilateral triangle is 28.5 inches, find the length of each side. Equilateral triangles are discussed in Appendix C.

2. Two angles are **complementary angles** if the sum of their measures is 90°. If angle A and angle B are complementary angles, and angle A is 21° more than twice angle B, find the measures of angle A and angle B. Complementary angles are discussed in Appendix C.

3. Two angles are **supplementary angles** if the sum of their measures is 180°. If angle A and angle B are supplementary angles, and angle B is 8° less than three times angle A, find the measures of angle A and angle B. Supplementary angles are discussed in Appendix C.

4. If one angle of a triangle is 20° larger than the smallest angle, and the third angle is six times as large as the smallest angle, find the measures of the three angles.

5. If one angle of a triangle is 10° greater than the smallest angle, and the third angle is 30° less than twice the smallest angle, find the measures of the three angles.

6. The length of a rectangle is to be 8 feet more than its width. What should the dimensions of the rectangle be if the perimeter is to be 48 feet?

7. In an **isosceles triangle** two sides are equal. The third side is 2 meters less than one of the other sides. Find the length of each side if the perimeter is 10 meters. Isosceles triangles are discussed in Appendix C.

8. The perimeter of a rectangle is 120 feet. Find the length and width of the rectangle if the length is twice the width.

9. The perimeter of the cement basement floor of a house is 240 feet. Find the length and width of the rectangular-shaped basement floor if the length is 24 feet less than twice the width.

10. If the two smaller angles of a parallelogram have equal measures and the two larger angles are each 30° larger than a smaller angle, find the measure of each angle.

11. If the two smaller angles of a parallelogram have equal measures and the two larger angles each measure 27° less than twice a smaller angle, find the measure of each angle.

12. The measure of one angle of a quadrilateral is 10° greater than the smallest angle, the third angle is 14° greater than twice the smallest angle, and the fourth angle is 21° greater than the smallest angle. Find the measures of the four angles of the quadrilateral.

13. A bookcase is to have four shelves as shown. The height of the bookcase is to be 2 feet more than the width, and only 20 feet of wood is available. What should be the dimensions of the bookcase?

14. What should be the dimensions of the bookcase in Exercise 13, if the height is to be twice its width?

15. Betty McKane plans to build storage shelves as shown. If she has only 45 feet of wood for the entire unit and

wishes the length to be 3 times the height, find the length and height of the unit.

16. An area is to be fenced in along a straight river bank as illustrated. If the length of the fenced-in area is to be 4 feet greater than the width, and the total amount of fencing used is 64 feet, find the width and length of the fenced-in area.

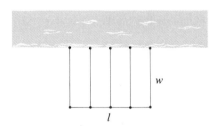

17. Consider the equation $A = l \cdot w$. What happens to the area if the length is doubled and the width is halved? Explain your answer.

18. Consider the equation $A = s^2$. What happens to the area if the length of a side, s, is doubled? Explain your answer.

19. Consider the equation $V = l \cdot w \cdot h$. What happens to the volume if the length, width, and height are all doubled? Explain your answer.

20. Consider the equation $V = \frac{4}{3}\pi r^3$. What happens to the volume if the radius is tripled? Explain your answer.

21. Make up your own realistic geometric word problem that can be solved using algebra. Write the problem as an equation and solve the equation. Answer the question asked in the original problem.

Cumulative Review Exercises

[1.3] *Insert either* $>$, $<$, *or* $=$ *in the shaded area to make the statement true.*

22. $-|-6|$ ▨ $|-4|$ **23.** $|-3|$ ▨ $-|3|$

[1.6] **24.** Evaluate $-6 - (-2) + (-4)$.

[2.1] **25.** Simplify $-6y + x - 3(x - 2) + 2y$.

JUST FOR FUN

1. Consider the accompanying figure. **(a)** Write a formula for determining the area of the colored region. **(b)** Find the area of the colored region when $S = 9$ inches and $s = 6$ inches.

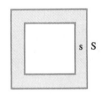

2. One way to express the area of the following figure is $(a + b)(c + d)$. Can you determine another expression, using the area of the four rectangles, to represent the area of the figure?

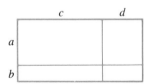

3. The total pressure, P, in pounds per square inch, exerted on an object x feet below sea level is given by the formula $P = 14.70 + 0.43x$. As shown in the accompany-

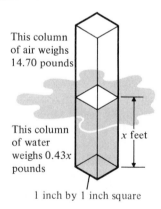

This column of air weighs 14.70 pounds

This column of water weighs $0.43x$ pounds

x feet

1 inch by 1 inch square

ing diagram, the 14.70 represents the weight in pounds of the column of air (from sea level to the top of the atmosphere) standing over a 1-inch by 1-inch square of seawater. The $0.43x$ represents the weight, in pounds, of a column of water 1 inch by 1 inch by x feet.

(a) A submarine can withstand a total pressure of 162 pounds per square inch. How deep can the submarine go?

(b) If the pressure gauge in the submarine registers a total pressure of 97.26 pounds per square inch, how deep is the submarine?

SUMMARY

GLOSSARY

Area *(113):* The total surface area within a figure's boundaries.

Circumference *(114):* The length of the curve that forms a circle.

Complementary angles *(137):* Two angles whose measures sum to 90°.

Diameter *(114):* A line segment through the center of a circle with both end points on the circle.

Equilateral triangle *(137):* A triangle with three sides of the same length.

Formula *(112):* An equation commonly used to express a specific physical concept mathematically.

Perimeter *(113):* The sum of the lengths of the sides of a figure.

Quadrilateral *(113):* A four-sided figure.

Radius *(114):* A line segment from the center of a circle to any point on the circle.

Supplementary angles *(137):* Two angles whose measures sum to 180°.

IMPORTANT FACTS

Simple interest formula: $i = prt$

The sum of the measures of the angles in any triangle is 180°.

The sum of the measures of the angles in any quadrilateral is 360°.

To Solve a Word Problem

1. Read the question carefully.
2. If possible, draw a sketch to help visualize the problem.

3. Determine which quantity you are being asked to find. Choose a letter to represent this unknown quantity; write down exactly what this letter represents. If there is more than one unknown quantity, express all unknown quantities in terms of the variable selected.
4. Write the word problem as an equation.
5. Solve the equation for the unknown quantity.
6. Answer the question or questions asked.
7. Check the solution in the original problem.

Review Exercises

[3.1] *Use the formula to find the value of the indicated variable for the values given.*

1. $C = \pi d$; find C when $d = 4$ and $\pi = 3.14$.

2. $A = \frac{1}{2}bh$; find A when $b = 12$ and $h = 8$.

3. $P = 2l + 2w$; find P when $l = 6$ and $w = 4$.

4. $i = prt$; find i when $p = 1000$, $r = 15\%$, and $t = 2$.

5. $E = IR$; find E when $I = 0.12$ and $R = 2000$.

6. $A = \pi r^2$; find A when $r = 3$ and $\pi = 3.14$.

7. $V = \frac{4}{3}\pi r^3$; find V when $r = 3$ and $\pi = 3.14$.

8. $Fd^2 = km$; find k when $F = 60$, $m = 12$, and $d = 2$.

9. $y = mx + b$; find b when $y = 15$, $m = 3$, and $x = -2$.

10. $2x + 3y = -9$; find y when $x = 12$.

11. $4x - 3y = 15 + x$; find y when $x = -3$.

12. $2x = y + 3z + 4$; find y when $x = 5$ and $z = -3$.

13. $IR = E + Rr$; find r when $I = 5$, $E = 100$, and $R = 200$.

Solve the given equation for y; then find the value of y for the given value of x.

14. $2x - y = 12$, $x = 10$

15. $3x - 2y = -4$, $x = 2$

16. $3x = 5 + 2y$, $x = -3$

17. $-6x - 2y = 20$, $x = 0$

18. $6 = -3x - 2y$, $x = -6$

19. $3y - 4x = -3$, $x = 2$

Solve for the variable indicated.

20. $F = ma$, for m

21. $A = \frac{1}{2}bh$, for h

22. $i = prt$, for t

23. $P = 2l + 2w$, for w

24. $2x - 3y = 6$, for y

25. $A = \dfrac{B + C}{2}$, for B

26. $V = \frac{4}{3}\pi r^2 h$, for h

Solve each problem.

27. How much interest will Karen pay if she borrows $600 for 2 years at 15% simple interest? (Use $i = prt$.)

[3.2, 3.3] *Solve each problem.*

29. One number is 4 more than the other. Find the two numbers if their sum is 62.

30. The sum of two consecutive integers is 255. Find the two integers.

31. The larger of two integers is 3 more than five times the smaller integer. Find the two numbers if the smaller subtracted from the larger is 31.

32. What is the cost of a car before tax if the total cost including a 5% tax is $8400?

33. In Paul's present position as a salesman he receives a base salary of $500 per week plus a 3% commission on all sales he makes. He is considering changing jobs and moving to another company where he would sell the same items. His base salary would be only $400 per week, but his commission would be 8% on all sales he makes. What weekly dollar sales would he have to make for the total salary of each company to be the same?

34. During the first week of a going-out-of-business sale, all items are reduced by 20%. During the second week of the sale, all items that still cost more than $100 are reduced by an additional $25. During the second week of the sale, Kathy purchased a camcorder for $495. What was the original price of the camcorder?

[3.4] **35.** If one angle of a triangle measures 10° greater than the smallest angle, and the third angle measures 10° less than twice the smallest angle, find the measures of the three angles.

36. One angle of a trapezoid measures 10° greater than the smallest angle. A third angle measures five times the smallest angle. The fourth angle measures 20° greater than four times the smallest angle. Find the measure of the four angles.

28. The perimeter of a rectangle is 16 inches. Find the length of the rectangle if the width is 2 inches.

37. Mrs. Appleby wants a garden whose length is 4 feet more than its width. The perimeter of the garden is to be 70 feet. What will be the dimensions of the garden?

[3.1–3.4]

38. The sum of two consecutive odd integers is 208. Find the two integers.

39. Mr. and Mrs. Lendel rent a car for $18 a day plus 16 cents a mile. If they plan to use the car for 2 days, how many miles can they drive if they have only $100?

40. What is the cost of a television before tax if the total cost including a 6% tax is $477?

41. Mr. McAdams sells water softeners. He receives a weekly salary of $300 plus a 5% commission on the sales he makes. If Mr. McAdams earned $900 last week, what was his dollar volume in sales?

42. One angle of a triangle is 8° greater than the smallest angle. The third angle is 4° greater than twice the smallest angle. Find the measures of the three angles of the triangle.

43. Dreyel Company plans to increase its number of employees by 25 per year. If the company presently has 427 employees, how long will it take before they reach 627 employees?

44. If the two larger angles of a parallelogram each measure 40° greater than the two smaller angles, find the measure of the four angles.

Practice Test

1. Use $P = 2l + 2w$ to find P when $l = 6$ feet and $w = 3$ feet.

2. Use $A = P + Prt$ to find A when $P = 100$, $r = 0.15$, and $t = 3$.

3. Use $V = \frac{1}{3}\pi r^2 h$ to find V when $r = 4$, $h = 6$, and $\pi = 3.14$.

Solve for the variable indicated.

4. $P = IR$, for R

5. $3x - 2y = 6$, for y

6. $A = \dfrac{a + b}{3}$, for a

7. $D = R(c + a)$, for c

8. The sum of two integers is 158. Find the two integers if the larger is 10 less than twice the smaller.

9. The sum of three consecutive integers is 42. Find the three integers.

10. Mr. Herron has only $20. If he wishes to leave a 15% tip and must pay 7% tax, find the price of the most expensive meal that he can order.

11. A triangle has a perimeter of 75 inches. Find the three sides if one side is 15 inches larger than the smallest side, and the third side is twice the smallest side.

12. The sum of the angles of a parallelogram is 360°. If the two smaller angles are equal and the two larger angles are each 30° greater than twice the smaller angles, find the measure of each angle.

4

Exponents, Polynomials, and Additional Applications

4.1 Exponents

4.2 Negative Exponents
(Optional)

4.3 Scientific Notation
(Optional)

4.4 Addition and Subtraction
of Polynomials

4.5 Multiplication
of Polynomials

4.6 Division of Polynomials

4.7 Rate and Mixture Problems

Summary

Review Exercises

Practice Test

Cumulative Review Test

See Section 4.7, Exercise 9.

4.1

Exponents

▶ **1** Review exponents.

▶ **2** Learn the rules of exponents.

▶ **1** To understand and work with polynomials, we need to expand our knowledge of exponents. Exponents were introduced in Chapter 1 (Section 1.8). Let us review the fundamental concepts. In the expression x^n, x is referred to as the **base** and n is called the **exponent.** x^n is read "x to the nth power."

$$x^2 = \underbrace{x \cdot x}_{\text{2 factors of } x}$$

$$x^4 = \underbrace{x \cdot x \cdot x \cdot x}_{\text{4 factors of } x}$$

$$x^m = \underbrace{x \cdot x \cdot x \cdot \cdots \cdot x}_{m \text{ factors of } x}$$

EXAMPLE 1   Write $xxxxyy$ using exponents.

Solution: $\underbrace{x \; x \; x \; x}_{\substack{\text{4 factors} \\ \text{of } x}} \; \underbrace{y \; y}_{\substack{\text{2 factors} \\ \text{of } y}} \; = x^4 y^2$ ■

Remember that, when a term containing a variable is given without a numerical coefficient, the numerical coefficient of the term is assumed to be 1. For example, $x = 1x$ and $x^2y = 1x^2y$.

Also recall that, when a variable or numerical value is given without an exponent, the exponent of that variable or numerical value is assumed to be 1. For example, $x = x^1$, $xy = x^1y^1$, $x^2y = x^2y^1$, and $2xy^2 = 2^1x^1y^2$.

Rules of Exponents   ▶ **2** Now we will learn the rules of exponents.

EXAMPLE 2   Multiply $x^4 \cdot x^3$.

Solution: $\overbrace{x^4}^{} \quad \cdot \quad \overbrace{x^3}^{}$

$\underbrace{x \cdot x \cdot x \cdot x \cdot x \cdot x \cdot x}_{} = x^7$ ■

Example 2 illustrates that when multiplying expressions with the same base we keep the base and *add* the exponents. This is the product rule of exponents.

Product Rule
_____

$$x^m \cdot x^n = x^{m+n}$$

In Example 2 we showed that $x^4 \cdot x^3 = x^7$. This problem could also be done using the product rule: $x^4 \cdot x^3 = x^{4+3} = x^7$.

EXAMPLE 3   Multiply using the product rule.

(a) $3^2 \cdot 3$ (b) $2^4 \cdot 2^2$ (c) $x \cdot x^4$ (d) $x^2 \cdot x^5$ (e) $y^4 \cdot y^7$

Solution: (a) $3^2 \cdot 3 = 3^2 \cdot 3^1 = 3^{2+1} = 3^3$ or 27
(b) $2^4 \cdot 2^2 = 2^{4+2} = 2^6$ or 64
(c) $x \cdot x^4 = x^1 \cdot x^4 = x^{1+4} = x^5$
(d) $x^2 \cdot x^5 = x^{2+5} = x^7$
(e) $y^4 \cdot y^7 = y^{4+7} = y^{11}$

COMMON STUDENT ERROR

Note in Example 3(a) that $3^2 \cdot 3^1$ is 3^3 and not 9^3. When multiplying powers of the same base, *do not multiply the bases.*

Correct	*Wrong*
$3^2 \cdot 3^1 = 3^3$	$\cancel{3^2 \cdot 3^1 = 9^3}$

Example 4 will be helpful in explaining the quotient rule of exponents.

EXAMPLE 4   Divide $x^5 \div x^3$.

Solution: $\dfrac{x^5}{x^3} = \dfrac{\cancel{x} \cdot \cancel{x} \cdot \cancel{x} \cdot x \cdot x}{\cancel{x} \cdot \cancel{x} \cdot \cancel{x}} = \dfrac{1 x^2}{1} = x^2$

When dividing expressions with the same base, keep the base and *subtract* the exponent in the denominator from the exponent in the numerator.

Quotient Rule

$$\frac{x^m}{x^n} = x^{m-n}, \qquad x \neq 0$$

In Example 4 we showed that $x^5/x^3 = x^2$. This problem could also be done using the quotient rule: $x^5/x^3 = x^{5-3} = x^2$.

EXAMPLE 5   Divide each expression.

(a) $\dfrac{3^5}{3^2}$ (b) $\dfrac{5^4}{5}$ (c) $\dfrac{x^{12}}{x^5}$ (d) $\dfrac{x^9}{x^5}$ (e) $\dfrac{x^7}{x}$

Solution: (a) $\dfrac{3^5}{3^2} = 3^{5-2} = 3^3$ or 27   (b) $\dfrac{5^4}{5} = \dfrac{5^4}{5^1} = 5^{4-1} = 5^3$ or 125

(c) $\dfrac{x^{12}}{x^5} = x^{12-5} = x^7$ (d) $\dfrac{x^9}{x^5} = x^{9-5} = x^4$

(e) $\dfrac{x^7}{x} = x^{7-1} = x^6$

COMMON STUDENT ERROR

Note in Example 5(a) that $3^5/3^2$ is 3^3 and not 1^3. When dividing powers of the same base, *do not divide out the bases*.

Correct	*Wrong*
$\dfrac{3^3}{3^1} = 3^2$ or 9	$\dfrac{3^3}{3^1} = 1^2$

The answer to Example 5(c), $\dfrac{x^{12}}{x^5}$, is x^7. We obtained this answer using the quotient rule. This answer could also be obtained by dividing out the common factor in both the numerator and denominator as follows. We divided out the product of five x's, which is x^5. We can indicate this process in shortened form as follows:

$$\frac{x^{12}}{x^5} = \frac{(x \cdot x \cdot x \cdot x \cdot x) \cdot x \cdot x \cdot x \cdot x \cdot x \cdot x \cdot x}{(x \cdot x \cdot x \cdot x \cdot x)} = x^7$$

$$\frac{x^{12}}{x^5} = \frac{x^5 \cdot x^7}{x^5} = x^7$$

Similarly, $\dfrac{x^5}{x^{12}}$ can be simplified by dividing out the common factor, x^5, as follows:

$$\frac{x^5}{x^{12}} = \frac{x^5}{x^5 \cdot x^7} = \frac{1}{x^7}$$

We will now simplify some expressions by dividing out common factors.

EXAMPLE 6 Simplify each of the following by dividing out a common factor in both the numerator and denominator.

(a) $\dfrac{x^9}{x^{12}}$ (b) $\dfrac{y^4}{y^9}$

Solution: (a) Write the denominator with a factor of x^9. Since $x^9 \cdot x^3 = x^{12}$, we break x^{12} up into $x^9 \cdot x^3$.

$$\frac{x^9}{x^{12}} = \frac{x^9}{x^9 \cdot x^3} = \frac{1}{x^3}$$

(b) $\dfrac{y^4}{y^9} = \dfrac{y^4}{y^4 \cdot y^5} = \dfrac{1}{y^5}$ ■

We will show another way to evaluate expressions like $\dfrac{x^9}{x^{12}}$ in the next section by using the negative exponent rule.

Example 7 will lead us into our next rule, the zero exponent rule.

EXAMPLE 7   Divide $\dfrac{x^3}{x^3}$.

Solution: By the quotient rule,

$$\frac{x^3}{x^3} = x^{3-3} = x^0$$

However,

$$\frac{x^3}{x^3} = \frac{\not{x} \cdot \not{x} \cdot \not{x}}{\not{x} \cdot \not{x} \cdot \not{x}} = 1$$

Since $x^3/x^3 = x^0$ and $x^3/x^3 = 1$, then x^0 must equal 1.

Example 7 illustrates the zero exponent rule.

Zero Exponent Rule

$$x^0 = 1, \qquad x \neq 0$$

EXAMPLE 8 Simplify each expression.

(a) 3^0 (b) x^0 (c) $3x^0$ (d) $(3x)^0$

Solution: (a) $3^0 = 1$
(b) $x^0 = 1$
(c) $3x^0 = 3(x^0)$ Remember, the exponent refers only to the immediately
$\qquad = 3 \cdot 1 = 3$ preceding symbol unless parentheses are used.
(d) $(3x)^0 = 1$

COMMON STUDENT ERROR

An expression raised to the zero power is not equal to 0; it is equal to 1.

Correct	*Wrong*
$x^0 = 1$	$x^0 = 0$
$5^0 = 1$	$5^0 = 0$

The zero exponent rule specifies that $x \neq 0$. Why can x not be 0? Consider the problem $0^2/0^2$.

$$\frac{0^2}{0^2} = \frac{0 \cdot 0}{0 \cdot 0} = \frac{0}{0} = \text{indeterminate} \qquad \text{(from Section 1.7)}$$

By the quotient rule,

$$\frac{0^2}{0^2} = 0^{2-2} = 0^0$$

Notice that $0^2/0^2 = 0^0$ and $0^2/0^2 = $ indeterminate. Therefore, 0^0 must be indeterminate.

The power rule will be explained with the aid of Example 9.

EXAMPLE 9 Simplify $(x^3)^2$.

Solution: $(x^3)^2 = \underbrace{x^3 \cdot x^3}_{\substack{2 \text{ factors} \\ \text{of } x^3}} = x^{3+3} = x^6$

Power Rule

$$(x^m)^n = x^{m \cdot n}$$

The power rule indicates that when we have an exponential expression raised to an exponent, we keep the base and *multiply* the exponents. Example 9 could also be solved using the power rule.

$$(x^3)^2 = x^{3 \cdot 2} = x^6$$

Note that the answers are the same.

EXAMPLE 10   Simplify each term.

(a) $(x^3)^4$ (b) $(3^4)^2$ (c) $(y^3)^8$

Solution: (a) $(x^3)^4 = x^{3 \cdot 4} = x^{12}$
(b) $(3^4)^2 = 3^{4 \cdot 2} = 3^8$
(c) $(y^3)^8 = y^{3 \cdot 8} = y^{24}$ ■

HELPFUL HINT

Students often confuse the product and power rules. Note the difference carefully.

Product Rule	*Power Rule*
$x^m \cdot x^n = x^{m+n}$	$(x^m)^n = x^{m \cdot n}$
$2^3 \cdot 2^5 = 2^{3+5} = 2^8$	$(2^3)^5 = 2^{3 \cdot 5} = 2^{15}$

Example 11 will help us in explaining the expanded power rule. As the name suggests, this rule is an expansion of the power rule.

EXAMPLE 11   Simplify $\left(\dfrac{ax}{by}\right)^4$.

Solution: $\left(\dfrac{ax}{by}\right)^4 = \dfrac{ax}{by} \cdot \dfrac{ax}{by} \cdot \dfrac{ax}{by} \cdot \dfrac{ax}{by}$

$$= \frac{a \cdot a \cdot a \cdot a \cdot x \cdot x \cdot x \cdot x}{b \cdot b \cdot b \cdot b \cdot y \cdot y \cdot y \cdot y} = \frac{a^4 \cdot x^4}{b^4 \cdot y^4} = \frac{a^4 x^4}{b^4 y^4}$$ ■

Example 11 illustrates the expanded power rule.

Expanded Power Rule

$$\left(\frac{ax}{by}\right)^m = \frac{a^m x^m}{b^m y^m}, \qquad b \neq 0, y \neq 0$$

The expanded power rule illustrates that every factor within parentheses is affected by an exponent outside the parentheses.

EXAMPLE 12   Simplify each expression.

(a) $(2x)^2$ (b) $(-x)^3$ (c) $(2xy)^3$ (d) $\left(\dfrac{-3x}{2y}\right)^2$

Solution: (a) $(2x)^2 = 2^2x^2 = 4x^2$ (b) $(-x)^3 = (-1x)^3 = (-1)^3x^3 = -1x^3 = -x^3$

(c) $(2xy)^3 = 2^3x^3y^3 = 8x^3y^3$ (d) $\left(\dfrac{-3x}{2y}\right)^2 = \dfrac{(-3)^2x^2}{2^2y^2} = \dfrac{9x^2}{4y^2}$ ∎

EXAMPLE 13   Simplify   (a) $\dfrac{8x^3y^2}{4xy^2}$ (b) $\left(\dfrac{8x^3y^2}{4xy^2}\right)^3$.

Solution: (a) We begin by dividing out common factors.

$$\frac{8x^3y^2}{4xy^2} = 2 \cdot \frac{x^3}{x} \cdot \frac{y^2}{y^2} = 2x^2$$

(b) To obtain the answer to part (b), we need only cube the answer to part (a), $2x^2$.

$$(2x^2)^3 = 2^3(x^2)^3 = 8x^6$$

Thus $\left(\dfrac{8x^3y^2}{4xy^2}\right)^3 = 8x^6$. ∎

Note that whenever we have an expression raised to a power, as in Example 13(b), we simplify the expression in parentheses as much as possible before using the expanded power rule.

EXAMPLE 14   Simplify $\left(\dfrac{25x^4y^3}{5x^2y^7}\right)^4$.

Solution: Begin by simplifying the expression within parentheses.

$$\left(\frac{25x^4y^3}{5x^2y^7}\right)^4 = \left(\frac{5x^2}{y^4}\right)^4 = \frac{5^4x^8}{y^{16}} = \frac{625x^8}{y^{16}}$$ ∎

COMMON STUDENT ERROR

Students often make errors in simplifying expressions containing exponents. It is very important that you have a thorough understanding of exponents. One of the most common errors made by students follows. Study this error carefully to make sure you do not make this error.

Correct	*Wrong*
$\dfrac{4}{2x} = \dfrac{\overset{2}{\cancel{4}}}{\underset{1}{\cancel{2}}x} = \dfrac{2}{x}$	$\dfrac{4}{x+2} = \dfrac{\overset{2}{\cancel{4}}}{x+2} = \dfrac{2}{x+1}$
$\dfrac{x^3y^2}{y^2} = \dfrac{x^3\cancel{y^2}}{\cancel{y^2}} = x^3$	$\dfrac{x^3+y^2}{y^2} = \dfrac{x^3+\cancel{y^2}}{\underset{1}{\cancel{y^2}}} = x^3+1$

The simplifications on the right are wrong because only common **factors** can be divided out (remember factors are multiplied together). In the denominator on the right, $x + 2$, the x and 2 are terms, and not factors, since they are being added. Also, in the numerator $x^3 + y^2$, the x^3 and y^2 are terms, not factors. No common factors can be divided out in the fractions on the right.

EXAMPLE 15 Simplify $(2x^2y^3)^4(xy^2)$.

Solution: Begin simplifying $(2x^2y^3)^4$ by using the expanded power rule. Then use the product rule to obtain the answer.

$$(2x^2y^3)^4(xy^2) = (16x^8y^{12})(x^1y^2)$$
$$= 16 \cdot x^8 \cdot x^1 \cdot y^{12} \cdot y^2$$
$$= 16x^9y^{14}$$

Summary of the Rules of Exponents Presented in This Section

1. $x^m \cdot x^n = x^{m+n}$ product rule

2. $\dfrac{x^m}{x^n} = x^{m-n}, \qquad x \neq 0$ quotient rule

3. $x^0 = 1, \qquad x \neq 0$ zero exponent rule

4. $(x^m)^n = x^{m \cdot n}$ power rule

5. $\left(\dfrac{ax}{by}\right)^m = \dfrac{a^m x^m}{b^m y^m}, \ b \neq 0, y \neq 0$ expanded power rule

---

Exercise Set 4.1

Simplify each of the following.

1. $x^4 \cdot x^3$
2. $x^5 \cdot x^2$
3. $y^2 \cdot y$
4. $4^2 \cdot 4$
5. $3^2 \cdot 3^3$
6. $x^4 \cdot x^2$
7. $y^3 \cdot y^2$
8. $x^3 \cdot x^4$
9. $y^4 \cdot y$
10. $\dfrac{x^4}{x^3}$
11. $\dfrac{x^{15}}{x^7}$
12. $\dfrac{y^3}{y}$
13. $\dfrac{5^4}{5^2}$
14. $\dfrac{3^5}{3^2}$
15. $\dfrac{x^9}{x^5}$
16. $\dfrac{x^3}{x^5}$
17. $\dfrac{y^2}{y}$
18. $\dfrac{x^{13}}{x^4}$
19. $\dfrac{x^2}{x^2}$
20. $\dfrac{3^4}{3^4}$
21. x^0
22. 5^0
23. $3x^0$
24. $-2x^0$
25. $(3x)^0$
26. $-(4x)^0$
27. $(-4x)^0$
28. $(x^2)^3$
29. $(x^5)^2$
30. $(x^2)^2$
31. $(x^5)^5$
32. $(x^4)^2$
33. $(x^3)^1$
34. $(x^3)^2$
35. $(x^3)^4$
36. $(x^5)^4$
37. $(x^4)^2$
38. $(2x)^2$
39. $(1.3x)^2$
40. $(-3x)^2$
41. $(-x)^2$
42. $(-x)^3$
43. $(4x^2)^3$
44. $(2.5x^3)^2$
45. $(-3x^3)^3$
46. $(xy)^4$
47. $(2x^2y)^3$
48. $(4x^3y^2)^3$
49. $(8.6x^2y^5)^2$
50. $(2xy^4)^3$
51. $(-6x^3y^2)^3$
52. $(9xy^4)^2$
53. $(-x^4y^5z^6)^3$
54. $(-2x^4y^2z)^3$
55. $\left(\dfrac{x}{y}\right)^2$
56. $\left(\dfrac{x}{3}\right)^2$
57. $\left(\dfrac{x}{5}\right)^3$
58. $\left(\dfrac{2}{x}\right)^3$
59. $\left(\dfrac{y}{x}\right)^5$
60. $\left(\dfrac{3}{y}\right)^4$
61. $\left(\dfrac{6}{x}\right)^3$
62. $\left(\dfrac{2x}{y}\right)^3$
63. $\left(\dfrac{3x}{y}\right)^3$
64. $\left(\dfrac{5x^2}{y}\right)^2$

65. $\left(\dfrac{2x}{5}\right)^2$

66. $\left(\dfrac{3x^4}{2}\right)^3$

67. $\left(\dfrac{4y^3}{x}\right)^3$

68. $\left(\dfrac{-4x^2}{5}\right)^2$

69. $\left(\dfrac{-3x^3}{4}\right)^3$

70. $\left(\dfrac{-x^5}{y^2}\right)^3$

Simplify each of the following.

71. $\dfrac{x^3y^2}{xy^5}$

72. $\dfrac{x^2y^6}{x^4y}$

73. $\dfrac{x^5y^7}{x^{12}y^3}$

74. $\dfrac{x^4y^5}{x^7y^{12}}$

75. $\dfrac{10x^3y^8}{2xy^{10}}$

76. $\dfrac{5x^{12}y^2}{10xy^9}$

77. $\dfrac{4xy}{16x^3y^2}$

78. $\dfrac{20x^4y^6}{5xy^9}$

79. $\dfrac{35x^4y^7}{10x^9y^{12}}$

80. $\dfrac{20x^8y^{12}}{5x^8y^7}$

81. $\dfrac{-36xy^9z}{12x^4y^5z^2}$

82. $\dfrac{4x^4y^7z^3}{32x^5y^4z^9}$

83. $\dfrac{-6x^2y^7z^5}{2x^5y^9z^6}$

84. $\dfrac{-25x^4y^{10}}{30x^3y^7z}$

85. $\left(\dfrac{4x^4}{2x^6}\right)^3$

86. $\left(\dfrac{5x^5}{10x^7}\right)^3$

87. $\left(\dfrac{8y^7}{2y^3}\right)^3$

88. $\left(\dfrac{125y^4}{25y^{10}}\right)^3$

89. $\left(\dfrac{27x^9}{30x^5}\right)^2$

90. $\left(\dfrac{18y^6}{24y^{10}}\right)^3$

91. $\left(\dfrac{x^4y^3}{x^2y^5}\right)^2$

92. $\left(\dfrac{2x^7y^2}{4xy^2}\right)^3$

93. $\left(\dfrac{9y^2z^7}{18y^7z}\right)^4$

94. $\left(\dfrac{y^7z^5}{y^8z^4}\right)^{10}$

95. $\left(\dfrac{3x^2y^5}{y^2}\right)^3$

96. $\left(\dfrac{-64xy^6}{32xy^9}\right)^4$

97. $\left(\dfrac{-x^4y^6}{x^2}\right)^2$

98. $\left(\dfrac{-x^3y^5}{xy^7}\right)^3$

99. $\left(\dfrac{-12x}{16x^7y^2}\right)^2$

100. $\left(\dfrac{-x^4z^7}{x^2z^5}\right)^4$

101. $x^6(3xy^4)$

102. $(2x^4y)(-y^5)$

103. $(-6xy^5)(3x^2y^4)$

104. $(-2xy)(3xy)$

105. $(3x^4y^2)(4xy^6)$

106. $(5x^2y)(3xy^5)$

107. $(5xy)(2xy^6)$

108. $(3x^2y)^2(xy)$

109. $(2xy)^2(3xy^2)$

110. $(3x^2)^4(2xy^5)$

111. $(x^4y^6)^3(3x^2y^5)$

112. $(4x^2y)(3xy^2)^3$

113. $(2x^2y^5)(3x^5y^4)^3$

114. $(5x^4y^7)(2x^3y)^3$

115. $(x^7y^5)(xy^2)^4$

116. $(xy^4)(xy^4)^3$

117. $(3x^4y^{10})^2(2x^2y^8)$

118. $(3x^6y)^2(4xy^8)$

Read the Common Student Error on page 147. Simplify the following expressions by dividing out common factors. If the expression cannot be simplified by dividing out common factors, state so.

119. $\dfrac{x+y}{x}$

120. $\dfrac{xy}{x}$

121. $\dfrac{x^2+2}{x}$

122. $\dfrac{2x^2}{2}$

123. $\dfrac{x+4}{2}$

124. $\dfrac{2x}{2}$

125. $\dfrac{x^2y^2}{x^2}$

126. $\dfrac{x^2+y^2}{x^2}$

127. $\dfrac{x}{x+1}$

128. $\dfrac{1}{x+1}$

129. $\dfrac{x^4}{x^2y}$

130. $\dfrac{y^2}{x^2+y}$

131. For what value of x is $x^0 \neq 1$?

132. Explain the difference between the product rule and power rule. Give an example of each.

133. Consider the expression $(-x^5y^7)^9$. When the power rule is used to simplify the expression, will the *sign* of the simplified expression be positive or negative? Explain how you determined your answer.

134. Consider the expression $(-9x^4y^6)^8$. When the power rule is used to simplify the expression, what will be the

sign of the simplified expression? Explain how you determined your answer.

135. Consider the expression $(-8x^5y^7)^6$. When simplified, what will be the *sign* of the simplified expression? Explain how you determined your answer.

136. In your own words discuss (a) the product rule, (b) the quotient rule, (c) the zero exponent rule, (d) the power rule, and (e) the expanded power rule.

Cumulative Review Exercises

[2.2] *Answer each question in your own words.*

✏️ **137.** What is a linear equation?

✏️ **138.** What is a conditional linear equation?

✏️ **139.** What is an identity?

[3.1] **140.** Find the circumference and area of the circle shown.

3 in.

141. Solve the equation $2x - 5y = 6$ for y.

JUST FOR FUN

Simplify each expression.

1. $\left(\dfrac{3x^4y^5}{6x^6y^8}\right)^3\left(\dfrac{9x^7y^8}{3x^3y^5}\right)^2$

2. $(2xy^4)^3\left(\dfrac{6x^2y^5}{3x^3y^4}\right)^3(3x^2y^4)^2$

4.2

Negative Exponents (Optional)

▶ **1**   Understand the negative exponent rule.

▶ **2**   Simplify expressions containing negative exponents.

▶ **1**   One additional rule that involves exponents is the negative exponent rule. You will need to understand negative exponents to be successful with scientific notation in the next section.

The negative exponent rule will be developed using the quotient rule illustrated in Example 1.

EXAMPLE 1   Simplify $\dfrac{x^3}{x^5}$ by (a) using the quotient rule, and (b) dividing out common factors.

Solution: **(a)** By the quotient rule,

$$\frac{x^3}{x^5} = x^{3-5} = x^{-2}$$

(b) By dividing out common factors,

$$\frac{x^3}{x^5} = \frac{\cancel{x} \cdot \cancel{x} \cdot \cancel{x}}{\cancel{x} \cdot \cancel{x} \cdot \cancel{x} \cdot x \cdot x} = \frac{1}{x^2}$$

In Example 1 we see that $\dfrac{x^3}{x^5}$ is equal to both x^{-2} and $\dfrac{1}{x^2}$. Therefore, x^{-2} must equal $\dfrac{1}{x^2}$. That is, $x^{-2} = \dfrac{1}{x^2}$. This is an example of the negative exponent rule.

Negative Exponent Rule

$$x^{-m} = \frac{1}{x^m}, \qquad x \neq 0$$

When a variable or number is raised to a negative exponent, the expression may be rewritten as 1 divided by the variable or number to that positive exponent.

Examples

$$x^{-3} = \frac{1}{x^3} \qquad 4^{-2} = \frac{1}{4^2} = \frac{1}{16}$$

$$y^{-7} = \frac{1}{y^7} \qquad 5^{-3} = \frac{1}{5^3} = \frac{1}{125}$$

COMMON STUDENT ERROR

Students often believe that a negative exponent automatically makes the entire expression negative. This is not true.

Expression	*Correct*	*Wrong*	*Also wrong*
3^{-2}	$\dfrac{1}{3^2}$	-3^2	$-\dfrac{1}{3^2}$
x^{-3}	$\dfrac{1}{x^3}$	$-x^3$	$-\dfrac{1}{x^3}$

To help you see that the negative exponent rule makes sense, consider the following sequence of exponential expressions and their corresponding values.

$$2^4 = 16$$
$$2^3 = 8 \qquad \text{(One-half of 16 is 8.)}$$
$$2^2 = 4 \qquad \text{(One-half of 8 is 4.)}$$
$$2^1 = 2 \qquad \text{(One-half of 4 is 2.)}$$
$$2^0 = 1 \qquad \text{(One-half of 2 is 1.)}$$
$$2^{-1} = \frac{1}{2^1} \text{ or } \frac{1}{2} \qquad \left(\text{One-half of 1 is } \tfrac{1}{2}.\right)$$
$$2^{-2} = \frac{1}{2^2} \text{ or } \frac{1}{4} \qquad \left(\text{One-half of } \tfrac{1}{2} \text{ is } \tfrac{1}{4}.\right)$$
$$2^{-3} = \frac{1}{2^3} \text{ or } \frac{1}{8} \qquad \left(\text{One-half of } \tfrac{1}{4} \text{ is } \tfrac{1}{8}.\right)$$
$$2^{-4} = \frac{1}{2^4} \text{ or } \frac{1}{16} \qquad \left(\text{One-half of } \tfrac{1}{8} \text{ is } \tfrac{1}{16}.\right)$$

Note that each time the exponent decreases by 1 the value of the expression is halved. For example, when we go from 2^4 to 2^3, the value of the expression goes from 16 to 8. If we continue decreasing the exponents beyond $2^0 = 1$, the next exponent in the pattern is -1. And if we take half of 1 we get $\frac{1}{2}$. This pattern illustrates that $x^{-m} = \dfrac{1}{x^m}$.

▶ **2** Generally, when you are asked to simplify an exponential expression **your final answer should contain no negative exponents.** You may simplify exponential expressions using the negative exponent rule and rules of exponents presented in the previous section. The following examples indicate how exponential expressions containing negative exponents may be simplified.

EXAMPLE 2   Use the negative exponent rule to write each expression with positive exponents.

(a) x^{-2} (b) y^{-4} (c) 3^{-2} (d) 5^{-1} (e) $\dfrac{1}{x^{-2}}$

Solution: (a) $x^{-2} = \dfrac{1}{x^2}$ (b) $y^{-4} = \dfrac{1}{y^4}$

(c) $3^{-2} = \dfrac{1}{3^2} = \dfrac{1}{9}$ (d) $5^{-1} = \dfrac{1}{5}$

(e) $\dfrac{1}{x^{-2}} = \dfrac{1}{1/x^2} = \dfrac{1}{1} \cdot \dfrac{x^2}{1} = x^2$ ∎

HELPFUL HINT

When a factor is moved from the denominator to the numerator or from the numerator to the denominator, the sign of the *exponent* changes.

$$x^{-4} = \frac{1}{x^4} \qquad \frac{1}{x^{-4}} = x^4$$

$$3^{-5} = \frac{1}{3^5} \qquad \frac{1}{3^{-5}} = 3^5$$

Now let's look at additional examples that combine two or more of the rules presented so far.

EXAMPLE 3   Simplify each term.

(a) $(y^{-3})^8$ (b) $(4^2)^{-3}$

Solution: (a) $(y^{-3})^8 = y^{(-3)(8)}$ by the power rule

$= y^{-24}$

$= \dfrac{1}{y^{24}}$ by the negative exponent rule

(b) $(4^2)^{-3} = 4^{(2)(-3)}$ by the power rule

$= 4^{-6}$

$= \dfrac{1}{4^6}$ by the negative exponent rule   ∎

EXAMPLE 4   Simplify each of the following.

(a) $x^3 \cdot x^{-5}$ (b) $3^{-4} \cdot 3^{-7}$

Solution: (a) $x^3 \cdot x^{-5} = x^{3+(-5)}$ by the product rule

$= x^{-2}$

$= \dfrac{1}{x^2}$ by the negative exponent rule

(b) $3^{-4} \cdot 3^{-7} = 3^{-4+(-7)}$ by the product rule

$= 3^{-11}$

$= \dfrac{1}{3^{11}}$ by the negative exponent rule   ∎

COMMON STUDENT ERROR

What is the sum of $3^2 + 3^{-2}$? Look carefully at the correct solution.

Correct	*Wrong*

$$3^2 + 3^{-2} = 9 + \frac{1}{9} \qquad \cancel{3^2 + 3^{-2} = 0}$$
$$= 9\frac{1}{9}$$

Note that $3^2 \cdot 3^{-2} = 3^{2+(-2)} = 3^0 = 1$.

EXAMPLE 5   Simplify each of the following.

(a) $\dfrac{x^7}{x^{10}}$ (b) $\dfrac{5^{-7}}{5^{-4}}$

Solution: (a) $\dfrac{x^7}{x^{10}} = x^{7-10}$ by the quotient rule

$\qquad\qquad = x^{-3}$

$\qquad\qquad = \dfrac{1}{x^3}$ by the negative exponent rule

(b) $\dfrac{5^{-7}}{5^{-4}} = 5^{-7-(-4)}$ by the quotient rule

$\qquad\quad = 5^{-7+4}$

$\qquad\quad = 5^{-3}$

$\qquad\quad = \dfrac{1}{5^3}$ or $\dfrac{1}{125}$ by the negative exponent rule   ■

HELPFUL HINT

Consider a division problem where a variable has a negative exponent in either its numerator or its denominator. To simplify the expression, we can move the variable with the negative exponent from the numerator to the denominator, or from the denominator to the numerator, and change the sign of the exponent. For example,

$$\frac{x^{-4}}{x^5} = \frac{1}{x^5 \cdot x^4} = \frac{1}{x^{5+4}} = \frac{1}{x^9}$$

$$\frac{y^3}{y^{-7}} = y^3 \cdot y^7 = y^{3+7} = y^{10}$$

Now consider a division problem where the variable has a negative exponent in both its numerator and denominator. To simplify such an expression, we move the variable with the lesser (more negative) exponent from the numerator to the denominator, or from the denominator to the numerator, and change the sign of the exponent from negative to positive. For example,

$$\frac{x^{-8}}{x^{-3}} = \frac{1}{x^8 \cdot x^{-3}} = \frac{1}{x^{8-3}} = \frac{1}{x^5} \qquad \text{Note that } -8 < -3.$$

$$\frac{y^{-4}}{y^{-7}} = y^7 \cdot y^{-4} = y^{7-4} = y^3 \qquad \text{Note that } -7 < -4.$$

EXAMPLE 6   Simplify each expression.

(a) $4x^2(5x^{-5})$ (b) $\dfrac{8x^3y^{-2}}{4xy^2}$ (c) $\dfrac{2x^2y^5}{8x^7y^{-3}}$

Solution: (a) $4x^2(5x^{-5}) = 4 \cdot 5 \cdot x^2 \cdot x^{-5} = 20x^{-3} = \dfrac{20}{x^3}$

(b) $\dfrac{\overset{2}{\cancel{8}}x^3y^{-2}}{\underset{1}{\cancel{4}}xy^2} = 2 \cdot \dfrac{x^3}{x} \cdot \dfrac{y^{-2}}{y^2}$

$\qquad = 2 \cdot x^2 \cdot \dfrac{1}{y^4}$

$\qquad = \dfrac{2x^2}{y^4}$

(c) $\dfrac{\overset{1}{\cancel{2}}x^2y^5}{\underset{4}{\cancel{8}}x^7y^{-3}} = \dfrac{1}{4} \cdot \dfrac{x^2}{x^7} \cdot \dfrac{y^5}{y^{-3}}$

$\qquad = \dfrac{1}{4} \cdot \dfrac{1}{x^5} \cdot y^8$

$\qquad = \dfrac{y^8}{4x^5}$

If you study Example 6(b), you can see that the variable with the negative exponent was moved from the numerator to the denominator. In 6(c), the variable with the negative exponent was moved from the denominator to the numerator. In each case, the sign of the exponent was changed from negative to positive when the variable factor was moved.

EXAMPLE 7   Simplify $(4x^{-3})^{-2}$.

Solution: Begin by using the expanded power rule.

$$(4x^{-3})^{-2} = 4^{-2}x^{(-3)(-2)}$$

$$= \dfrac{1}{4^2}x^6$$

$$= \dfrac{x^6}{16}$$

COMMON STUDENT ERROR

Consider the following simplifications. Can you explain why the simplification on the right is wrong?

Correct   *Wrong*

$\dfrac{x^3y^{-2}}{w} = \dfrac{x^3}{wy^2}$ $\cancel{\dfrac{x^3 + y^{-2}}{w}} \;\; \cancel{\dfrac{x^3}{w + y^2}}$

The reason the simplification on the right is wrong is because in the numerator $x^3 + y^{-2}$ the y^{-2} *is not a factor;* it is a term. We will learn how to simplify expressions like this when we study complex fractions in Section 6.6.

Summary of Rules of Exponents

1. $x^m \cdot x^n = x^{m+n}$ product rule

2. $\dfrac{x^m}{x^n} = x^{m-n}$, $x \neq 0$ quotient rule

3. $x^0 = 1$, $x \neq 0$ zero exponent rule

4. $(x^m)^n = x^{m \cdot n}$ power rule

5. $\left(\dfrac{ax}{by}\right)^m = \dfrac{a^m x^m}{b^m y^m}$, $b \neq 0$, $y \neq 0$ expanded power rule

6. $x^{-m} = \dfrac{1}{x^m}$, $x \neq 0$ negative exponent rule

Exercise Set 4.2

Simplify each of the following.

1. x^{-2} **2.** y^{-5} **3.** 5^{-1} **4.** 6^{-2}

5. $\dfrac{1}{x^{-4}}$ **6.** $\dfrac{1}{x^{-2}}$ **7.** $\dfrac{1}{x^{-1}}$ **8.** $\dfrac{1}{y^{-3}}$

9. $\dfrac{1}{5^{-2}}$ **10.** $\dfrac{1}{6^{-3}}$ **11.** $(x^{-2})^3$ **12.** $(x^{-4})^2$

13. $(y^{-7})^3$ **14.** $(y^3)^{-8}$ **15.** $(x^5)^{-2}$ **16.** $(x^{-9})^{-2}$

17. $(2^{-3})^{-2}$ **18.** $(2^{-3})^2$ **19.** $x^4 \cdot x^{-1}$ **20.** $x^{-3} \cdot x^1$

21. $x^7 \cdot x^{-5}$ **22.** $x^{-3} \cdot x^{-2}$ **23.** $3^{-2} \cdot 3^4$ **24.** $5^3 \cdot 5^{-4}$

25. $\dfrac{x^9}{x^{12}}$ **26.** $\dfrac{x^2}{x^{-1}}$ **27.** $\dfrac{y^6}{y^{-3}}$ **28.** $\dfrac{x^{-2}}{x^5}$

29. $\dfrac{x^{-7}}{x^{-3}}$ **30.** $\dfrac{x^{-8}}{x^{-3}}$ **31.** $\dfrac{3^2}{3^{-1}}$ **32.** $\dfrac{2^6}{2^{-1}}$

33. 3^{-3} **34.** x^{-7} **35.** $\dfrac{1}{z^{-9}}$ **36.** $\dfrac{1}{4^{-3}}$

37. $(x^5)^{-5}$ **38.** $(x^{-3})^{-4}$ **39.** $(y^{-2})^{-3}$ **40.** $x^9 \cdot x^{-12}$

41. $x^5 \cdot x^{-9}$ **42.** $x^{-3} \cdot x^{-5}$ **43.** $x^{-12} \cdot x^{-7}$ **44.** $4^{-3} \cdot 4^3$

45. $\dfrac{x^{-3}}{x^5}$ **46.** $\dfrac{y^6}{y^{-8}}$ **47.** $\dfrac{y^9}{y^{-1}}$ **48.** $\dfrac{3^{-4}}{3}$

49. $\dfrac{2^{-3}}{2^{-3}}$ **50.** y^{-1} **51.** z^{-7} **52.** $\dfrac{1}{2^{-5}}$

53. $\dfrac{1}{1^{-7}}$ **54.** $(z^{-7})^{-3}$ **55.** $(x^{-4})^{-1}$ **56.** $(x^{-3})^0$

57. $(x^0)^{-3}$ **58.** $(2^{-2})^{-1}$ **59.** $2^{-3} \cdot 2$ **60.** $6^4 \cdot 6^{-2}$

61. $6^{-4} \cdot 6^2$

62. $\dfrac{x^{-5}}{x^{-9}}$

63. $\dfrac{x^{-1}}{x^{-4}}$

64. $\dfrac{x^{-4}}{x^{-1}}$

65. $(3^2)^{-1}$

66. $(5^{-2})^{-2}$

67. $\dfrac{5}{5^{-2}}$

68. $\dfrac{x^6}{x^7}$

69. $\dfrac{2^{-4}}{2^{-2}}$

70. $x^{-12} \cdot x^8$

71. $\dfrac{7^{-1}}{7^{-1}}$

Simplify each of the following.

72. $4xy^{-1}$

73. $5x^{-1}y$

74. $(6x^2)^{-2}$

75. $(3x^3)^{-1}$

76. $3x^{-2}y^2$

77. $5x^4y^{-1}$

78. $5x^{-5}y^{-2}$

79. $(3x^2y^3)^{-2}$

80. $(4x^2y^{-3})^{-2}$

81. $(x^5y^{-3})^{-3}$

82. $(3x^{-3}y^4)^{-1}$

83. $3x(5x^{-4})$

84. $(4x^{-2})(5x^{-3})$

85. $2x^5(3x^{-6})$

86. $6x^4(-2x^{-2})$

87. $(9x^5)(-3x^{-7})$

88. $(4x^2y)(3x^3y^{-1})$

89. $(2x^{-3}y^{-2})(x^4y)$

90. $(5x^{-7})(4x^{-2}y)$

91. $(3y^{-2})(5x^{-1}y^3)$

92. $\dfrac{8x^4}{4x^{-1}}$

93. $\dfrac{3x^5}{6x^{-2}}$

94. $\dfrac{12x^{-2}}{3x^5}$

95. $\dfrac{2y^{-6}}{6y^4}$

96. $\dfrac{5x^{-2}}{25x^{-5}}$

97. $\dfrac{36x^{-4}}{9x^{-2}}$

98. $\dfrac{12x^{-2}y}{2x^3y^2}$

99. $\dfrac{3x^4y^{-2}}{6y^3}$

100. $\dfrac{16x^{-7}y^{-2}}{4x^5y^2}$

101. $\dfrac{32x^4y^{-2}}{4x^{-2}y^{-3}}$

102. $\dfrac{9x^4y^{-7}}{18x^{-1}y^{-1}}$

 103. (a) Does $a^{-1}b^{-1} = \dfrac{1}{ab}$?
Explain your answer.

(b) Does $a^{-1} + b^{-1} = \dfrac{1}{a+b}$?
Explain your answer.

 104. (a) Does $\dfrac{x^{-1}y^2}{z} = \dfrac{y^2}{xz}$?
Explain your answer.

(b) Does $\dfrac{x^{-1} + y^2}{z} = \dfrac{y^2}{x+z}$?
Explain your answer.

105. In your own words describe the negative exponent rule.

Cumulative Review Exercises

[1.9] **106.** Evaluate $2[6 - (4 - 5)] \div 2 - 5^2$.

107. Evaluate $\dfrac{-3^2 \cdot 4 \div 2}{\sqrt{9} - 2^2}$.

[2.6] **108.** The instructions on a bottle of concentrated household cleaner say to mix 8 ounces of the cleaner with 3 gallons of water. If your bucket holds only 2.5 gallons of water, how much cleaner should you use?

[3.3] **109.** The larger of two integers is one more than three times the smaller. If the sum of the two integers is 37, find the two integers.

JUST FOR FUN

1. Often problems involving exponents can be done in more than one way. Simplify

$$\left(\dfrac{3x^2y^3}{z}\right)^{-2}$$

(a) by first using the expanded power rule.

(b) by first using the negative exponent rule.

4.3

Scientific Notation (Optional)

▶ **1** Convert a decimal number to and from scientific notation.

▶ **2** Do calculations with numbers in scientific notation form.

▶ **1**   When working with scientific problems, we often deal with very large and very small numbers. For example, the distance from Earth to the sun is about 93,000,000 miles. The wavelength of a yellow color of light is about 0.0000006 meter. Because it is difficult to work with many zeros, scientists often express such numbers with exponents. For example, the number 93,000,000 might be written 9.3×10^7 and the number 0.0000006 might be written 6.0×10^{-7}. Numbers such as 9.3×10^7 and 6.0×10^{-7} are in a form called **scientific notation.** Each number written in scientific notation is written as a number greater than or equal to 1 and less than 10 ($1 \leq a < 10$) multiplied by some power of 10.

Examples of Numbers in Scientific Notation

$$1.2 \times 10^6$$
$$3.762 \times 10^3$$
$$8.07 \times 10^{-2}$$
$$1 \times 10^{-5}$$

Consider the number 68,400.

$$68,400 = 6.84 \times 10,000$$
$$= 6.84 \times 10^4 \qquad \text{Note that } 10,000 = 10 \cdot 10 \cdot 10 \cdot 10 = 10^4.$$

Therefore, $68,400 = 6.84 \times 10^4$. Note that to go from 68,400 to 6.84 the decimal point was moved 4 places to the left. Also note that the exponent on the 10, the 4, is the same as the number of places the decimal point was moved to the left. Here is a simplified procedure for writing a number in scientific notation:

To Write a Number in Scientific Notation

1. Move the decimal in the original number to the right of the first non-zero digit. This will give a number greater than or equal to 1 and less than 10.
2. Count the number of places you have moved the decimal to obtain the number in step 1. If the original number was 10 or greater, the count is to be considered positive. If the original number was less than 1, the count is to be considered negative.
3. Multiply the number obtained in step 1 by 10 raised to the count (power) found in step 2.

EXAMPLE 1   Write the following numbers using scientific notation.
(a) 10,700   (b) 0.000386
(c) 972,000   (d) 0.0083

Solution: (a) 10,700 means 10,700.

$$10{,}700. = 1.07 \times 10^4$$

4 places

The original number is greater than 10; therefore, the exponent is positive.

(b) $0.000386 = 3.86 \times 10^{-4}$

4 places

The original number is less than 1; therefore, the exponent is negative.

(c) $972{,}000. = 9.72 \times 10^5$

5 places

(d) $0.0083 = 8.3 \times 10^{-3}$

3 places

To Convert from a Number Given in Scientific Notation

1. Observe the exponent of the power of 10.

2. (a) If the exponent is positive, move the decimal in the number (greater than or equal to 1 and less than 10) to the right the same number of places as the exponent. It may be necessary to add zeros to the number. This will result in a number greater than or equal to 10.

   (b) If the exponent is negative, move the decimal in the number to the left the same number of places as the exponent (dropping the negative sign). It may be necessary to add zeros. This will result in a number less than 1.

EXAMPLE 2   Write each number without exponents.
(a) 3.2×10^4 (b) 6.28×10^{-3} (c) 7.95×10^8

Solution: (a) Moving the decimal four places to the right gives

$$3.2 \times 10^4 = 3.2 \times 10{,}000 = 32{,}000$$

(b) Move the decimal three places to the left.

$$6.28 \times 10^{-3} = 0.00628$$

(c) Move the decimal eight places to the right.

$$7.95 \times 10^8 = 795{,}000{,}000$$

▶ **2**   We can use the rules of exponents presented in Section 4.1 when working with numbers written in scientific notation.

EXAMPLE 3   Multiply $(4.2 \times 10^6)(2 \times 10^{-4})$.

Solution: $(4.2 \times 10^6)(2 \times 10^{-4}) = (4.2 \times 2)(10^6 \times 10^{-4})$

$$= 8.4 \times 10^2$$

$$= 840$$

EXAMPLE 4 Divide $\dfrac{6.2 \times 10^{-5}}{2 \times 10^{-3}}$.

Solution: $\dfrac{6.2 \times 10^{-5}}{2 \times 10^{-3}} = \left(\dfrac{6.2}{2}\right)\left(\dfrac{10^{-5}}{10^{-3}}\right)$

$$= 3.1 \times 10^{-5-(-3)}$$
$$= 3.1 \times 10^{-5+3}$$
$$= 3.1 \times 10^{-2}$$
$$= 0.031$$

■

EXAMPLE 5 Multiply $(42,100,000)(0.008)$.

Solution: Change each number to scientific notation form.

$$(42,100,000)(0.008) = (4.21 \times 10^7)(8 \times 10^{-3})$$
$$= (4.21 \times 8)(10^7 \times 10^{-3})$$
$$= 33.68 \times 10^4$$
$$= 336,800$$

■

 **Calculator Corner**

What will your calculator show when you multiply very large or very small numbers? The answer depends on whether your calculator has the ability to display an answer in scientific notation form. On calculators without the ability to express numbers in scientific notation, you will probably get an error message because the answer will be too large or too small for the display.

Example: On a calculator without scientific notation:

$\boxed{\text{C}}$ 8000000 $\boxed{\times}$ 600000 $\boxed{=}$ Error

Example: On a calculator that uses scientific notation form:

$\boxed{\text{C}}$ 8000000 $\boxed{\times}$ 600000 $\boxed{=}$ 4.8   12

This 4.8   12 means 4.8×10^{12}.

Example: On a calculator that uses scientific notation form:

$\boxed{\text{C}}$ .0000003 $\boxed{\times}$ .004 $\boxed{=}$ 1.2   −9

This 1.2   −9 means 1.2×10^{-9}.

Exercise Set 4.3

Express each number in scientific notation form.

1. 42,000 **2.** 3,610,000 **3.** 900 **4.** 0.00062

5. 0.053 **6.** 0.0000462 **7.** 19,000 **8.** 5,260,000,000

9. 0.00000186 **10.** 0.0003 **11.** 0.00000914 **12.** 37,000

13. 107 **14.** 0.02 **15.** 0.153 **16.** 416,000

Express each number without exponents.

17. 4.2×10^3

18. 1.63×10^{-4}

19. 4×10^7

20. 6.15×10^5

21. 2.13×10^{-5}

22. 9.64×10^{-7}

23. 3.12×10^{-1}

24. 4.6×10^1

25. 9×10^6

26. 7.3×10^4

27. 5.35×10^2

28. 1.04×10^{-2}

29. 3.5×10^4

30. 2.17×10^{-6}

31. 1×10^4

32. 1×10^{-3}

Perform the indicated operation and express each number without exponents.

33. $(4 \times 10^2)(3 \times 10^5)$

34. $(2 \times 10^{-3})(3 \times 10^2)$

35. $(5.1 \times 10^1)(3 \times 10^{-4})$

36. $(1.6 \times 10^{-2})(4 \times 10^{-3})$

37. $\dfrac{6.4 \times 10^5}{2 \times 10^3}$

38. $\dfrac{8 \times 10^{-3}}{2 \times 10^1}$

39. $\dfrac{8.4 \times 10^{-6}}{4 \times 10^{-3}}$

40. $\dfrac{25 \times 10^3}{5 \times 10^{-2}}$

41. $\dfrac{4 \times 10^5}{2 \times 10^4}$

42. $\dfrac{16 \times 10^3}{8 \times 10^{-3}}$

Perform the indicated operation by first converting each number to scientific notation form. Write the answer in scientific notation form.

43. $(700{,}000)(6{,}000{,}000)$

44. $(0.0006)(5{,}000{,}000)$

45. $(0.003)(0.00015)$

46. $(230{,}000)(3000)$

47. $\dfrac{1{,}400{,}000}{700}$

48. $\dfrac{20{,}000}{0.0005}$

49. $\dfrac{0.00004}{200}$

50. $\dfrac{0.0012}{0.000006}$

51. $\dfrac{150{,}000}{0.0005}$

52. List the following numbers from smallest to largest: 4.8×10^5, 3.2×10^{-1}, 4.6, 8.3×10^{-4}

53. List the following numbers from smallest to largest: 9.2×10^{-5}, 8.4×10^3, 1.3×10^{-1}, 6.2×10^4

54. The distance from Earth to the planet Jupiter is approximately 4.5×10^8 miles. If a spacecraft travels at a speed of 25,000 miles per hour, how long, in hours, would it take the spacecraft to travel from Earth to Jupiter? Use distance = rate × time.

See Exercise 54.

55. If a computer can do a calculation in 0.000004 second, how long, in seconds, would it take the computer to do 8 trillion (8,000,000,000,000) calculations?

56. The half-life of a radioactive isotope is the time required for half the quantity of the isotope to decompose. The half-life of uranium 238 is 4.5×10^9 years, and the half-life of uranium 234 is 2.5×10^5 years. How many times greater is the half-life of uranium 238 than uranium 234?

57. A treaty between the United States and Canada requires that during the tourist season a minimum of 100,000 cubic feet of water per second flows over Niagara Falls (another 130,000 to 160,000 cubic feet/sec is diverted for power generation). Find the minimum amount of water that will flow over the falls in a 24-hour period during the tourist season.

58. In Example 2(b) we showed that $6.28 \times 10^{-3} = 0.00628$. Show that you obtain the same results by using the negative exponent rule on 10^{-3}.

59. Consider the problem $\dfrac{(4 \cdot 10^3)(6 \cdot 10^x)}{24 \cdot 10^{-5}} = 1$. Determine the value of x that makes this a true statement. Explain how you determined your answer.

60. Consider the problem $\dfrac{25 \cdot 10^8}{(5 \cdot 10^{-3})(5 \cdot 10^x)} = 1$. Determine the value of x that makes this a true statement. Explain how you determined your answer.

Cumulative Review Exercises

[1.9] **61.** Evaluate $4x^2 + 3x + \dfrac{x}{2}$ when $x = 0$.

[2.3] **62. (a)** If $-x = -\dfrac{3}{2}$, what is the value of x?
   (b) If $5x = 0$, what is the value of x?

[2.5] **63.** Solve the equation $2x - 3(x - 2) = x + 2$.

[4.1] **64.** Simplify $\left(\dfrac{-2x^5 y^7}{8x^8 y^3}\right)^3$.

JUST FOR FUN

1. (a) Light travels at a speed of 1.86×10^5 miles per second. A *light year* is the distance the light travels in one year. Determine the number of miles in a light year.

(b) Earth is approximately 93,000,000 miles from the sun. How long does it take light from the sun to reach Earth?

4.4

Addition and Subtraction of Polynomials

► **1** Identify polynomials.

► **2** Add polynomials.

► **3** Subtract polynomials.

► **1** A **polynomial in x** is an expression containing the sum of a finite number of terms of the form ax^n, for any real number a and any whole number n.

Examples of Polynomials	*Not Polynomials*	
$2x$	$4x^{1/2}$	(fractional exponent)
$\dfrac{1}{3}x - 4$	$3x^2 + 4x^{-1} + 5$	(negative exponent)
$x^2 - 2x + 1$	$4 + \dfrac{1}{x}$	$\left(\dfrac{1}{x} = x^{-1},\ \text{negative exponent}\right)$

A polynomial is written in **descending order** (or **descending powers**) **of the variable** when the exponents on the variable decrease from left to right.

Example of Polynomial in Descending Order

$$2x^3 + 4x^2 - 6x + 3$$

Note in the example that the constant, 3, can be written as $3x^0$ since x^0 is 1.

A polynomial of one term is called a **monomial.** A **binomial** is a two-termed polynomial. A **trinomial** is a three-termed polynomial. Polynomials containing more than three terms are not given special names. The term "poly" is a prefix meaning "many." The chart that follows summarizes this information.

Type of Polynomial	*Number of Terms*	*Examples*		
Monomial	One	$5,$	$4x,$	$-6x^2$
Binomial	Two	$x + 4,$	$x^2 - 6,$	$2x^2 - 5x$
Trinomial	Three	$x^2 - 2x + 3,$	$5x^2 - 6x + 7$	

The **degree of a term** of a polynomial in one variable is the exponent on the variable in that term.

Term	*Degree of Term*	
$4x^2$	Second	
$2y^5$	Fifth	
$-5x$	First	$(-5x$ can be written $-5x^1$.)
3	Zero	(3 can be written $3x^0$.)

The **degree of a polynomial** in one variable is the same as that of its highest-degree term.

Polynomial	*Degree of Polynomial*	
$8x^3 + 2x^2 - 3x + 4$	Third	$(8x^3$ is highest-degree term.)
$x^2 - 4$	Second	$(x^2$ is highest-degree term.)
$2x - 1$	First	$(2x$ or $2x^1$ is highest-degree term.)
4	Zero	(4 or $4x^0$ is highest-degree term.)

Addition of Polynomials

▶ **2**   In Section 2.1 we stated that like terms are terms having the same variables and the same exponents. That is, like terms differ only in their numerical coefficients.

Examples of Like Terms

$$3, \quad -5$$
$$2x, \quad x$$
$$-2x^2, \quad 4x^2$$
$$3y^2, \quad 5y^2$$
$$3xy^2, \quad 5xy^2$$

> **To add polynomials,** combine the like terms of the polynomials.

EXAMPLE 1   Simplify $(4x^2 + 6x + 3) + (2x^2 + 5x - 1)$.

Solution:

$$(4x^2 + 6x + 3) + (2x^2 + 5x - 1)$$

$= 4x^2 + 6x + 3 + 2x^2 + 5x - 1$ Remove parentheses.

$= \underline{4x^2 + 2x^2} \; \underline{+ 6x + 5x} \; \underline{+ 3 - 1}$ Rearrange terms.

$= \qquad 6x^2 \quad + 11x \quad + 2$ Combine like terms.   ■

EXAMPLE 2   Simplify $(4x^2 + 3x + y) + (x^2 - 6x + 3)$.

Solution:

$$(4x^2 + 3x + y) + (x^2 - 6x + 3)$$

$= 4x^2 + 3x + y + x^2 - 6x + 3$ Remove parentheses.

$= \underline{4x^2 + x^2} \; \underline{+ 3x - 6x} + y + 3$ Rearrange terms.

$= \qquad 5x^2 \quad - \quad 3x \quad + y + 3$ Combine like terms.   ■

EXAMPLE 3 Simplify $(3x^2y - 4xy + y) + (x^2y + 2xy + 3y)$.

Solution: $(3x^2y - 4xy + y) + (x^2y + 2xy + 3y)$

$= 3x^2y - 4xy + y + x^2y + 2xy + 3y$ Remove parentheses.

$= \underbrace{3x^2y + x^2y}\ \underbrace{- 4xy + 2xy}\ \underbrace{+ y + 3y}$ Rearrange terms.

$= \qquad 4x^2y \qquad - \quad 2xy \qquad + \quad 4y$ Combine like terms.   ■

Addition of Polynomials in Columns

In most of the text, when we add polynomials we will do so as in Examples 1 through 3. That is, we will list horizontally the polynomials being added. However, in Section 4.6, when we divide polynomials, there will be steps where we add polynomials in columns.

To Add Polynomials in Columns

1. Arrange polynomials in descending order one under the other with like terms in the same columns.
2. Find the sum of the terms in each column.

EXAMPLE 4 Add $4x^2 - 2x + 2$ and $-2x^2 - x + 4$ using columns.

Solution:
$$
\begin{array}{r}
4x^2 - 2x + 2 \\
-2x^2 - x + 4 \\
\hline
2x^2 - 3x + 6
\end{array}
$$
■

EXAMPLE 5 Add $(3x^3 + 2x - 4)$ and $(2x^2 - 6x - 3)$ using columns.

Solution: Since the polynomial $3x^3 + 2x - 4$ does not have an x^2 term, we will add the term $0x^2$ to the polynomial. This procedure sometimes helps in aligning like terms.

$$
\begin{array}{r}
3x^3 + 0x^2 + 2x - 4 \\
2x^2 - 6x - 3 \\
\hline
3x^3 + 2x^2 - 4x - 7
\end{array}
$$
■

Subtraction of Polynomials

▶ 3

To Subtract Polynomials

1. Remove parentheses. (This will have the effect of changing the sign of *every* term within the parentheses of the polynomial being subtracted.)
2. Combine like terms.

EXAMPLE 6 Simplify $(3x^2 - 2x + 5) - (x^2 - 3x + 4)$.

Solution: $(3x^2 - 2x + 5) - (x^2 - 3x + 4)$

$= 3x^2 - 2x + 5 - x^2 + 3x - 4$ Remove parentheses (change the sign of each term being subtracted).

$= \underbrace{3x^2 - x^2}\ \underbrace{- 2x + 3x}\ \underbrace{+ 5 - 4}$ Rearrange terms.

$= \qquad 2x^2 \qquad + \quad x \qquad + \quad 1$ Combine like terms.   ■

EXAMPLE 7   Subtract $(-x^2 - 2x + 3)$ from $(x^3 + 4x + 6)$.

Solution:
$$(x^3 + 4x + 6) - (-x^2 - 2x + 3)$$
$$= x^3 + 4x + 6 + x^2 + 2x - 3 \qquad \text{Remove parentheses.}$$
$$= x^3 + x^2 + 4x + 2x + 6 - 3 \qquad \text{Rearrange terms.}$$
$$= x^3 + x^2 + 6x + 3 \qquad \text{Combine like terms.} \quad \blacksquare$$

COMMON STUDENT ERROR

One of the most common mistakes made by students occurs when subtracting polynomials. When subtracting one polynomial from another, **the sign of each term in the polynomial being subtracted must change, not just the sign of the first term.**

Correct	*Wrong*
$6x^2 - 4x + 3 - (2x - 3x + 4)$	$6x^2 - 4x + 3 - (2x^2 - 3x + 4)$
$= 6x^2 - 4x + 3 - 2x^2 + 3x - 4$	$= 6x^2 - 4x + 3 - 2x^2 - 3x + 4$
$= 4x^2 - x - 1$	$= 4x^2 - 7x + 7$

Do not make this mistake!

Subtraction of Polynomials in Columns

To Subtract Polynomials in Columns

1. Write *the polynomial being subtracted* below the polynomial from which it is being subtracted. List like terms in the same column.
2. **Change the sign of each term** in the polynomial being subtracted. (This step can be done mentally, if you like.)
3. Find the sum of the terms in each column.

EXAMPLE 8   Subtract $(x^2 - 6x + 5)$ from $(3x^2 + 5x + 7)$ using columns.

Solution: Align like terms in columns (step 1).

$$\begin{array}{r} 3x^2 + 5x + 7 \\ -(x^2 - 6x + 5) \\ \hline \end{array} \qquad \text{Align like terms.}$$

Change *all* signs in the second row (step 2); then add (step 3).

$$\begin{array}{r} 3x^2 + 5x + 7 \\ - x^2 + 6x - 5 \\ \hline 2x^2 + 11x + 2 \end{array} \qquad \begin{array}{l} \text{Change all signs.} \\[6pt] \text{Add.} \end{array} \quad \blacksquare$$

EXAMPLE 9   Using columns, subtract $(2x^2 - 6)$ from $(-3x^3 + 4x - 3)$.

Solution: To help with aligning like terms, we will write each expression with descending powers of x. If a given power of x is missing, we will write that term with a numerical coefficient of 0.

$$-3x^3 + 4x - 3 = -3x^3 + 0x^2 + 4x - 3$$
$$2x^2 - 6 = 2x^2 + 0x - 6$$

Align like terms.

$$-3x^3 + 0x^2 + 4x - 3$$
$$\underline{-(2x^2 + 0x - 6)}$$

Change all signs in the second row; then add.

$$-3x^3 + 0x^2 + 4x - 3$$
$$\underline{-\ 2x^2 - 0x + 6}$$
$$-3x^3 - 2x^2 + 4x + 3$$

∎

Note: Many of you will find that you can change the signs mentally and can therefore align and change the signs in one step.

Exercise Set 4.4

Indicate the expressions that are polynomials. If the polynomial has a specific name—for example, monomial or binomial—give that name.

1. $4x$	**2.** $3x^2 - 6x + 7$	**3.** -12	**4.** $3x^{-2}$
5. $-4x - 6x^2$	**6.** $4x^3 - 8$	**7.** $6x^2 - 2x + 8$	**8.** $x - 3$
9. $3x^{1/2} + 2x$	**10.** $-2x^2 + 5x^{-1}$	**11.** $2x + 5$	**12.** $4 - 3x$
13. 6	**14.** $x^{1/3} + x^{2/3}$	**15.** $3x^3 - 2x^2 + 4x - 7$	**16.** $x^2 - 3$
17. $4 - x^2 - 6x$	**18.** $-3x$	**19.** $2x^{-2}$	**20.** $6x^2 + 3x - 5$

Express each polynomial in descending order. If the polynomial is already in descending order, so state. Give the degree of each polynomial.

21. $3x$	**22.** 6	**23.** $2x^2 - 6 + x$
24. $-4 + x^2 - 2x$	**25.** $-8 - 4x - x^2$	**26.** $2x + 4 - x^2$
27. $x^3 - 6$	**28.** $-x - 1$	**29.** $2x^2 + 5x - 8$
30. $3x^3 - x + 4$	**31.** $4 - 6x^3 + x^2 - 3x$	**32.** $-4 + x - 3x^2 + 4x^3$
33. $-2x + 5x^2 - 4$	**34.** $1 - x^3 + 3x$	**35.** $5x + 3x^2 - 6 - 2x^3$
36. $4 - 2x - 3x^2 + 5x^4$		

Add as indicated.

37. $(2x + 3) + (4x - 2)$	**38.** $(3x - 6) + (2x - 3)$
39. $(-4x + 8) + (2x + 3)$	**40.** $(-5x - 3) + (-2x + 3)$
41. $(5x + 8) + (-6x - 10)$	**42.** $(-8x + 4) + (3x - 12)$
43. $(9x - 12) + (12x - 9)$	**44.** $(3x - 8) + (-8x + 5)$
45. $(x^2 + 2x - 3) + (4x + 3.8)$	**46.** $(-2x^2 + 3x - 9) + (-2x - 3)$
47. $(5x - 7) + (2x^2 + 3x + 12)$	**48.** $(-3x + 8) + (-2x^2 - 3x - 5)$
49. $(3x^2 - 4x + 8) + (2x^2 + 5x + 12)$	**50.** $(x^2 - 6x + 7) + (-x^2 + 3x + 5)$
51. $(-3x^2 - 4x + 8) + (5x - 2x^2 + \frac{1}{2})$	**52.** $(9x^2 + 3x - 12) + (5x^2 - \frac{1}{3}x - 3)$
53. $(8x^2 + 4) + (-2.6x^2 - 5x)$	**54.** $(8x^3 + 4x^2 + 6) + (0.2x^2 + 5x)$
55. $(-7x^3 - 3x^2 + 4) + (4x + 5x^3 - 7)$	**56.** $(9x^3 - 2x^2 + 4x - 7) + (2x^3 - 6x^2 - 4x + 3)$
57. $(x^2 + xy - y^2) + (2x^2 - 3xy + y^2)$	**58.** $(x^2y + 6x^2 - 3xy^2) + (-x^2y - 12x^2 + 4xy^2)$
59. $(4x^2y + 2x - 3) + (3x^2y - 5x + 5)$	**60.** $(x^2y + x - y) + (2x^2y + 2x - 6y + 3)$

Add in columns.

61. Add $3x - 6$ and $4x + 5$	**62.** Add $-2x + 5$ and $-3x - 5$
63. Add $x^2 - 2x + 4$ and $3x + 12$	**64.** Add $4x^2 - 6x + 5$ and $-2x - 8$

65. Add $-2x^2 + 4x - 12$ and $-x^2 - 2x$

66. Add $5x^2 + x + 9$ and $2x^2 - 12$

67. Add $3x^2 + 4x - 5$ and $4x^2 + 3x - 8$

68. Add $-5x^2 - 3$ and $x^2 + 2x - 9$

69. Add $2x^3 + 3x^2 + 6x - 9$ and $7 - 4x^2$

70. Add $-3x^3 + 3x + 9$ and $2x^2 - 4$

71. Add $6x^3 - 4x^2 + x - 9$ and $-x^3 - 3x^2 - x + 7$

72. Add $4x^3 + 7$ and $-2x^3 - 4x - 1$

73. Add $xy + 6x + 4$ and $2xy - 3x - 1$

74. Add $x^2y - 6x + 3$ and $-2x^2y - 4x - 8$

Subtract as indicated.

75. $(3x - 4) - (2x + 2)$

76. $(6x + 3) - (4x - 2)$

77. $(-2x - 3) - (-5x - 7)$

78. $(12x - 3) - (-2x + 7)$

79. $(-x + 4) - (-x + 9)$

80. $(4x + 8) - (3x + 9)$

81. $(6 - 12x) - (3 - 5x)$

82. $(4x^2 - 6x + 3) - (3x + 7)$

83. $(9x^2 + 7x - 5) - (3x^2 + 3.5)$

84. $(-2x^2 + 4x - 5.2) - (5x^2 + 3x + 7.5)$

85. $(5x^2 - x - 1) - (-3x^2 - 2x - 5)$

86. $(5x^2 - 7) - (4x - 3)$

87. $(5x^2 - x + 12) - (5 + x)$

88. $(-5x^2 - 2x) - (2x^2 - 7x + 9)$

89. $(9x - 6) - (-2x^2 + 4x - 8)$

90. $(8x^3 + 5x^2 - 4) - (4x - 3 + 6x^2)$

91. $(4x^3 - 6x^2 + 5x - 7) - (6x + \frac{2}{3}x^2 - 3)$

92. $(-3x^2 + 4x - 7) - (x^3 + 4x^2 - 8x + 5)$

93. $(9x^3 - \frac{1}{5}) - (x^2 + 5x)$

94. $(3x^3 - 6x^2 + 5x) - (4x^3 - 2x^2 + 5)$

95. Subtract $(4x - 6)$ from $(3x + 5)$

96. Subtract $(-4x + 7)$ from $(-3x - 9)$

97. Subtract $(5x - 6)$ from $(2x^2 - 4x + 8)$

98. Subtract $(2x^2 - 6x + 4)$ from $(5x^2 + 6x + 8)$

99. Subtract $(4x^3 - 6x^2)$ from $(3x^3 + 5x^2 + 9x - 7)$

100. Subtract $(-4x^2 + 8x - 7)$ from $(-5x^3 - 6x^2 + 7)$

Perform each subtraction using columns.

101. Subtract $(2x - 7)$ from $(5x + 10)$

102. Subtract $(6x + 8)$ from $(2x - 5)$

103. Subtract $(-9x - 4)$ from $(-5x + 3)$

104. Subtract $(-3x + 8)$ from $(6x^2 - 5x + 3)$

105. Subtract $(4x^2 - 7)$ from $(9x^2 + 7x - 9)$

106. Subtract $(4x^2 + 7x - 9)$ from $(x^2 - 6x + 3)$

107. Subtract $(-4x^2 + 6x)$ from $(x - 6)$

108. Subtract $(x^2 - 6)$ from $(x^2 + 4x)$

109. Subtract $(x^2 + 6x - 7)$ from $(4x^3 - 6x^2 + 7x - 9)$

110. Subtract $(2x^3 + 4x^2 - 9x)$ from $(-5x^3 + 4x - 12)$

111. In your own words describe a polynomial.

112. (a) What is a monomonial? Make up three examples.

(b) What is a binomial? Make up three examples.

(c) What is a trinomial? Make up three examples.

113. (a) Explain how to find the degree of a term in one variable.

(b) Explain how to find the degree of a polynomial in one variable.

114. Make up your own fifth degree polynomial with three terms. Explain why it is a fifth degree polynomial with three terms.

115. Explain how to write a polynomial in descending order of the variable.

116. Make up your own addition problem where the sum of two binomials is $-2x + 4$.

117. Make up your own addition problem where the sum of two trinomials is $2x^2 + 5x - 6$.

118. Make up your own subtraction problem where the difference of two trinomials is $x - 2$.

119. Make up your own subtraction problem where the difference of two trinomials is $-x^2 + 4x - 5$.

Cumulative Review Exercises

[1.4] **120.** Insert either $>$, $<$, or $=$ in the shaded area to make the statement true. $|-4|$ ▭ $|-6|$.

[1.7] *Answer true or false.*

121. The product of two negative numbers is always a negative number.

122. The sum of two negative numbers is always a negative number.

123. The difference of two negative numbers is always a negative number.

124. The quotient of two negative numbers is always a negative number.

[4.1] **125.** Simplify $\left(\dfrac{3x^4y^5}{6x^7y^4}\right)^3$.

JUST FOR FUN

Simplify each expression.

1. $(3x^2 - 6x + 3) - (2x^2 - x - 6) - (x^2 + 7x - 9)$

2. $3x^2y - 6xy - 2xy + 9xy^2 - 5xy + 3x$

3. $4(x^2 + 2x - 3) - 6(2 - 4x - x^2) - 2x(x + 2)$

4.5

Multiplication of Polynomials

▶ **1** Multiply a monomial by a monomial.

▶ **2** Multiply a monomial by a polynomial.

▶ **3** Multiply two binomials using the distributive property.

▶ **4** Multiply two binomials using the FOIL method.

▶ **5** Identify and multiply special products.

▶ **6** Multiply any two polynomials.

Multiplying a Monomial by a Monomial

▶ **1** We begin our discussion of multiplication of polynomials by multiplying a monomial by a monomial. To multiply two monomials, multiply their coefficients and use the product rule of exponents to determine the exponents on the variables. Problems of this type were done in Section 4.1.

EXAMPLE 1 Multiply $(3x^2)(5x^5)$.

Solution: $(3x^2)(5x^5) = 3 \cdot 5 \cdot x^2 \cdot x^5 = 15x^{2+5} = 15x^7$.

EXAMPLE 2 Multiply $(-2x^6)(3x^4)$.

Solution: $(-2x^6)(3x^4) = (-2)(3) \cdot x^6 \cdot x^4 = -6x^{6+4} = -6x^{10}$.

EXAMPLE 3 Multiply $(5x^2y)(8x^5y^4)$.

Solution: Remember that when a variable is given without an exponent we assume that the exponent on the variable is 1.

$(5x^2y)(8x^5y^4) = 40x^{2+5}y^{1+4} = 40x^7y^5$

EXAMPLE 4 Multiply $6xy^2z^5(-3x^4y^7z)$.

Solution: $6xy^2z^5(-3x^4y^7z) = -18x^5y^9z^6$

EXAMPLE 5 Multiply $(-4x^4z^9)(-3xy^7z^3)$.

Solution: $(-4x^4z^9)(-3xy^7z^3) = 12x^5y^7z^{12}$.

**Multiplying
a Polynomial
by a Monomial**

▶ **2**   To multiply a polynomial by a monomial, we use the distributive property presented earlier.

$$a(b + c) = ab + ac$$

The distributive property can be expanded to

$$a(b + c + d + \cdots + n) = ab + ac + ad + \cdots + an$$

EXAMPLE 6   Multiply $2x(3x^2 + 4)$.

Solution: $2x(3x^2 + 4) = (2x)(3x^2) + (2x)(4)$
$$= 6x^3 + 8x$$ ■

Notice that the use of the distributive property results in monomials being multiplied by monomials. If we study Example 6, we see that the $2x$ and $3x^2$ are both monomials, as are the $2x$ and 4.

EXAMPLE 7   Multiply $-3x(4x^2 - 2x - 1)$.

Solution: $-3x(4x^2 - 2x - 1) = (-3x)(4x^2) + (-3x)(-2x) + (-3x)(-1)$
$$= -12x^3 + 6x^2 + 3x$$ ■

EXAMPLE 8   Multiply $3x^2(4x^3 - 2x + 7)$.

Solution: $3x^2(4x^3 - 2x + 7) = (3x^2)(4x^3) + (3x^2)(-2x) + (3x^2)(7)$
$$= 12x^5 - 6x^3 + 21x^2$$ ■

EXAMPLE 9   Multiply $2x(3x^2y - 6xy + 5)$.

Solution: $2x(3x^2y - 6xy + 5) = (2x)(3x^2y) + (2x)(-6xy) + (2x)(5)$
$$= 6x^3y - 12x^2y + 10x$$ ■

EXAMPLE 10   Multiply $(3x^2 - 2xy + 3)4x$.

Solution: $(3x^2 - 2xy + 3)4x = (3x^2)(4x) + (-2xy)(4x) + (3)(4x)$
$$= 12x^3 - 8x^2y + 12x$$

This problem could be written as $4x(3x^2 - 2xy + 3)$ by the commutative property of multiplication, and then solved as in Examples 6 through 9.   ■

**Multiplying
a Binomial
by a Binomial**

▶ **3**   Now we will discuss multiplying a binomial by a binomial. Before we explain how to do this, consider the multiplication problem $43 \cdot 12$.

$$
\begin{array}{r}
43 \longleftarrow \text{multiplicand} \\
\underline{12} \longleftarrow \text{multiplier}
\end{array}
$$

$$
\begin{array}{r}
2(4) \longrightarrow \quad 86 \longleftarrow 2(3) \\
1(4) \longrightarrow \underline{43} \longleftarrow 1(3) \\
516
\end{array}
$$

Note how the 2 multiplies both the 3 and the 4 and the 1 also multiplies both the 3 and the 4. That is, every number in the multiplier multiplies every number in the multiplicand. The same thing must happen whenever any two polynomials are multiplied. That is, **every term in one polynomial must multiply every term in the other polynomial.**

Consider multiplying $(a + b)(c + d)$. Treating $(a + b)$ as a single term and using the distributive property, we get

$$(a + b)(c + d) = (a + b)c + (a + b)d$$

Using the distributive property a second time gives

$$= ac + bc + ad + bd$$

Notice how each term of the first polynomial was multiplied by each term of the second polynomial, and all the products were added to obtain the answer.

EXAMPLE 11 Mulitply $(3x + 2)(x - 5)$.

Solution:
$$\begin{aligned}
(3x + 2)(x - 5) &= (3x + 2)x + (3x + 2)(-5) \\
&= 3x(x) + 2(x) + 3x(-5) + 2(-5) \\
&= 3x^2 + 2x - 15x - 10 \\
&= 3x^2 - 13x - 10
\end{aligned}$$

Note that after performing the multiplication like terms must be combined. ∎

EXAMPLE 12 Multiply $(x - 4)(y + 3)$.

Solution:
$$\begin{aligned}
(x - 4)(y + 3) &= (x - 4)y + (x - 4)3 \\
&= xy - 4y + 3x - 12
\end{aligned}$$ ∎

FOIL Method

▶ **4** A common method used to multiply two binomials is the *FOIL method*. This procedure also results in each term of one binomial being multiplied by each term in the other binomial. Students often prefer to use this method when multiplying two binomials.

The FOIL Method

Consider
$$(a + b)(c + d)$$

F stands for **first**—multiply the first terms of each binomial together:

$$\overset{\text{F}}{\overbrace{(a + b)}}(c + d) \qquad \text{product } ac$$

O stands for **outer**—multiply the two outer terms together:

$$\overset{\text{O}}{\overbrace{(a + b)(c + d)}} \qquad \text{product } ad$$

I stands for **inner**—multiply the two inner terms together:

$$(a + \overset{\text{I}}{\overbrace{b)(c}} + d) \qquad \text{product } bc$$

L stands for **last**—multiply the last terms together:

$$(a + \overset{\text{L}}{\overbrace{b)(c + d}}) \qquad \text{product } bd$$

The answer will be the sum of the products.

$$(a + b)(c + d) = ac + ad + bc + bd$$

The FOIL method is not actually a different method used to multiply binomials, but rather an acronym to help students remember to correctly apply the distributive property. We could have used IFOL or any arrangement of the four letters. However, FOIL is easier to remember than the other arrangements.

EXAMPLE 13   Using the FOIL method, multiply $(2x - 3)(x + 4)$.

Solution:

$$(2x - 3)(x + 4)$$

$$\overset{F}{(2x)(x)} + \overset{O}{(2x)(4)} + \overset{I}{(-3)(x)} + \overset{L}{(-3)(4)}$$

$$= \quad 2x^2 \quad + \quad 8x \quad - \quad 3x \quad - \quad 12$$

$$= 2x^2 + 5x - 12$$

Thus $(2x - 3)(x + 4) = 2x^2 + 5x - 12$.   ■

EXAMPLE 14   Multiply $(4 - 2x)(6 - 5x)$.

Solution:

$$(4 - 2x)(6 - 5x)$$

$$\overset{F}{4(6)} + \overset{O}{4(-5x)} + \overset{I}{(-2x)(6)} + \overset{L}{(-2x)(-5x)}$$

$$= \quad 24 \quad - \quad 20x \quad - \quad 12x \quad + \quad 10x^2$$

$$= 10x^2 - 32x + 24$$

Thus $(4 - 2x)(6 - 5x) = 10x^2 - 32x + 24$.   ■

EXAMPLE 15   Multiply $(x - 5)(x + 5)$.

Solution:

$$\overset{F}{(x)(x)} + \overset{O}{(x)(5)} + \overset{I}{(-5)(x)} + \overset{L}{(-5)(5)}$$

$$= \quad x^2 \quad + \quad 5x \quad - \quad 5x \quad - \quad 25$$

$$= x^2 - 25$$

Thus $(x - 5)(x + 5) = x^2 - 25$.   ■

EXAMPLE 16   Multiply $(2x + 3)(2x - 3)$.

Solution:

$$\overset{F}{(2x)(2x)} + \overset{O}{(2x)(-3)} + \overset{I}{(3)(2x)} + \overset{L}{(3)(-3)}$$

$$= \quad 4x^2 \quad - \quad 6x \quad + \quad 6x \quad - \quad 9$$

$$= 4x^2 - 9$$

Thus $(2x + 3)(2x - 3) = 4x^2 - 9$.   ■

Special Products

▶ **5** Examples 15 and 16 are examples of the special product of the sum and difference of two quantities.

Product of Sum and Difference of Two Quantities

$$(a + b)(a - b) = a^2 - b^2$$

The preceding product is also referred to as the **difference-of-squares formula** because the expression on the right side of the equal sign is the difference of two squares.

EXAMPLE 17   Use the rule for finding the product of the sum and difference of two quantities to multiply each expression.

(a) $(x + 3)(x - 3)$ (b) $(2x + 4)(2x - 4)$ (c) $(3x + 2y)(3x - 2y)$

Solution: (a) If we let $x = a$ and $3 = b$, then

$$(a + b)(a - b) = a^2 - b^2$$
$$\downarrow \quad \downarrow \ \downarrow \quad \downarrow \quad \downarrow \quad \downarrow$$
$$(x + 3)(x - 3) = (x)^2 - (3)^2$$
$$= x^2 - 9$$

(b)
$$(a + b)(a - b) = a^2 - b^2$$
$$\downarrow \quad \downarrow \ \downarrow \quad \downarrow \quad \downarrow \quad \downarrow$$
$$(2x + 4)(2x - 4) = (2x)^2 - (4)^2$$
$$= 4x^2 - 16$$

(c)
$$(a + b)(a - b) = a^2 - b^2$$
$$\downarrow \quad \downarrow \ \downarrow \quad \downarrow \quad \downarrow \quad \downarrow$$
$$(3x + 2y)(3x - 2y) = (3x)^2 - (2y)^2$$
$$= 9x^2 - 4y^2$$

This problem could also be done using the FOIL method. ■

EXAMPLE 18   Using the FOIL method, multiply $(x + 3)^2$.

Solution: $(x + 3)^2 = (x + 3)(x + 3)$

$$\begin{array}{cccc} \text{F} & \text{O} & \text{I} & \text{L} \end{array}$$
$$x(x) + x(3) + 3(x) + (3)(3)$$
$$= x^2 + 3x + 3x + 9$$
$$= x^2 + 6x + 9$$
■

Example 18 is an example of the square of a binomial, another special product.

Square of Binomial Formulas

$$(a + b)^2 = (a + b)(a + b) = a^2 + 2ab + b^2$$
$$(a - b)^2 = (a - b)(a - b) = a^2 - 2ab + b^2$$

EXAMPLE 19   Use the square of the binomial formula to multiply each expression.

(a) $(x + 5)^2$ (b) $(2x + 4)^2$
(c) $(3x + 2y)(3x + 2y)$ (d) $(x - 3)(x - 3)$

Solution: (a) If we let $x = a$ and $5 = b$, then

$$(a + b)(a + b) = a^2 + 2a\,b + b^2$$
$$\downarrow \quad \downarrow\downarrow \quad \downarrow \qquad \downarrow \qquad \downarrow\downarrow \qquad \downarrow$$
$$(x + 5)(x + 5) = (x)^2 + 2(x)(5) + (5)^2$$
$$= x^2 + 10x + 25$$

(b) $$(a + b)(a + b) = a^2 + 2a\,b + b^2$$
$$\downarrow \quad \downarrow\downarrow \quad \downarrow \qquad \downarrow \qquad \downarrow\downarrow \qquad \downarrow$$
$$(2x + 4)(2x + 4) = (2x)^2 + 2(2x)(4) + (4)^2$$
$$= 4x^2 + 16x + 16$$

(c) $$(a + b)(a + b) = a^2 + 2a\,b + b^2$$
$$\downarrow \quad \downarrow\downarrow \quad \downarrow \qquad \downarrow \qquad \downarrow\downarrow \qquad \downarrow$$
$$(3x + 2y)(3x + 2y) = (3x)^2 + 2(3x)(2y) + (2y)^2$$
$$= 9x^2 + 12xy + 4y^2$$

(d) $$(a - b)(a - b) = a^2 - 2a\,b + b^2$$
$$\downarrow \quad \downarrow\downarrow \quad \downarrow \qquad \downarrow \qquad \downarrow\downarrow \qquad \downarrow$$
$$(x - 3)(x - 3) = (x)^2 - 2(x)(3) + (3)^2$$
$$= x^2 - 6x + 9$$

This problem could also be done using the FOIL method.   ■

COMMON STUDENT ERROR

	Correct	*Wrong*
	$(a + b)^2 = a^2 + 2ab + b^2$	~~$(a + b)^2 = a^2 + b^2$~~
	$(a - b)^2 = a^2 - 2ab + b^2$	~~$(a - b)^2 = a^2 - b^2$~~

Do not forget middle term when you square a binomial.

$$(x + 2)^2 \neq x^2 + 4$$
$$(x + 2)^2 = (x + 2)(x + 2)$$
$$= x^2 + 4x + 4$$

Multiplying
a Polynomial
by a Polynomial

▶ **6**   When multiplying a binomial by a binomial, we saw that every term in the first binomial was multiplied by every term in the second binomial. When multiplying any two polynomials, each term of one polynomial must be multiplied by each term of the other polynomial. We can accomplish this with use of the distributive property. Consider the multiplication $(3x + 2)(4x^2 - 5x - 3)$. We can find the answer using the distributive property as follows.

$$(3x + 2)(4x^2 - 5x - 3)$$
$$= 3x\,(4x^2 - 5x - 3) + 2\,(4x^2 - 5x - 3)$$
$$= 12x^3 - 15x^2 - 9x + 8x^2 - 10x - 6$$
$$= 12x^3 - 7x^2 - 19x - 6$$

Thus $(3x + 2)(4x^2 - 5x - 3) = 12x^3 - 7x^2 - 19x - 6$.

Multiplication problems can be performed by using the distributive property, as was just illustrated. However, many students prefer to multiply a polynomial by a polynomial using a different procedure. On page 168 we showed that when multiplying the number 43 by the number 12, we multiply each digit in the number 43 by each digit in the number 12. Review that multiplication at this time. We can follow a similar procedure when multiplying a polynomial by a polynomial, as illustrated in the following examples. We must be careful, however, to align like terms in the same columns when performing the individual multiplications.

EXAMPLE 20 Multiply $(3x + 4)(2x + 5)$.

Solution: First write the polynomials one beneath the other.

$$\begin{array}{r} 3x + 4 \\ \underline{2x + 5} \end{array}$$

Next, multiply each term in $(3x + 4)$ by 5.

$$\begin{array}{r} 3x + 4 \\ \underline{2x + 5} \\ 5(3x + 4) \longrightarrow 15x + 20 \end{array}$$

Next, multiply each term in $(3x + 4)$ by $2x$ and align like terms.

$$\begin{array}{r} 3x + 4 \\ \underline{2x + 5} \\ 15x + 20 \\ 2x(3x + 4) \longrightarrow \underline{6x^2 + 8x } \\ 6x^2 + 23x + 20 \end{array}$$ Add like terms in columns.

The same answer is obtained using the FOIL method. ◼

EXAMPLE 21 Multiply $(5x - 2)(2x^2 + 3x - 4)$.

Solution: For convenience we place the shorter expression on the bottom, as illustrated.

$$\begin{array}{r} 2x^2 + 3x - 4 \\ \underline{5x - 2} \\ -4x^2 - 6x + 8 \\ \underline{10x^3 + 15x^2 - 20x } \\ 10x^3 + 11x^2 - 26x + 8 \end{array}$$

Multiply top polynomial by -2.
Multiply top polynomial by $5x$ and align like terms.
Add like terms in columns. ◼

EXAMPLE 22 Multiply $x^2 - 3x + 2$ by $2x^2 - 3$.

Solution:

$$\begin{array}{r} x^2 - 3x + 2 \\ \underline{2x^2 - 3} \\ -3x^2 + 9x - 6 \\ \underline{2x^4 - 6x^3 + 4x^2 } \\ 2x^4 - 6x^3 + x^2 + 9x - 6 \end{array}$$

Multiply top polynomial by -3.
Multiply top polynomial by $2x^2$ and align like terms.
Add like terms in columns. ◼

EXAMPLE 23   Multiply $(3x^3 - 2x^2 + 4x + 6)(x^2 - 5x)$.

Solution:

$$3x^3 - 2x^2 + 4x + 6$$
$$x^2 - 5x$$
$$\overline{-15x^4 + 10x^3 - 20x^2 - 30x}$$ Multiply top polynomial by $-5x$.
$$\underline{3x^5 - 2x^4 + 4x^3 + 6x^2}$$ Multiply top polynomial by x^2.
$$3x^5 - 17x^4 + 14x^3 - 14x^2 - 30x$$ Add like terms in columns.

Exercise Set 4.5

Multiply.

1. $x^2 \cdot 3xy$
2. $6xy^2 \cdot 3xy^4$
3. $5x^4y^5(6xy^2)$
4. $-5x^2y^3(6x^2y^5)$
5. $4x^4y^6(-7x^2y^9)$
6. $12x^6y^2(2x^9y^7)$
7. $9xy^6 \cdot 6x^5y^8$
8. $(3x^5y^6)(5y^7)$
9. $(6x^2y)(\frac{1}{2}x^4)$
10. $\frac{2}{3}x(6x^2y^3)$

Multiply.

11. $3(x + 4)$
12. $3(x - 4)$
13. $2x(x - 3)$
14. $-5x(x + 2)$
15. $-4x(-2x + 6)$
16. $-x(3x - 5)$
17. $2x(x^2 + 3x - 1)$
18. $-x(2x^2 - 6x + 5)$
19. $-2x(x^2 - 2x + 5)$
20. $-3x(-2x^2 + 5x - 6)$
21. $5x(-4x^2 + 6x - 4)$
22. $x(x^2 - x + 1)$
23. $(3x^2 + 4x - 5)8x$
24. $1.2x(3x^2 + y)$
25. $0.3x(2xy + 5x - 6y)$
26. $-\frac{1}{2}x(2x^2 + 4x - 6y^2)$
27. $(x - y - 3)y$
28. $\frac{1}{3}y(y^2 - 6y + 3x)$

Multiply.

29. $(x + 3)(x + 4)$
30. $(2x - 3)(x + 5)$
31. $(2x + 5)(3x - 6)$
32. $(x + 5)(x - 5)$
33. $(2x - 4)(2x + 4)$
34. $(4 + 3x)(2 - x)$
35. $(5 - 3x)(6 + 2x)$
36. $(4 + 6x)(x - 3)$
37. $(-x + 3)(2x + 5)$
38. $(6x - 1)(-2x + 5)$
39. $(x + 4)(x + 3)$
40. $(x - 3)(x + 5)$
41. $(x + 4)(x - 2)$
42. $(2x + 3)(x + 5)$
43. $(3x + 4)(2x + 5)$
44. $(3x - 6)(4x - 2)$
45. $(3x + 4)(2x - 3)$
46. $(4x + 4)(x + 1)$
47. $(x - 1)(x + 1)$
48. $(3x - 8)(2x + 3)$
49. $(2x - 3)(2x - 3)$
50. $(6x - 1)(2x + 1)$
51. $(4 - x)(3 + 2x)$
52. $(6 - 2x)(5x - 3)$
53. $(2x + 3)(4 - 2x)$
54. $(5 - 6x)(2x - 7)$
55. $(x + y)(x - y)$
56. $(x + 2y)(2x - 3)$
57. $(2x - 3y)(3x + 2y)$
58. $(x + 3)(2y - 5)$
59. $(4x - 3y)(2y - 3)$
60. $(2x - 0.1)(x + 2.4)$
61. $(x + 0.6)(x + 0.3)$
62. $(3x - 6)(x + \frac{1}{3})$
63. $(2x + 4)(x + \frac{1}{2})$
64. $(x + 4)(x - \frac{1}{2})$

Multiply using a special product formula.

65. $(x + 4)(x - 4)$
66. $(x + 3)^2$
67. $(2x - 1)(2x + 1)$
68. $(x + 2)(x + 2)$
69. $(x + y)^2$
70. $(2x - 3)(2x + 3)$
71. $(x - 0.2)^2$
72. $(x + 3y)(x + 3y)$
73. $(3x + 5)(3x - 5)$
74. $(5x + 4)(5x - 4)$

Multiply.

75. $(x + 3)(2x^2 + 4x - 1)$
76. $(2x + 3)(4x^2 - 5x + 6)$
77. $(5x + 4)(x^2 - x + 4)$
78. $(2x - 5)(3x^2 - 4x + 7)$

79. $(-2x^2 - 4x + 1)(7x - 3)$
81. $(-3x + 9)(-6x^2 + 5x - 3)$
83. $(3x^2 - 2x + 4)(2x^2 + 3x + 1)$
85. $(x^2 - x + 3)(x^2 - 2x)$
87. $(3x^3 + 2x^2 - x)(x - 3)$
89. $(a + b)(a^2 - ab + b^2)$

80. $(4x^2 + 9x - 2)(x - 2)$
82. $(4x^2 + 1)(2x + 1)$
84. $(x^2 - 2x + 3)(x^2 - 4)$
86. $(-3x^2 - 2x + 4)(2x^2 - 4x + 3)$
88. $(-x^3 + x^2 - 6x + 3)(x^2 + 2x)$
90. $(a - b)(a^2 + ab + b^2)$

91. Will the product of a monomial and a monomial always be a monomial? Explain your answer.
92. Will the product of a monomial and a binomial ever be a trinomial? Explain your answer.
93. Will the product of two binomials after like terms are combined always be a trinomial? Explain your answer.

94. Consider the multiplication $3x^2(2x^{} - 5x^{} + 3x^{}) = 6x^8 - 15x^5 + 9x^3$. Determine the exponents to be placed in the shaded area to make a true statement. Explain how you determined your answer.

Cumulative Review Exercises

[3.3] **95.** The cost of a taxi ride is $2.00 for the first mile and $1.50 for each additional mile or part thereof. Find the maximum distance Heidi can ride in the taxi if she has only $20.

[4.1] **96.** Simplify $\left(\dfrac{4x^8y^5}{8x^8y^6}\right)^4$.

[4.1–4.2] **97.** Evaluate each of the following:
(a) -6^3 (b) 6^{-3}

[4.4] **98.** Subtract $5x^2 - 4x - 3$ from $-x^2 - 6x + 5$.

JUST FOR FUN

Perform each polynomial multiplication.

1. $\sqrt{5}x\left(2x^2 + \sqrt{5}x - \dfrac{1}{2}\right)$ *Hint:* $(\sqrt{5})2 = 2\sqrt{5}$ and $\sqrt{5} \cdot \sqrt{5} = \sqrt{25} = 5$

2. $\left(\dfrac{x}{2} + \dfrac{2}{3}\right)\left(\dfrac{2x}{3} - \dfrac{2}{5}\right)$
3. $(2x^3 - 6x^2 + 5x - 3)(3x^3 - 6x + 4)$

4.6

Division of Polynomials

▶ **1** Divide a polynomial by a monomial.
▶ **2** Divide a polynomial by a binomial.
▶ **3** Check division of polynomial problems.
▶ **4** Write polynomials in descending order when dividing.

Dividing a Polynomial by a Monomial

▶ **1**

To divide a polynomial by a monomial, divide each term of the polynomial by the monomial.

EXAMPLE 1   Divide: $\dfrac{2x + 16}{2}$.

Solution: $\dfrac{2x + 16}{2} = \dfrac{2x}{2} + \dfrac{16}{2}$

$= x + 8.$ ■

EXAMPLE 2   Divide: $\dfrac{4x^2 - 8x}{2x}$

Solution: $\dfrac{4x^2 - 8x}{2x} = \dfrac{4x^2}{2x} - \dfrac{8x}{2x}$

$= 2x - 4.$ ■

EXAMPLE 3   Divide: $\dfrac{4x^3 - 6x^2 + 8x - 3}{2x}$

Solution: $\dfrac{4x^3 - 6x^2 + 8x - 3}{2x} = \dfrac{4x^3}{2x} - \dfrac{6x^2}{2x} + \dfrac{8x}{2x} - \dfrac{3}{2x}$

$= 2x^2 - 3x + 4 - \dfrac{3}{2x}$ ■

EXAMPLE 4   Divide: $\dfrac{3x^3 - 6x^2 + 4x - 1}{-3x}$

Solution: In this problem a negative sign appears in the denominator. One method to simplify this problem is to multiply both numerator and denominator by -1.

$$\frac{(-1)(3x^3 - 6x^2 + 4x - 1)}{(-1)(-3x)} = \frac{-3x^3 + 6x^2 - 4x + 1}{3x}$$

$$= \frac{-3x^3}{3x} + \frac{6x^2}{3x} - \frac{4x}{3x} + \frac{1}{3x}$$

$$= -x^2 + 2x - \frac{4}{3} + \frac{1}{3x}$$ ■

COMMON STUDENT ERROR

Correct	*Wrong*
$\dfrac{x + 2}{2} = \dfrac{x}{2} + \dfrac{2}{2} = \dfrac{x}{2} + 1$	$\dfrac{x + \overset{1}{\cancel{2}}}{\underset{1}{\cancel{2}}} = \dfrac{x + 1}{1} = x + 1$
$\dfrac{x + 2}{x} = \dfrac{x}{x} + \dfrac{2}{x} = 1 + \dfrac{2}{x}$	$\dfrac{\overset{1}{\cancel{x}} + 2}{\underset{1}{\cancel{x}}} = \dfrac{1 + 2}{1} = 3$

Can you explain why the procedures on the right are wrong?

Dividing a Polynomial by a Binomial

▶ **2** We divide a polynomial by a binomial in much the same way as we perform long division. This procedure will be explained in Example 5.

EXAMPLE 5 Divide: $\dfrac{x^2 + 6x + 8}{x + 2}$ ⟵ dividend
⟵ divisor

Solution: Rewrite the divison problem as

$$x + 2 \overline{)x^2 + 6x + 8}$$

Divide x^2 (the first term in the dividend) by x (the first term in the divisor).

$$\frac{x^2}{x} = x$$

Place the quotient, x, above the like term containing x in the dividend.

$$x + 2 \overline{)\overset{\textstyle x}{x^2 + 6x + 8}}$$

Next, multiply the x by $x + 2$ as you would do in long division and place the terms of the product under their like terms.

Now subtract $x^2 + 2x$ from $x^2 + 6x$. When subtracting, remember to change the sign of the terms being subtracted and then add the like terms.

$$
\begin{array}{r}
x \\
x + 2 \overline{)\; x^2 + 6x + 8} \\
\underline{-\; x^2 \mp 2x} \\
4x
\end{array}
$$

Next, bring down the 8.

$$
\begin{array}{r}
x \\
x + 2 \overline{)x^2 + 6x + 8} \\
\underline{x^2 + 2x} \\
4x + 8
\end{array}
$$

Determine the quotient of $4x$ divided by x.

$$\frac{4x}{x} = +4$$

Place the $+4$ above the constant in the dividend.

$$
\begin{array}{r}
x + 4 \\
x + 2 \overline{)x^2 + 6x + 8} \\
\underline{x^2 + 2x} \\
4x + 8
\end{array}
$$

Multiply the $x + 2$ by 4 and place the terms of the product under their like terms.

Now subtract.

$$
\begin{array}{r}
x + 4 \quad\longleftarrow\text{ quotient}\\
x + 2 \overline{\smash{)}\,x^2 + 6x + 8}\\
\underline{x^2 + 2x}\\
4x + 8\\
\underline{\,4x \not{+} 8}\\
0 \quad\longleftarrow\text{ remainder}
\end{array}
$$

Thus

$$
\frac{x^2 + 6x + 8}{x + 2} = x + 4
$$

There is no remainder.

EXAMPLE 6   Divide: $\dfrac{6x^2 - 5x + 5}{2x + 3}$

Solution:

$$
\begin{array}{c}
\dfrac{6x^2}{2x} \quad \dfrac{-14x}{2x}\\
\downarrow \qquad \swarrow\\
3x - 7\\
2x + 3 \overline{\smash{)}\,6x^2 - 5x + 5}\\
\underline{\,6x^2 \not{+} 9x} \quad\longleftarrow 3x(2x + 3)\\
-14x + 5\\
\underline{\not{+}\,14x \not{+} 21} \quad\longleftarrow -7(2x + 3)\\
26 \quad\longleftarrow\text{ remainder}
\end{array}
$$

When there is a remainder, as in this example, list the quotient, plus the remainder above the divisor. Thus

$$
\frac{6x^2 - 5x + 5}{2x + 3} = 3x - 7 + \frac{26}{2x + 3}
$$

Checking Polynomial Division

▶ **3**   The answer to a division problem can be checked. Consider the division problem $13 \div 5$.

$$
\begin{array}{r}
2\\
5 \overline{\smash{)}\,13}\\
\underline{10}\\
3
\end{array}
$$

Note that the divisor times the quotient, plus the remainder, equals the dividend:

$$
(\text{Divisor} \times \text{quotient}) + \text{remainder} = \text{dividend}
$$
$$
(5 \cdot 2) + 3 = 13
$$
$$
10 + 3 = 13
$$
$$
13 = 13 \qquad \text{true}
$$

This same procedure can be used to check all division problems.

To Check Division of Polynomials

$$
(\text{Divisor} \cdot \text{quotient}) + \text{remainder} = \text{dividend}
$$

Let us check the answer to Example 6. The divisior is $2x + 3$, the quotient is $3x - 7$, the remainder is 26, and the dividend is $6x^2 - 5x + 5$.

Check: (divisor $\times$ quotient) + remainder = dividend

$$(2x + 3)(3x - 7) + 26 = 6x^2 - 5x + 5$$
$$(6x^2 - 5x - 21) + 26 = 6x^2 - 5x + 5$$
$$6x^2 - 5x + 5 = 6x^2 - 5x + 5 \qquad \text{true}$$

▸ **4 When dividing a polynomial by a binomial, both the polynomial and binomial should be listed in descending order. If a given power term is missing, it is often helpful to include that term with a numerical coefficient of 0.** For example, when dividing $(6x^2 + x^3 - 4)/(x - 2)$, we rewrite the problem as $(x^3 + 6x^2 + 0x - 4)/(x - 2)$ before beginning the division.

EXAMPLE 7 $(9x^3 - x - 28) \div (3x - 4)$.

Solution: Since there is no x^2 term in the dividend, we will add $0x^2$ to help align like terms.

$$\frac{9x^3}{3x}, \; \frac{12x^2}{3x}, \; \frac{15x}{3x}$$

$$
\begin{array}{r}
3x^2 + 4x + 5 \\
3x - 4 \overline{\smash{)}9x^3 + 0x^2 - x - 28} \\
\underline{9x^3 - 12x^2} \qquad\qquad\qquad 3x^2(3x-4) \\
12x^2 - x \\
\underline{12x^2 - 16x} \qquad\qquad 4x(3x-4) \\
15x - 28 \\
\underline{15x - 20} \qquad 5(3x-4) \\
-8 \qquad \text{remainder}
\end{array}
$$

$$\frac{9x^3 - x - 28}{3x - 4} = 3x^2 + 4x + 5 - \frac{8}{3x - 4}$$

Exercise Set 4.6

Divide as indicated.

1. $\dfrac{2x + 4}{2}$ **2.** $\dfrac{4x - 6}{2}$ **3.** $\dfrac{2x + 6}{2}$ **4.** $\dfrac{4x + 3}{2}$

5. $\dfrac{3x + 8}{2}$ **6.** $\dfrac{5x - 12}{6}$ **7.** $\dfrac{-6x + 4}{2}$ **8.** $\dfrac{-4x + 5}{-3}$

9. $\dfrac{-9x - 3}{-3}$ **10.** $\dfrac{5x - 4}{-5}$ **11.** $\dfrac{3x + 6}{x}$ **12.** $\dfrac{4x - 3}{2x}$

13. $\dfrac{9 - 3x}{-3x}$ **14.** $\dfrac{6 - 5x}{-3x}$ **15.** $\dfrac{3x^2 + 6x - 9}{3}$

16. $\dfrac{12x^2 - 6x + 3}{3}$ **17.** $\dfrac{-4x^2 + 6x + 8}{2}$ **18.** $\dfrac{5x^2 + 4x - 8}{2}$

19. $\dfrac{x^2 + 4x - 3}{x}$ **20.** $\dfrac{4x^2 - 6x + 7}{x}$ **21.** $\dfrac{6x^2 - 4x + 12}{2x}$

22. $\dfrac{8x^2 - 5x + 10}{-2x}$

23. $\dfrac{4x^3 + 6x^2 - 8}{-4x}$

24. $\dfrac{-12x^3 + 6x^2 - 15x}{-3x}$

25. $\dfrac{9x^3 + 3x^2 - 12}{3x^2}$

26. $\dfrac{-10x^3 - 6x^2 + 15}{-5x}$

Divide as indicated

27. $\dfrac{x^2 + 4x + 3}{x + 1}$

28. $\dfrac{x^2 + 7x + 10}{x + 5}$

29. $\dfrac{2x^2 + 13x + 15}{x + 5}$

30. $\dfrac{2x^2 + x - 10}{x - 2}$

31. $\dfrac{6x^2 + 16x + 8}{3x + 2}$

32. $\dfrac{2x^2 + 13x + 15}{2x + 3}$

33. $\dfrac{2x^2 + x - 10}{2x + 5}$

34. $\dfrac{8x^2 - 26x + 15}{4x - 3}$

35. $\dfrac{2x^2 + 7x - 18}{2x - 3}$

36. $\dfrac{x^2 - 25}{x - 5}$

37. $\dfrac{4x^2 - 9}{2x - 3}$

38. $\dfrac{9x^2 - 4}{3x + 2}$

39. $\dfrac{6x + 8x^2 - 25}{4x + 9}$

40. $\dfrac{10x + 3x^2 + 6}{x + 2}$

41. $\dfrac{6x + 8x^2 - 12}{2x + 3}$

42. $\dfrac{6x^2 - 13 - 11x}{2x - 5}$

43. $\dfrac{x^3 + 3x^2 + 5x + 3}{x + 1}$

44. $\dfrac{4x^3 + 12x^2 + 7x - 3}{2x + 3}$

45. $\dfrac{2x^3 - 3x^2 - 3x + 6}{x - 1}$

46. $\dfrac{9x^3 - 3x^2 - 9x + 4}{3x + 2}$

47. $\dfrac{2x^3 + 6x - 4}{x + 4}$

48. $\dfrac{x^3 - 8}{x - 3}$

49. $\dfrac{x^3 + 8}{x + 2}$

50. $\dfrac{x^3 - 27}{x - 3}$

51. $\dfrac{x^3 + 27}{x + 3}$

52. $\dfrac{4x^3 - 5x}{2x - 1}$

53. $\dfrac{9x^3 - x + 3}{3x - 2}$

54. $\dfrac{-x^3 - 6x^2 + 2x - 3}{x - 1}$

✎ **55.** Consider the division problem

$$\dfrac{15x^\blacksquare + 25x^\blacksquare + 5x^\blacksquare + 10x^\blacksquare}{5x^2} = 3x^5 + 5x^4 + x^2 + 2.$$

Determine the exponents to be placed in the shaded areas to make a true statement. Explain how you obtained your answer.

Cumulative Review Exercises

[1.3] **56.** Consider the set of numbers $\{2, -5, 0, \sqrt{7},$ $\frac{2}{5}, -6.3, \sqrt{3}, -23/34\}$. List those that are (a) natural numbers, (b) whole numbers, (c) rational numbers, (d) irrational numbers, and (e) real numbers.

[1.7] **57.** **(a)** To what is $0/1$ equal?
  (b) How do we refer to an expression like $1/0$?

[1.9] **58.** Give the order of operations to be followed when evaluating a mathematical problem.

[2.5] **59.** Solve the equation $2(x + 3) + 2x = x + 4$. $-\frac{2}{3}$

JUST FOR FUN

1. $\dfrac{4x^3 - 4x + 6}{2x + 3}$ *Hint:* The quotient will contain fractions.

2. $\dfrac{3x^3 - 5}{3x - 2}$

4.7

Rate and Mixture Problems

▶ **1** Set up and solve rate (or motion) problems involving only one rate.

▶ **2** Learn the distance formula.

▶ **3** Set up and solve rate problems involving two rates.

▶ **4** Set up and solve mixture problems.

In Chapter 2 we discussed applications of algebra that could be solved using proportions. In Chapter 3 we give a general procedure for setting up and solving application problems. Now we will use the procedure discussed in Chapter 3 to solve two other types of application problems, rate and mixture application problems.

Rate and mixture problems are grouped in the same section because, as you will learn shortly, you will use the same general multiplication procedure to solve them. We will begin by discussing rate problems.

Rate Problems with Only One Rate

▶ **1** A rate (for motion) problem is one in which an item is moving at a specified rate for a specified period of time. A car traveling at a constant speed, a swimming pool being filled or drained (the water is moving at a specified rate), and spaghetti being cut on a conveyor belt (conveyor belt moving at a specified speed) are all rate problems.

The formula often used to solve rate problems follows.

Rate Problems

$$\text{Amount} = \text{rate} \cdot \text{time}$$

The amount can be a measure of many different quantities, depending on the rate. For example, if the rate is measuring *distance* per unit time, the amount will be *distance*. If the rate is measuring *volume* per unit time, the amount will be *volume;* and so on.

The distance traveled by a car moving at a constant rate for a specific time can be found by the formula distance = rate · time. When a swimming pool is being filled at a constant rate for a specific period of time, the volume of water in the pool can be found by the formula volume = rate · time.

EXAMPLE 1 A swimming pool is being filled at a rate of 10 gallons per minute. How many gallons have been added after 25 minutes?

Solution: Since we are discussing gallons, which measure volume, the formula we will use is volume = rate · time. We are given the rate, 10 gallons per minute, and the time, 25 minutes. We are asked to find the volume.

$$\text{Volume} = \text{rate} \cdot \text{time}$$
$$= 10 \cdot 25 = 250$$

Thus the volume after 25 minutes is 250 gallons. ∎

Let us look at the units of measurement in Example 1. The rate is given in gallons per minute and the time is given in minutes. If we analyze the units, we see that the volume is measured in gallons.

$$\text{Volume} = \text{rate} \cdot \text{time}$$
$$= \frac{\text{gallons}}{\text{minute}} \cdot \text{minutes}$$
$$= \text{gallons}$$

EXAMPLE 2   A patient is to receive 1200 cubic centimeters of fluid intravenously over an 8-hour period. What should be the average intravenous flow rate?

Solution: Here we are given the volume and the length of time the fluid is to be administered. We are asked to find the rate.

$$\text{Volume} = \text{rate} \cdot \text{time}$$
$$1200 = r \cdot 8$$
$$\frac{1200}{8} = r$$
$$150 = r$$

The fluid should be administered at a rate of 150 cubic centimeters per hour.   ■

Can you explain why the rate in Example 2 must be in cubic centimeters per hour?

The Distance Formula

▶ **2**   When the "*amount*" in the rate formula is "*distance*" we often refer to the formula as the **distance formula,**

$$\text{distance} = \text{rate} \cdot \text{time} \quad \text{or} \quad d = r \cdot t$$

Examples 3 and 4 illustrate the use of the distance formula.

EXAMPLE 3   An oil-well-drilling device can drill 3 feet per hour. How long will it take to drill to a depth of 1870 feet?

Solution: Since we are given a distance of 1870 feet we will use the distance formula. We are given the distance and the rate and need to solve for the time, t.

$$\text{distance} = \text{rate} \cdot \text{time}$$
$$1870 = 3t$$
$$\frac{1870}{3} = t$$
$$623.33 = t$$

or $$t = 623.33 \text{ hours}$$ ■

HELPFUL HINT   When working rate problems, the units must be consistent with each other. If you are given a problem where the units are not consistent with each other, you will need to change one of the quantities so that the units will be consistent before you substitute the values into the formula. Example 4 illustrates how this is done.

EXAMPLE 4   A conveyor belt transporting uncut spaghetti moves at a rate of 1.5 feet per second. A cutting blade is activated at regular intervals to cut the spaghetti into proper lengths. At what time intervals should the blade be activated if the spaghetti is to be in 9-inch lengths?

Solution: Since we are working with lengths, which are distances, we will use the distance formula. We are asked to find the time, t, at which the blade should be activated. Since the rate is given in *feet* per second and the length is given in *inches*, one of these units must be changed. One foot equals 12 inches; thus, to change from feet per second to inches per second, we multiply the rate by 12.

$$1.5 \text{ feet per second} = (1.5)(12) = 18 \text{ inches per second}$$
$$\text{distance} = \text{rate} \cdot \text{time}$$
$$9 = 18 \cdot t$$
$$\frac{9}{18} = t$$
$$\frac{1}{2} = t$$

The blade should cut at $\frac{1}{2}$-second intervals. ∎

Rate Problems with Two Rates

▶ **3**   Now we will look at some rate problems that involve *two rates*. For example, we may be considering two trains traveling at two different speeds. In these problems, we generally begin by letting the variable represent one of the unknown quantities, and then we represent the second unknown quantity in terms of the first unknown quantity. For example, suppose we are talking about two trains, and we are told one train travels 20 miles per hour faster than another train. We might let r represent the rate of one of the trains and $r + 20$ the rate of the second train.

To solve problems of this type that use the distance formula, we generally add the two distances, or subtract the smaller distance from the larger, or set the two distances equal to each other, depending on the information given in the problem.

Often, when working problems involving two different rates, we construct a table to help organize the information given. Examples 5 through 8 illustrate the procedures used.

EXAMPLE 5   Two trains leave the same station along parallel tracks going in opposite directions. The train traveling east has a speed of 40 miles per hour. The train traveling west has a speed of 60 miles per hour. In how many hours will they be 500 miles apart?

Solution: We are asked to find the time, t, when the trains are 500 miles apart.

Let t = time for the trains to travel 500 miles apart

When the trains reach a distance 500 miles apart, each train has traveled for the same amount of time, t. We set up a table to help analyze the problem. The distance is found by multiplying the rate by the time.

West ◄————— 500 miles —————► East

◄— 60 mph   40 mph —►

Train	Rate	Time	Distance
East	40	t	$40t$
West	60	t	$60t$

Since the trains are traveling in opposite directions, the sum of their distances must be 500 miles.

$$\left(\begin{array}{c}\text{distance traveled}\\\text{by train 1}\end{array}\right) + \left(\begin{array}{c}\text{distance traveled}\\\text{by train 2}\end{array}\right) = 500 \text{ miles}$$

$$40t + 60t = 500$$
$$100t = 500$$
$$t = 5$$

The two trains will be 500 miles apart in 5 hours. ∎

EXAMPLE 6   Two cross-country skiers start skiing at the same time on the same trail going in the same direction. The more advanced skier averages 6 miles per hour, while the beginning skier averages 2 miles per hour. After how many hours of skiing will the two skiers be 10 miles apart?

Solution: We are asked to find the time it takes for the skiers to become separated by 10 miles. We will construct a table to aid us in setting up the problem.

Let t = time until skiers are 10 miles apart

When the two skiers are 10 miles apart, each has skied for the same number of hours, t.

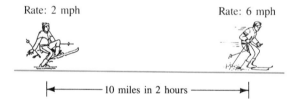

Rate: 2 mph   Rate: 6 mph

◄————— 10 miles in 2 hours —————►

Skier	Rate	Time	Distance
Slower	2	t	$2t$
Faster	6	t	$6t$

Since the skiers are traveling in the same direction, the distance between the skiers can be found by subtracting the distance traveled by the slower skier from the distance traveled by the faster skier.

$$\left(\begin{array}{c}\text{distance traveled}\\\text{by faster skier}\end{array}\right) - \left(\begin{array}{c}\text{distance traveled}\\\text{by slower skier}\end{array}\right) = 10 \text{ miles}$$

$$6t - 2t = 10$$
$$4t = 10$$
$$t = 2.5$$

Thus after 2.5 hours of skiing the two skiers will be 10 miles apart. ∎

EXAMPLE 7   Two construction crews are 20 miles apart working toward each other. Both are laying sewer pipe in a straight line that will eventually be connected together. Both crews will work the same hours. One crew has more modern equipment and more workers and can lay a greater length of pipe per day. If the faster crew lays 0.4 mile of pipe per day more than the slower crew, and the two pipes are connected after 10 days, find the rate at which each crew lays pipe.

Solution: We are asked to find the rate, r. We are told that both crews work for 10 days.

$$\text{Let } r = \text{rate of slower crew}$$
$$\text{then } r + 0.4 = \text{rate of faster crew}$$

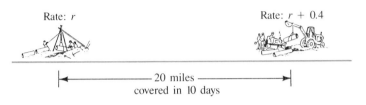

Rate: r

Rate: $r + 0.4$

20 miles
covered in 10 days

Crew	Rate	Time	Distance
Slower	r	10	$10r$
Faster	$r + 0.4$	10	$10(r + 0.4)$

The total distance covered by both crews must be 20 miles.

$$\left(\begin{array}{c}\text{distance covered} \\ \text{by slower crew}\end{array}\right) + \left(\begin{array}{c}\text{distance covered} \\ \text{by faster crew}\end{array}\right) = 20 \text{ miles}$$

$$10r + 10(r + 0.4) = 20$$
$$10r + 10r + 4 = 20$$
$$20r + 4 = 20$$
$$20r = 16$$
$$\frac{20r}{20} = \frac{16}{20}$$
$$r = 0.8$$

Thus the slower crew lays 0.8 mile of pipe per day and the faster crew lays $x + 0.4$ or $0.8 + 0.4 = 1.2$ miles of pipe per day. ∎

EXAMPLE 8   A mother and daughter plan to go hiking together on the Appalachian Trail. The mother hikes at an average of 4 miles per hour, the daughter at 5 miles per hour. If the daughter begins hiking $\frac{1}{2}$ hour after the mother, and they plan to meet on the trail:

(a) How long will it take for the mother and daughter to meet?
(b) How far from the starting point will they be when they meet?

Solution: (a) Since the daughter is the faster hiker, she will cover the same distance in less time. When they meet, they have both traveled the same distance. Since the rate is given in miles per hour, the time will be in hours. The daughter begins $\frac{1}{2}$ hour later; therefore, her time will be $\frac{1}{2}$ hour less than her mother's. We are asked to find the time for the mother and daughter to meet.

$$\text{Let } t = \text{time mother is hiking}$$
$$\text{then } t - \frac{1}{2} = \text{time daughter is hiking}$$

Daughter

Mother

Rate: 5 mph   Rate: 4 mph
Time: $t - \frac{1}{2}$ Time: t

Hiker	Rate	Time	Distance
Mother	4	t	$4t$
Daughter	5	$t - \frac{1}{2}$	$5(t - \frac{1}{2})$

$$\text{Distance mother} = \text{distance daughter}$$

$$4t = 5\left(t - \frac{1}{2}\right)$$

$$4t = 5t - \frac{5}{2}$$

$$4t - 5t = 5t - 5t - \frac{5}{2}$$

$$-t = -\frac{5}{2}$$

$$t = \frac{5}{2}$$

Thus they will meet in $\frac{5}{2}$ or $2\frac{1}{2}$ hours.

(b) The distance can be found using either the mother or daughter. We will use the mother.

$$d = r \cdot t$$

$$= 4 \cdot \frac{5}{2} = \frac{20}{2} = 10 \text{ miles}$$

The mother and daughter will meet 10 miles from the starting point.   ■

Mixture Problems

▶**4**   Now we will work some mixture problems. Any problem in which two or more quantities are combined to produce a different quantity or a single quantity is separated into two or more different quantities may be considered a mixture problem. Mixture problems are familiar to everyone, as we can see in the everyday examples that follow.

When solving mixture problems, we often let the variable represent one unknown quantity, and then we represent a second unknown quantity in terms of the first unknown quantity. For example, if we know that when two solutions are mixed they make a total of 80 liters, we may represent the number of liters of one of the solutions as x and the number of liters of the second solution as $80 - x$. Note that when we add x and $80 - x$ we get the sum of 80.

We generally solve mixture problems by using the fact that the amount (or value) of one part of the mixture plus the amount (or value) of the second part of the mixture is equal to the total amount (or value) of the total mixture.

Again, as we did with rate problems involving two rates, we will use tables to help analyze the problem.

EXAMPLE 9   Deborah wishes to mix coffee worth $7 per pound with 12 pounds of coffee worth $4 per pound.

(a) How many pounds of coffee worth $7 per pound must be mixed to obtain a mixture worth $6 per pound?

(b) How much of the mixture will be produced?

Solution: (a) We are asked to find the number of pounds of $7 coffee.

Let x = number of pounds of $7 coffee

Often it is helpful to make a sketch of the situation. After we draw a sketch, we will construct a table. In our sketch we will use a cup to put the coffee in.

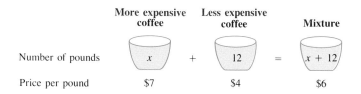

The value of the coffee is found by multiplying the number of pounds by the price per pound.

Coffee	Price	Number of Pounds	Value of Coffee
More expensive	7	x	$7x$
Less expensive	4	12	4(12)
Mixture	6	$x + 12$	$6(x + 12)$

value of $7 coffee + value of $4 coffee = value of $6 mixture

$$7x + 4(12) = 6(x + 12)$$
$$7x + 48 = 6x + 72$$
$$x + 48 = 72$$
$$x = 24 \text{ pounds}$$

Thus 24 pounds of the $7 coffee must be mixed with 12 pounds of the $4 coffee to obtain a mixture worth $6 per pound.

(b) The number of pounds of the mixture is

$$x + 12 = 24 + 12 = 36 \text{ pounds} \qquad \blacksquare$$

EXAMPLE 10 Tsong invests $15,000, part at 8% simple interest and the rest at 6% simple interest for a period of 1 year. How much did Tsong invest at each rate if his total interest for the year was $1100?

Solution: We use the simple interest formula that was introduced in Section 3.1 to solve this problem: interest = principal · rate · time.

$$\text{Let } x = \text{amount invested in the 6\% account}$$
$$\text{then } 15{,}000 - x = \text{amount invested in the 8\% account}$$

Account	Principal	Rate	Time	Interest
6%	x	0.06	1	0.06x
8%	$15{,}000 - x$	0.08	1	$0.08(15{,}000 - x)$

Since the sum of the interest from the two accounts is $1100, we write the equation

$$\left(\begin{array}{c}\text{interest from}\\ \text{6\% account}\end{array}\right) + \left(\begin{array}{c}\text{interest from}\\ \text{8\% account}\end{array}\right) = \text{total interest}$$

$$0.06x + 0.08(15{,}000 - x) = 1100$$
$$0.06x + 0.08(15{,}000) - 0.08(x) = 1100$$
$$0.06x + 1200 - 0.08x = 1100$$
$$-0.02x + 1200 = 1100$$
$$-0.02x = -100$$
$$x = \frac{-100}{-0.02} = 5000$$

Thus $5000 was invested at 6% interest. The amount invested at 8% was

$$15{,}000 - x = 15{,}000 - 5000 = 10{,}000$$

Therefore, $5000 was invested at 6% and $10,000 was invested at 8%. The total amount invested is $15,000, which checks with the information given. ■

EXAMPLE 11   Debby Sunderland has a total of 30 dimes and quarters. The total value of these coins is $4.50. How many of each coin does she have?

Solution: We are asked to find the number of each type of coin.

$$\text{Let } x = \text{number of dimes}$$
$$\text{then } 30 - x = \text{number of quarters}$$

The total value of the coins is found by multiplying the value of the coin by the number of coins.

Coin	Value of Coin	Number of Coins	Total Value of Coins
Dime	0.10	x	$0.10x$
Quarter	0.25	$30 - x$	$0.25(30 - x)$

$$\text{value of dimes} + \text{value of quarters} = \text{total value}$$
$$0.10x + 0.25(30 - x) = 4.50$$
$$0.10x + 7.5 - 0.25x = 4.50$$
$$-0.15x + 7.5 = 4.50$$
$$-0.15x = -3.0$$
$$x = \frac{-3.0}{-0.15} = 20$$

Thus there are 20 dimes and $30 - 20 = 10$ quarters.

Check: 20 dimes = $2.00
10 quarters = $2.50
Total = $4.50   true   ■

EXAMPLE 12   How many liters of a 25% salt solution must be added to 80 liters of a 40% salt solution to get a solution that is 30% salt?

Solution: We are asked to find the number of liters of the 25% salt solution.

Let x = number of liters of 25% salt solution

Let us draw a sketch of the situation.

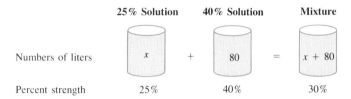

	25% Solution	40% Solution	Mixture
Numbers of liters	x	80	$x + 80$
Percent strength	25%	40%	30%

The amount of pure salt in a given solution is found by multiplying the percent strength by the number of liters.

Solution	Strength	Liters	Amount of Pure Salt
25%	0.25	x	$0.25x$
40%	0.40	80	$0.40(80)$
mixture	0.30	$x + 80$	$0.30(x + 80)$

$$\left(\begin{array}{c}\text{amount of pure salt} \\ \text{in 25\% solution}\end{array}\right) + \left(\begin{array}{c}\text{amount of pure salt} \\ \text{in 40\% solution}\end{array}\right) = \left(\begin{array}{c}\text{amount of pure salt} \\ \text{in 30\% mixture}\end{array}\right)$$

$$0.25x + 0.40(80) = 0.30(x + 80)$$
$$0.25x + 32 = 0.30x + 24$$
$$0.25x + 8 = 0.30x$$
$$8 = 0.05x$$
$$\frac{8}{0.05} = x$$
$$160 = x$$

Therefore, 160 liters of 25% salt solution must be added to the 80 liters of 40% solution to get a 30% salt solution. The total number of liters that will be obtained is 160 + 80 or 240. ∎

Exercise Set 4.7

Set up an algebraic equation that can be used to solve each problem. Solve the equation, and find the values desired.

1. How fast must a car travel to cover 150 miles in 3 hours?

2. Maria's small aboveground swimming pool has a capacity of 13,500 gallons of water. If a hose can supply 9 gallons per minute, how long will it take Maria's son to fill the pool?

3. Fred can lay 42 bricks per hour. How long will it take him to lay 546 bricks?

4. At what rate must a photocopying machine copy to make 100 copies in 2.5 minutes?

5. A certain laser can cut through steel at a rate of 0.2 centimeter per minute. How thick is a steel door if it requires 12 minutes to cut through it?

6. *Apollo 11* took approximately 87 hours to reach the moon, a distance of about 238,000 miles. Find the average rate of speed of the Apollo. Give the answer rounded to the nearest mile per hour.

7. A patient is to receive 1500 cubic centimeters of an intravenous fluid over a period of 6 hours. What should be the intravenous flow rate?

8. On a sunny day a work crew pouring cement on a new highway pours a volume of 1.4 tons per hour. Find the time it will take for the crew to pour 6.3 tons.

9. Kilauea Volcano's lava is flowing slowly toward the ocean. If the lava flows a distance of 57 meters in 9.5 hours, find the average rate of flow of the lava.

10. Morton can check 12 parts *per minute* on the assembly line. How many parts will he be able to check in an 8-*hour* day?

11. If John walks at a speed of 3 miles per hour, how far will he walk in 20 minutes?

12. A standard T120 VHS video cassette tape contains 246 meters of tape. When played on standard play speed (SP), the tape will play for 2 hours. The tape will play for 4 hours on long play speed (LP) and will play for 6 hours on super long play speed (SLP). Determine the rate of play of the tape at (a) SP, (b) LP, and (c) SLP.

Solve the following rate problems. See Examples 5 through 8.

13. Two planes leave an airport at the same time. One plane flies north at 500 miles per hour. The other flies south at 650 miles per hour. In how many hours will they be 4025 miles apart?

14. Two joggers, Sonya and Jeri, start from the same point and run in the same direction. Sonya jogs at 8 miles per hour, and Jeri jogs at 11 miles per hour. In how many hours will Sonya and Jeri be 9 miles apart?

15. Two trains, an Amtrak train and the Santa Fe Special, are 804 miles apart. Both start at the same time and travel toward each other. They meet 6 hours later. If the speed of the Amtrak train is 30 miles per hour faster than the Santa Fe Special, find the speed of each train.

16. Two trains, a Conrail train and the Durango Special, leave South Street Station at the same time traveling in the same direction along parallel tracks. The Conrail train travels 80 miles per hour, and the Durango Special travels 62 miles per hour. In how many hours will the two trains be 144 miles apart?

17. Two sailboats are 9.8 miles apart and sailing toward each other. The larger boat, the *Lorelei*, is traveling 4 miles per hour faster than the smaller boat, the *Brease Along*. If the two boats pass each other after 0.7 hour, find the speed of each boat.

18. Marie has two pet snails, Speedy and Slowpoke. The two snails are 12.4 inches apart and crawling toward each other. Speedy is crawling at an average rate that is 0.2 inch per hour faster than Slowpoke. Find the speed of the two snails if they meet after 2 hours.

19. A VHS T120 video cassette tape plays at a rate of 6.72 feet per minute for 2 hours (120 minutes) on standard

play speed (SP). What is the rate of the tape on super long play speed (SLP) if the same tape plays for 6 hours on this speed?

20. Two construction crews are laying blacktop on a road. They start at the same time at opposite ends of a 12-mile road and work toward one another. One crew lays blacktop at an average rate of 0.75 mile a day greater than the other crew. If the two crews meet after 3.2 days, find the rate of each crew.

21. Two ranchers are 16.5 miles apart at opposite ends of the Flying W Ranch in Montana. Both start at the same time and travel toward each other on horseback. Chet is walking his horse Chestnut while Annie is trotting on her horse Midnight. If Midnight is traveling at a rate of 3 miles per hour faster than Chestnut, and they meet after 1.5 hours, at what speed is each horse is traveling?

22. Mrs. Weber and her son belong to a health club and exercise together regularly. They start two treadmills at the same time. The son's machine is set for 6 miles per hour and the mother's machine is set for 4 miles per hour. When they finish, they compare the distances and find that together they had traveled a total of 11 miles. For how long had they worked out?

23. Serge and Francine go mountain climbing together. Francine begins climbing the mountain 30 minutes before Serge and averages 18 feet per minute. When Serge begins climbing, he averages 20 feet per minute. How far up the mountain will they meet?

24. Two rockets are to be launched in the same direction one hour apart. The first rocket is launched at noon and travels at 12,000 miles per hour. If the second rocket travels at 14,400 miles per hour, at what time will the rockets be the same distance from Earth?

25. The moving walkway in the United Airlines terminal at Chicago's O'Hare International Airport (also referred to as a "travelator" by United Airlines) is like a flat escalator moving along the ground. It has the effect of increasing the overall speed, with respect to the ground, of the people walking on it. When Marquerita Vela walked the floor along side the moving walkway at her normal speed of 100 feet per minute it took her 2.75 minutes to walk the length of the moving walkway. When she returned the same distance, walking at her normal speed on the moving walkway, it took her only 1.25 minutes. (a) Find the rate of speed of the moving walkway. (b) Find the length of the moving walkway.

26. Tanya starts for school walking at an average rate of 120 feet per minute. A short while later she realizes that she forgot her lunch and turns around and jogs back to the house at 360 feet per minute. If it took her a total of 10 minutes to go and return home, how far had Tanya walked before she turned around?

Solve the following mixture problems. See Examples 9 through 12.

27. Mr. Ellis invested $8900, part at 8% simple interest and the rest at 11% simple interest for a period of 1 year. How much does he invest at each rate if his total annual interest from both investments is $874? Use interest = principal · rate · time.

28. Linda invested $6000, part at 9% simple interest and the rest at 10% simple interest for a period of 1 year. If she received a total annual interest of $562 from both investments, how much did she invest at each rate?

29. Julio invested $5000, part at 10% simple interest and part at 15% simple interest for a period of 1 year. How much did he invest at each rate if the same amount of interest was received from each account?

30. The Clars invested $12,500, part at 8% simple interest and part at 12% simple interest for a period of 1 year. How much was invested at each rate if the same amount of interest was received from each account?

31. Charles has a total of 28 dimes and quarters. The total value of the coins is $3.55. How many of each coin does he have?

32. Kathleen has a total of 62 dimes and nickels. The total value of all the coins is $3.75. How many of each coin does she have?

33. Phil has a total of 12 bills in his wallet. Some are $1 bills and some are $10 bills. The total value of the 12 bills is $39. How many of each type does he have?

34. Mari has a total of 25 quarters and half-dollars. The total value of the coins is $9.00. How many of each coin does she have?

35. Casey holds two part-time jobs. One job pays $6.00 per hour, and the other pays $6.50 per hour. Last week Casey worked a total of 18 hours and earned $114.00. How many hours did he work on each job?

36. Almonds cost $6.00 per pound. Walnuts cost $6.40 per pound. How many pounds of each should Bridget mix to produce a 30-pound mixture that costs $6.25 per pound?

37. How many pounds of coffee costing $6.20 per pound must Jack mix with 18 pounds of coffee costing $5.60 per pound to produce a mixture that costs $5.80 per pound?

38. There were 600 people at a movie. Adult admission was $6. Children's admission was $5. How many adults and how many children attended if the total receipts were $3250 for the day?

39. In chemistry class, Ramon has 1 liter of a 20% sulfuric acid solution. How much of a 12% sulfuric acid solution must be mixed with the 1 liter of 20% solution to make a 15% sulfuric acid solution?

40. Doug Fisher, a pharmacist, has a 60% solution of the drug sodium iodite. He also has a 25% solution of the same drug. He gets a prescription calling for a 40% solution of the drug. How much of each solution should he mix to make 0.5 liter of the 40% solution?

41. Six quarts of the orange juice punch for the class party contains 12% orange juice. If $\frac{1}{2}$ quart of water is added to the punch, find the percent of orange juice in the new mixture.

42. The label on a 12-ounce can of frozen concentrate Hawaiian Punch indicates that when the can of concentrate is mixed with 3 cans of cold water the resulting mixture is 10% juice. Find the percentage of pure juice in the concentrate.

43. Scott's Family grass seed sells for $2.25 per pound and Scott's Spot Filler grass seed sells for $1.90 per pound. How many pounds of each should be mixed to get a 10-pound mixture that sells for $2.00 per pound?

44. Avon stock is selling at $37 a share. Coca-Cola stock is selling at $75 a share. Mr. Abelard has a maximum of $7800 to invest. He wishes to purchase five times as many shares of Avon as of Coca-Cola. How many shares of each will he purchase?

45. United Airlines stock is selling at $140 per share. Getty Oil stock is selling at $22 a share. An investor has a total of $8000 to invest. She wishes to purchase four times as many shares of United Airlines as Getty Oil. If only whole shares of stock can be purchased:

(a) How many shares of each will she purchase?

(b) How much money will be left over?

Cumulative Review Exercises

[1.2] **46. (a)** Divide $2\frac{3}{4} \div 1\frac{5}{8}$.
   (b) Add $2\frac{3}{4} + 1\frac{5}{8}$.

[2.5] **47.** Solve the equation $6(x - 3) = 4x - 18 + 2x$.

[2.6] **48.** Solve the proportion $\dfrac{6}{x} = \dfrac{72}{9}$.

[2.7] **49.** Solve the inequality $3x - 4 \le -4x + 3(x - 1)$.

JUST FOR FUN

1. Every 20 minutes Americans dump enough cars into junkyards so that the cars, if stacked one on top of the other, would reach the top of the Empire State Building, 1472 feet tall.
   (a) Write an equation using distance = rate · time, for t in hours, for determining how long it would take for the stack of cars to reach a height of 10 miles. There are 5,280 feet in a mile.
   (b) Find how long it would take for the stack of cars to reach a height of 10 miles.
2. Sixty-six lines of type will fit on a standard $8\frac{1}{2} \times 11$-inch sheet of paper. A Panasonic printer types 10 characters per inch at a speed of 100 characters per second. How long will it take the printer to type a full page of 66 lines if the page is set for a 1-inch margin on the left and a $\frac{1}{2}$-inch margin on the right (disregad the carriage return time).
3. An automatic garage door opener is designed to begin to open when a car is 100 feet from the garage. At what rate will the garage door have to open if it is to raise 6 feet by the time a car traveling at 4 miles per hour reaches it? (1 mile per hour = 1.47 feet per second.)
4. The radiator of Mark's 1968 Dodge Challenger has a capacity of 16 quarts. If it is presently filled with a 20% antifreeze solution, how many quarts must Mark drain and replace with pure antifreeze to make the radiator contain a 50% antifreeze solution?

SUMMARY

GLOSSARY

Binomial *(161):* A two-termed polynomial.
Degree of a polynomial *(162):* The same as the highest-degree term in the polynomial.
Degree of a term *(162):* The exponent on the variable when the polynomial is in one variable.
Descending order, or power, of the variable *(161):* Polynomial written so that the exponents decrease from left to right.

Monomial *(161):* A one-term polynomial.
Polynomial in x *(161):* An expression containing the sum of only a finite number of terms of the form ax^n, for any real number a and any whole number n.
Scientific notation form *(157):* A number greater than or equal to 1 and less than 10 multiplied by some power of 10.
Trinomial *(161):* A three-termed polynomial.

IMPORTANT FACTS

Rules of Exponents

1. $x^m x^n = x^{m+n}$ product rule
2. $\dfrac{x^m}{x^n} = x^{m-n}$ quotient rule
3. $x^0 = 1, x \neq 0$ zero exponent rule
4. $(x^m)^n = x^{mn}$ power rule
5. $\left(\dfrac{ax}{by}\right)^m = \dfrac{a^m x^m}{b^m y^m}, b \neq 0, y \neq 0$ expanded power rule
6. $x^{-m} = \dfrac{1}{x^m}, x \neq 0$ negative exponent rule

Product of sum and difference of two quantities (also called the **difference of two squares**):
$$(a + b)(a - b) = a^2 - b^2$$

FOIL   Method to Multiply Two Binomials
(**First, Outer, Inner, Last**)

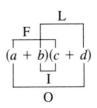

Square of a binomial:
$$(a + b)^2 = a^2 + 2ab + b^2$$
$$(a - b)^2 = a^2 - 2ab + b^2$$

Rate formula: Amount = rate · time
Distance formula: Distance = rate · time

Review Exercises

[4.1] *Simplify each of the following expressions.*

1. $x^4 \cdot x^2$

2. $x^3 \cdot x^5$

3. $3^2 \cdot 3^3$

4. $2^4 \cdot 2$

5. $\dfrac{x^4}{x}$

6. $\dfrac{x^6}{x^6}$

7. $\dfrac{3^5}{3^3}$

8. $\dfrac{4^5}{4^3}$

9. $\dfrac{x^6}{x^8}$

10. $\dfrac{x^7}{x^2}$

11. x^0

12. $3x^0$

13. $(3x)^0$

14. 4^0

15. $(2x)^2$

16. $(3x)^3$

17. $(-2x)^2$

18. $(-3x)^3$

19. $(2x^2)^4$

20. $(-x^4)^3$

21. $(-x^3)^4$

22. $\left(\dfrac{2x^3}{y}\right)^2$

23. $\left(\dfrac{3x^4}{2y}\right)^3$

24. $6x^2 \cdot 4x^3$

25. $\dfrac{16x^2y}{4xy^2}$

26. $(2x^2y)^2 \cdot 3x$

27. $\left(\dfrac{9x^2y}{3xy}\right)^2$

28. $(2x^2y)^3(3xy^4)$

29. $4x^2y^3(2x^3y^4)^2$

30. $6x^4(2x^3y^4)^2$

31. $\left(\dfrac{8x^4y^3}{2xy^5}\right)^2$

32. $\left(\dfrac{5x^4y^7}{10xy^{10}}\right)^3$

[4.2] *Simplify each of the following expressions.*

33. x^{-3}

34. y^{-7}

35. 5^{-2}

36. $\dfrac{1}{x^{-3}}$

37. $\dfrac{1}{x^{-7}}$

38. $\dfrac{1}{3^{-2}}$

39. $x^3 \cdot x^{-5}$

40. $x^{-2} \cdot x^{-3}$

41. $x^4 \cdot x^{-7}$

42. $x^{13} \cdot x^{-5}$

43. $\dfrac{x^2}{x^{-3}}$

44. $\dfrac{x^5}{x^{-2}}$

45. $\dfrac{x^{-3}}{x^3}$

46. $(3x^4)^{-2}$

47. $(4x^{-3}y)^{-3}$

48. $(-5x^{-2})^3$

49. $-2x^3 \cdot 4x^5$

50. $(3x^{-2}y)^3$

51. $(4x^{-2}y^3)^{-2}$

52. $2x(3x^{-2})$

53. $(5x^{-2}y)(2x^4y)$

54. $4x^5(6x^{-7}y^2)$

55. $2x^{-4}(3x^{-2}y^{-1})$

56. $\dfrac{6xy^4}{2xy^{-1}}$

57. $\dfrac{9x^{-2}y^3}{3xy^2}$

58. $\dfrac{25xy^{-6}}{5y^{-2}}$

59. $\dfrac{36x^4y^7}{9x^5y^{-3}}$

60. $\dfrac{4x^5y^{-2}}{8x^7y^3}$

[4.3] *Express each number in scientific notation form.*

61. 364,000

62. 1,640,000

63. 0.00763

64. 0.176

65. 2080

66. 0.000314

Express each number without exponents.

67. 4.2×10^{-3}

68. 1.65×10^4

69. 9.7×10^5

70. 4.38×10^{-6}

71. 9.14×10^{-1}

72. 5.36×10^2

Perform the indicated operation and write each answer without exponents.

73. $(2.3 \times 10^2)(2 \times 10^4)$

74. $(4.2 \times 10^{-3})(3 \times 10^5)$

75. $(6.4 \times 10^{-3})(3.1 \times 10^3)$

76. $\dfrac{6.8 \times 10^3}{2 \times 10^{-2}}$

77. $\dfrac{36 \times 10^4}{4 \times 10^6}$

78. $\dfrac{15 \times 10^{-3}}{5 \times 10^2}$

Perform the indicated operation by first converting each number to scientific notation form. Write the answer in scientific notation form.

79. $(60,000)(20,000)$

80. $(0.00004)(600,000)$

81. $(0.00023)(40,000)$

82. $\dfrac{40,000}{0.0002}$

83. $\dfrac{0.000068}{0.02}$

84. $\dfrac{1,500,000}{0.003}$

[4.4] *Indicate if each expression is a polynomial. If the polynomial has a specific name, give that name. If the polynomial is not written in descending order, rewrite it in descending order. State the degree of each polynomial.*

85. $x + 3$

86. -2

87. $x^2 - 4 + 3x$

88. $-3 - x + 4x^2$

89. $-5x^2 + 3$

90. $4x^{1/2} - 6$

91. $x - 4x^2$

92. $x^3 + x^{-2} + 3$

93. $x^3 - 2x - 6 + 4x^2$

[4.4–4.6] *Perform the operations indicated.*

94. $(x + 3) + (2x + 4)$

95. $(5x - 5) + (4x + 6)$

96. $(-3x + 4) + (5x - 9)$

97. $(4x^2 + 6x + 5) + (-6x + 9)$

98. $(-x^2 + 6x - 7) + (-2x^2 + 4x - 8)$

99. $(12x^2 + 4x - 8) + (-x^2 - 6x + 5)$

100. $(2x - 4.3) - (x + 2.4)$

101. $(-4x + 8) - (-2x + 6)$

102. $(9x^2 - \frac{3}{4}x) - (\frac{1}{2}x - 4)$

103. $(6x^2 - 6x + 1) - (12x + 5)$

104. $(-2x^2 + 8x - 7) - (3x^2 + 12)$

105. $(x^2 + 7x - 3) - (x^2 + 3x - 5)$

106. $2(x + 5)$

107. $x(2x - 4)$

108. $4.5x(x^2 - 3x)$

109. $3x(2x^2 - 4x + 7)$

110. $-x(3x^2 - 6x - 1)$

111. $-4x(-6x^2 + 4x - 2)$

112. $(x + 4)(x + 5)$

113. $(2x + 4)(x - 3)$

114. $(4x + 6)^2$

115. $(6 - 2x)(2 + 3x)$

116. $(x + 4)(x - 4)$

117. $(3x + 1)(x^2 + 2x + 4)$

118. $(x - 1)(3x^2 + 4x - 6)$

119. $(-5x + 2)(-2x^2 + 3x - 6)$

120. $\dfrac{2x + 4}{2}$

121. $\dfrac{4x - 8}{4}$

122. $\dfrac{8x^2 + 4x}{x}$

123. $\dfrac{6x^2 + 9x - 4}{3}$

124. $\dfrac{8x^2 + 6x - 4}{x}$

125. $\dfrac{8x^2 - 4x}{2x}$

126. $\dfrac{16x - 4}{-2}$

127. $\dfrac{12 + 6x}{-3}$

128. $\dfrac{5x^2 + 10x + 2}{2x}$

129. $\dfrac{x^2 + x - 12}{x - 3}$

130. $\dfrac{6x^2 - 11x + 3}{3x - 1}$

131. $\dfrac{5x^2 + 28x - 10}{x + 6}$

132. $\dfrac{4x^3 + 12x^2 + x - 12}{2x + 3}$

133. $\dfrac{4x^3 - 5x + 4}{2x - 1}$

[4.7] *Solve the following rate problems.*

134. A train travels at 70 miles per hour. How long will it take the train to travel 280 miles?

135. How fast must a plane fly to travel 3500 miles in 6.5 hours?

136. Two joggers follow the same route. Marty jogs at 8 kilometers per hour and Nick at 6 kilometers per hour. If they leave at the same time, how long will it take for them to be 4 kilometers apart?

Solve the following mixture problems.

137. Kathy Platico wishes to place part of her $12,000 into a savings account earning 8% simple interest and part into a savings account earning $7\frac{1}{4}\%$ simple interest. How much should she invest in each if she wishes her interest for the year to be $900?

138. A chemist wishes to make 2 liters of an 8% acid solution by mixing a 10% acid solution and a 5% acid solution. How many liters of each should the chemist use?

Solve the following problems

139. Marty completed the 26-mile Boston Marathon in 4 hours. Find his average rate of speed.

140. Two trains leave from the same station on parallel tracks going in opposite directions. One train travels at 50 miles per hour and the other at 60 miles per hour. How long will it take for the trains to be 440 miles apart?

141. A butcher combined hamburger that cost $3.50 per pound with hamburger that cost $4.10 per pound. How many pounds of each were used to make 80 pounds of a mixture that sells for $3.65 per pound?

142. Joan has a total of 32 stamps. Some are 29-cent stamps and some are 19-cent stamps. How many of each type does she have if the total value of her stamps is $8.08?

143. Two brothers who are 230 miles apart start driving toward each other. If the younger brother travels 5 miles per hour faster than the older brother and the brothers meet after 2 hours, find the speed traveled by each brother.

144. How many liters of a 30% acid solution must be mixed with 2 liters of a 12% acid solution to obtain a 15% acid solution?

Practice Test

Simplify each of the following expressions.

1. $2x^2 \cdot 3x^4$

2. $(3x^2)^3$

3. $\dfrac{8x^4}{2x}$

4. $\left(\dfrac{3x^2y}{6xy^3}\right)^3$

5. $(2x^3y^{-2})^{-2}$

6. $\dfrac{2x^4y^{-2}}{10x^7y^4}$

In Problems 7 through 9, determine whether each expression is a polynomial. If the polynomial has a specific name, give that name.

7. $x^2 - 4 + 6x$

8. -3

9. $x^{-2} + 4$

10. Write the polynomial $-5 + 6x^3 - 2x^2 + 5x$ in descending order, and give its degree.

Perform the operations indicated.

11. $(2x + 4) + (3x^2 - 5x - 3)$

12. $(x^2 - 4x + 7) - (3x^2 - 8x + 7)$

13. $(4x^2 - 5) - (x^2 + x - 8)$

14. $3x(4x^2 - 2x + 5)$

15. $(4x + 7)(2x - 3)$

16. $(6 - 4x)(5 + 3x)$

17. $(2x - 4)(3x^2 + 4x - 6)$

18. $\dfrac{16x^2 + 8x - 4}{4}$

19. $\dfrac{3x^2 - 6x + 5}{-3x}$

20. $\dfrac{8x^2 - 2x - 15}{2x - 3}$

21. Madison can fertilize 0.7 acre per hour on his farm. How long will it take him to fertilize a 40-acre farm?

22. Train A travels 60 miles per hour for 4 hours. If train B

is to travel the same distance in 3 hours, find the speed of train B.

23. How many liters of 20% salt solution must be added to 60 liters of 40% salt solution to get a solution that is 35% salt?

* Perform the indicated operation by first converting each number to scientific notation form. Write the answer in scientific notation form.

24. $(42,000)(30,000)$

25. $\dfrac{0.0008}{4000}$

* From optional section.

Cumulative Review Test

1. Evaluate $16 \div (4 - 6) \cdot 5$.
2. Solve the equation $2x + 5 = 3(x - 5)$.
3. Solve the equation $3(x - 2) - (x + 4) = 2x - 10$.
4. Solve the inequality $2x - 14 > 5x + 1$. Graph the solution on the number line.
5. Solve for w. $v = lwh$.

6. Solve for y. $4x - 3y = 6$.
7. Simplify $(3x^4)(2x^5)$.
8. Simplify $(3x^2y^4)^3(5x^2y)$.
9. Write the polynomial $-2x + 3x^2 - 5$ with exponents in descending order, and give its degree.
10. Subtract $6x^2 - 3x + 4$ from $2x^2 - 9x - 7$.

Perform the indicated operations.

11. $(2x^2 + 4x - 3) + (6x^2 - 7x + 12)$.
12. $(4x^2 - 5x - 2) - (3x^2 - 2x + 5)$.
13. $(2x - 3)(3x - 5)$.
14. $(2x^2 + 4x + 8)(x - 5)$.
15. $\dfrac{9x^2 - 6x + 8}{3x}$.
16. $\dfrac{2x^2 + x - 6}{x + 2}$.

17. At Graham's Grocery Store, 3 cans of soup are selling for $1.25. Find the cost of 8 cans of soup.
18. Eleven increased by twice a number is nineteen. Find the number.
19. The length of a rectangle is four more than twice the width. Find the dimensions of the rectangle if its perimeter is 26 feet.
20. Two runners start at the same point and run in opposite directions. One runs at 6 mph and the other runs at 8 mph. In how many hours will they be 28 miles apart?

5 Factoring

5.1 Factoring a Monomial from a Polynomial

5.2 Factoring by Grouping

5.3 Factoring Trinomials with $a = 1$

5.4 Factoring Trinomials with $a \neq 1$

5.5 Special Factoring Formulas and a General Review of Factoring

5.6 Solving Quadratic Equations Using Factoring

Summary

Review Exercises

Practice Test

See Section 5.3, Exercise 71.

5.1

**Factoring
a Monomial
from a Polynomial**

▶**1** Identify factors.

▶**2** Find the greatest common factor of two or more numbers.

▶**3** Find the greatest common factor of two or more terms.

▶**4** Factor a monomial from a polynomial.

▶**1** In Chapter 4 we learned how to multiply polynomials. In this chapter we focus on factoring, the reverse process of multiplication. In Section 4.5 we showed that $2x(3x^2 + 4) = 6x^3 + 8x$. In this chapter we start with an expression like $6x^3 + 8x$ and determine that its factors are $2x$ and $3x^2 + 4$, and write $6x^3 + 8x = 2x(3x^2 + 4)$. To **factor an expression** means to write the expression as a product of its factors. Factoring is important because it can be used to solve equations.

> **If $a \cdot b = c$, then a and b are said to be *factors* of c.**

$3 \cdot 5 = 15$; thus 3 and 5 are factors of 15.

$x^3 \cdot x^4 = x^7$; thus x^3 and x^4 are factors of x^7.

$x(x + 2) = x^2 + 2x$; thus x and $(x + 2)$ are factors of $x^2 + 2x$.

$(x - 1)(x + 3) = x^2 + 2x - 3$; thus $(x - 1)$ and $(x + 3)$ are factors of $x^2 + 2x - 3$.

A given number or expression may have many factors. Consider the number 30.

$$1 \cdot 30 = 30, \qquad 2 \cdot 15 = 30, \qquad 3 \cdot 10 = 30, \qquad 5 \cdot 6 = 30$$

Thus the positive factors of 30 are 1, 2, 3, 5, 6, 10, 15, and 30. Factors can also be negative. Since $(-1)(-30) = 30$, -1 and -30 are also factors of 30. In fact, for each factor a of an expression, $-a$ must also be a factor. Other factors of 30 are therefore -1, -2, -3, -5, -6, -10, -15, and -30. When asked to list the factors of an expression with a positive numerical coefficient that contains a variable, we generally list only positive factors.

EXAMPLE 1   List the factors of $6x^3$.

Solution: factors   factors

$\overbrace{1 \cdot 6x^3} = 6x^3$ $\overbrace{x \cdot 6x^2} = 6x^3$

$2 \cdot 3x^3 = 6x^3$ $2x \cdot 3x^2 = 6x^3$

$3 \cdot 2x^3 = 6x^3$ $3x \cdot 2x^2 = 6x^3$

$6 \cdot x^3 \ = 6x^3$ $6x \cdot x^2 \ = 6x^3$

The factors of $6x^3$ are 1, 2, 3, 6, x, $2x$, $3x$, $6x$, x^2, $2x^2$, $3x^2$, $6x^2$, x^3, $2x^3$, $3x^3$, and $6x^3$. The opposite (or negative) of each of these factors is also a factor, but these opposites are generally not listed unless specifically asked for.   ■

Here are examples of multiplying and factoring:

Multiplying	*Factoring*
$2(3x + 4) = 6x + 8$	$6x + 8 = 2(3x + 4)$
$5x(x + 4) = 5x^2 + 20x$	$5x^2 + 20x = 5x(x + 4)$
$(x + 1)(x + 3) = x^2 + 4x + 3$	$x^2 + 4x + 3 = (x + 1)(x + 3)$

▶ **2** To factor a monomial from a polynomial, we make use of the *greatest common factor (GCF)*. If after reading the following material you wish to see additional material on obtaining the GCF, you may read Appendix B where one of the topics discussed is finding the GCF.

Recall from Section 1.2 that the **greatest common factor** of two or more numbers is the greatest number that divides all the numbers. The greatest common factor of the numbers 6 and 8 is 2. Two is the greatest number that divides both 6 and 8. What is the GCF of 48 and 60? When the GCF of two or more numbers is not easily found, we can determine the GCF by writing each number as a product of prime numbers. A **prime number** is an integer greater than 1 that has exactly two factors, itself and one. The first 13 prime numbers are

$$2, 3, 5, 7, 11, 13, 17, 19, 23, 29, 31, 37, 41.$$

A positive integer (other than 1) that is not prime is called **composite.** The number 1 is neither prime nor composite, it is called a **unit.**

To write a number as a product of prime numbers, follow the procedure illustrated in Examples 2 and 3.

EXAMPLE 2 Write 48 as a product of prime numbers.

Solution: Select any two numbers whose product is 48. Two possibilities are $6 \cdot 8$ and $4 \cdot 12$, but there are other choices. Continue breaking down the factors until all the factors are prime, as illustrated in Fig. 5.1.

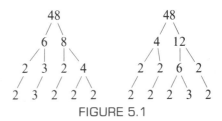

FIGURE 5.1

Note that no matter how you select your initial factors,

$$48 = 2 \cdot 2 \cdot 2 \cdot 2 \cdot 3 = 2^4 \cdot 3$$

■

EXAMPLE 3 Write 60 as a product of prime numbers.

Solution: One breakdown is shown in Fig. 5.2.

Therefore, $60 = 2 \cdot 2 \cdot 3 \cdot 5 = 2^2 \cdot 3 \cdot 5.$ FIGURE 5.2

■

To Determine the GCF of Two or More Numbers

1. Write each number as a product of prime numbers.
2. Determine the prime factors common to all the numbers.
3. Multiply the common factors found in step 2. The product of these factors will be the GCF.

EXAMPLE 4   Determine the greatest common factor of 48 and 60.

Solution: In Examples 1 and 2 we found that

$$48 = 2 \cdot 2 \cdot 2 \cdot 2 \cdot 3 = 2^4 \cdot 3$$
$$60 = 2 \cdot 2 \cdot 3 \cdot 5 = 2^2 \cdot 3 \cdot 5$$

There are two factors of 2 and one factor of 3 common to both numbers. The product of these factors is the GCF of 48 and 60:

$$\text{GCF} = 2 \cdot 2 \cdot 3 = 12$$

The GCF of 48 and 60 is 12. Twelve is the greatest number that divides both 48 and 60. ∎

EXAMPLE 5   Find the GCF of 18 and 24.

Solution: $18 = 2 \cdot 3 \cdot 3 = 2 \cdot 3^2$
$24 = 2 \cdot 2 \cdot 2 \cdot 3 = 2^3 \cdot 3$

One factor of 2 and one factor of 3 are common to both numbers.

$$\text{GCF} = 2 \cdot 3 = 6$$ ∎

▶ **3**   The GCF of a collection of terms containing variables is easily found. Consider the terms x^3, x^4, x^5, and x^6. The GCF of these terms is x^3, since x^3 is the highest power of x that divides all four terms. We can illustrate this by writing the terms in factored form.

$$x^3 = x^3 \cdot 1$$
$$x^4 = x^3 \cdot x$$
$$x^5 = x^3 \cdot x^2$$
$$x^6 = x^3 \cdot x^3$$

GCF of all four terms

EXAMPLE 6   Find the GCF of the terms p^6, p^2, p^7, and p^4.

Solution: The GCF is p^2 since p^2 is the highest power of p that divides each term. ∎

EXAMPLE 7   Find the GCF of the terms x^2y^3, x^3y^2, and xy^4.

Solution: The highest power of x common to all three terms is x^1 or x. The highest power of y common to all three terms is y^2. So the GCF of the three terms is xy^2. ∎

Greatest Common Factor of Two or More Terms

To find the GCF of two or more terms, take each factor the *fewest* number of times that it appears in any of the terms.

EXAMPLE 8   Find the GCF of the terms xy, x^2y^2, and x^3.

Solution: The GCF is x. The smallest power of x that appears in any of the terms is x. Since the x^3 term does not contain a power of y, the GCF does not contain a y. ∎

EXAMPLE 9 Find the GCF of each pair of terms.

(a) $x(x + 3)$ and $2(x + 3)$ (b) $x(x - 2)$ and $x - 2$
(c) $2(x + y)$ and $3x(x + y)$

Solution: (a) The GCF is $(x + 3)$.
(b) $x - 2$ can be written as $1(x - 2)$. The GCF of $x(x - 2)$ and $1(x - 2)$ is therefore $x - 2$.
(c) The GCF is $(x + y)$. ∎

EXAMPLE 10 Find the GCF of each set of terms.

(a) $18y^2$, $15y^3$, $21y^5$ (b) $-20x^2$, $8x$, $40x^3$ (c) $4x^2$, x^2, x^3

Solution: (a) The GCF of 18, 15, and 21 is 3. The GCF of the three terms is $3y^2$.
(b) The GCF of -20, 8, and 40 is 4. The GCF of the three terms is $4x$.
(c) The GCF is x^2. ∎

▶ **4**

To Factor a Monomial from a Polynomial
1. Determine the greatest common factor of all terms in the polynomial.
2. Write each term as the product of the GCF and its other factor.
3. Use the distributive property to factor out the GCF.

EXAMPLE 11 Factor $6x + 12$.

Solution: The GCF is 6.

$$6x + 12 = 6 \cdot x + 6 \cdot 2 \qquad \text{Write each term as a product of the GCF and some other factor.}$$

$$= 6(x + 2) \qquad \text{Distributive property}$$

To check the factoring process, multiply the factors using the distributive property.

Check: $6(x + 2) = 6x + 12$ ∎

EXAMPLE 12 Factor $8x - 10$.

Solution: The GCF of $8x$ and -10 is 2.

$$8x - 10 = 2 \cdot 4x - 2 \cdot 5$$
$$= 2(4x - 5)$$ ∎

EXAMPLE 13 Factor $6y^2 + 9y^5$.

Solution: The GCF is $3y^2$.

$$6y^2 + 9y^5 = 3y^2 \cdot 2 + 3y^2 \cdot 3y^3$$
$$= 3y^2(2 + 3y^3)$$ ∎

EXAMPLE 14　Factor $8x^3 + 12x^2 - 16x$.

Solution:　The GCF is $4x$.

$$8x^3 + 12x^2 - 16x = 4x \cdot 2x^2 + 4x \cdot 3x - 4x \cdot 4$$
$$= 4x(2x^2 + 3x - 4)$$

Check: $4x(2x^2 + 3x - 4) = 8x^3 + 12x^2 - 16x$ ∎

EXAMPLE 15　Factor $45x^2 - 30x + 5$.

Solution:　The GCF is 5.

$$45x^2 - 30x + 5 = 5 \cdot 9x^2 - 5 \cdot 6x + 5 \cdot 1$$
$$= 5(9x^2 - 6x + 1)$$ ∎

EXAMPLE 16　Factor $4x^3 + x^2 + 8x^2 y$.

Solution:　The GCF is x^2.

$$4x^3 + x^2 + 8x^2 y = x^2 \cdot 4x + x^2 \cdot 1 + x^2 \cdot 8y$$
$$= x^2(4x + 1 + 8y)$$ ∎

Notice in Examples 15 and 16 that, when one of the terms is itself the GCF, we express it in factored form as the product of the term itself and 1.

EXAMPLE 17　Factor $3x(2x - 1) + 4(2x - 1)$.

Solution:　The GCF is $(2x - 1)$. Factoring out the GCF gives

$$3x(2x - 1) + 4(2x - 1) = (3x + 4)(2x - 1)$$

The answer may also be expressed as $(2x - 1)(3x + 4)$ by the commutative property of multiplication. ∎

EXAMPLE 18　Factor $x(x + 3) - 5(x + 3)$.

Solution:　The GCF is $(x + 3)$. Factoring out the GCF gives

$$x(x + 3) - 5(x + 3) = (x - 5)(x + 3)$$ ∎

EXAMPLE 19　Factor $3x^2 + 5xy + 7y^2$.

Solution:　The only factor common to all three terms is 1. Whenever the only factor common to all the terms in a polynomial is 1, the polynomial cannot be factored by the method presented in this section. ∎

Whenever you are factoring a polynomial by any of the methods presented in this chapter, the first step will always be to see if there is a common factor (other than 1) to all the terms in the polynomial. If so, factor the greatest common factor from each term using the distributive property.

HELPFUL HINT **Checking a Factoring Problem**

A factoring problem may be checked by multiplying the factors together. The product of the factors should be identical to the expression that was originally factored. You should check all factoring problems.

Exercise Set 5.1

Write each number as a product of prime numbers.

1. 40
2. 70
3. 90
4. 180
5. 200
6. 96

Find the greatest common factor for the two given numbers.

7. 36, 20
8. 45, 27
9. 60, 84
10. 120, 96
11. 72, 90
12. 76, 68

Determine the greatest common factor for each set of terms.

13. x^2, x, x^3
14. x^2, x^5, x^7
15. $3x$, $6x^2$, $9x^3$
16. $6p$, $4p^2$, $8p^3$
17. x, y, z
18. xy, x, x^2
19. xy, xy^2, xy^3
20. x^2y, x^3y, $4x$
21. x^3y^7, x^7y^{12}, x^5y^5
22. $4x^3$, $12x$, $8y$
23. -8, $24x$, $48x^2$
24. $24p^3$, $18p^2q$, $9q^2$
25. $12x^4y^7$, $6x^3y^9$, $9x^{12}y^7$
26. $14x^8y^2$, $16x^7y^7$, $12xy^5$
27. $36x^2y$, $15x^3$, $14x^2y$
28. $-8x^2$, $-9x^3$, $12xy^3$
29. $2(x + 3)$, $3(x + 3)$
30. $4(x - 5)$, $3x(x - 5)$
31. $x^2(2x - 3)$, $5(2x - 3)$
32. $x(2x + 5)$, $2x + 5$
33. $3x - 4$, $y(3x - 4)$
34. $x(x + 5)$, $x + 5$

Factor the GCF from each term in the expression. If an expression cannot be factored, so state.

35. $2x + 4$
36. $4x + 2$
37. $15x - 5$
38. $12x + 15$
39. $13x + 5$
40. $6x^2 + 3x$
41. $16x^2 - 12x$
42. $27y - 9y^2$
43. $20p - 18p^2$
44. $9x + 18x^2$
45. $6x^3 - 8x$
46. $7x^5 - 9x^4$
47. $36x^{12} - 24x^8$
48. $45y^{12} + 30y^{10}$
49. $24y^{15} - 9y^3$
50. $38x^4 - 16x^5$
51. $x + 3xy^2$
52. $2x^2y - 6x$

53. $6x + 5y$
54. $3x^2y + 6x^2y^2$
55. $16xy^2z + 4x^3y$
56. $80x^5y^3z^4 - 36x^2yz^3$
57. $34x^2y^2 + 16xy^4$
58. $42xy^6z^{12} - 18y^4z^2$
59. $36xy^2z^3 + 36x^3y^2z$
60. $19x^4y^{12}z^{13} - 8x^5y^3z^9$
61. $14y^3z^5 - 9xy^3z^5$
62. $7x^4y^9 - 21x^3y^7z^5$
63. $3x^2 + 6x + 9$
64. $x^3 + 6x^2 - 4x$
65. $9x^2 + 18x + 3$
66. $4x^2 - 16x + 24$
67. $3x^3 - 6x^2 + 12x$
68. $15x^2 + 9x - 9$
69. $45x^2 - 16x + 10$
70. $5x^3 - 6x^2 + x$
71. $15p^2 - 6p + 9$
72. $35y^3 - 7y^2 + 14y$
73. $24x^6 + 8x^4 - 4x^3$
74. $44x^5y + 11x^3y + 22x^2$
75. $48x^2y + 16xy^2 + 33xy$
76. $52x^2y^2 + 16xy^3 + 26z$

77. $x(x + 2) + 3(x + 2)$

78. $5x(2x - 5) + 3(2x - 5)$

79. $7x(4x - 3) - 4(4x - 3)$

80. $3x(7x + 1) - 2(7x + 1)$

81. $4x(2x + 1) + 1(2x + 1)$

82. $3x(4x - 5) + 1(4x - 5)$

83. $4x(2x + 1) + 2x + 1$

84. $3x(4x - 5) + 4x - 5$

85. What is a factored expression?

86. What is the greatest common factor of two or more numbers?

87. In your own words explain how to factor a monomial from a polynomial.

88. How may any factoring problem be checked?

Cumulative Review Exercises

[2.1] **89.** Simplify $3x - (x - 6) + 4(3 - x)$.

[2.5] **90.** Solve the equation $2(x + 3) - x = 5x + 2$.

[3.1] **91.** If $A = P(1 + rt)$, find r when $A = 1000$, $t = 2$, and $P = 500$.

92. Solve the formula $A = \frac{1}{2}bh$ for h.

JUST FOR FUN

1. Factor $4x^2(x - 3)^3 - 6x(x - 3)^2 + 4(x - 3)$.

2. Factor $6x^5(2x + 7) + 4x^3(2x + 7) - 2x^2(2x + 7)$.

3. Consider the expression

$$1 + 2 - 3 + 4 + 5 - 6 + 7 + 8 - 9$$
$$+ 10 + 11 - 12 + 13 + 14 - 15$$

(a) Construct groups of three terms (for example, the first group is $1 + 2 - 3$), and write the sum of each

group as a product of 3 and another factor [for example, the first group would be 3(0)].

(b) Factor out the common factor of 3.

(c) Find the sum of the numbers.

(d) Use the procedure above to find the sum of the numbers if the process above was continued until
$. . . + 31 + 32 - 33$.

5.2

Factoring by Grouping

▶ **1** Factor a polynomial containing four terms by grouping.

▶ **1** It may be possible to factor a polynomial containing four or more terms by removing common factors from groups of terms. This process is called *factoring by grouping*. In Section 5.4 we discuss factoring trinomials. One of the methods we will use requires a knowledge of factoring by grouping. Example 1 illustrates the procedure for factoring by grouping.

EXAMPLE 1 Factor $ax + ay + bx + by$.

Solution: There is no factor (other than 1) common to all four terms. However, a is common to the first two terms and b is common to the last two terms. Factor a from the first two terms and b from the last two terms.

$$ax + ay + bx + by = a(x + y) + b(x + y)$$

Now $(x + y)$ is common to both terms. Factor out $(x + y)$.

$$a(x + y) + b(x + y) = (a + b)(x + y)$$

Thus $ax + ay + bx + by = (a + b)(x + y)$. ∎

> **To Factor a Four-term Polynomial Using Grouping**
>
> 1. Determine if there are any factors common to all four terms. If so, factor the greatest common factor from each of the four terms.
> 2. If necessary, arrange the four terms so that the first two terms have a common factor and the last two have a common factor.
> 3. Use the distributive property to factor each group of two terms.
> 4. Factor the greatest common factor from the results of step 3.

EXAMPLE 2   Factor $x^2 + 3x + 4x + 12$ by grouping.

Solution: Factor an x from the first two terms and a 4 from the last two terms.

$$x^2 + 3x + 4x + 12 = x(x + 3) + 4(x + 3)$$

Now factor the common factor $(x + 3)$.

$$x(x + 3) + 4(x + 3) = (x + 4)(x + 3)$$

Thus $x^2 + 3x + 4x + 12 = (x + 4)(x + 3)$.

EXAMPLE 3   Factor $6x^2 + 9x + 8x + 12$ by grouping.

Solution: $6x^2 + 9x + 8x + 12 = 3x(2x + 3) + 4(2x + 3)$
$$= (3x + 4)(2x + 3)$$

A factoring by grouping problem can be checked by multiplying the factors using the FOIL method. If you have not made a mistake your result will be the polynomial you began with. Following is a check of Example 3.

$$(3x + 4)(2x + 3) = (3x)(2x) + (3x)(3) + 4(2x) + 4(3)$$
$$= 6x^2 + 9x + 8x + 12$$

Note that this is the polynomial we started with. Therefore, the factoring is correct.

EXAMPLE 4   Factor $6x^2 + 8x + 9x + 12$ by grouping.

Solution: $6x^2 + 8x + 9x + 12 = 2x(3x + 4) + 3(3x + 4)$
$$= (2x + 3)(3x + 4)$$

Notice that Example 4 is the same as Example 3 with the two middle terms switched. The answers to Examples 3 and 4 are the same. When factoring by grouping, the two like terms may be switched and the answer will remain the same.

EXAMPLE 5   Factor $x^2 + 3x + x + 3$ by grouping.

Solution: x is the common factor of the first two terms. Is there a common factor of the last two terms? Yes; remember that 1 is a factor of every term. Factor a 1 from the last two terms.

$$x^2 + 3x + x + 3 = x^2 + 3x + 1 \cdot x + 1 \cdot 3$$
$$= x(x + 3) + 1(x + 3)$$
$$= (x + 1)(x + 3)$$

Note that $x + 3$ was expressed as $1(x + 3)$.

EXAMPLE 6   Factor $4x^2 - 2x - 2x + 1$ by grouping.

Solution: When $2x$ is factored from the first two terms, we get

$$4x^2 - 2x - 2x + 1 = 2x(2x - 1) - 2x + 1$$

What should we factor from the last two terms? We wish to factor $-2x + 1$ in such a manner that we end up with an expression that is a multiple of $(2x - 1)$. **Whenever we wish to change the sign of *each* term of an expression, we can factor out a negative number from each term.** In this case we factor out a negative 1.

$$-2x + 1 = -1(2x - 1)$$

Now rewrite $-2x + 1$ as $-1(2x - 1)$.

$$2x(2x - 1)\ -2x + 1 = 2x(2x - 1)\ -1(2x - 1)$$

Now factor out the common factor $(2x - 1)$.

$$2x(2x - 1) - 1(2x - 1) = (2x - 1)(2x - 1) \quad \text{or} \quad (2x - 1)^2 \qquad ■$$

EXAMPLE 7   Factor $x^2 + 3x - x - 3$ by grouping.

Solution: $x^2 + 3x - x - 3 = x(x + 3) - x - 3$
$$= x(x + 3) - 1(x + 3)$$
$$= (x - 1)(x + 3)$$

Note: $-x - 3 = -1(x + 3)$.   ■

EXAMPLE 8   Factor $3x^2 - 6x - 4x + 8$ by grouping.

Solution: $3x^2 - 6x - 4x + 8 = 3x(x - 2) - 4(x - 2)$
$$= (3x - 4)(x - 2)$$

Note: $-4x + 8 = -4(x - 2)$.   ■

HELPFUL HINT	When factoring four terms by grouping, if the coefficient of the third term is positive, as in Examples 2 through 5, you will generally factor out a positive coefficient from the last two terms. If the coefficient of the third term is negative, as in Examples 6 through 8, you will generally factor out a negative coefficient from the last two terms.

In the examples illustrated so far, the two middle terms have been like terms. This need not be the case, as illustrated in Example 9.

EXAMPLE 9 Factor by grouping $xy + 3x - 2y - 6$.

Solution: This problem contains two variables, x and y. The procedure to factor here is basically the same as before. Factor an x from the first two terms and a -2 from the last two terms.

$$yx + 3x - 2y - 6 = x(y + 3) - 2(y + 3)$$
$$= (x - 2)(y + 3)$$

EXAMPLE 10 Factor $2x^2 + 4xy + 3xy + 6y^2$.

Solution: We will factor out a $2x$ from the first two terms and a $3y$ from the last two terms.

$$2x^2 + 4xy + 3xy + 6y^2 = 2x(x + 2y) + 3y(x + 2y)$$

Now factor out the common factor $(x + 2y)$.

$$2x(x + 2y) + 3y(x + 2y) = (2x + 3y)(x + 2y)$$

If Example 10 were given as $2x^2 + 3xy + 4xy + 6y^2$, would the results be the same? Try it and see.

EXAMPLE 11 Factor $6r^2 - 9rs + 8rs - 12s^2$.

Solution: $6r^2 - 9rs + 8rs - 12s^2 = 3r(2r - 3s) + 4s(2r - 3s)$
$$= (3r + 4s)(2r - 3s)$$

EXAMPLE 12 Factor $3x^2 - 15x + 6x - 30$.

Solution: The first step in any factoring problem is to determine if all the terms have a common factor. If so, we factor out that common factor. In this polynomial, a 3 is common to every term. We therefore begin by factoring out the 3.

$$3x^2 - 15x + 6x - 30 = 3(x^2 - 5x + 2x - 10)$$

Now we factor the remaining expression by grouping.

$$3(x^2 - 5x + 2x - 10) = 3[x(x - 5) + 2(x - 5)]$$
$$= 3[(x + 2)(x - 5)]$$
$$= 3(x + 2)(x - 5)$$

Thus $3x^2 - 15x + 6x - 30 = 3(x + 2)(x - 5)$.

Exercise Set 5.2

Factor by grouping.

1. $x^2 + 4x + 3x + 12$

2. $x^2 + 5x + 2x + 10$

3. $x^2 + 2x + 5x + 10$

4. $x^2 - 2x + 3x - 6$

5. $x^2 + 3x + 2x + 6$

6. $x^2 + 2x + 3x + 6$

7. $x^2 + 3x - 5x - 15$

8. $x^2 + 3x - 2x - 6$

9. $4x^2 + 6x - 6x - 9$

10. $4x^2 - 6x + 6x - 9$

11. $3x^2 + 9x + x + 3$

12. $x^2 + 4x + x + 4$

13. $4x^2 - 2x - 2x + 1$

14. $2x^2 + 6x - x - 3$

15. $8x^2 + 32x + x + 4$

16. $8x^2 - 4x - 2x + 1$

17. $3x^2 - 2x + 3x - 2$

18. $35x^2 + 21x - 40x - 24$

19. $3x^2 - 2x - 3x + 2$

20. $35x^2 - 40x + 21x - 24$

21. $15x^2 - 18x - 20x + 24$

22. $15x^2 - 20x - 18x + 24$

23. $x^2 + 2xy - 3xy - 6y^2$

24. $x^2 - 3xy + 2xy - 6y^2$

25. $6x^2 - 9xy + 2xy - 3y^2$

26. $3x^2 - 18xy + 4xy - 24y^2$

27. $10x^2 - 12xy - 25xy + 30y^2$

28. $12x^2 - 9xy + 4xy - 3y^2$

29. $x^2 + bx + ax + ab$

30. $xy + 3x + 2y + 6$

31. $xy + 4x - 2y - 8$

32. $x^2 - 2x + ax - 2a$

33. $a^2 + 2a + ab + 2b$

34. $2x^2 - 8x + 3xy - 12y$

35. $xy - x + 5y - 5$

36. $y^2 - yb + ya - ab$

37. $6 + 4y - 3x - 2xy$

38. $3y - 6 - xy + 2x$

39. $a^3 + 2a^2 + a + 2$

40. $x^3 - 3x^2 + 2x - 6$

41. $x^3 + 4x^2 - 3x - 12$

42. $y^3 + 2y^2 - 4y - 8$

43. $2x^2 - 12x + 8x - 48$

44. $3x^2 - 3x - 3x + 3$

45. $4x^2 + 8x + 8x + 16$

46. $2x^3 - 5x^2 - 6x^2 + 15x$

47. $6x^3 + 9x^2 - 2x^2 - 3x$

48. $9x^3 + 6x^2 - 45x^2 - 30x$

49. $2x^2 - 4xy + 8xy - 16y^2$

50. $18x^2 + 27xy + 12xy + 18y^2$

✎ 51. What is the first step in any factoring by grouping problem?

✎ 52. How can you check the solution to a factoring by grouping problem?

✎ 53. A 4-term factoring by grouping problem results in the factors $(x - 2)(x + 4)$. List the expression that was factored, and explain how you determined the answer.

✎ 54. A 4-term factoring by grouping problem results in the factors $(x - 2y)(x - 3)$. List the expression that was factored, and explain how you determined the answer.

Cumulative Review Exercises

[3.1] **55.** The diameter of a willow tree grows at an average rate of 3.5 inches per year. What is the approximate age of a willow tree that has a diameter of 25 inches?

[3.3] **56.** Steve takes the Transit Authority bus to and from work each day. The one-way fare is 85 cents. The Transit Authority offers a monthly bus pass for $30, which provides unlimited bus transportation. How many days would Steve have to travel to and from work in a given month to make it worthwhile for him to purchase the monthly bus pass?

[4.6] **57.** Divide $\dfrac{15x^3 - 6x^2 - 9x + 5}{3x}$

58. Divide $\dfrac{x^2 - 9}{x - 3}$

JUST FOR FUN

Factor by grouping.

1. $3x^5 - 15x^3 + 2x^3 - 10x$

2. $x^3 + xy - x^2y - y^2$

3. $18a^2 + 3ax^2 - 6ax - x^3$

5.3

Factoring Trinomials with $a = 1$

▶**1**   Factor trinomials of the form $ax^2 + bx + c$ where $a = 1$.

▶**2**   Understand the trial and error method of factoring trinomials.

▶**3**   Remove a common factor from the trinomial before factoring a trinomial.

AN IMPORTANT NOTE REGARDING FACTORING TRINOMIALS

Factoring trinomials is very important in algebra, higher-level mathematics, physics, and other science courses. Because it is so important, you need to study and learn Sections 5.3 and 5.4 well. *To be successful in Chapter 6, Rational Expressions and Equations, you must be able to factor trinomials. You should study and learn Sections 5.3 and 5.4 so well that when you need to factor trinomials in Chapter 6 you will not have to keep looking back to this chapter to remember how to factor them.*

In this section we will learn to factor trinomials of the form $ax^2 + bx + c$, where a, the numerical coefficient of the squared term, is 1. That is, we will be factoring trinomials of the form $x^2 + bx + c$. One example of this type of trinomial is $x^2 + 5x + 6$. Note that x^2 means $1x^2$.

In Section 5.4, we will learn to factor trinomials of the form $ax^2 + bx + c$, where $a \neq 1$. One example of this type of trinomial is $2x^2 + 7x + 3$.

▶**1**   Now we will discuss how to factor trinomials of the form $ax^2 + bx + c$, where a, the numerical coefficient of the squared term, is equal to 1. Examples of such trinomials are

$$x^2 + 7x + 12 \qquad\qquad x^2 - 2x - 24$$
$$a = 1, b = 7, c = 12 \qquad a = 1, b = -2, c = -24$$

Recall that factoring is the reverse process of multiplication. We can show with the FOIL method that

$$(x + 3)(x + 4) = x^2 + 7x + 12 \quad \text{and} \quad (x - 6)(x + 4) = x^2 - 2x - 24$$

Therefore, $x^2 + 7x + 12$ and $x^2 - 2x - 24$ factor as follows:

$$x^2 + 7x + 12 = (x + 3)(x + 4) \quad \text{and} \quad x^2 - 2x - 24 = (x - 6)(x + 4)$$

Notice that each of these trinomials when factored results in the product of two binomials in which the first term of each binomial is x and the second term is a number (including its sign). In general, when we factor a trinomial of the form $x^2 + bx + c$ we will get a pair of binomial factors as follows:

$$x^2 + bx + c = (x + \blacksquare)(x + \blacksquare)$$

numbers
go here

If, for example, we find that the numbers that go in the shaded areas of the factors are 4 and -6, the factors are written $(x + 4)$ and $(x - 6)$. Notice that instead of listing the second factor as $(x + (-6))$, we list it as $(x - 6)$.

The procedure that follows is one method to determine the numbers to place in the shaded area when factoring a given trinomial of the form $x^2 + bx + c$.

In Section 4.5 we illustrated how the FOIL method is used to multiply two binomials. Let us multiply $(x + 3)(x + 4)$ using the FOIL method.

$$\begin{array}{l} \overset{\text{F}}{(x} + 3)\overset{\text{L}}{(x} + 4) = x^2 + \underbrace{4x + 3x}_{} + 12 \\ \qquad\qquad\quad = x^2 + 7x + 12 \end{array}$$

We see that $(x + 3)(x + 4) = x^2 + 7x + 12$.

Note that the **sum** of the outer and inner terms **is 7x** and the **product** of the last term **is 12.** To factor $x^2 + 7x + 12$, we look for two numbers whose product is 12 and whose sum is 7.

Factors of 12	Sum of Factors
$(1)(12) = 12$	$1 + 12 = 13$
$(2)(6) = 12$	$2 + 6 = 8$
$(3)(4) = 12$	$3 + 4 = 7$
$(-1)(-12) = 12$	$-1 + (-12) = -13$
$(-2)(-6) = 12$	$-2 + (-6) = -8$
$(-3)(-4) = 12$	$-3 + (-4) = -7$

The only factors of 12 whose sum is a positive 7 are 3 and 4. The factors of $x^2 + 7x + 12$ will therefore be $(x + 3)$ and $(x + 4)$.

$$x^2 + 7x + 12 = (x + 3)(x + 4)$$

In the previous illustration all the possible factors of 12 were listed so that you could see them. However, when working a problem, once you find the specific factors you are seeking you need go no further.

To Factor Trinomials of the Form $ax^2 + bx + c$, Where $a = 1$

1. Determine two numbers whose product equals the constant, c, and whose sum equals b.
2. Use the two numbers found in step 1, including their signs, to write the trinomial in factored form. The trinomial in factored form will be

$$(x + \text{one number})(x + \text{second number})$$

How do we determine the two numbers mentioned in steps 1 and 2? The sign of the constant, c, is a key in finding the two numbers. *The Helpful Hint that follows is very important and useful. Read it carefully.*

HELPFUL HINT

When asked to factor a trinomial of the form $x^2 + bx + c$, the first thing you should do is to observe the sign of the constant.

(a) **If the constant, c, is positive, then both numbers in the factors will have the same sign, either both positive or both negative. Furthermore, that common sign will be the same as b, the sign of the coefficient of the x term of the trinomial being factored.**

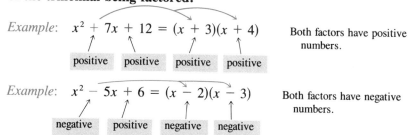

Example: $x^2 + 7x + 12 = (x + 3)(x + 4)$ Both factors have positive
                                                   numbers.

              positive   positive   positive   positive

Example: $x^2 - 5x + 6 = (x - 2)(x - 3)$ Both factors have negative
                                                   numbers.

              negative   positive   negative   negative

(b) **If the constant is negative, then the two numbers will have opposite signs. That is, one number will be positive and the other number will be negative.**

Example: $x^2 + x - 6 = (x + 3)(x - 2)$

                            negative   positive   negative

Example: $x^2 - 2x - 8 = (x + 2)(x - 4)$

                            negative   positive   negative

We will use this information as a starting point when factoring trinomials.

EXAMPLE 1   Complete the last column of the chart for parts (a) through (d).

	Sign of Coefficient of x Term	*Sign of Constant of Trinomial*	*Signs of Numbers in Factors*
(a)	−	+	
(b)	−	−	
(c)	+	−	
(d)	+	+	

Solution: (a) Since the constant is positive, both numbers must have the same sign. Since the coefficient of the x term is negative, both factors will contain negative numbers.
(b) Since the constant is negative, one factor will contain a positive number and the other will contain a negative number.
(c) Since the constant is negative, one factor will contain a positive number and the other will contain a negative number.
(d) Since the constant is positive, both numbers must have the same sign. Since the coefficient of the x term is positive, both factors will contain positive numbers.   ■

EXAMPLE 2   Factor $x^2 + x - 6$.

Solution: We must find two numbers whose product is the constant, -6, and whose sum is the coefficient of the x term, 1. Remember that x means $1x$. Since the constant is negative, one number must be positive and the other negative. Recall that the product of two numbers with unlike signs is a negative number. We now list the factors of -6 and look for the two factors whose sum is 1.

Factors of -6	Sum of Factors
$1(-6) = -6$	$1 + (-6) = -5$
$2(-3) = -6$	$2 + (-3) = -1$
$3(-2) = -6$	$3 + (-2) = 1$
$6(-1) = -6$	$6 + (-1) = 5$

Note that the factors 1 and -6 are different from the factors -1 and 6.

The numbers 3 and -2 have a product of -6 and a sum of 1. Thus the factors are $(x + 3)$ and $(x - 2)$.

$$x^2 + x - 6 = (x + 3)(x - 2)$$

The order of the factors is not crucial. Therefore, $x^2 + x - 6 = (x - 2)(x + 3)$ is also an acceptable answer.   ∎

Trinomial factoring problems can be checked by multiplying the factors using the FOIL method. If the factoring is correct, the product obtained using the FOIL method will be identical to the original trinomial. Let us check the factors obtained in Example 2.

Check: $(x + 3)(x - 2) = x^2 - 2x + 3x - 6 = x^2 + x - 6$

Since the product of the factors is the original trinomial, the factoring was correct.

EXAMPLE 3   Factor $x^2 - x - 6$.

Solution: The factors of -6 are illustrated in Example 2. The factors whose product is -6 and whose sum is -1 are 2 and -3.

Factors	Sum of Factors
$(2)(-3) = -6$	$2 + (-3) = -1$

$$x^2 - x - 6 = (x + 2)(x - 3)$$ ∎

EXAMPLE 4   Factor $x^2 - 5x + 6$.

Solution: We must find two numbers whose product is 6 and whose sum is -5. Since the constant, 6, is positive, both factors must have the same sign. Since the coefficient of the x term, -5, is negative, both numbers must be negative. Note that a negative number times a negative number gives a positive product. We now list the negative factors of 6 and look for the pair whose sum is -5.

Factors of 6	Sum of Factors
$(-1)(-6)$	$-1 + (-6) = -7$
$(-2)(-3)$	$-2 + (-3) = -5$

The factors of 6 whose sum is -5 are -2 and -3.

$$x^2 - 5x + 6 = (x - 2)(x - 3)$$ ∎

EXAMPLE 5   Factor $x^2 + 2x - 8$.

Solution: We must find the factors of -8 whose sum is 2. Since the constant is negative, one factor will be positive and the other factor will be negative.

Factors of -8	Sum of Factors
(1)(-8)	$1 + (-8) = -7$
(2)(-4)	$2 + (-4) = -2$
(4)(-2)	$4 + (-2) = 2$
(8)(-1)	

Since we have found the two numbers, 4 and -2, whose product is -8 and whose sum is 2, we need go no further.

$$x^2 + 2x - 8 = (x + 4)(x - 2)$$ ■

EXAMPLE 6   Factor $x^2 - 6x + 9$.

Solution: We must find the factors of 9 whose sum is -6. Both factors must be negative. Can you explain why? The two factors whose product is 9 and whose sum is -6 are -3 and -3.

$$x^2 - 6x + 9 = (x - 3)(x - 3)$$
$$= (x - 3)^2$$ ■

EXAMPLE 7   Factor $x^2 - 4x - 60$.

Solution: We must find two numbers whose product is -60 and whose sum is -4. Since the constant is negative, one factor must be positive and the other negative. The desired factors are -10 and 6 because $(-10)(6) = -60$ and $-10 + 6 = -4$.

$$x^2 - 4x - 60 = (x - 10)(x + 6)$$ ■

EXAMPLE 8   Factor $x^2 + 4x + 12$.

Solution: Let us first find the two numbers whose product is 12 and whose sum is 4. Since both the constant and x term are positive, the two numbers must also be positive.

Factors of 12	Sum of Factors
(1)(12)	$1 + 12 = 13$
(2)(6)	$2 + 6 = 8$
(3)(4)	$3 + 4 = 7$

Note that there are no two numbers whose product is 12 and whose sum is 4. When two numbers cannot be found to satisfy the given conditions, the trinomial cannot be factored by the method presented in this section. Therefore, we write the answer *"cannot be factored."*   ■

There is at most one pair of numbers that satisfies the two specific conditions of the problem. For example, when factoring $x^2 - 2x - 24$, the two numbers whose product is -24 and whose sum is -2 are -6 and 4. No other pair of numbers will satisfy these specific conditions. Thus the only factors of $x^2 - 2x - 24$ are $(x - 6)(x + 4)$.

A slightly different type of problem is illustrated in Example 9.

EXAMPLE 9   Factor $x^2 + 2xy + y^2$ completely.

Solution: In this problem the second term contains two variables, x and y, and the last term is not a constant. The procedure used to solve this problem is similar to that outlined previously. You should realize, however, that the product of the first terms of the factors we are looking for must be x^2, and the product of the last terms of the factors must be y^2.

We must find two numbers whose product is 1 (from $1y^2$) and whose sum is 2 (from $2xy$). The two numbers are 1 and 1. Thus

$$x^2 + 2xy + y^2 = (x + 1y)(x + 1y) = (x + y)(x + y) = (x + y)^2 \qquad \blacksquare$$

EXAMPLE 10   Factor $x^2 - xy - 6y^2$.

Solution: Determine two numbers whose product is -6 and whose sum is -1. The numbers are -3 and 2. The last terms must be $-3y$ and $2y$ to obtain the $-6y^2$.

$$x^2 - xy - 6y^2 = (x - 3y)(x + 2y) \qquad \blacksquare$$

Trial and Error Method

▶ **2**   Another method that can be used to factor trinomials of the form $x^2 + bx + c$ is called the **trial and error method.** With this method we write down factors of the form $(x + \ \ )(x + \ \ )$ and then try different sets of factors of the constant, c, in the shaded areas of the parentheses. We try each set of factors by multiplying using the FOIL method, and continue until we find the set whose sum of the products of the outer and inner terms is the same as the x term in the trinomial. For example, to factor the trinomial $x^2 - 6x - 16$, we determine the possible factors of -16. Then we try each set of factors until we obtain a set whose product from the FOIL method contains $-6x$, the same x term as in the trinomial. We now illustrate how to factor $x^2 - 6x - 16$ using trial and error. Begin by listing the possible factors of -16.

Factor $x^2 - 6x - 16$

Factors of -16	Possible Factors	Product of Factors	Is Middle Term of Product of Factors $-6x$?
$(16)(-1)$	$(x + 16)(x - 1)$	$x^2 + 15x - 16$	No
$(8)(-2)$	$(x + 8)(x - 2)$	$x^2 + 6x - 16$	No
$(4)(-4)$	$(x + 4)(x - 4)$	$x^2 - 16$	No
$(2)(-8)$	$(x + 2)(x - 8)$	$x^2 \ \ -6x \ - 16$	Yes
$(1)(-16)$	$(x + 1)(x - 16)$	$x^2 - 15x - 16$	No

We see that $x^2 - 6x - 16$ factors as follows: $x^2 - 6x - 16 = (x + 2)(x - 8)$

Once we found the correct factors we could have stopped and answered the question. All the possible factors were listed here for your benefit.

If you use this method you should still make use of the information about signs given to you earlier in this section. For example, if the constant is positive, then both numbers in the factors will have the same sign, and they will have the sign of the coefficient of the x term. If the constant is negative, then one factor will contain a positive number and the other factor will contain a negative number.

When the set of possible factors of the constant is small, you may wish to use the trial and error method. The trial and error method will be discussed further in the next section.

EXAMPLE 11   Factor $x^2 - 14x + 48$ using the trial and error method.

Solution: Since the constant, 48, is positive and the x term, $-14x$, is negative, both factors must be negative. Therefore, we will list only the negative factors of 48.

$$x^2 \; \boxed{-14x} \; + \; 48$$

Factors of 48	Possible Factors	Product of Factors	Is Middle Term of Product of Factors −14x?
$-1(-48)$	$(x - 1)(x - 48)$	$x^2 - 49x + 48$	No
$-2(-24)$	$(x - 2)(x - 24)$	$x^2 - 26x + 48$	No
$-3(-16)$	$(x - 3)(x - 16)$	$x^2 - 19x + 48$	No
$-4(-12)$	$(x - 4)(x - 12)$	$x^2 - 16x + 48$	No
$-6(-8)$	$(x - 6)(x - 8)$	$x^2 \; \boxed{-14x} \; + \; 48$	Yes

Thus $x^2 - 14x + 48 = (x - 6)(x - 8)$. ■

EXAMPLE 12   Factor $x^2 + 3xy - 18y^2$ by trial and error.

Solution: Since the last term of the trinomial is negative, one factor will contain a positive number and the other factor will contain a negative number. We try sets of factors of -18 until we find the set that gives the correct middle coefficient of 3. We may disregard the second variable, y, at this time and just work with $x^2 + 3x - 18$. Once we obtain the set of factors we are seeking we include the variable y with the factors.

$$x^2 + 3x - 18 = (x + 6)(x - 3)$$

Thus, $$x^2 + 3xy - 18y^2 = (x + 6y)(x - 3y)$$

Note that the sum of the products of the outer and inner terms of $(x + 6y)(x - 3y)$ is the $+3xy$ we are seeking. ■

▶ **3**   There will be times when each term of a trinomial has a common factor. When this occurs, factor out the common factor as explained in Section 5.1 before factoring the trinomial. **Whenever the numerical coefficient of the highest-powered term is not 1, you should check for a common factor.** After factoring out any common factor, you should factor the remaining trinomial by one of the methods presented in this section.

EXAMPLE 13   Factor $2x^2 + 2x - 12$.

Solution: Since the numerical coefficient of the squared term is not 1, we check for a common factor. Note that 2 is common to each term of the polynomial, so we factor it out.

$$2x^2 + 2x - 12 = 2(x^2 + x - 6) \qquad \text{Factor out the common factor.}$$

Now factor the remaining trinomial $x^2 + x - 6$ into $(x + 3)(x - 2)$.

$$2x^2 + 2x - 12 = 2(x + 3)(x - 2)$$

Note that the 2 that is factored out *is a part of the answer*. After the 2 has been factored out, it plays no part in the factoring of the remaining trinomial. ■

EXAMPLE 14   Factor $3x^3 + 24x^2 - 60x$.

Solution: $3x$ divides each term of the polynomial and therefore is a common factor.

$$3x^3 + 24x^2 - 60x = 3x(x^2 + 8x - 20) \qquad \text{Factor out the common factor.}$$
$$= 3x(x + 10)(x - 2) \qquad \text{Factor the remaining trinomial.} \qquad \blacksquare$$

Exercise Set 5.3

Factor each expression. If an expression cannot be factored by the method presented in this section, so state.

1. $x^2 + 7x + 10$

2. $x^2 - 8x + 12$

3. $x^2 + 5x + 6$

4. $x^2 + 7x + 6$

5. $x^2 + 7x + 12$

6. $x^2 - x - 6$

7. $x^2 - 7x + 9$

8. $y^2 - 6y + 8$

9. $y^2 - 16y + 15$

10. $x^2 + 8x - 20$

11. $x^2 + x - 6$

12. $p^2 - 3p - 10$

13. $k^2 - 2k - 15$

14. $x^2 - 6x + 10$

15. $b^2 - 11b + 18$

16. $x^2 + 11x - 30$

17. $x^2 - 8x - 15$

18. $x^2 - 10x + 21$

19. $a^2 + 12a + 11$

20. $x^2 + 16x + 64$

21. $x^2 + 13x - 30$

22. $x^2 - 30x - 64$

23. $x^2 + 4x + 4$

24. $x^2 - 4x + 4$

25. $x^2 + 6x + 9$

26. $x^2 - 6x + 9$

27. $x^2 + 10x + 25$

28. $x^2 - 10x - 25$

29. $w^2 - 18w + 45$

30. $x^2 - 11x + 10$

31. $x^2 + 22x - 48$

32. $x^2 - 2x + 8$

33. $x^2 - x - 20$

34. $x^2 - 17x - 60$

35. $y^2 - 9y + 14$

36. $x^2 + 15x + 56$

37. $x^2 + 12x - 64$

38. $x^2 - 18x + 80$

39. $x^2 - 14x + 24$

40. $x^2 - 13x + 36$

41. $x^2 - 2x - 80$

42. $x^2 + 18x + 32$

43. $x^2 - 17x + 60$

44. $x^2 - 15x - 16$

45. $x^2 + 30x + 56$

46. $x^2 - 2xy + y^2$

47. $x^2 - 4xy + 4y^2$

48. $x^2 - 6xy + 8y^2$

49. $x^2 + 8xy + 15y^2$

50. $x^2 - 5xy - 14y^2$

Factor completely.

51. $2x^2 - 14x + 12$

52. $3x^2 - 9x - 30$

53. $5x^2 + 20x + 15$

54. $4x^2 + 12x - 16$

55. $2x^2 - 14x + 24$

56. $3y^2 - 33y + 54$

57. $x^3 - 3x^2 - 18x$

58. $x^3 + 11x^2 - 42x$

59. $2x^3 + 6x^2 - 56x$

60. $3x^3 - 36x^2 + 33x$

61. $x^3 + 4x^2 + 4x$

62. $2x^3 - 12x^2 + 10x$

63. How can a trinomial factoring problem be checked?   **64.** Explain how to determine the factors when factoring a trinomial of the form $x^2 + bx + c$.

Write the trinomial whose factors are listed. Explain how you determined your answer.

65. $(x - 3)(x - 8)$

66. $(x - 3y)(x + 6y)$

67. $2(x - 5y)(x + y)$

68. $3(x + y)(x - y)$

Cumulative Review Exercises

[4.5] **69.** Multiply $(2x^2 + 5x - 6)(x - 2)$.

[4.6] **70.** Divide $3x^2 - 10x - 10$ by $x - 4$.

[4.7] **71.** Dr. Kaufman, a chemist, mixes 4 liters of an 18% acid solution with 1 liter of a 26% acid solution. Find the strength of the mixture.

[5.2] **72.** Factor by grouping: $3x^2 + 5x - 6x - 10$.

JUST FOR FUN

Factor each expression.

1. $x^2 + 0.6x + 0.08$

2. $x^2 - 0.5x - 0.06$

3. $x^2 + \frac{2}{5}x + \frac{1}{25}$

4. $x^2 - \frac{2}{3}x + \frac{1}{9}$

5.4

Factoring Trinomials with $a \neq 1$

▸**1**   Factor trinomials of the form $ax^2 + bx + c$, $a \neq 1$, by trial and error.

▸**2**   Factor trinomials of the form $ax^2 + bx + c$, $a \neq 1$, by grouping.

AN IMPORTANT NOTE TO STUDENTS

In this section we discuss two methods of factoring trinomials of the form $ax^2 + bx + c$, $a \neq 1$. That is, we will be factoring trinomials whose squared term has a numerical coefficient not equal to 1, after removing any common factors. Examples of trinomials with $a \neq 1$ are

$$2x^2 + 11x + 12 \quad (a = 2) \qquad 4x^2 - 3x + 1 \quad (a = 4)$$

The methods we will discuss are (1) **factoring by trial and error** and (2) **factoring by grouping.** We present two different methods for factoring these trinomials because some students, and some instructors, prefer one method, while others prefer the second method. You may use either method unless your instructor asks you to use a specific method. *We will use the same examples to illustrate both methods so that you can make a comparison. Each method is treated independently of the other. So if your teacher asks you to use a specific method, either factoring by grouping or factoring by trial and error, you need only read the material related to that specific method.* Factoring by trial and error is covered on pages 217–223, and factoring by grouping is covered on pages 224–228.

Method 1: Trial and Error

▸**1**   Let us now discuss factoring trinomials of the form $ax^2 + bx + c$, $a \neq 1$, by the trial and error method. The trial and error method was introduced in Section 5.3. It may be helpful for you to reread that material before going any further.

Recall that factoring is the reverse process of multiplying. Consider the product of the following two binomials:

$$
\begin{array}{cccc}
\textbf{F} & \textbf{O} & \textbf{I} & \textbf{L} \\
\end{array}
$$

$$
\begin{aligned}
(2x + 3)(x + 5) &= 2x(x) + (2x)(5) + 3(x) + 3(5) \\
&= 2x^2 + 10x + 3x + 15 \\
&= 2x^2 + 13x + 15
\end{aligned}
$$

Notice that the product of the first terms of the binomials gives the x-squared term of the trinomial, $2x^2$. Also notice that the product of the last terms of the binomials gives the last term, or constant, of the trinomial, $+15$. Finally, notice that the sum of the products of the outer terms and inner terms of the binomials gives the middle term of the trinomial, $+13x$. When we factor a trinomial using trial and error, we make use of these important facts. Note that $2x^2 + 13x + 15$ in factored form is $(2x + 3)(x + 5)$.

$$2x^2 + 13x + 15 = (2x + 3)(x + 5)$$

When factoring a trinomial of the form $ax^2 + bx + c$ by the trial and error method, you must keep in mind that when you obtain your factors the product of the first terms in the factors must equal the first term of the trinomial, ax^2. Also, the product of the constants in the factors, including their signs, must equal the constant, c, of the trinomial.

Product of constants in factors must equal c.

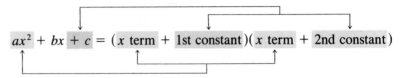

Product of x terms in factors must equal ax^2.

For example, when factoring the trinomial $2x^2 + 7x + 6$, each of the following pairs of factors has a product of the first terms equal to $2x^2$ and a product of the last terms equal to 6.

Trinomial	Possible Factors	Product of First Terms	Product of Last Terms
$2x^2 + 7x + 6$	$(2x + 1)(x + 6)$	$2x(x) = 2x^2$	$1(6) = 6$
	$(2x + 2)(x + 3)$	$2x(x) = 2x^2$	$2(3) = 6$
	$(2x + 3)(x + 2)$	$2x(x) = 2x^2$	$3(2) = 6$
	$(2x + 6)(x + 1)$	$2x(x) = 2x^2$	$6(1) = 6$

Each of these sets of factors is a possible answer, but only one has the correct factors. How do we determine which is the correct factoring of the trinomial $2x^2 + 7x + 6$? The key lies in the bx term. We know that when we multiply two binomials using the FOIL method the sum of the products of the outer and inner terms gives us the bx term of the trinomial. We use this concept in reverse to determine the correct set of factors. We need to find the set of factors whose sum of the products of the outer and inner terms is equal to the bx term of the trinomial.

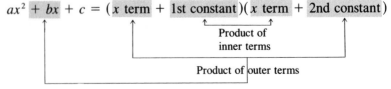

Sum of products of outer and inner terms must equal bx.

Now look at the possible sets of factors we obtained for $2x^2 + 7x + 6$ to see if any of the factors yield the correct x term, $7x$.

Trinomial	Possible Factors	Product of the First Terms	Product of the Last Terms	Sum of the Products of Outer and Inner Terms
$2x^2 + 7x + 6$	$(2x + 1)(x + 6)$	$2x^2$	6	$2x(6) + 1(x) = 13x$
	$(2x + 2)(x + 3)$	$2x^2$	6	$2x(3) + 2(x) = 8x$
	$(2x + 3)(x + 2)$	$2x^2$	6	$2x(2) + 3(x) = 7x$
	$(2x + 6)(x + 1)$	$2x^2$	6	$2x(1) + 6(x) = 8x$

Since $(2x + 3)(x + 2)$ yields the correct x term, $7x$, the trinomial $2x^2 + 7x + 6$ factors into $(2x + 3)(x + 2)$.

$$2x^2 + 7x + 6 = (2x + 3)(x + 2)$$

We can check this factoring using the FOIL method.

$$\textbf{F} \qquad \textbf{O} \qquad \textbf{I} \qquad \textbf{L}$$

Check: $(2x + 3)(x + 2) = 2x(x) + 2x(2) + 3(x) + 3(2)$
$$= 2x^2 + 4x + 3x + 6$$
$$= 2x^2 + 7x + 6$$

Since we obtained the original trinomial, our factoring is correct.

Note in the preceding illustration that $(2x + 1)(x + 6)$ are different factors than $(2x + 6)(x + 1)$, because in one case the 1 is paired with the $2x$ and in the second case the 1 is paired with the x. The factors $(2x + 1)(x + 6)$ and $(x + 6)(2x + 1)$ are, however, the same set of factors with their order reversed.

HELPFUL HINT

When factoring a trinomial of the form $ax^2 + bx + c$, remember that the sign of the constant, c, and the sign of the x term, bx, offer valuable information. When factoring a trinomial by trial and error, first check the sign of the constant. If it is positive, then the signs in both factors will be the same as the sign of the bx term. If the constant is negative, then one factor will contain a plus sign and the other a negative sign.

Now we outline the procedure to factor trinomials of the form $ax^2 + bx + c$, $a \neq 1$, by trial and error. Keep in mind that the more you practice, the better you will become at factoring.

To Factor Trinomials of the Form $ax^2 + bx + c$, $a \neq 1$, Using Trial and Error

1. Determine if there is any factor common to all three terms. If so, factor it out.
2. Write all pairs of factors of the coefficient of the squared term, a.
3. Write all pairs of factors of the constant term, c.
4. Try various combinations of these factors until the correct middle term, bx, is found.

When factoring using this procedure, if there is more than one pair of numbers whose product is a, we generally begin with the middle-sized pair. We will illustrate the procedure with Examples 1 through 8.

EXAMPLE 1   Factor $2x^2 + 11x + 12$.

Solution: We first determine that there are no common factors other than 1 to all three terms. Since the first term is $2x^2$, one factor must contain a $2x$ and the other an x. The product of the last terms in the factors must be 12. Since the constant and the coefficient of the x term are both positive, only the positive factors of 12 need be considered. We will list the positive factors of 12, the possible factors of the trinomial, and the sum of the products of the outer and inner terms. Once we find the factors of 12 that yield the proper sum of the products of the outer and inner terms, $11x$, we can write the answer.

Factors of 12	Possible Factors of Trinomial	Sum of the Products of the Outer and Inner Terms
1(12)	$(2x + 1)(x + 12)$	$25x$
2(6)	$(2x + 2)(x + 6)$	$14x$
3(4)	$(2x + 3)(x + 4)$	$11x$
4(3)	$(2x + 4)(x + 3)$	$10x$
6(2)	$(2x + 6)(x + 2)$	$10x$
12(1)	$(2x + 12)(x + 1)$	$14x$

Since the product of $(2x + 3)$ and $(x + 4)$ yields the correct bx term, $11x$, they are the correct factors.

$$2x^2 + 11x + 12 = (2x + 3)(x + 4)$$

In Example 1, our first factor could have been written with an x and the second with a $2x$. Had we done this we still would have obtained the correct answer: $(x + 4)(2x + 3)$. In Example 1, we could have stopped once we found the pair of factors that yielded the $11x$. Instead, we listed all the factors so that you could study them.

EXAMPLE 2   Factor $5x^2 - 7x - 6$.

Solution: The one factor must contain a $5x$ and the other an x. We now list the factors of -6 and look for the set of factors that yield $-7x$.

Factors of -6	Possible Factors	Sum of the Products of the Outer and Inner Terms
$-1(6)$	$(5x - 1)(x + 6)$	$29x$
$-2(3)$	$(5x - 2)(x + 3)$	$13x$
$-3(2)$	$(5x - 3)(x + 2)$	$7x$
$-6(1)$	$(5x - 6)(x + 1)$	$-x$

Since we did not obtain the desired quantity, $-7x$, by writing the negative factor with the $5x$, we will now try listing the negative factor with the x.

$1(-6)$	$(5x + 1)(x - 6)$	$-29x$
$2(-3)$	$(5x + 2)(x - 3)$	$-13x$
$3(-2)$	$(5x + 3)(x - 2)$	$-7x$
$6(-1)$	$(5x + 6)(x - 1)$	x

We see that $(5x + 3)(x - 2)$ gives the $-7x$ we are looking for. Thus

$$5x^2 - 7x - 6 = (5x + 3)(x - 2)$$

Again we listed all the possible combinations for you to study.

HELPFUL HINT

In Example 2, we were asked to factor $5x^2 - 7x - 6$. When we considered the product of $-3(2)$ in the first set of possible factors, we obtained

Factors of -6	Possible Factors	Sum of the Products of the Outer and Inner Terms
$-3(2)$	$(5x - 3)(x + 2)$	$7x$

Later in the problem we tried the factors $3(-2)$ and obtained the correct answer.

$3(-2)$	$(5x + 3)(x - 2)$	$-7x$

When factoring a trinomial with a *negative constant*, if you obtain the correct *bx* term, except the sign is the opposite of the one you are seeking, *reverse the signs on the constants* in the factors. This should give you the set of factors you are seeking.

EXAMPLE 3   Factor $8x^2 + 33x + 4$.

Solution: There are no factors common to all three terms. Since the first term is $8x^2$, there are a number of possible combinations for the first terms in the factors. Since $8 = 8 \cdot 1$ and $8 = 4 \cdot 2$, the possible factors may be of the form $(8x\ \ )(x\ \ )$ or $(4x\ \ )(2x\ \ )$. When this situation occurs, we generally start with the middle-sized pair of factors. Thus we begin with $(4x\ \ )(2x\ \ )$. If this pair does not lead to the solution, we will then try $(8x\ \ )(x\ \ )$. We now list the factors of the constant, 4. Since all signs are positive, we list only the positive factors of 4.

Factors of 4	Possible Factors	Sum of the Products of the Outer and Inner Terms
$1(4)$	$(4x + 1)(2x + 4)$	$18x$
$2(2)$	$(4x + 2)(2x + 2)$	$12x$
$4(1)$	$(4x + 4)(2x + 1)$	$12x$

Since we did not obtain the factors with $(4x\ \ )(2x\ \ )$, we now try $(8x\ \ )(x\ \ )$.

Factors of 4	Possible Factors	Sum of the Products of the Outer and Inner Terms
1(4)	$(8x + 1)(x + 4)$	$33x$
2(2)	$(8x + 2)(x + 2)$	$18x$
4(1)	$(8x + 4)(x + 1)$	$12x$

Since the product of $(8x + 1)$ and $(x + 4)$ yields the correct x term, $33x$, they are the correct factors.

$$8x^2 + 33x + 4 = (8x + 1)(x + 4)$$

■

EXAMPLE 4   Factor $4x^2 - 4x + 1$.

Solution: The factors must be of the form $(4x\quad)(x\quad)$ or $(2x\quad)(2x\quad)$. We start with the middle-sized factors $(2x\quad)(2x\quad)$. Since the constant is positive and the coefficient of the x term is negative, both factors must be negative.

Factors of 1	Possible Factors	Sum of the Products of the Outer and Inner Terms
$(-1)(-1)$	$(2x - 1)(2x - 1)$	$-4x$

Since we found the correct factors, we can stop.

$$4x^2 - 4x + 1 = (2x - 1)(2x - 1) = (2x - 1)^2$$

■

EXAMPLE 5   Factor $2x^2 + 3x + 5$.

Solution: The factors will be of the form $(2x\quad)(x\quad)$. We need only consider the positive factors of 5. Can you explain why?

Factors of 5	Possible Factors	Sum of the Products of the Outer and Inner Terms
1(5)	$(2x + 1)(x + 5)$	$11x$
5(1)	$(2x + 5)(x + 1)$	$7x$

Since we have tried all possible combinations and we have not obtained the x term, $3x$, this trinomial cannot be factored.

■

If you come across a trinomial that cannot be factored, as in Example 5, do not leave the answer blank. Instead, write "cannot be factored." However, before you write the answer "cannot be factored," recheck the problem and make sure you have tried every possible combination.

EXAMPLE 6   Factor $4x^2 + 7xy + 3y^2$.

Solution: This trinomial is different from the other trinomials in that the last term is not a constant but contains a y^2. Don't let this scare you. The factoring process is the same, except that the second term of both factors will contain a y. We begin by considering factors of the form $(2x \quad)(2x \quad)$. If we cannot find the factors, then we try factors of the form $(4x \quad)(x \quad)$.

Factors of 3	Possible Factors	Sum of the Products of the Outer and Inner Terms
1(3)	$(2x + y)(2x + 3y)$	$8xy$
3(1)	$(2x + 3y)(2x + y)$	$8xy$
1(3)	$(4x + y)(x + 3y)$	$13xy$
3(1)	$(4x + 3y)(x + y)$	$\boxed{7xy}$

$$4x^2 + 7xy + 3y^2 = (4x + 3y)(x + y)$$

Check: $(4x + 3y)(x + y) = 4x^2 + 4xy + 3xy + 3y^2 = 4x^2 + 7xy + 3y^2$ ■

EXAMPLE 7   Factor $6x^2 - 13xy - 8y^2$.

Solution: We begin by looking at factors of the form $(3x \quad)(2x \quad)$. If we cannot find the solution from factors of this form, we will try factors of the form $(6x \quad)(x \quad)$. Since the last term, $-8y^2$, is negative, one factor will contain a plus sign and the other will contain a minus sign.

Factors of -8	Possible Factors	Sum of the Products of the Outer and Inner Terms
1(−8)	$(3x + y)(2x - 8y)$	$-22xy$
2(−4)	$(3x + 2y)(2x - 4y)$	$-8xy$
4(−2)	$(3x + 4y)(2x - 2y)$	$2xy$
8(−1)	$(3x + 8y)(2x - y)$	$13xy$ ⟵

We are looking for $-13xy$. When we considered 8(−1), we obtained $13xy$. As explained in the Helpful Hint on page 221, if we reverse the signs of the numbers in the factors, we will obtain the factors we are seeking.

$$(3x + 8y)(2x - y) \qquad \text{Gives } 13xy.$$
$$(3x - 8y)(2x + y) \qquad \text{Gives } -13xy.$$

Therefore, $6x^2 - 13xy - 8y^2 = (3x - 8y)(2x + y)$.   ■

Now we will look at an example in which all three terms of the trinomial have a common factor.

EXAMPLE 8   Factor $4x^3 + 10x^2 + 6x$.

Solution: The first step in *any factoring problem* is to determine if all the terms contain a common factor. If so, factor out that common factor immediately. In this example, $2x$ is common to all three terms. We begin by factoring out the $2x$. Then we continue the factoring process using the trial and error method.

$$4x^3 + 10x^2 + 6x = 2x(2x^2 + 5x + 3)$$
$$= 2x(2x + 3)(x + 1)$$ ■

Method 2:
Factoring
by Grouping

▶ **2** We will now discuss the factoring by grouping method. The steps in the box that follow give the procedure for factoring trinomials by grouping.

To Factor Trinomials of the Form $ax^2 + bx + c$, $a \neq 1$, by Grouping

---

1. Determine if there is a factor common to all three terms. If so, factor it out.
2. Find two numbers whose product is equal to the product of a times c, and whose sum is equal to b.
3. Rewrite the bx term as the sum or difference of two terms using the factors found in step 2.
4. Factor by grouping as explained in Section 5.2.

This process will be made clear in Example 9. We will rework examples 1 through 8 here using the factoring by grouping method. Example 9 that follows is the same trinomial given in Example 1. After you study this method and try some exercises, you will gain a feel for which method you prefer using.

EXAMPLE 9 Factor $2x^2 + 11x + 12$.

Solution: First determine if there is a common factor to all the terms of the polynomial. There are no common factors (other than 1) to the three terms.

$$a = 2, \qquad b = 11, \qquad c = 12$$

1. We must find two numbers whose product is $a \cdot c$ and whose sum is b. We must therefore find two numbers whose product equals $2 \cdot 12$ or 24 and whose sum equals 11. Only the positive factors of 24 need be considered since all signs of the trinomial are positive.

Factors of 24	Sum of Factors
(1)(24)	$1 + 24 = 25$
(2)(12)	$2 + 12 = 14$
(3)(8)	$3 + 8 = 11$
(4)(6)	$4 + 6 = 10$

   The desired factors are 3 and 8.

2. Rewrite the $11x$ term as the sum or difference of two terms using the values found in step 1. Therefore, we rewrite $11x$ as $3x + 8x$.
   $$2x^2 + 11x + 12$$
   $$= 2x^2 + \overset{\frown}{3x + 8x} + 12$$

3. Now factor by grouping. Start by factoring out a common factor from the first two terms and a common factor from the last two terms. This procedure was discussed in Section 5.2.

   $$\underset{\substack{x \text{ is common} \\ \text{factor.}}}{\underbrace{2x^2 + 3x}} + \underset{\substack{4 \text{ is common} \\ \text{factor.}}}{\underbrace{8x + 12}}$$
   $$= x(2x + 3) + 4(2x + 3)$$
   $$= (x + 4)(2x + 3)$$

   ■

Note that in step 2 of Example 9 we rewrote $11x$ as $3x + 8x$. Would it have made a difference if we had written $11x$ as $8x + 3x$? Let us work it out and see.

$$2x^2 + 11x + 12$$
$$= 2x^2 + \overbrace{8x + 3x} + 12$$

$$\underbrace{2x \text{ is common}}_{\text{factor.}} \qquad \underbrace{3 \text{ is common}}_{\text{factor.}}$$

$$\overbrace{2x^2 + 8x} + \overbrace{3x + 12}$$
$$= 2x(x + 4) + 3(x + 4)$$
$$= (2x + 3)(x + 4)$$

Since $(2x + 3)(x + 4) = (x + 4)(2x + 3)$, the answers are the same. We obtained the same answer by writing the $11x$ as either $3x + 8x$ or $8x + 3x$. *In general, when rewriting the bx term of the trinomial using the specific factors found, the terms may be listed in either order.* You should, however, check after you list the two terms to make sure that the sum of the terms you listed equals the *bx* term.

EXAMPLE 10 Factor $5x^2 - 7x - 6$.

Solution: There are no common factors other than 1.

$$a = 5, \qquad b = -7, \qquad c = -6$$

The product of a times c is $5(-6) = -30$. We must find two numbers whose product is -30 and whose sum is -7.

Factors of -30	*Sum of Factors*
$(-1)(30)$	$-1 + 30 = 29$
$(-2)(15)$	$-2 + 15 = 13$
$(-3)(10)$	$-3 + 10 = 7$
$(-5)(6)$	$-5 + 6 = 1$
$(-6)(5)$	$-6 + 5 = -1$
$(-10)(3)$	$-10 + 3 = -7$
$(-15)(2)$	$-15 + 2 = -13$
$(-30)(1)$	$-30 + 1 = -29$

Rewrite the $-7x$ as $-10x + 3x$.

$$5x^2 - 7x - 6$$
$$= 5x^2 - \overbrace{10x + 3x} - 6 \qquad \text{Now factor by grouping.}$$
$$= 5x(x - 2) + 3(x - 2)$$
$$= (5x + 3)(x - 2)$$

In Example 10, we could have expressed the $-7x$ as $3x - 10x$ and obtained the same answer. Try working Example 10 by rewriting $-7x$ as $3x - 10x$.

HELPFUL HINT

Notice in Example 10 that we were looking for two factors of -30 whose sum was -7. When we considered the factors $(-3)(10)$ we obtained a sum of 7. The factors we eventually obtained that gave a sum of -7 were $(3)(-10)$. Note that, when the *constant of the trinomial is negative,* if we switch the signs of the constant in the factors, the sign of the sum of the factors changes. Thus, when trying pairs of factors to obtain the *bx* term, if you obtain the coefficient you are seeking, except that it has the opposite sign, reverse the signs in the factors. This should give you the coefficient you are seeking.

EXAMPLE 11   Factor $8x^2 + 33x + 4$.

Solution: There are no common factors other than 1. We must find two numbers whose product is $8 \cdot 4$ or 32 and whose sum is 33. The numbers are 1 and 32.

Factors of 32	Sum of Factors
(1)(32)	1 + 32 = 33

Rewrite $33x$ as $32x + x$.

$$8x^2 + 33x + 4$$
$$= 8x^2 + 32x + x + 4$$
$$= 8x(x + 4) + 1(x + 4)$$
$$= (8x + 1)(x + 4)$$ ■

EXAMPLE 12   Factor $4x^2 - 4x + 1$.

Solution: There are no common factors other than 1. We must find two numbers whose product is $4 \cdot 1$ or 4 and whose sum is -4. Since the product of a times c is positive and the b term is negative, both numerical factors must be negative.

Factors of 4	Sum of Factors
(−1)(−4)	−1 + (−4) = −5
(−2)(−2)	−2 + (−2) = −4

The desired factors are -2 and -2.

$$4x^2 - 4x + 1$$
$$= 4x^2 - 2x - 2x + 1 \qquad \text{Rewrite } -4x \text{ as } -2x - 2x.$$
$$= 2x(2x - 1) - 2x + 1$$
$$= 2x(2x - 1) - 1(2x - 1) \qquad \text{Rewrite } -2x + 1 \text{ as } -1(2x - 1).$$
$$= (2x - 1)(2x - 1) \text{ or } (2x - 1)^2$$ ■

When attempting to factor a trinomial, if there are no two integers whose product equals $a \cdot c$ and whose sum equals b, the trinomial cannot be factored by this method.

EXAMPLE 13   Factor $2x^2 + 3x + 5$.

Solution: There are no common factors other than 1. We must find two numbers whose product is 10 and whose sum is 3.

Factors of 10	*Sum of Factors*
(1)(10)	$1 + 10 = 11$
(2)(5)	$2 + 5 = 7$

Since there are no two factors of 10 whose sum is 3, we conclude that this trinomial cannot be factored by this method. ▪

EXAMPLE 14   Factor $4x^2 + 7xy + 3y^2$.

Solution: There are no common factors other than 1. This trinomial contains two variables. It is factored in basically the same manner as the previous examples. Determine two numbers whose product is $4 \cdot 3$ or 12 and whose sum is 7. The two numbers are 4 and 3.

$$4x^2 + 7xy + 3y^2$$
$$= 4x^2 + 4xy + 3xy + 3y^2$$
$$= 4x(x + y) + 3y(x + y)$$
$$= (4x + 3y)(x + y)$$ ▪

EXAMPLE 15   Factor $6x^2 - 13xy - 8y^2$.

Solution: There are no common factors other than 1. Determine two numbers whose product is $6(-8)$ or -48 and whose sum is -13. Since the product is negative, one factor must be positive and the other negative. Some factors are:

Product of Factors	*Sum of Factors*
(1)(-48)	$1 + (-48) = -47$
(2)(-24)	$2 + (-24) = -22$
(3)(-16)	$3 + (-16) = -13$

There are many other factors, but we have found the ones we are looking for. The two numbers whose product is -48 and whose sum is -13 are -16 and 3.

$$6x^2 - 13xy - 8y^2$$
$$= 6x^2 - 16xy + 3xy - 8y^2$$
$$= 2x(3x - 8y) + y(3x - 8y)$$
$$= (2x + y)(3x - 8y)$$

Check: $(2x + y)(3x - 8y)$

$$\begin{array}{cccc} \text{F} & \text{O} & \text{I} & \text{L} \\ (2x)(3x) + & (2x)(-8y) + & (y)(3x) + & (y)(-8y) \\ 6x^2 \quad - & 16xy \quad + & 3xy \quad - & 8y^2 \end{array}$$
$$6x^2 - 13xy - 8y^2$$ ▪

Remember that in any factoring problem our first step is to determine if all terms in the polynomial have a common factor other than 1. If so, we use the distributive property to factor the GCF from each term. We then continue to factor by one of the methods discussed in this chapter, if possible.

EXAMPLE 16　Factor $4x^3 + 10x^2 + 6x$.

Solution:　The factor $2x$ is common to all three terms. Factor the $2x$ from each term of the polynomial.

$$4x^3 + 10x^2 + 6x = 2x(2x^2 + 5x + 3)$$

Now continue by factoring $2x^2 + 5x + 3$. The two numbers whose product is $2 \cdot 3$ or 6 and whose sum is 5 are 2 and 3.

$$2x[2x^2 + 5x + 3]$$
$$= 2x[2x^2 + 2x + 3x + 3]$$
$$= 2x[2x(x + 1) + 3(x + 1)]$$
$$= 2x(2x + 3)(x + 1)$$ ∎

HELPFUL HINT

Which Method Should You Use to Factor a Trinomial?

If your instructor asks you to use a specific method, then you should use that method. If your instructor does not require the use of a specific method, you should use the method you feel most comfortable with. You may wish to start with the trial and error method if there are only a few possible factors to try. If you cannot find the factors by trial and error or if there are many possible factors to consider, you may wish to use the grouping procedure. With time and practice you will learn which method you feel most comfortable with and which method gives you greater success.

Exercise Set 5.4

Factor completely. If an expression cannot be factored, so state.

1. $3x^2 + 5x + 2$

2. $3x^2 + 4x + 1$

3. $6x^2 + 13x + 6$

4. $5x^2 + 13x + 6$

5. $2x^2 + 5x + 3$

6. $4x^2 + 4x - 3$

7. $2x^2 + 11x + 15$

8. $3x^2 - 2x - 8$

9. $3x^2 - 10x - 8$

10. $4x^2 - 11x + 7$

11. $5y^2 - 8y + 3$

12. $5m^2 - 16m + 3$

13. $5a^2 - 12a + 6$

14. $2x^2 - x - 1$

15. $4x^2 + 13x + 3$

16. $6y^2 - 19y + 15$

17. $5x^2 + 11x + 4$

18. $3x^2 - 2x - 5$

19. $5y^2 - 16y + 3$

20. $5x^2 + 2x + 7$

21. $3x^2 + 14x - 5$

22. $7x^2 + 43x + 6$

23. $7x^2 - 16x + 4$

24. $15x^2 - 19x + 6$

25. $3x^2 - 10x + 7$

26. $3y^2 - 22y + 7$

27. $5z^2 - 33z - 14$

28. $3z^2 - 11z - 6$

29. $8x^2 + 2x - 3$

30. $8x^2 + 6x - 9$

31. $10x^2 - 27x + 5$

32. $6x^2 + 7x - 10$

33. $8x^2 - 2x - 15$

34. $8x^2 + 13x - 6$

35. $6x^2 + 33x + 15$

36. $18x^2 - 3x - 10$

37. $6x^2 + 4x - 10$

38. $12z^2 + 32z + 20$

39. $6x^3 + 5x^2 - 4x$

40. $8x^2 + 2x - 20$

41. $4x^3 + 2x^2 - 6x$

42. $18x^3 - 21x^2 - 9x$

43. $6x^3 + 4x^2 - 10x$

44. $300x^2 - 400x - 400$

45. $60x^2 + 40x + 5$

46. $36x^2 - 36x + 9$

47. $2x^2 + 5xy + 2y^2$

48. $8x^2 - 8xy - 6y^2$

49. $2x^2 - 7xy + 3y^2$

50. $15x^2 - xy - 6y^2$

51. $18x^2 + 18xy - 8y^2$

52. $12a^2 - 34ab + 24b^2$

53. What is the first step in factoring any trinomial?

54. How may any trinomial factoring problem be checked?

55. Explain in your own words the procedure used to factor a trinomial of the form $ax^2 + bx + c$, $a \neq 1$.

Cumulative Review Exercises

[2.5] **56.** Solve the equation $3x + 4 = -(x - 6)$.

[3.4] **57.** The perimeter of a rectangle is 22 feet. Find the dimensions of the rectangle if the length is two more than twice the width.

[5.1] **58.** Factor $36x^4y^3 - 12xy^2 + 24x^5y^6$.

[5.3] **59.** Factor $x^2 - 15x + 54$.

JUST FOR FUN

Factor each trinomial.

1. $18x^2 + 9x - 20$.

2. $8x^2 - 99x + 36$.

5.5

Special Factoring Formulas and a General Review of Factoring

▸ **1** Factor the difference of two squares.

▸ **2** Factor the sum and difference of two cubes.

▸ **3** Learn the general procedure for factoring a polynomial.

There are special formulas for certain types of factoring problems that are used very often. The special formulas we focus on in this section are the *difference of two squares* and the *sum and difference of two cubes*. There is no special formula for the sum of two squares; this is because the sum of two squares cannot be factored over the set of real numbers. You will need to memorize the three formulas in this section so that you can use them automatically whenever you need them.

Difference
of Two Squares

▶ **1** Let us begin with the difference of two squares. Consider the binomial $x^2 - 9$. Note that each term of the binomial can be expressed as the square of some expression.

$$x^2 - 9 = x^2 - 3^2$$

This is an example of a difference of two squares problem. To factor the difference of two squares, it is convenient to use the difference of two squares formula first introduced in Section 4.5.

Difference of Two Squares
_____

$$a^2 - b^2 = (a + b)(a - b)$$

EXAMPLE 1 Factor $x^2 - 9$.

Solution: If we write $x^2 - 9$ as a difference of two squares, we have $x^2 - 3^2$. Using the difference of two squares formula, where a is replaced by x and b is replaced by 3, we obtain the following:

$$a^2 - b^2 = (a + b)(a - b)$$
$$x^2 - 3^2 = (x + 3)(x - 3)$$ ∎

EXAMPLE 2 Factor each of the following using the difference of two squares formula.

(a) $x^2 - 16$ (b) $4x^2 - 9$ (c) $16x^2 - 9y^2$

Solution: (a) $x^2 - 16 = (x)^2 - (4)^2$
$$= (x + 4)(x - 4)$$

(b) $4x^2 - 9 = (2x)^2 - (3)^2$
$$= (2x + 3)(2x - 3)$$

(c) $16x^2 - 9y^2 = (4x)^2 - (3y)^2$
$$= (4x + 3y)(4x - 3y)$$ ∎

EXAMPLE 3 Factor each of the following differences of squares.

Solution: (a) $16x^4 - 9y^4$ (b) $x^6 - y^4$

(a) Rewrite $16x^4$ as $(4x^2)^2$ and $9y^4$ as $(3y^2)^2$, then use the difference of two squares formula.

$$16x^4 - 9y^4 = (4x^2)^2 - (3y^2)^2$$
$$= (4x^2 + 3y^2)(4x^2 - 3y^2)$$

(b) Rewrite x^6 as $(x^3)^2$ and y^4 as $(y^2)^2$, then use the difference of two squares formula.

$$x^6 - y^4 = (x^3)^2 - (y^2)^2$$
$$= (x^3 + y^2)(x^3 - y^2)$$ ∎

EXAMPLE 4 Factor $4x^2 - 16y^2$ using the difference of two squares formula.

Solution: First remove the common factor, 4.

$$4x^2 - 16y^2 = 4(x^2 - 4y^2)$$

Now use the formula for the difference of two squares.

$$4(x^2 - 4y^2) = 4[(x)^2 - (2y)^2]$$
$$= 4(x + 2y)(x - 2y)$$

COMMON STUDENT ERROR

The difference of two squares can be factored. However, it is not possible to factor the sum of two squares over the set of real numbers.

Correct	*Wrong*
$a^2 - b^2 = (a + b)(a - b)$	$a^2 + b^2 = (a + b)(a + b)$

Sum and Difference of Two Cubes

▶ **2**   We begin our discussion of the sum and difference of two cubes with a multiplication of polynomials problem. Consider the product of $(a + b)(a^2 - ab + b^2)$.

$$
\begin{array}{r}
a^2 - ab + b^2 \\
a + b \\
\hline
a^2b - ab^2 + b^3 \\
a^3 - a^2b + ab^2 \\
\hline
a^3 \qquad\qquad + b^3
\end{array}
$$

Thus $(a + b)(a^2 - ab + b^2) = a^3 + b^3$. Since factoring is the opposite of multiplying, we may factor $a^3 + b^3$ as follows:

$$a^3 + b^3 = (a + b)(a^2 - ab + b^2)$$

We see, using the same procedure, that $a^3 - b^3 = (a - b)(a^2 + ab + b^2)$. The expression $a^3 + b^3$ is a sum of two cubes and the expression $a^3 - b^3$ is a difference of two cubes. The formulas for factoring the sum and difference of two cubes follow.

Sum of Two Cubes

$$a^3 + b^3 = (a + b)(a^2 - ab + b^2)$$

Difference of Two Cubes

$$a^3 - b^3 = (a - b)(a^2 + ab + b^2)$$

Note that the trinomials $a^2 - ab + b^2$ and $a^2 + ab + b^2$ cannot be factored further. Now let us do some factoring problems using the sum and difference of two cubes.

EXAMPLE 5   Factor $x^3 + 8$.

Solution: Rewrite $x^3 + 8$ as a sum of two cubes: $x^3 + 8 = (x)^3 + (2)^3$

If we let x be a and 2 be b, then, using the sum of cubes formula, we get

$$x^3 + 8 = (x)^3 + (2)^3$$
$$= (x + 2)[x^2 - x(2) + 2^2]$$
$$= (x + 2)(x^2 - 2x + 4) \qquad \blacksquare$$

You can check this problem by multiplying $(x + 2)(x^2 - 2x + 4)$. If factored correctly, the product of the factors will equal the original expression, $x^3 + 8$. Try it and see.

EXAMPLE 6   Factor $y^3 - 27$.

Solution: Rewrite $y^3 - 27$ as a difference of two cubes.

$$y^3 - 27 = (y)^3 - (3)^3$$
$$= (y - 3)[y^2 + y(3) + 3^2]$$
$$= (y - 3)(y^2 + 3y + 9) \qquad \blacksquare$$

EXAMPLE 7   Factor $8a^3 - b^3$.

Solution: Rewrite $8a^3 - b^3$ as a difference of two cubes. Note that $(2a)^3 = 8a^3$; thus we can write

$$8a^3 - b^3 = (2a)^3 - (b)^3$$
$$= (2a - b)[(2a)^2 + (2a)(b) + b^2]$$
$$= (2a - b)(4a^2 + 2ab + b^2) \qquad \blacksquare$$

EXAMPLE 8   Factor $8r^3 + 27s^3$.

Solution: Rewrite $8r^3 + 27s^3$ as a sum of two cubes. Since $8r^3 = (2r)^3$ and $27s^3 = (3s)^3$, we write

$$8r^3 + 27s^3 = (2r)^3 + (3s)^3$$
$$= (2r + 3s)[(2r)^2 - (2r)(3s) + (3s)^2]$$
$$= (2r + 3s)(4r^2 - 6rs + 9s^2) \qquad \blacksquare$$

A General Review of Factoring

▶ **3**   In this chapter we presented a number of different methods of factoring. We will now combine problems and techniques from this and previous sections.

Here is a general procedure for factoring any polynomial:

General Procedure to Factor a Polynomial

---

1. Determine if the polynomial has a greatest common factor other than 1. If so, factor out the GCF from every term in the polynomial.
2. If the polynomial has two terms (or is a binomial), determine if it is a difference of two squares or a sum or difference of two cubes. If so, factor using the appropriate formula.
3. If the polynomial has three terms, factor the trinomial using the methods discussed in Sections 5.3 and 5.4.
4. If the polynomial has more than three terms, then try factoring by grouping.
5. As a final step, examine your factored polynomial to see if any factors listed have a common factor and can be factored further. If you find a common factor, factor it out at this point.

EXAMPLE 9   Factor $3x^4 - 27x^2$.

Solution: First determine if there is a greatest common factor other than 1. Since $3x^2$ is common to both terms, factor it out.

$$3x^4 - 27x^2 = 3x^2(x^2 - 9)$$
$$= 3x^2(x + 3)(x - 3)$$

Note that $x^2 - 9$ is a difference of two squares.   ■

EXAMPLE 10   Factor $3x^2y^2 - 6xy^2 - 24y^2$.

Solution: Begin by factoring the GCF, $3y^2$, from each term. Then factor the remaining trinomial.

$$3x^2y^2 - 6xy^2 - 24y^2 = 3y^2(x^2 - 2x - 8)$$
$$= 3y^2(x - 4)(x + 2)$$ ■

EXAMPLE 11   Factor $10a^2b - 15ab + 20b$.

Solution: $10a^2b - 15ab + 20b = 5b(2a^2 - 3a + 4)$

Since $2a^2 - 3a + 4$ cannot be factored we stop here.   ■

EXAMPLE 12   Factor $2xy + 4x + 2y + 4$.

Solution: Always begin by determining if the polynomial has a common factor. In this example, 2 is the GCF. Factor a 2 from each term.

$$2xy + 4x + 2y + 4 = 2(xy + 2x + y + 2)$$

Now factor by grouping.

$$= 2[x(y + 2) + 1(y + 2)]$$
$$= 2(x + 1)(y + 2)$$ ■

In Example 12, what would happen if we forgot to factor out the GCF, 2? Let's rework the problem without first factoring out the GCF, 2, and see what happens. Factor $2x$ from the first two terms, and 2 from the last two terms.

$$2xy + 4x + 2y + 4 = 2x(y + 2) + 2(y + 2)$$
$$= (2x + 2)(y + 2)$$

In step 5 of the general procedure, we are told to examine the factored polynomial to see if any factor listed has a common factor. If we study the factors, we see that the factor $2x + 2$ has a common factor of 2. If we factor out the 2 from $2x + 2$, we will obtain the same answer obtained in Example 12.

$$(2x + 2)(y + 2) = 2(x + 1)(y + 2)$$

EXAMPLE 13   Factor $12x^2 + 12x - 9$.

Solution: First factor out the common factor 3. Then factor the remaining trinomial by one of the methods discussed in Section 5.4 (either by grouping or trial and error).

$$12x^2 + 12x - 9 = 3(4x^2 + 4x - 3)$$
$$= 3(2x + 3)(2x - 1)$$ ■

EXAMPLE 14   Factor $2x^4y + 54xy$.

Solution: First factor out the common factor $2xy$.

$$2x^4y + 54xy = 2xy(x^3 + 27)$$
$$= 2xy(x + 3)(x^2 - 3x + 9)$$

Note that $x^3 + 27$ is a sum of two cubes.   ■

Exercise Set 5.5

Factor the difference of two squares.

1. $x^2 - 4$
2. $x^2 - 9$
3. $y^2 - 25$
4. $z^2 - 64$
5. $x^2 - 49$
6. $x^2 - a^2$
7. $x^2 - y^2$
8. $4x^2 - 9$
9. $9y^2 - 16$
10. $16x^2 - 9y^2$
11. $64a^2 - 36b^2$
12. $100x^2 - 81y^2$
13. $25x^2 - 16$
14. $y^4 - 4x^2$
15. $z^4 - 81x^2$
16. $9x^4 - 16y^4$
17. $9x^4 - 81y^2$
18. $4x^4 - 25y^4$
19. $49m^4 - 16n^2$
20. $2x^4 - 50y^2$
21. $20x^2 - 180$
22. $4x^3 - xy^2$

Factor the sum or difference of two cubes.

23. $x^3 + y^3$
24. $x^3 - y^3$
25. $a^3 - b^3$
26. $a^3 + b^3$
27. $x^3 + 8$
28. $x^3 - 8$
29. $x^3 - 27$
30. $a^3 + 27$
31. $a^3 + 1$
32. $a^3 - 1$
33. $8x^3 + 27$
34. $27y^3 - 8$
35. $27a^3 - 64$
36. $64 - x^3$
37. $27 - 8y^3$
38. $1 + 27y^3$
39. $8x^3 - 27y^3$
40. $64x^3 - 27y^3$

Factor each of the following completely.

41. $2x^2 - 2x - 12$
42. $3x^2 - 9x - 12$
43. $x^2y - 16y$
44. $2x^2 - 8$
45. $3x^2 + 6x + 3$
46. $3x^2 - 9x - 12x + 36$
47. $5x^2 + 10x - 15$
48. $4x^2 - 100$
49. $3xy - 6x + 9y - 18$
50. $x^2y + 2xy - 6xy - 12y$
51. $2x^2 - 72$
52. $4ya^2 - 36y$
53. $3x^2y - 27y$
54. $2x^3 - 50x$
55. $3x^3y^2 + 3y^2$
56. $x^4 - 8x$
57. $2x^3 - 16$
58. $x^3 - 64x$
59. $6x^2 - 4x + 24x - 16$
60. $4x^2y - 6xy - 20xy + 30y$
61. $3x^3 - 10x^2 - 8x$
62. $4x^3 - 22x^2 + 30x$
63. $4x^2 + 5x - 6$
64. $12a^2 - 36a + 27$
65. $25b^2 - 100$
66. $3b^2 - 75c^2$
67. $a^5b^2 - 4a^3b^4$
68. $12x^2 + 8x - 18x - 12$
69. $3x^4 - 18x^3 + 27x^2$
70. $a^6 + 4a^4b^2$

71. $x^3 + 25x$
72. $8y^2 + 23y - 3$
73. $y^4 - 16$
74. $16m^3 + 250$
75. $10a^2 + 25ab - 60b^2$
76. $ac + 2a + bc + 2b$
77. $9x^2 + 12x - 5$
78. $2ab + 4a - 3b - 6$
79. $x^3 - 25x$
80. $9 - 9y^4$

81. **(a)** Write the formula for factoring the difference of two squares.

   (b) In your own words, explain how to factor the difference of two squares.

82. **(a)** Write the formula for factoring the sum of two cubes.

   (b) In your own words, explain how to factor the sum of two cubes.

83. **(a)** Write the formula for factoring the difference of two cubes.

   (b) In your own words, explain how to factor the difference of two cubes.

Cumulative Review Exercises

[2.7] **84.** Solve the inequality and graph the solution on the number line: $6(x - 2) < 4x - 3 + 2x$.

[3.1] **85.** Solve the equation $2x - 5y = 6$ for y.

[4.1] **86.** Simplify $\left(\dfrac{4x^4y}{6xy^5}\right)^3$.

[4.2] **87.** Simplify $x^{-2}x^{-3}$.

JUST FOR FUN

1. Factor $x^6 + 1$.
2. Factor $x^6 - 27y^9$.
3. Have you ever seen the proof that 1 is equal to 2? Here it is.

   Let $a = b$, then square both sides of the equation:

   $$a^2 = b^2$$
   $$a^2 = b \cdot b$$
   $$a^2 = ab \qquad \text{Substitute } a = b.$$
   $$a^2 - b^2 = ab - b^2 \qquad \begin{array}{l}\text{Subtract } b^2 \text{ from both} \\ \text{sides of the equation.}\end{array}$$
   $$(a + b)(a - b) = b(a - b) \qquad \begin{array}{l}\text{Factor both sides of the} \\ \text{equation.}\end{array}$$

   $$\frac{(a + b)(a - b)}{(a - b)} = \frac{b(a - b)}{(a - b)} \qquad \begin{array}{l}\text{Divide both sides of the} \\ \text{equation by } (a - b) \\ \text{and divide out} \\ \text{common factors.}\end{array}$$

   $$a + b = b$$
   $$b + b = b \qquad \text{Substitute } a = b.$$
   $$2b = b$$
   $$\frac{2b}{b} = \frac{b}{b} \qquad \begin{array}{l}\text{Divide both sides of the} \\ \text{equation by } b.\end{array}$$
   $$2 = 1$$

   Obviously, $2 \neq 1$. Therefore, we must have made an error somewhere. Can you find it? Cannot divide both sides of equation by $a - b$ because it equals 0.

5.6

Solving Quadratic Equations Using Factoring

▶ **1** Recognize quadratic equations.

▶ **2** Solve quadratic equations using factoring.

▶ **1** In this section we introduce **quadratic equations,** which are equations that contain a second-degree term and no term of a higher degree.

> **Quadratic Equation**
>
> Quadratic equations have the form
> $$ax^2 + bx + c = 0$$
> where a, b, and c are real numbers, $a \neq 0$.

Examples of Quadratic Equations

$$x^2 + 2x - 3 = 0$$
$$3x^2 - 4x = 0$$
$$2x^2 - 3 = 0$$

Quadratic equations like those given, where one side of the equation is written in descending order of the variable and the other side of the equation is equal to zero, are said to be in **standard form.**

Some quadratic equations can be solved by factoring. Two methods that can be used to solve quadratic equations that cannot be solved by factoring are given in Chapter 10. To solve a quadratic equation by factoring, we use the zero-factor property.

Zero-factor Property

If $ab = 0$, then $a = 0$ or $b = 0$.

In other words, if the product of two factors is 0, then at least one of the factors must be 0.

EXAMPLE 1 Solve the equation $(x + 3)(x + 4) = 0$.

Solution: Since the product of the factors equals 0, according to the preceding rule, one or both factors must equal zero. Set each factor equal to 0, and solve each resulting equation.

$$(x + 3) = 0 \qquad \text{or} \qquad (x + 4) = 0$$
$$x + 3 = 0 \qquad\qquad x + 4 = 0$$
$$x + 3 - 3 = 0 - 3 \qquad\qquad x + 4 - 4 = 0 - 4$$
$$x = -3 \qquad \text{or} \qquad x = -4$$

Thus, if x is either -3 or -4, the product of the factors is 0. The solutions to the equation are -3 and -4.

Check:
$$x = -3 \qquad\qquad x = -4$$
$$(x + 3)(x + 4) = 0 \qquad (x + 3)(x + 4) = 0$$
$$(-3 + 3)(-3 + 4) = 0 \qquad (-4 + 3)(-4 + 4) = 0$$
$$0(1) = 0 \qquad\qquad -1(0) = 0$$
$$0 = 0 \qquad\qquad 0 = 0$$ ■

EXAMPLE 2 Solve the equation $(4x - 3)(2x + 4) = 0$.

Solution: Set each factor equal to 0 and solve for x.

$$(4x - 3) = 0 \qquad \text{or} \qquad (2x + 4) = 0$$
$$4x - 3 = 0 \qquad\qquad 2x + 4 = 0$$
$$4x = 3 \qquad\qquad 2x = -4$$
$$x = \frac{3}{4} \qquad \text{or} \qquad x = -2$$

The solutions to the equation are $\frac{3}{4}$ and -2. ■

▶ **2**

To Solve a Quadratic Equation Using Factoring

1. Write the equation in standard form with the squared term positive. This will result in the one side of the equation being equal to 0.
2. Factor the side of the equation that is not 0.
3. Set each factor *containing a variable* equal to zero and find the solution.

EXAMPLE 3   Solve the equation $2x^2 = 12x$.

Solution: To make the right side of the equation equal to 0, we subtract $12x$ from both sides of the equation. Then we factor out a $2x$ from both terms.

$$2x^2 = 12x$$
$$2x^2 - 12x = 12x - 12x$$
$$2x^2 - 12x = 0$$
$$2x(x - 6) = 0$$

Now set each factor equal to zero.

$$2x = 0 \quad \text{or} \quad x - 6 = 0$$
$$x = \frac{0}{2} \qquad\qquad x = 6$$
$$x = 0$$

The solutions to the equation are 0 and 6.

EXAMPLE 4   Solve the equation $x^2 + 10x + 28 = 4$.

Solution: To make the right side of the equation equal to 0, we subtract 4 from both sides.

$$x^2 + 10x + 24 = 0$$

Now factor:

$$(x + 4)(x + 6) = 0$$

And solve:

$$x + 4 = 0 \quad \text{or} \quad x + 6 = 0$$
$$x = -4 \qquad\qquad x = -6$$

The solutions are -4 and -6.

EXAMPLE 5   Solve the equation $3x^2 + 2x - 12 = -7x$.

Solution: Since all terms are not on the same side of the equation, add $7x$ to both sides of the equation.

$$3x^2 + 9x - 12 = 0$$

Factor out the common factor.

$$3(x^2 + 3x - 4) = 0$$

Factor the remaining trinomial.

$$3(x + 4)(x - 1) = 0$$

Now solve for x.

$$x + 4 = 0 \qquad \text{or} \qquad x - 1 = 0$$
$$x = -4 \qquad\qquad\qquad x = 1$$

Since the 3 that was factored does not contain a variable, we do not have to set it equal to zero. The solutions to the equation are -4 and 1. ■

EXAMPLE 6 Solve the equation $-x^2 + 5x + 6 = 0$.

Solution: When the squared term is negative, we generally make it positive by multiplying both sides of the equation by -1.

$$-1(-x^2 + 5x + 6) = -1 \cdot 0$$
$$x^2 - 5x - 6 = 0$$

Note that the sign of each term on the left side of the equation changed and that the right side of the equation remained zero. Now proceed as before.

$$x^2 - 5x - 6 = 0$$
$$(x - 6)(x + 1) = 0$$
$$x - 6 = 0 \qquad \text{or} \qquad x + 1 = 0$$
$$x = 6 \qquad\qquad\qquad x = -1$$

A check using the original equation will show that the solutions are 6 and -1. ■

COMMON STUDENT ERROR

Be careful not to confuse factoring a polynomial with solving an equation.

Correct	*Wrong*
Factor: $x^2 + 3x + 2$	Factor: $x^2 + 3x + 2$
$(x + 2)(x + 1)$	$(x + 2)(x + 1)$
	$x + 2 = 0 \quad \text{or} \quad x + 1 = 0$
	$x = -2 \quad \text{or} \quad x = -1$

Do you know what is wrong with the example on the right? The expression $x^2 + 3x + 2$ is a polynomial (a trinomial), and not an equation. Since it is not an equation, it cannot be solved. When you are given a polynomial, you cannot just add "= 0" to the polynomial to change it to an equation.

Correct

Solve: $x^2 + 3x + 2 = 0$
$$(x + 2)(x + 1) = 0$$
$$x + 2 = 0 \quad \text{or} \quad x + 1 = 0$$
$$x = -2 \quad \text{or} \quad x = -1$$

EXAMPLE 7   Solve the equation $x^2 = 9$.

Solution: Subtract 9 from both sides of the equation; then factor using the difference of two squares formula.

$$x^2 - 9 = 0$$
$$(x + 3)(x - 3) = 0$$

$x + 3 = 0$ or $x - 3 = 0$

$x = -3$ $x = 3$ ∎

EXAMPLE 8   The product of two numbers is 66. Find the two numbers if one number is 5 more than the other.

Solution: Let x = smaller number

$x + 5$ = large number

$$x(x + 5) = 66$$
$$x^2 + 5x = 66$$
$$x^2 + 5x - 66 = 0$$
$$(x - 6)(x + 11) = 0$$
$$x = 6, x = -11$$

Remember that x represents the smaller of the two numbers. This problem has two possible solutions.

	Solution 1	*Solution 2*
smaller number	6	−11
larger number	$x + 5 = 6 + 5 = 11$	$x + 5 = -11 + 5 = -6$

One solution is: smaller number 6, larger number 11. A second solution is: smaller number −11, larger number −6.

Check: Product of the two numbers is 66. $6 \cdot 11 = 66$ $(-11)(-6) = 66$

One number is 5 more than the other number.   11 is 5 more than 6   −6 is 5 more than −11

If the question had stated "the product of two *positive* numbers is 66," the only solution would be 6 and 11.   ∎

EXAMPLE 9   Find the length and width of a rectangle if its length is 3 more than the width and its area is 54.

Solution: Let x = width

$x + 3$ = length (see Fig. 5.3.)

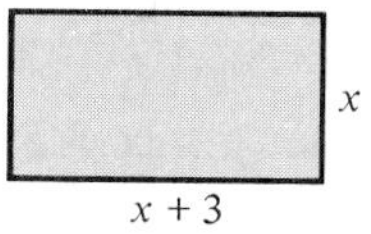

$$\text{area} = \text{length} \cdot \text{width}$$
$$54 = (x + 3)x$$
$$54 = x^2 + 3x$$
$$0 = x^2 + 3x - 54$$

or $x^2 + 3x - 54 = 0$

$$(x - 6)(x + 9) = 0$$

$x - 6 = 0$ or $x + 9 = 0$

$x = 6$ $x = -9$

$x + 3$

FIGURE 5.3

Since the length of a geometric figure cannot be a negative number, the only solution is

$$\text{width} = 6, \qquad \text{length} = x + 3 = 6 + 3 = 9$$

Check: $A = l \cdot w$

$A = 9 \cdot 6$

$54 = 54 \qquad$ true   ∎

Exercise Set 5.6

Solve each equation.

1. $x(x + 3) = 0$

2. $4x(x - 7) = 0$

3. $5x(x - 9) = 0$

4. $(x + 3)(x - 5) = 0$

5. $(2x + 5)(x - 3) = 0$

6. $(2x + 3)(x - 5) = 0$

7. $x^2 - 16 = 0$

8. $x^2 - 25 = 0$

9. $x^2 - 12x = 0$

10. $x^2 + 4x = 0$

11. $9x^2 + 18x = 0$

12. $x^2 + 6x + 5 = 0$

13. $x^2 + x - 12 = 0$

14. $x^2 + 6x + 9 = 0$

15. $x^2 - 12x = -20$

16. $3y^2 - 2 = -y$

17. $z^2 + 3z = 18$

18. $3x^2 = -21x - 18$

19. $3x^2 - 6x - 72 = 0$

20. $x^2 = 3x + 18$

21. $x^2 + 19x = 42$

22. $3x^2 - 9x - 30 = 0$

23. $2y^2 + 22y + 60 = 0$

24. $w^2 + 45 + 18w = 0$

25. $-2x - 8 = -x^2$

26. $-9x + 20 = -x^2$

27. $-x^2 + 30x + 64 = 0$

28. $-y^2 + 12y - 11 = 0$

29. $x^2 - 3x - 18 = 0$

30. $z^2 + 16z = -64$

31. $3p^2 = 22p - 7$

32. $5w^2 - 16w = -3$

33. $3r^2 + r = 2$

34. $3x^2 = 7x + 20$

35. $4x^2 + 4x - 48 = 0$

36. $6x^2 + 13x + 6 = 0$

37. $6x^2 - 5x = 4$

38. $2x^2 - 4x - 6 = 0$

39. $2x^2 - 10x = -12$

40. $x^2 - 25 = 0$

41. $2x^2 = 32x$

42. $4x^2 - 9 = 0$

43. $x^2 = 36$

44. $2x^2 - 32 = 0$

45. $x^2 = 9$

46. $x^2 = 4$

Express each problem as an equation, and solve.

47. The product of two consecutive positive even integers is 80. Find the two integers.

48. The product of two consecutive positive even integers is 120. Find the two integers.

49. The product of two consecutive positive odd integers is 63. Find the two integers.

50. The product of two positive integers is 108. Find the two numbers if one is 3 more than the other.

51. The product of two positive integers is 35. Find the two numbers if the larger number is 3 less than twice the smaller number.

52. The product of two positive integers is 64. Find the two integers if one number is four times the other.

53. The area of a rectangle is 36 square feet. Find the length and width if the length is four times the width.

54. The area of a rectangle is 54 square inches. Find the length and width if the length is 3 inches less than twice the width.

55. If each side of a square is increased by 6 meters, the area becomes 64 square meters. Find the length of a side of the original square.

56. If each side of a square is increased by 4 meters, the area becomes 121 square meters. Find the length of a side of the original square.

The sum of the first n *even numbers is given by the formula* s = n² + n. *Find* n *for the given sums.*

57. $s = 12$

58. $s = 30$

59. **(a)** When solving the equation $(x + 1)(x - 2) = 4$, explain why we **cannot** solve the equation by writing $x + 1 = 4$ or $x - 2 = 4$, and then solving for x.

(b) Solve the equation $(x + 1)(x - 2) = 4$.

60. When solving an equation like $3(x - 4)(x + 5) = 0$, we set the factors $x - 4$ and $x + 5$ equal to 0, but we do not set the 3 equal to 0. Can you explain why?

Cumulative Review Exercises

[4.4] **61.** Subtract $x^2 - 4x + 6$ from $3x + 2$.

[4.5] **62.** Multiply $(3x^2 + 2x - 4)(2x - 1)$

[4.6] **63.** Divide $\dfrac{6x^2 - 19x + 15}{3x - 5}$ by dividing the numerator by the denominator.

[5.4] **64.** Divide $\dfrac{6x^2 - 19x + 15}{3x - 5}$ by factoring the numerator and dividing out common factors.

JUST FOR FUN

1. When a cannon is fired, under certain specific conditions, the height of the cannonball from the ground, in feet, at any time, t, can be found by using the formula $h = -16t^2 + 128t$.

   (a) Find the height of the cannonball at 2 seconds.

   (b) Find the time it takes for the cannonball to hit the ground.

   Hint: What is the value of h at impact?

2. Solve the equation $x^3 + 3x^2 - 10x = 0$.

SUMMARY

GLOSSARY

Composite number *(199):* A positive integer, other than 1, that is not prime.

Factor an expression *(198):* To factor an expression means to write the expression as a product of its factors.

Factors *(198):* If $a \cdot b = c$, then a and b are factors of c.

Greatest common factor (GCF) *(199):* The greatest factor that divides each of the terms in an expression.

Prime number *(199):* An integer greater than 1 that has exactly two factors, itself and one.

Quadratic equation *(235):* An equation of the form $ax^2 + bx + c = 0$, $a \neq 0$.

IMPORTANT FACTS

Difference of two squares:
$$a^2 - b^2 = (a + b)(a - b)$$

Sum of two cubes:
$$a^3 + b^3 = (a + b)(a^2 - ab + b^2)$$

Difference of two cubes:
$$a^3 - b^3 = (a - b)(a^2 + ab + b^2)$$

Note: The sum of two squares, $a^2 + b^2$, cannot be factored over the set of real numbers.

Zero-factor property: If $a \cdot b = 0$, then $a = 0$ or $b = 0$.

General Procedure to Factor a Polynomial

1. Determine if the polynomial has a greatest common factor other than 1. If so, factor out the GCF from every term in the polynomial.

2. If the polynomial has two terms (or is a binomial), determine if it is a difference of two squares or a sum or difference of two cubes. If so, factor using the appropriate formula.

3. If the polynomial has three terms, factor the trinomial using the methods discussed in Sections 5.3 and 5.4.

4. If the polynomial has more than three terms, try factoring by grouping.

5. As a final step, examine your factored polynomial to see if any factors listed have a common factor and can be factored further. If you find a common factor, factor it out at this point.

Review Exercises

[5.1] *Find the greatest common factor for each set of terms.*

1. $x^3, x^5, 2x^2$

2. $3p, 6p^2, 9p^3$

3. $18x, 24, 36y^2$

4. $40x^2y^3, 36x^3y^4, 16x^5y^2z$

5. $9xyz, 12xz, 36, x^2y$

6. $-32x^5, 16x^2, 24x^2y$

7. $x(2x - 5), 3(2x - 5)$

8. $x(x + 5), x + 5$

Factor each expression. If an expression cannot be factored, so state.

9. $5x - 20$

10. $9x + 33$

11. $16y^2 - 12y$

12. $55p^3 - 20p^2$

13. $24x^2y + 18x^3y^2$

14. $18x^2y - 9xy$

15. $2x^2 + 4x - 8$

16. $60x^4y^4 + 6x^9y^3 - 18x^5y^2$

17. $24x^2 - 13y^2 + 6xy$

18. $x(5x + 3) - 2(5x + 3)$

19. $3x(x - 1) - 2(x - 1)$

20. $2x(4x - 3) + 4x - 3$

[5.2] *Factor by grouping.*

21. $x^2 + 3x + 2x + 6$

22. $x^2 - 5x + 3x - 15$

23. $x^2 - 7x + 7x - 49$

24. $2a^2 - 2ab - a + b$

25. $3xy + 3x + 2y + 2$

26. $x^2 + 3x - 2xy - 6y$

27. $5x^2 + 20x - x - 4$

28. $5x^2 - xy + 20xy - 4y^2$

29. $12x^2 - 8xy + 15xy - 10y^2$

30. $12x^2 + 15xy - 8xy - 10y^2$

31. $ab - a + b - 1$

32. $3x^2 - 9xy + 2xy - 6y^2$

33. $20x^2 - 12x + 15x - 9$

34. $6x^2 + 9x - 2x - 3$

[5.3] *Factor completely.*

35. $x^2 + 6x + 8$

36. $x^2 - 8x + 15$

37. $x^2 - x - 20$

38. $x^2 + x - 20$

39. $x^2 - 3x - 18$

40. $x^2 - 9x + 14$

41. $x^2 - 12x - 45$

42. $x^2 + 11x + 24$

43. $x^3 + 5x^2 + 4x$

44. $x^3 - 3x^2 - 40x$

45. $x^2 - 2xy - 15y^2$

46. $4x^3 + 32x^2y + 60xy^2$

[5.4] *Factor completely.*

47. $2x^2 + 7x - 4$

48. $3x^2 + 13x + 4$

49. $4x^2 - 9x + 5$

50. $3x^2 - 5x - 12$

51. $4x^2 + 4x - 15$

52. $5x^2 - 32x + 12$

53. $3x^2 + 13x + 12$

54. $6x^2 + 31x + 5$

55. $2x^2 + 9x - 35$

56. $6x^2 + 11x - 10$

57. $8x^2 - 18x - 35$

58. $4x^2 + 20x + 25$

59. $9x^3 - 12x^2 + 4x$

60. $18x^3 - 24x^2 - 10x$

61. $4x^2 - 16xy + 15y^2$

62. $16x^2 - 22xy - 3y^2$

[5.5] *Factor completely.*

63. $x^2 - 25$

64. $x^2 - 64$

65. $4x^2 - 16$

66. $81x^2 - 9y^2$

67. $64x^4 - 81y^4$

68. $16 - 25y^2$

69. $4x^4 - 9y^4$

70. $100x^4 - 121y^4$

71. $x^3 - y^3$

72. $x^3 + y^3$

73. $a^3 + 8$

74. $a^3 - 1$

75. $a^3 + 27$

76. $x^3 - 8$

77. $8x^3 - y^3$

78. $27 - 8y^3$

[5.1–5.5] *Factor completely.*

79. $8x^2 + 16x - 24$

80. $2x^2 - 16x + 32$

81. $4x^2 - 36$

82. $3y^2 - 27$

83. $x^2 - 10x + 24$

84. $x^2 - 6x - 27$

85. $4x^2 - 4x - 15$

86. $6x^2 - 33x + 36$

87. $8x^3 - 8$

88. $x^3y - 27y$

89. $x^2y - xy + 4xy - 4y$

90. $6x^3 + 30x^2 + 9x^2 + 45x$

91. $x^2 + 5xy + 6y^2$

92. $2x^2 - xy - 10y^2$

93. $4x^2 - 20xy + 25y^2$

94. $16y^2 - 49z^2$

95. $ab + 7a + 6b + 42$

96. $16y^5 - 25y^7$

97. $2x^3 + 12x^2y + 16xy^2$

98. $6x^2 + 5xy - 21y^2$

99. $32x^3 + 32x^2 + 6x$

100. $y^4 - 1$

[5.6] *Solve each equation.*

101. $x(x - 4) = 0$

102. $(x + 3)(x + 4) = 0$

103. $(x - 5)(3x + 2) = 0$

104. $x^2 - 3x = 0$

105. $5x^2 + 20x = 0$

106. $x^2 - 2x - 24 = 0$

107. $x^2 + 8x + 15 = 0$

108. $x^2 = -2x + 8$

109. $x^2 - 12 = -x$

110. $3x^2 + 21x + 30 = 0$

111. $x^2 - 6x + 8 = 0$

112. $6x^2 + 6x - 12 = 0$

113. $8x^2 - 3 = -10x$

114. $2x^2 + 15x = 8$

115. $4x^2 - 16 = 0$

116. $36x^2 - 49 = 0$

[5.6] *Express each problem as an equation and solve.*

117. The product of two consecutive positive integers is 110. Find the two integers.

118. The product of two consecutive positive even integers is 48. Find the two integers.

119. The product of two positive integers is 40. Find the integers if the larger is 2 less than twice the smaller.

120. The area of a rectangle is 63 square feet. Find the length and width of the rectangle if the length is 2 feet greater than the width.

121. One square has a side 4 inches longer than the side of a second square. If the area of the larger square is 81 square inches, find the length of a side of each square.

Practice Test

1. Find the greatest common factor of $4x^4$, $12x^5$, and $10x^2$.

2. Find the greatest common factor of $6x^2y^3$, $9xy^2$, and $12xy^5$.

Factor completely.

3. $4x^2y - 8xy$

4. $24x^2y - 6xy + 9x$

5. $x^2 - 3x + 2x - 6$

6. $3x^2 - 12x + x - 4$

7. $5x^2 - 15xy - 3xy + 9y^2$

8. $x^2 + 12x + 32$

9. $x^2 + 5x - 24$

10. $x^2 - 9xy + 20y^2$

11. $2x^2 - 22x + 60$

12. $2x^3 - 3x^2 + x$

13. $12x^2 - xy - 6y^2$

14. $x^2 - 9y^2$

15. $x^3 + 27$

Solve each equation.

16. $(x - 2)(2x - 5) = 0$

17. $x^2 + 6 = -5x$

18. $x^2 + 4x - 5 = 0$

Solve each problem.

19. The product of two positive integers is 36. Find the two integers if the larger is 1 more than twice the smaller.

20. The area of a rectangle is 24 square meters. Find the length and width of the rectangle if its length is 2 meters greater than its width.

6

Rational Expressions and Equations

6.1 Reducing Rational Expressions

6.2 Multiplication and Division of Rational Expressions

6.3 Addition and Subtraction of Rational Expressions with a Common Denominator

6.4 Finding the Least Common Denominator

6.5 Addition and Subtraction of Rational Expressions

6.6 Complex Fractions (Optional)

6.7 Solving Equations Containing Rational Expressions

6.8 Applications of Rational Equations

Summary

Review Exercises

Practice Test

Cumulative Review Test

See Section 6.8, Exercise 13.

6.1

Reducing Rational Expressions

▶ **1**   Identify and define rational expressions.

▶ **2**   Determine the values for which a rational expression is defined.

▶ **3**   Recognize the three signs of a fraction.

▶ **4**   Reduce rational expressions.

▶ **5**   Factor a negative 1 from a polynomial.

▶ **1**   In this chapter we focus on rational expressions. To be successful with this material, you need a thorough understanding of the factoring techniques discussed in Chapter 5.

A **rational expression** (also called an **algebraic fraction**) is an algebraic expression of the form p/q, where p and q are polynomials and $q \neq 0$. Examples of rational expressions are

$$\frac{4}{5}, \qquad \frac{x-6}{x}, \qquad \frac{x^2+2x}{x-3}, \qquad \frac{x}{x^2-4}$$

▶ **2**   Note that the denominator of a rational expression cannot equal 0 since division by 0 is not permitted. In the expression $(x+3)/x$, x cannot have a value of 0 since the denominator would then equal 0. We say that the expression $(x+3)/x$ is *defined* when x is not 0. The expression $(x+3)/x$ is defined for all real numbers except 0. It is *undefined* when x is 0. In $(x^2+4x)/(x-3)$, x cannot have a value of 3 for that would result in the denominator having a value of 0. What values of x cannot be used in the expression $x/(x^2-4)$? If you answered 2 and -2, you answered correctly. **Whenever we list a rational expression containing a variable in the denominator, we always assume that the value or values of the variable that make the denominator 0 are excluded.**

One method that can be used to determine the value or values of the variable that are excluded is to set the denominator equal to 0 and then solve the resulting equation for the variable.

EXAMPLE 1   Determine the value or values of the variable for which the rational expression is defined.

(a) $\dfrac{x+3}{2x-5}$ (b) $\dfrac{x+3}{x^2+3x-10}$

Solution: (a) We need to determine the value or values of x that make $2x-5$ equal to 0 and exclude these. We can do this by setting $2x-5$ equal to 0 and solving the resulting equation for x.

$$2x - 5 = 0$$
$$2x = 5$$
$$x = \frac{5}{2}$$

Thus we do not consider $x = \dfrac{5}{2}$ when we consider the rational expression $\dfrac{x + 3}{2x - 5}$. The expression is defined for all real numbers except $x = \dfrac{5}{2}$. We will sometimes shorten our answer and write, $x \neq \dfrac{5}{2}$.

(b) To determine the value or values that are excluded, we set the denominator equal to zero and solve the resulting equation for the variable.

$$x^2 + 3x - 10 = 0$$
$$(x + 5)(x - 2) = 0$$
$$x + 5 = 0 \quad \text{or} \quad x - 2 = 0$$
$$x = -5 \quad \text{or} \quad x = 2$$

Therefore, we do not consider the values $x = -5$ or $x = 2$ when we consider the rational expression $(x + 3)/(x^2 + 3x - 10)$, for both $x = -5$ and $x = 2$ result in the denominator having a value of zero. This expression is defined for all real numbers except $x = -5$ and $x = 2$. Thus, $x \neq -5$ and $x \neq 2$. ■

▶ **3**   Three signs are associated with any fraction: the sign of the numerator, the sign of the denominator, and the sign of the fraction.

$$\text{sign of fraction} \longrightarrow + \frac{-a}{+b}$$

sign of numerator

sign of denominator

Whenever any of the three signs is omitted, we assume it to be positive. For example,

$$\frac{a}{b} \quad \text{means} \quad + \frac{+a}{+b}$$

$$\frac{-a}{b} \quad \text{means} \quad + \frac{-a}{+b}$$

$$-\frac{a}{b} \quad \text{means} \quad - \frac{+a}{+b}$$

Changing any two of the three signs of a fraction does not change the value of a fraction. Thus

$$\frac{-a}{b} = -\frac{a}{b} = \frac{a}{-b}$$

Generally, we do not write a fraction with a negative denominator. For example, the expression $\dfrac{2}{-5}$ would be written as either $\dfrac{-2}{5}$ or $-\dfrac{2}{5}$. The expression $\dfrac{x}{-(4 - x)}$ can be written $\dfrac{x}{x - 4}$ since $-(4 - x) = -4 + x$ or $x - 4$.

▶**4** A rational expression is **reduced to its lowest terms** when the numerator and denominator have no common factors other than 1. The fraction $\frac{9}{12}$ is not in reduced form because the 9 and 12 both contain the common factor of 3. When the 3 is factored out, the reduced fraction is $\frac{3}{4}$.

$$\frac{9}{12} = \frac{\overset{1}{\cancel{3}} \cdot 3}{\underset{1}{\cancel{3}} \cdot 4} = \frac{3}{4}$$

The rational expression $\dfrac{ab - b^2}{2b}$ is not in reduced form because both the numerator and denominator have a common factor of b. To reduce this expression, factor the b from each term in the numerator; then divide out the common factor b.

$$\frac{ab - b^2}{2b} = \frac{\cancel{b}(a - b)}{2\cancel{b}} = \frac{a - b}{2}$$

$\dfrac{ab - b^2}{2b}$ becomes $\dfrac{a - b}{2}$ when reduced to its lowest terms.

> **To Reduce Rational Expressions**
>
> **1.** Factor both the numerator and denominator as completely as possible.
> **2.** Divide both the numerator and denominator by any common factors.

EXAMPLE 2   Reduce $\dfrac{5x^3 + 10x^2 - 25x}{5x^2}$ to its lowest terms.

Solution: Factor the numerator. The greatest common factor of each term in the numerator is $5x$.

$$\frac{5x^3 + 10x^2 - 25x}{5x^2} = \frac{5x(x^2 + 2x - 5)}{5x^2}$$
$$= \frac{x^2 + 2x - 5}{x}$$

EXAMPLE 3   Reduce $\dfrac{x^2 + 2x - 3}{x + 3}$ to its lowest terms.

Solution: Factor the numerator; then divide out the common factor.

$$\frac{x^2 + 2x - 3}{x + 3} = \frac{\cancel{(x + 3)}(x - 1)}{\cancel{x + 3}} = x - 1$$

The expression reduces to $x - 1$.

EXAMPLE 4   Reduce $\dfrac{x^2 - 16}{x - 4}$ to its lowest terms.

Solution: Factor the numerator; then divide out common factors.

$$\frac{x^2 - 16}{x - 4} = \frac{(x + 4)\cancel{(x - 4)}}{\cancel{x - 4}} = x + 4$$

COMMON STUDENT ERROR

Remember: Only common *factors* can be divided out when *multiplying* expressions.

$$\text{Correct} \qquad\qquad \text{Wrong}$$

$$\frac{\overset{4}{\cancel{16}}\overset{x}{\cancel{x^2}}}{\underset{1}{\cancel{4}}\underset{1}{\cancel{x}}} = 4x \qquad\qquad \frac{\cancel{x^2} - \cancel{16}}{\cancel{x} - \cancel{4}}$$

In the denominator of the example on the left, $4x$, the 4 and x are factors since they are *multiplied* together. The 4 and the x are also both factors of the numerator $16x^2$, since $16x^2$ can be written $4 \cdot x \cdot 4x$.

Many students divide out *terms* incorrectly. In the expression $\dfrac{x^2 - 16}{x - 4}$, neither x nor 4 is a factor of either the numerator or denominator. The x and -4 are *terms* of the denominator.

EXAMPLE 5   Reduce $\dfrac{2x^2 + 7x + 6}{x^2 - x - 6}$ to lowest terms.

Solution: $\dfrac{2x^2 + 7x + 6}{x^2 - x - 6} = \dfrac{(2x + 3)\cancel{(x + 2)}}{(x - 3)\cancel{(x + 2)}} = \dfrac{2x + 3}{x - 3}.$

Note that $\dfrac{2x + 3}{x - 3}$ cannot be simplified any further.   ■

▶ **5   When −1 is factored from a polynomial, the sign of each term in the polynomial changes.**

Examples

$$-3x + 5 = -1(3x - 5) = -(3x - 5)$$
$$6 - 2x = -1(-6 + 2x) = -(2x - 6)$$
$$-2x^2 + 3x - 4 = -1(2x^2 - 3x + 4) = -(2x^2 - 3x + 4)$$

Whenever the terms in a numerator and denominator differ only in their signs (one is the opposite or additive inverse of the other), we can factor out −1 from either the numerator or denominator and then divide out the common factor. This procedure is illustrated in Examples 6 and 7.

EXAMPLE 6   Reduce $\dfrac{3x - 7}{7 - 3x}$ to lowest terms.

Solution: Since each term in the numerator differs only in sign from its like term in the denominator, we will factor −1 from each term in the denominator.

$$\frac{3x - 7}{7 - 3x} = \frac{3x - 7}{-1(-7 + 3x)}$$

$$= \frac{\overset{1}{\cancel{3x - 7}}}{-\underset{1}{\cancel{(3x - 7)}}} = -1$$ ■

EXAMPLE 7 Reduce $\dfrac{4x^2 - 23x - 6}{6 - x}$ to lowest terms.

Solution:
$$\frac{4x^2 - 23x - 6}{6 - x} = \frac{(4x + 1)(x - 6)}{6 - x}$$
$$= \frac{(4x + 1)\cancel{(x - 6)}}{-1\cancel{(x - 6)}}$$
$$= \frac{4x + 1}{-1} = -(4x + 1)$$

Exercise Set 6.1

Determine the value or values where the expression is defined. (See Example 1.)

1. $\dfrac{x + 4}{x}$

2. $\dfrac{2}{x + 5}$

3. $\dfrac{4}{x - 6}$

4. $\dfrac{5}{2x - 3}$

5. $\dfrac{x + 4}{x^2 - 4}$

6. $\dfrac{7}{x^2 - 7x + 12}$

7. $\dfrac{x - 3}{x^2 + 6x - 16}$

8. $\dfrac{x^2 + 3}{2x^2 - 13x + 15}$

Reduce each expression to its lowest terms.

9. $\dfrac{x}{x + xy}$

10. $\dfrac{3x}{6x + 9}$

11. $\dfrac{4x + 12}{x + 3}$

12. $\dfrac{3x^2 + 6x}{3x^2 + 9x}$

13. $\dfrac{x^3 + 6x^2 + 3x}{2x}$

14. $\dfrac{x^2y^2 - xy + 3y}{y}$

15. $\dfrac{x^2 + 2x + 1}{x + 1}$

16. $\dfrac{x - 1}{x^2 + 2x - 3}$

17. $\dfrac{x^2 - 2x}{x^2 - 4x + 4}$

18. $\dfrac{x^2 + 3x - 18}{2x - 6}$

19. $\dfrac{x^2 - x - 6}{x^2 - 4}$

20. $\dfrac{x^2 + 6x + 9}{x^2 - 9}$

21. $\dfrac{2x^2 - 4x - 6}{x - 3}$

22. $\dfrac{4x^2 - 12x - 40}{2x^2 - 16x + 30}$

23. $\dfrac{2x - 3}{3 - 2x}$

24. $\dfrac{4x - 8}{4 - 2x}$

25. $\dfrac{x^2 - 2x - 8}{4 - x}$

26. $\dfrac{5 - x}{x^2 - 2x - 15}$

27. $\dfrac{x^2 + 3x - 18}{-2x^2 + 6x}$

28. $\dfrac{2x^2 + 5x - 12}{2x - 3}$

29. $\dfrac{2x^2 + 5x - 3}{1 - 2x}$

30. $\dfrac{x^2 - 9}{x^2 - 2x - 15}$

31. $\dfrac{6x^2 + x - 2}{2x - 1}$

32. $\dfrac{4x^2 - 13x + 10}{(x - 2)^2}$

33. $\dfrac{6x^2 + 7x - 20}{2x + 5}$

34. $\dfrac{16x^2 + 24x + 9}{4x + 3}$

35. $\dfrac{6x^2 - 13x + 6}{3x - 2}$

36. $\dfrac{x^2 - 25}{(x - 5)^2}$

37. $\dfrac{x^2 - 3x + 4x - 12}{x - 3}$

38. $\dfrac{x^2 - 2x + 4x - 8}{2x^2 + 3x + 8x + 12}$

39. $\dfrac{2x^2 - 8x + 3x - 12}{2x^2 + 8x + 3x + 12}$

40. $\dfrac{x^3 + 1}{x^2 - x + 1}$

41. $\dfrac{x^3 - 8}{x - 2}$

42. In any rational expression where there is a variable in the denominator, what do we always assume about the variable?

43. Explain why x can represent any real number in the expression $(x + 3)/(x^2 + 4)$.

44. Consider the rational expression $\dfrac{x - 3}{x - 3}$. What value, if any, can x not represent? Explain your answer.

45. Consider the expression $\dfrac{x}{(x - 4)^2}$. What value, if any, can x not represent? Explain your answer.

46. Is $-\dfrac{x + 4}{4 - x}$ equal to -1? Explain how you determined your answer.

47. Is $-\dfrac{3x + 2}{-3x - 2}$ equal to 1? Explain how you determined your answer.

48. In your own words, explain how to reduce a rational expression to lowest terms.

49. Determine the denominator that will make the statement $\dfrac{x^2 - x - 6}{\rule{1cm}{0.4pt}} = x - 3$ true. Explain how you determined your answer.

50. Determine the numerator that will make the statement $\dfrac{\rule{1cm}{0.4pt}}{x + 4} = x + 3$ true. Explain how you determined your answer.

Cumulative Review Exercises

[3.2] **51.** Solve the formula $z = \dfrac{x - y}{2}$ for y.

[3.4] **52.** Find the measure of the three angles of a triangle if one angle is $30°$ greater than the smallest angle, and the third angle is $10°$ greater than three times the smallest angle.

[4.1] **53.** Simplify $\left(\dfrac{3x^6 y^2}{9x^4 y^3}\right)^2$.

[4.4] **54.** Subtract $6x^2 - 4x - 8 - (-3x^2 + 6x + 9)$.

6.2

Multiplication and Division of Rational Expressions

▶ **1** Multiply rational expressions.

▶ **2** Divide rational expressions.

▶ **1** In Section 1.2 we reviewed multiplication of numerical fractions. Recall that to multiply two fractions we multiply their numerators together and multiply their denominators together.

Multiplication

$$\frac{a}{b} \cdot \frac{c}{d} = \frac{a \cdot c}{b \cdot d}, \qquad b \neq 0 \quad \text{and} \quad d \neq 0$$

EXAMPLE 1   Multiply $\left(\dfrac{3}{5}\right)\left(\dfrac{-2}{9}\right)$.

Solution: First divide out common factors; then multiply.

$$\frac{\overset{1}{\cancel{3}}}{5} \cdot \frac{-2}{\underset{3}{\cancel{9}}} = \frac{1 \cdot (-2)}{5 \cdot 3} = -\frac{2}{15}$$

The same principles apply when multiplying rational expressions containing variables. Before multiplying, you should first divide out any factors common to both a numerator and a denominator.

To Multiply Rational Expressions

---

1. Factor all numerators and denominators as completely as possible.
2. Divide out common factors.
3. Multiply numerators together and multiply denominators together.

EXAMPLE 2   Multiply $\dfrac{3x^2}{2y} \cdot \dfrac{4y^3}{3x}$.

Solution: This problem can be represented as

$$\frac{3xx}{2y} \cdot \frac{4yyy}{3x}$$

$$\frac{\overset{1\,1}{\cancel{3}\cancel{x}x}}{2y} \cdot \frac{4yyy}{\underset{1\,1}{\cancel{3}\cancel{x}}} \qquad \text{Divide out the 3's and } x\text{'s.}$$

$$\frac{\overset{1\,1}{\cancel{3}\cancel{x}x}}{\underset{1\,1}{2\cancel{y}}} \cdot \frac{\overset{2\,1}{\cancel{4}\cancel{y}yy}}{\underset{1\,1}{\cancel{3}\cancel{x}}} \qquad \text{Divide both the 4 and the 2 by 2, and divide out the } y\text{'s.}$$

Now multiply the remaining numerators together and the remaining denominators together.

$$\frac{2xy^2}{1} \quad \text{or} \quad 2xy^2$$

Rather than illustrating this entire process when multiplying rational expressions, we will often proceed as follows:

$$\frac{3x^2}{2y} \cdot \frac{4y^3}{3x}$$

$$= \frac{\overset{1\ x}{\cancel{3}\cancel{x}^2}}{\underset{1\,1}{2\cancel{y}}} \cdot \frac{\overset{2\ y^2}{\cancel{4}\cancel{y}^3}}{\underset{1\,1}{\cancel{3}\cancel{x}}} = 2xy^2 \qquad \blacksquare$$

EXAMPLE 3   Multiply $\dfrac{-2a^3b^2}{3x^3y} \cdot \dfrac{4a^2x}{5b^2y^3}$.

Solution: $\dfrac{-2a^3\cancel{b^2}}{3x^3y} \cdot \dfrac{4a^2x}{5\cancel{b^2}y^3} = \dfrac{-8a^5}{15x^2y^4}$. $\blacksquare$

EXAMPLE 4   Multiply $(x - 2) \cdot \dfrac{3}{x^2 - 2x}$.

Solution: $\dfrac{\cancel{(x-2)}}{1} \cdot \dfrac{3}{x\cancel{(x-2)}} = \dfrac{3}{x}$. $\blacksquare$

EXAMPLE 5   Multiply $\dfrac{(x+2)^2}{6x^2} \cdot \dfrac{3x}{x^2-4}$.

Solution: $\dfrac{(x+2)(x+2)}{6x^2} \cdot \dfrac{3x}{(x+2)(x-2)}$

$$= \frac{\cancel{(x+2)}(x+2)}{\cancel{6}x^{\cancel{2}}} \cdot \frac{\overset{1}{\cancel{3}}\overset{1}{\cancel{x}}}{\cancel{(x+2)}(x-2)} = \frac{x+2}{2x(x-2)}.$$

This answer cannot be simplified further.   ■

EXAMPLE 6   Multiply $\dfrac{x-3}{2x} \cdot \dfrac{4x}{3-x}$.

Solution: $\dfrac{x-3}{\underset{1}{\cancel{2x}}} \cdot \dfrac{\overset{2}{\cancel{4x}}}{3-x} = \dfrac{2(x-3)}{3-x}$.

This problem is still not complete. In Section 6.1 we showed that $3-x$ is $-1(-3+x)$ or $-1(x-3)$. Thus

$$\frac{2(x-3)}{3-x} = \frac{2\cancel{(x-3)}}{-1\cancel{(x-3)}} = -2$$ ■

HELPFUL HINT

When only the signs differ in a numerator and denominator in a multiplication problem, factor out -1 *from either the numerator or denominator;* then divide out the common factor.

$$\frac{a-b}{x} \cdot \frac{y}{b-a} = \frac{\cancel{a-b}}{x} \cdot \frac{y}{-1\cancel{(a-b)}} = \frac{-y}{x}$$

EXAMPLE 7   Multiply $\dfrac{3x+2}{2x-1} \cdot \dfrac{4-8x}{3x+2}$.

Solution: $\dfrac{3x+2}{2x-1} \cdot \dfrac{4-8x}{3x+2} = \dfrac{3x+2}{2x-1} \cdot \dfrac{4(1-2x)}{3x+2} = \dfrac{(3x+2)}{2x-1} \cdot \dfrac{4(1-2x)}{(3x+2)}$.

Note that the factor $(1-2x)$ in the numerator of the second fraction differs only in sign from the denominator of the first fraction $(2x-1)$. We will therefore factor a -1 from the numerator.

$$= \frac{\cancel{(3x+2)}}{(2x-1)} \cdot \frac{4(-1)(2x-1)}{\cancel{(3x+2)}}$$

$$= \frac{\cancel{(3x+2)}}{\cancel{(2x-1)}} \cdot \frac{-4\cancel{(2x-1)}}{\cancel{(3x+2)}} = \frac{-4}{1} = -4$$ ■

EXAMPLE 8   Multiply $\dfrac{2x^2+7x-15}{4x^2-8x+3} \cdot \dfrac{2x^2+x-1}{x^2+6x+5}$.

Solution:
$$\frac{(2x - 3)(x + 5)}{(2x - 3)(2x - 1)} \cdot \frac{(2x - 1)(x + 1)}{(x + 1)(x + 5)}$$

$$= \frac{\cancel{(2x - 3)}\cancel{(x + 5)}}{\cancel{(2x - 3)}\cancel{(2x - 1)}} \cdot \frac{\cancel{(2x - 1)}\cancel{(x + 1)}}{\cancel{(x + 1)}\cancel{(x + 5)}} = 1$$

EXAMPLE 9 Multiply $\dfrac{2x^3 - 14x^2 + 12x}{6y^2} \cdot \dfrac{-2y}{3x^2 - 3x}$.

Solution:
$$\frac{2x(x^2 - 7x + 6)}{6y^2} \cdot \frac{-2y}{3x(x - 1)}$$

$$= \frac{2x(x - 6)(x - 1)}{6y^2} \cdot \frac{-2y}{3x(x - 1)}$$

$$= \frac{\cancel{2x}(x - 6)\cancel{(x - 1)}}{\underset{3\ y}{\cancel{6y^2}}} \cdot \frac{-2\cancel{y}}{3\cancel{x}\cancel{(x - 1)}}$$

$$= \frac{-2(x - 6)}{9y}$$

Note that $\dfrac{-2(x - 6)}{9y}$ and $\dfrac{-2x + 12}{9y}$ are both acceptable answers.

EXAMPLE 10 Multiply $\dfrac{x^2 - y^2}{x + y} \cdot \dfrac{x + 2y}{2x^2 - xy - y^2}$.

Solution:
$$\frac{(x + y)(x - y)}{x + y} \cdot \frac{x + 2y}{(2x + y)(x - y)}$$

$$= \frac{\cancel{(x + y)}\cancel{(x - y)}}{\cancel{(x + y)}} \cdot \frac{x + 2y}{(2x + y)\cancel{(x - y)}}$$

$$= \frac{x + 2y}{2x + y}$$

Division of Rational Expressions

▶ **2** In Chapter 1 we learned that to divide one fraction by a second we invert the divisor and multiply.

Division

$$\frac{a}{b} \div \frac{c}{d} = \frac{a}{b} \cdot \frac{d}{c} = \frac{ad}{bc}, \qquad b \neq 0, \quad d \neq 0, \quad \text{and} \quad c \neq 0$$

EXAMPLE 11 Divide as indicated.

(a) $\dfrac{3}{5} \div \dfrac{4}{5}$ (b) $\dfrac{2}{3} \div \dfrac{5}{6}$

Solution: (a) $\dfrac{3}{\cancel{5}} \cdot \dfrac{\overset{1}{\cancel{5}}}{4} = \dfrac{3 \cdot 1}{1 \cdot 4} = \dfrac{3}{4}$ (b) $\dfrac{2}{\cancel{3}} \cdot \dfrac{\overset{2}{\cancel{6}}}{5} = \dfrac{2 \cdot 2}{1 \cdot 5} = \dfrac{4}{5}$

The same principles are used when dividing rational expressions.

To Divide Rational Expressions

Invert the divisor (the second fraction) and multiply.

EXAMPLE 12 $\dfrac{5x^2}{z} \div \dfrac{4z^3}{3}$.

Solution: $\dfrac{5x^2}{z} \cdot \dfrac{3}{4z^3} = \dfrac{15x^2}{4z^4}$.

EXAMPLE 13 $\dfrac{x^2 - 9}{x + 4} \div \dfrac{x - 3}{x + 4}$.

Solution: $\dfrac{x^2 - 9}{x + 4} \cdot \dfrac{x + 4}{x - 3} = \dfrac{(x + 3)(x - 3)}{(x + 4)} \cdot \dfrac{(x + 4)}{(x - 3)}$

$$= x + 3.$$

EXAMPLE 14 $\dfrac{-1}{2x - 3} \div \dfrac{3}{3 - 2x}$.

Solution: $\dfrac{-1}{2x - 3} \cdot \dfrac{3 - 2x}{3} = \dfrac{-1}{(2x - 3)} \cdot \dfrac{-1(2x - 3)}{3}$

$$= \dfrac{(-1)(-1)}{(1)(3)} = \dfrac{1}{3}.$$

EXAMPLE 15 $\dfrac{x^2 + 8x + 15}{x^2} \div (x + 3)^2$.

Solution: $\dfrac{x^2 + 8x + 15}{x^2} \cdot \dfrac{1}{(x + 3)^2} = \dfrac{(x + 5)(x + 3)}{x^2} \cdot \dfrac{1}{(x + 3)(x + 3)}$

$$= \dfrac{x + 5}{x^2(x + 3)}.$$

EXAMPLE 16 $\dfrac{12x^2 - 22x + 8}{3x} \div \dfrac{3x^2 + 2x - 8}{2x^2 + 4x}$.

Solution: $\dfrac{12x^2 - 22x + 8}{3x} \cdot \dfrac{2x^2 + 4x}{3x^2 + 2x - 8} = \dfrac{2(6x^2 - 11x + 4)}{3x} \cdot \dfrac{2x(x + 2)}{(3x - 4)(x + 2)}$

$$= \dfrac{2(3x - 4)(2x - 1)}{3x} \cdot \dfrac{2x(x + 2)}{(3x - 4)(x + 2)}$$

$$= \dfrac{4(2x - 1)}{3}$$

Exercise Set 6.2

Multiply as indicated.

1. $\dfrac{3x}{2y} \cdot \dfrac{y^2}{6}$

2. $\dfrac{15x^3y^2}{z} \cdot \dfrac{z}{5xy^3}$

3. $\dfrac{16x^2}{y^4} \cdot \dfrac{5x^2}{y^2}$

4. $\dfrac{32m}{5n^3} \cdot \dfrac{-15m^2n^3}{4}$

5. $\dfrac{6x^5y^3}{5z^3} \cdot \dfrac{6x^4}{5yz^4}$

6. $\dfrac{x^2 - 4}{x^2 - 9} \cdot \dfrac{x + 3}{x - 2}$

7. $\dfrac{3x - 2}{3x + 2} \cdot \dfrac{4x - 1}{1 - 4x}$

8. $\dfrac{x - 6}{2x + 5} \cdot \dfrac{2x}{-x + 6}$

9. $\dfrac{x^2 + 7x + 12}{x + 4} \cdot \dfrac{1}{x + 3}$

10. $\dfrac{x^2 + 3x - 10}{2x} \cdot \dfrac{x^2 - 3x}{x^2 - 5x + 6}$

11. $\dfrac{a^2 - b^2}{a} \cdot \dfrac{a^2 + ab}{a + b}$

12. $\dfrac{x^2 - 25}{x^2 - 3x - 10} \cdot \dfrac{x + 2}{x}$

13. $\dfrac{6x^2 - 14x - 12}{6x + 4} \cdot \dfrac{x + 3}{2x^2 - 2x - 12}$

14. $\dfrac{2x^2 - 9x + 9}{8x - 12} \cdot \dfrac{2x}{x^2 - 3x}$

15. $\dfrac{x + 3}{x - 3} \cdot \dfrac{x^3 - 27}{x^2 + 3x + 9}$

16. $\dfrac{x^3 + 8}{x^2 - x - 6} \cdot \dfrac{x + 3}{x^2 - 2x + 4}$

Divide as indicated.

17. $\dfrac{6x^3}{y} \div \dfrac{2x}{y^2}$

18. $\dfrac{9x^3}{4} \div \dfrac{1}{16y^2}$

19. $\dfrac{25xy^2}{7z} \div \dfrac{5x^2y^2}{14z^2}$

20. $\dfrac{36y}{7z^2} \div \dfrac{3xy}{2z}$

21. $\dfrac{7a^2b}{xy} \div \dfrac{7}{6xy}$

22. $2xz \div \dfrac{4xy}{z}$

23. $\dfrac{3x^2 + 6x}{x} \div \dfrac{2x + 4}{x^2}$

24. $\dfrac{x - 3}{4y^2} \div \dfrac{x^2 - 9}{2xy}$

25. $(x - 3) \div \dfrac{x^2 + 3x - 18}{x}$

26. $\dfrac{1}{x^2 - 17x + 30} \div \dfrac{1}{x^2 + 7x - 18}$

27. $\dfrac{x^2 - 12x + 32}{x^2 - 6x - 16} \div \dfrac{x^2 - x - 12}{x^2 - 5x - 24}$

28. $\dfrac{a - b}{9a + 9b} \div \dfrac{a^2 - b^2}{a^2 + 2a + 1}$

29. $\dfrac{2x^2 + 9x + 4}{x^2 + 7x + 12} \div \dfrac{2x^2 - x - 1}{(x + 3)^2}$

30. $\dfrac{a^2 - b^2}{9} \div \dfrac{3a - 3b}{27x^2}$

31. $\dfrac{x^2 - y^2}{x^2 - 2xy + y^2} \div \dfrac{x + y}{x - y}$

32. $\dfrac{9x^2 - 9y^2}{6x^2y^2} \div \dfrac{3x + 3y}{12x^2y^5}$

Perform the indicated operation.

33. $\dfrac{12x^2}{6y^2} \cdot \dfrac{36xy^5}{12}$

34. $\dfrac{y^3}{8} \cdot \dfrac{9x^2}{y^3}$

35. $\dfrac{45a^2b^3}{12c^3} \cdot \dfrac{4c}{9a^3b^5}$

36. $\dfrac{-2xw}{y^5} \div \dfrac{6x^2}{y^6}$

37. $\dfrac{-xy}{a} \div \dfrac{-2ax}{6y}$

38. $\dfrac{27x}{5y^2} \div 3x^2y^2$

39. $\dfrac{80m^4}{49x^5y^7} \cdot \dfrac{14x^{12}y^5}{25m^5}$

40. $\dfrac{-18x^2y}{11z^2} \cdot \dfrac{22z^3}{x^2y^5}$

41. $(2x + 5) \cdot \dfrac{1}{4x + 10}$

42. $\dfrac{1}{4x - 3} \cdot (20x - 15)$

43. $\dfrac{1}{7x^2y} \cdot \dfrac{1}{21x^3y}$

44. $\dfrac{x^2y^5}{3z} \div \dfrac{3z}{2x}$

45. $\dfrac{12a^2}{4bc} \div \dfrac{3a^2}{bc}$

46. $\dfrac{4 - x}{x - 4} \cdot \dfrac{x - 3}{3 - x}$

47. $\dfrac{5 - 2x}{x + 8} \cdot \dfrac{-x - 8}{2x - 5}$

48. $\dfrac{x - 3}{x + 5} \cdot \dfrac{2x^2 + 10x}{2x - 6}$

49. $\dfrac{6x + 6y}{a} \div \dfrac{12x + 12y}{a^2}$

50. $\dfrac{2a + 2b}{3} \div \dfrac{a^2 - b^2}{a - b}$

51. $\dfrac{a^2b^2}{6x + 6y} \div \dfrac{ab}{x^2 - y^2}$

52. $\dfrac{1}{-x - 4} \div \dfrac{x^2 - 7x}{x^2 - 3x - 28}$

53. $\dfrac{x^2 - 5x - 24}{x^2 - x - 12} \cdot \dfrac{x^2 + x - 6}{x^2 - 10x + 16}$

54. $\dfrac{4x + 4y}{xy^2} \cdot \dfrac{x^2y}{3x + 3y}$

55. $\dfrac{a^2 + 6a + 9}{a^2 - 4} \cdot \dfrac{a - 2}{a + 3}$

56. $\dfrac{x^2}{x^2 - 4} \cdot \dfrac{x^2 - 5x + 6}{x^2 - 3x}$

57. $\dfrac{x^2 + 10x + 21}{x + 7} \div (x + 3)$

58. $\dfrac{x^2 - 9x + 14}{x^2 - 5x + 6} \div \dfrac{x^2 - 5x - 14}{x + 2}$

59. $\dfrac{3x^2 - x - 2}{x + 7} \div \dfrac{x - 1}{4x^2 + 25x - 21}$

60. $\dfrac{(x + 2)^2}{x - 2} \div \dfrac{x^2 - 4}{2x - 4}$

61. $\dfrac{9x^2 + 6x - 8}{x - 3} \cdot \dfrac{(x - 3)^2}{3x + 4}$

62. $\dfrac{5x^2 + 17x + 6}{x + 3} \cdot \dfrac{x - 1}{5x^2 + 7x + 2}$

63. $\dfrac{2x + 4y}{x^2 + 4xy + 4y^2} \cdot \dfrac{x + 2y}{2}$

64. $\dfrac{x^2 - y^2}{x^2 + xy} \cdot \dfrac{3x^2 + 6x}{3x^2 - 2xy - y^2}$

65. $\dfrac{x^2 - 4}{2y} \div \dfrac{2 - x}{6xy}$

66. $\dfrac{2x^2 + 7x - 15}{1 - x} \div \dfrac{3x^2 + 13x - 10}{x - 1}$

67. $\dfrac{x^2 - y^2}{8x^2 - 16xy + 8y^2} \cdot \dfrac{4x - 4y}{x + y}$

68. $\dfrac{x^2 - 4y^2}{x^2 + 3xy + 2y^2} \cdot \dfrac{x + y}{x^2 - 4xy + 4y^2}$

69. $\dfrac{x^3 - 64}{x + 4} \div \dfrac{x^2 + 4x + 16}{x^2 + 8x + 16}$

Consider the multiplication problems that follow. What polynomial should be in the shaded area of the second fraction to make the statement true? Explain how you determined your answer.

70. $\dfrac{x + 3}{x - 4} \cdot \dfrac{\boxed{}}{x + 3} = x + 2$

71. $\dfrac{x - 5}{x + 2} \cdot \dfrac{\boxed{}}{x - 5} = 2x - 3$

72. In your own words, explain how to multiply rational expressions.

73. In your own words, explain how to divide rational expressions.

Cumulative Review Exercises

[4.5] **74.** Multiply $(4x^3y^2z^4)(5xy^3z^7)$.

[4.6] **75.** Divide $\dfrac{4x^3 - 5x}{2x - 1}$.

[5.4] **76.** Factor $3x^2 - 9x - 30$.

[5.6] **77.** Solve $3x^2 - 9x - 30 = 0$.

JUST FOR FUN

Simplify each of the following:

1. $\left(\dfrac{x + 2}{x^2 - 4x - 12} \cdot \dfrac{x^2 - 9x + 18}{x - 2} \right) \div \dfrac{x^2 + 5x + 6}{x^2 - 4}$

2. $\left(\dfrac{x^2 - x - 6}{2x^2 - 9x + 9} \div \dfrac{x^2 + x - 12}{x^2 + 3x - 4} \right) \cdot \dfrac{2x^2 - 5x + 3}{x^2 + x - 2}$

6.3

Addition and Subtraction of Rational Expressions with a Common Denominator

▶ **1**   Add and subtract rational expressions with a common denominator.

▶ **1**   Recall that when adding (or subtracting) two arithmetic fractions with a common denominator we add (or subtract) the numerators while keeping the common denominator.

Addition and Subtraction

$$\frac{a}{c} + \frac{b}{c} = \frac{a + b}{c}, c \neq 0 \qquad \frac{a}{c} - \frac{b}{c} = \frac{a - b}{c}, c \neq 0$$

EXAMPLE 1   Add $\dfrac{3}{8} + \dfrac{2}{8}$.

Solution: $\dfrac{3}{8} + \dfrac{2}{8} = \dfrac{3 + 2}{8} = \dfrac{5}{8}$

Note that we did not reduce $\frac{2}{8}$ to $\frac{1}{4}$ because the common denominator is 8. Remember that we do not add the denominators; *only the numerators are added*.   ∎

EXAMPLE 2 $\dfrac{5}{7} - \dfrac{1}{7}$.

Solution: $\dfrac{5}{7} - \dfrac{1}{7} = \dfrac{5 - 1}{7} = \dfrac{4}{7}$ ∎

To add or subtract rational expressions, we use the same principle.

To Add or Subtract Expressions with a Common Denominator

1. Add or subtract the numerators.
2. Place the sum or difference of the numerators found in step 1 over the common denominator.
3. Reduce the fraction if possible.

EXAMPLE 3 $\dfrac{3}{x + 2} + \dfrac{x - 4}{x + 2}$.

Solution: $\dfrac{3}{x + 2} + \dfrac{x - 4}{x + 2} = \dfrac{3 + (x - 4)}{x + 2} = \dfrac{x - 1}{x + 2}$.   ∎

EXAMPLE 4 $\dfrac{3x + 5}{x - 3} - \dfrac{2x}{x - 3}$.

Solution: $\dfrac{3x + 5}{x - 3} - \dfrac{2x}{x - 3} = \dfrac{3x + 5 - 2x}{x - 3} = \dfrac{x + 5}{x - 3}$.   ∎

EXAMPLE 5 $\dfrac{2x^2 + 5}{x + 3} + \dfrac{6x - 5}{x + 3}$.

Solution: $\dfrac{2x^2 + 5}{x + 3} + \dfrac{6x - 5}{x + 3} = \dfrac{2x^2 + 5 + (6x - 5)}{x + 3} = \dfrac{2x^2 + 6x}{x + 3}$.

Now factor $2x$ from each term in the numerator and reduce.

$$= \frac{2x\cancel{(x + 3)}}{\cancel{(x + 3)}} = 2x \qquad \blacksquare$$

EXAMPLE 6 $\dfrac{x^2 + 3x - 2}{(x + 5)(x - 2)} + \dfrac{4x + 12}{(x + 5)(x - 2)}$.

Solution: $\dfrac{x^2 + 3x - 2}{(x + 5)(x - 2)} + \dfrac{4x + 12}{(x + 5)(x - 2)} = \dfrac{x^2 + 3x - 2 + (4x + 12)}{(x + 5)(x - 2)}$

$$= \frac{x^2 + 7x + 10}{(x + 5)(x - 2)}$$

$$= \frac{\cancel{(x + 5)}(x + 2)}{\cancel{(x + 5)}(x - 2)}$$

$$= \frac{x + 2}{x - 2} \qquad \blacksquare$$

When subtracting rational expressions, be sure to subtract the entire numerator of the fraction being subtracted. Study the following common student error very carefully.

COMMON STUDENT ERROR

Consider the problem

$$\frac{4x}{x - 2} - \frac{2x + 1}{x - 2}$$

Many students begin problems of this type incorrectly. Here are the correct and incorrect ways of working this subtraction problem.

Correct	*Wrong*
$\dfrac{4x}{x - 2} - \dfrac{2x + 1}{x - 2} = \dfrac{4x - (2x + 1)}{x - 2}$	$\dfrac{\cancel{4x}}{\cancel{x - 2}} - \dfrac{2x + 1}{x - 2} = \dfrac{4x - 2x + 1}{x - 2}$
$= \dfrac{4x - 2x - 1}{x - 2}$	
$= \dfrac{2x - 1}{x - 2}$	

Note that the entire numerator of the second fraction (not just the first term) **must be subtracted.** Also note that the sign of *each* term of the numerator being subtracted will change when the parentheses are removed.

EXAMPLE 7 $\dfrac{x^2 - 2x + 3}{x^2 + 7x + 12} - \dfrac{x^2 - 4x - 5}{x^2 + 7x + 12}.$

Solution: $\dfrac{x^2 - 2x + 3}{x^2 + 7x + 12} - \dfrac{x^2 - 4x - 5}{x^2 + 7x + 12} = \dfrac{x^2 - 2x + 3 - (x^2 - 4x - 5)}{x^2 + 7x + 12}$

$$= \dfrac{x^2 - 2x + 3 - x^2 + 4x + 5}{x^2 + 7x + 12}$$

$$= \dfrac{2x + 8}{x^2 + 7x + 12}$$

$$= \dfrac{2(x + 4)}{(x + 3)(x + 4)}$$

$$= \dfrac{2}{x + 3}$$

EXAMPLE 8 $\dfrac{5x}{x - 6} - \dfrac{4x^2 - 16x - 18}{x - 6}.$

Solution: $\dfrac{5x}{x - 6} - \dfrac{4x^2 - 16x - 18}{x - 6} = \dfrac{5x - (4x^2 - 16x - 18)}{x - 6}$

$$= \dfrac{5x - 4x^2 + 16x + 18}{x - 6}$$

$$= \dfrac{-4x^2 + 21x + 18}{x - 6}$$

$$= \dfrac{-(4x^2 - 21x - 18)}{x - 6}$$

$$= \dfrac{-(4x + 3)(x - 6)}{(x - 6)}$$

$$= -(4x + 3)$$

Exercise Set 6.3

Add or subtract as indicated.

1. $\dfrac{x - 1}{6} + \dfrac{x}{6}$

2. $\dfrac{x + 11}{8} + \dfrac{2x + 5}{8}$

3. $\dfrac{x - 7}{3} - \dfrac{4}{3}$

4. $\dfrac{2x + 3}{5} - \dfrac{x}{5}$

5. $\dfrac{x + 2}{x} - \dfrac{5}{x}$

6. $\dfrac{3x + 6}{2} - \dfrac{x}{2}$

7. $\dfrac{1}{x} + \dfrac{x + 2}{x}$

8. $\dfrac{3x + 4}{x + 1} + \dfrac{6x + 5}{x + 1}$

9. $\dfrac{4}{x + 2} + \dfrac{x + 3}{x + 2}$

10. $\dfrac{x - 3}{x} + \dfrac{x + 3}{x}$

11. $\dfrac{x - 4}{x} - \dfrac{x + 4}{x}$

12. $\dfrac{x}{x - 2} + \dfrac{2x + 3}{x - 2}$

13. $\dfrac{4x - 3}{x - 7} - \dfrac{2x + 8}{x - 7}$

14. $\dfrac{4x - 5}{3x^2} + \dfrac{2x + 5}{3x^2}$

15. $\dfrac{9x + 7}{6x^2} - \dfrac{3x + 4}{6x^2}$

16. $\dfrac{4x - 6}{x^2 + 2x} + \dfrac{7x + 5}{x^2 + 2x}$

17. $\dfrac{-2x - 4}{x^2 + 2x + 1} + \dfrac{3x + 5}{x^2 + 2x + 1}$

18. $\dfrac{-2x + 6}{x^2 + x - 6} + \dfrac{3x - 3}{x^2 + x - 6}$

19. $\dfrac{4}{x^2 - 2x - 3} + \dfrac{x - 3}{x^2 - 2x - 3}$

20. $\dfrac{-x - 4}{x^2 - 16} + \dfrac{2(x + 4)}{x^2 - 16}$

21. $\dfrac{x + 4}{3x + 2} - \dfrac{x + 4}{3x + 2}$

22. $\dfrac{2x + 4}{(x + 2)(x - 3)} - \dfrac{x + 7}{(x + 2)(x - 3)}$

23. $\dfrac{2x + 4}{x - 7} - \dfrac{6x + 5}{x - 7}$

24. $\dfrac{x^2 + 2x}{3x} - \dfrac{x^2 + 5x + 6}{3x}$

25. $\dfrac{x^2 + 4x + 3}{x + 2} - \dfrac{5x + 9}{x + 2}$

26. $\dfrac{3x^2}{x^2 + 2x} - \dfrac{4x}{x^2 + 2x}$

27. $\dfrac{4}{2x + 3} + \dfrac{6x + 5}{2x + 3}$

28. $\dfrac{-2x + 5}{5x - 10} + \dfrac{2(x - 5)}{5x - 10}$

29. $\dfrac{x^2}{x + 3} + \dfrac{9}{x + 3}$

30. $\dfrac{x^2 - 2x - 3}{x^2 - x - 6} + \dfrac{x - 3}{x^2 - x - 6}$

31. $\dfrac{4x + 12}{3 - x} - \dfrac{3x + 15}{3 - x}$

32. $\dfrac{-x - 7}{2x - 9} - \dfrac{-3x - 16}{2x - 9}$

33. $\dfrac{x^2 - 2}{x^2 + 6x - 7} - \dfrac{-4x + 19}{x^2 + 6x - 7}$

34. $\dfrac{x^2 + 6x}{(x + 9)(x + 5)} - \dfrac{27}{(x + 9)(x + 5)}$

35. $\dfrac{x^2 - 13}{x + 4} - \dfrac{3}{x + 4}$

36. $\dfrac{x^2 - 6}{2x + 3} - \dfrac{-3x^2 + 3}{2x + 3}$

37. $\dfrac{3x^2 - 7x}{4x^2 - 8x} + \dfrac{x}{4x^2 - 8x}$

38. $\dfrac{3x^2 + 15x}{x^3 + 2x^2 - 8x} + \dfrac{2x^2 + 5x}{x^3 + 2x^2 - 8x}$

39. $\dfrac{2x^2 - 6x + 5}{2x^2 + 18x + 16} - \dfrac{8x + 21}{2x^2 + 18x + 16}$

40. $\dfrac{x^3 - 10x^2 + 35x}{x(x - 6)} - \dfrac{x^2 + 5x}{x(x - 6)}$

41. $\dfrac{x^2 + 3x - 6}{x^2 - 5x + 4} - \dfrac{-2x^2 + 4x - 4}{x^2 - 5x + 4}$

42. $\dfrac{4x^2 + 5}{9x^2 - 64} - \dfrac{x^2 - x + 29}{9x^2 - 64}$

43. $\dfrac{5x^2 + 40x + 8}{x^2 - 64} + \dfrac{x^2 + 9x}{x^2 - 64}$

44. $\dfrac{20x^2 + 5x + 1}{6x^2 + x - 2} - \dfrac{8x^2 - 12x - 5}{6x^2 + x - 2}$

45. When subtracting rational expressions, what must happen to the sign of each term of the numerator being subtracted?

46. Explain why

$$\dfrac{4x - 3}{5x + 4} - \dfrac{2x - 7}{5x + 4} \neq \dfrac{4x - 3 - 2x - 7}{5x + 4}$$

Show what the next step should be.

47. Explain why

$$\dfrac{6x - 2}{x^2 - 4x + 3} - \dfrac{3x^2 - 4x + 5}{x^2 - 4x + 3} \neq \dfrac{6x - 2 - 3x^2 - 4x + 5}{x^2 - 4x + 3}$$

Show what the next step should be.

48. Explain why

$$\dfrac{4x + 5}{x^2 - 6x} - \dfrac{-x^2 + 3x + 6}{x^2 - 6x} \neq \dfrac{4x + 5 + x^2 + 3x + 6}{x^2 - 6x}$$

Show what the next step should be.

In Exercises 49 through 52, list the polynomial to be placed to the shaded area to make a true statement. Explain how you determined your answer.

49. $\dfrac{x^2 - 6x + 3}{x + 3} + \dfrac{\rule{1.2em}{0.6em}}{x + 3} = \dfrac{2x^2 - 5x - 6}{x + 3}$

50. $\dfrac{-x^2 - 4x + 3}{2x + 5} + \dfrac{\rule{1.2em}{0.6em}}{2x + 5} = \dfrac{5x - 7}{2x + 5}$

51. $\dfrac{4x^2 - 6x - 7}{x^2 - 4} - \dfrac{\rule{1.2em}{0.6em}}{x^2 - 4} = \dfrac{2x^2 + x - 3}{x^2 - 4}$

52. $\dfrac{-3x^2 - 9}{(x + 4)(x - 2)} - \dfrac{\rule{3em}{0.6em}}{(x + 4)(x - 2)} = \dfrac{x^2 + 3x}{(x + 4)(x - 2)}$

53. In your own words, explain how to add or subtract rational expressions that have the same denominator.

Cumulative Review Exercises

[2.5] **54.** Solve the equation:
$6x + 4 = -(x + 2) - 3x + 4$.

[2.6] **55.** The instructions on a bottle of concentrated hummingbird food indicate that 6 ounces of the concentrate should be mixed with one gallon (128 ounces) of water. If you wish to mix the concentrate with only 24 ounces of water, how much concentrate should be used?

[3.3] **56.** The American Health Racquet Club has two payment plans. Plan 1 is a yearly membership fee of $100 plus $2 per hour for use of the court. Plan 2 is an annual membership fee of $250 with no charge for court time.

(a) How many hours would Shamo have to play in a year to make the cost of plan 1 equal to the cost of plan 2?

(b) If Shamo plans to play an average of 4 hours per week for the year, which plan should he use?

See Exercise 56.

[6.2] **57.** Divide $\dfrac{x^2 + x - 6}{2x^2 + 7x + 3} \div \dfrac{x^2 + 5x + 6}{x^2 - 4}$.

JUST FOR FUN

Add or subtract as indicated.

1. $\dfrac{3x - 2}{x^2 - 9} - \dfrac{4x^2 - 6}{x^2 - 9} + \dfrac{5x - 1}{x^2 - 9}$.

2. $\dfrac{x^2 - 6x + 3}{x + 2} + \dfrac{x^2 - 2x}{x + 2} - \dfrac{2x^2 - 3x + 5}{x + 2}$

6.4

Finding the Least Common Denominator

▶**1** Find the least common denominator for rational expressions.

▶**1** To add two numerical fractions with unlike denominators, we must first obtain a common denominator.

EXAMPLE 1 Add $\dfrac{3}{5} + \dfrac{4}{7}$.

Solution: The least common denominator (LCD) [or least common multiple (LCM)] of 5 and 7 is 35. Thirty-five is the smallest number that is divisible by both 5 and 7. Rewrite each fraction so that it has a denominator equal to the LCD, 35.

$$\frac{3}{5} + \frac{4}{7} = \frac{3}{5} \cdot \frac{7}{7} + \frac{4}{7} \cdot \frac{5}{5}$$

$$= \frac{21}{35} + \frac{20}{35} = \frac{41}{35} \quad \text{or} \quad 1\frac{6}{35} \qquad \blacksquare$$

To add or subtract rational expressions, we must also write each expression with a common denominator.

> **To Find the Least Common Denominator of Rational Expressions**
>
> 1. Factor each denominator completely. Factors in any given denominator that occur more than once should be expressed as powers. For example, $(x + 5)(x + 5)$ should be expressed as $(x + 5)^2$.
> 2. List all different factors (other than 1) that appear in any of the denominators. When the same factor appears in more than one denominator, write the factor with the highest power that appears.
> 3. The least common denominator is the product of all the factors in step 2.

EXAMPLE 2 Find the least common denominator.

$$\frac{1}{3} + \frac{1}{x}$$

Solution: The only factor (other than 1) of the first denominator is 3. The only factor (other than 1) of the second denominator is x. The LCD is therefore $3 \cdot x = 3x$. ∎

EXAMPLE 3 Find the LCD.

$$\frac{3}{5x} - \frac{2}{x^2}$$

Solution: The factors that appear in the denominators are 5 and x. List each factor with its highest power. The LCD is the product of these factors.

$$\text{LCD} = 5 \cdot \overset{\text{highest power of } x}{x^2} = 5x^2$$ ∎

EXAMPLE 4 Find the LCD.

$$\frac{1}{18x^3y} + \frac{5}{27x^2y^3}$$

Solution: Write both 18 and 27 as products of prime numbers. $18 = 2 \cdot 3^2$ and $27 = 3^3$. *If you have forgotten how to write a number as a product of prime numbers, read Section 5.1 or Appendix B now.*

$$\frac{1}{18x^3y} + \frac{5}{27x^2y^3} = \frac{1}{2 \cdot 3^2 x^3 y} + \frac{5}{3^3 x^2 y^3}$$

The factors that appear are 2, 3, x, and y. List the highest powers of each of these factors.

$$\text{LCD} = 2 \cdot 3^3 \cdot x^3 \cdot y^3 = 54x^3y^3$$ ∎

EXAMPLE 5 Find the LCD.

$$\frac{3}{x} - \frac{2y}{x + 5}$$

Solution: The factors that appear are x and $(x + 5)$. *Note that the x in the second denominator, $x + 5$, is not a factor of that denominator since the operation is addition rather than multiplication.*

$$\text{LCD} = x(x + 5)$$ ∎

EXAMPLE 6   Find the LCD.

$$\frac{3}{2x^2 - 4x} + \frac{x^2}{x^2 - 4x + 4}$$

Solution: Factor both denominators.

$$\frac{3}{2x(x - 2)} + \frac{x^2}{(x - 2)(x - 2)} = \frac{3}{2x(x - 2)} + \frac{x^2}{(x - 2)^2}$$

The factors that appear are 2, x, and $x - 2$. List the highest powers of each of these factors.

$$\text{LCD} = 2 \cdot x \cdot (x - 2)^2 = 2x(x - 2)^2$$

EXAMPLE 7   Find the LCD.

$$\frac{5x}{x^2 - x - 12} - \frac{6x^2}{x^2 - 7x + 12}$$

Solution: Factor both denominators.

$$\frac{5x}{(x + 3)(x - 4)} - \frac{6x^2}{(x - 3)(x - 4)}$$

$$\text{LCD} = (x + 3)(x - 4)(x - 3)$$

Although $(x - 4)$ is a common factor of each denominator the highest power of that factor that appears in each denominator is 1.

EXAMPLE 8   Find the LCD.

$$\frac{3x}{x^2 - 14x + 48} + x + 9$$

Solution: Factor the denominator of the first term.

$$\frac{3x}{(x - 6)(x - 8)} + x + 9$$

The denominator of $x + 9$ is 1. The expression can be rewritten as

$$\frac{3x}{(x - 6)(x - 8)} + \frac{x + 9}{1}$$

The LCD is therefore $1(x - 6)(x - 8)$ or simply $(x - 6)(x - 8)$.

Exercise Set 6.4

Find the least common denominator for each expression.

1. $\dfrac{x}{3} + \dfrac{x - 1}{3}$

2. $\dfrac{4 - x}{5} - \dfrac{12}{5}$

3. $\dfrac{1}{2x} + \dfrac{1}{3}$

4. $\dfrac{1}{x + 2} - \dfrac{3}{5}$

5. $\dfrac{3}{5x} + \dfrac{7}{2}$

6. $\dfrac{2x}{3x} + 1$

7. $\dfrac{2}{x^2} + \dfrac{3}{x}$

8. $\dfrac{5x}{x + 1} + \dfrac{6}{x + 2}$

9. $\dfrac{x + 4}{2x + 3} + x$

10. $\dfrac{x + 4}{2x} + \dfrac{3}{7x}$

11. $\dfrac{x}{x + 1} + \dfrac{4}{x^2}$

12. $\dfrac{x}{3x^2} + \dfrac{9}{15x^4}$

13. $\dfrac{x + 3}{16x^2 y} - \dfrac{5}{9x^3}$

14. $\dfrac{-4}{8x^2 y^2} + \dfrac{7}{5x^4 y^5}$

15. $\dfrac{x^2 + 3}{18x} - \dfrac{x - 7}{12(x + 5)}$

16. $\dfrac{x - 7}{3x + 5} - \dfrac{6}{x + 5}$

17. $\dfrac{2x - 7}{x^2 + x} - \dfrac{x^2}{x + 1}$

18. $\dfrac{9}{(x-4)(x+2)} - \dfrac{x+8}{x+2}$

19. $\dfrac{15}{36x^2y} + \dfrac{x+3}{15xy^3}$

20. $\dfrac{x^2-4}{x^2-16} + \dfrac{3}{x+4}$

21. $\dfrac{6}{2x+8} + \dfrac{6x+3}{3x-9}$

22. $6x^2 + \dfrac{9x}{x-3}$

23. $\dfrac{9x+4}{x+6} - \dfrac{3x-6}{x+5}$

24. $\dfrac{x+2}{x^2+11x+18} - \dfrac{x^2-4}{x^2-3x-10}$

25. $\dfrac{x-2}{x^2-5x-24} + \dfrac{3}{x^2+11x+24}$

26. $\dfrac{6x+5}{x^2-4} - \dfrac{3x}{x^2-5x-14}$

27. $\dfrac{6}{x+3} - \dfrac{x+5}{x^2-4x+3}$

28. $\dfrac{3x-8}{x^2-1} + \dfrac{x^2+5}{x+1}$

29. $\dfrac{2x}{x^2-x-2} - \dfrac{3}{x^2+4x+3}$

30. $\dfrac{6x+5}{x+2} + \dfrac{4x}{(x+2)^2}$

31. $\dfrac{3x-5}{x^2+4x+4} + \dfrac{3}{x+2}$

32. $\dfrac{9x+7}{(x+3)(x+2)} - \dfrac{4x}{(x-3)(x+2)}$

33. $\dfrac{x}{3x^2+16x-12} + \dfrac{6}{3x^2+17x-6}$

34. $\dfrac{2x-7}{2x^2+5x+2} - \dfrac{x^2}{3x^2+4x-4}$

35. $\dfrac{2x-3}{4x^2+4x+1} + \dfrac{x^2-4}{8x^2+10x+3}$

36. In your own words, explain how to find the least common denominator of two rational expressions.

Cumulative Review Exercises

[1.2] **37.** Subtract $4\dfrac{3}{5} - 2\dfrac{5}{9}$.

[1.10] *Name each of the properties illustrated.*

38. $x + 3 = 3 + x$.

39. $4(x + y) = 4x + 4y$.

40. $(x + y) + z = x + (y + z)$.

[2.4] **41. (a)** Explain in a step-by-step manner how you would solve the equation $3x + 5 = 0$ for x.

(b) Use the procedure outlined in part (a) to solve the equation $3x + 5 = 0$.

[5.6] **42. (a)** Explain in a step-by-step manner how you would solve the equation $x^2 + x - 2 = 0$ for x using factoring.

(b) Use the procedure in part (a) to solve the equation $x^2 + x - 2 = 0$.

JUST FOR FUN

Find the least common denominator.

1. $\dfrac{3}{2x^3y^6} - \dfrac{5}{6x^5y^9} + \dfrac{1}{5x^{12}y^2}$.

2. $\dfrac{x}{x-2} - \dfrac{4}{x^2-4} + \dfrac{3}{x+2}$.

3. $\dfrac{4}{x^2-x-12} + \dfrac{3}{x^2-6x+8} + \dfrac{5}{x^2+x-6}$

6.5

Addition and Subtraction of Rational Expressions

▶ **1**   Add and subtract rational expressions.

In Section 6.3, we discussed how to add or subtract rational expressions with a common denominator. Now we will discuss the procedure for adding and subtracting rational expressions that are not given with a common denominator.

▶ **1**   The method used to add or subtract rational expressions with unlike denominators is outlined in Example 1.

EXAMPLE 1 Add $\dfrac{3}{x} + \dfrac{5}{y}$.

Solution: First, determine the LCD as outlined in Section 6.4.

$$\text{LCD} = xy$$

Now write each fraction with the LCD. We do this by multiplying **both** the numerator and denominator of each fraction by any factors needed to obtain the LCD.

In this problem the fraction on the left must be multiplied by y/y and the fraction on the right must be multiplied by x/x.

$$\frac{y}{y} \cdot \frac{3}{x} + \frac{5}{y} \cdot \frac{x}{x} = \frac{3y}{xy} + \frac{5x}{xy}$$

By multiplying both the numerator and denominator by the same factor, we are in effect multiplying by 1, which does not change the value of the fraction, only its appearance. Thus the new fraction is equivalent to the original fraction.

Now add the numerators, while leaving the LCD alone.

$$\frac{3y}{xy} + \frac{5x}{xy} = \frac{3y + 5x}{xy} \quad \text{or} \quad \frac{5x + 3y}{xy} \qquad \blacksquare$$

To Add or Subtract Two Rational Expressions with Unlike Denominators

1. Determine the LCD.
2. Rewrite each fraction as an equivalent fraction with the LCD. This is done by multiplying both the numerator and denominator of each fraction by any factors needed to obtain the LCD.
3. Add or subtract the numerators while maintaining the LCD.
4. When possible, factor the remaining numerator and reduce the fraction.

EXAMPLE 2 Add $\dfrac{5}{4x^2y} + \dfrac{3}{14xy^3}$.

Solution: The LCD is $28x^2y^3$. We must write each fraction with the denominator $28x^2y^3$. To do this, multiply the left fraction by $7y^2/7y^2$ and the right fraction by $2x/2x$.

$$\frac{7y^2}{7y^2} \cdot \frac{5}{4x^2y} + \frac{3}{14xy^3} \cdot \frac{2x}{2x} = \frac{35y^2}{28x^2y^3} + \frac{6x}{28x^2y^3}$$

$$= \frac{35y^2 + 6x}{28x^2y^3} \quad \text{or} \quad \frac{6x + 35y^2}{28x^2y^3} \qquad \blacksquare$$

EXAMPLE 3 Add $\dfrac{3}{x + 2} + \dfrac{4}{x}$.

Solution: We must write each fraction with the LCD $x(x + 2)$. To do this, multiply the fraction on the left by x/x and the fraction on the right by $(x + 2)/(x + 2)$.

$$\frac{x}{x} \cdot \frac{3}{(x + 2)} + \frac{4}{x} \cdot \frac{(x + 2)}{(x + 2)} = \frac{3x}{x(x + 2)} + \frac{4(x + 2)}{x(x + 2)}$$

$$= \frac{3x + 4(x + 2)}{x(x + 2)}$$

$$= \frac{3x + 4x + 8}{x(x + 2)} = \frac{7x + 8}{x(x + 2)} \qquad \blacksquare$$

EXAMPLE 4   Subtract $\dfrac{x}{x + 5} - \dfrac{2}{x - 3}$.

Solution: The LCD is $(x + 5)(x - 3)$. The fraction on the left must be multiplied by $(x - 3)/(x - 3)$ to obtain the LCD. The fraction on the right must be multiplied by $(x + 5)/(x + 5)$ to obtain the LCD.

$$\frac{(x - 3)}{(x - 3)} \cdot \frac{x}{(x + 5)} - \frac{2}{(x - 3)} \cdot \frac{(x + 5)}{(x + 5)} = \frac{x(x - 3)}{(x - 3)(x + 5)} - \frac{2(x + 5)}{(x - 3)(x + 5)}$$

$$= \frac{x^2 - 3x}{(x - 3)(x + 5)} - \frac{2x + 10}{(x - 3)(x + 5)}$$

$$= \frac{x^2 - 3x - (2x + 10)}{(x - 3)(x + 5)}$$

$$= \frac{x^2 - 3x - 2x - 10}{(x - 3)(x + 5)}$$

$$= \frac{x^2 - 5x - 10}{(x - 3)(x + 5)}$$

The numerator and denominator have no common factors, so we can leave the answer in this form. If you wish, you can multiply the factors in the denominator and give the answer as $(x^2 - 5x - 10)/(x^2 + 2x - 15)$. Generally, we will leave the answers in factored form.   ■

EXAMPLE 5   Subtract $\dfrac{x + 2}{x - 4} - \dfrac{x + 3}{x + 4}$.

Solution: The LCD is $(x - 4)(x + 4)$.

$$\frac{(x + 4)}{(x + 4)} \cdot \frac{(x + 2)}{(x - 4)} - \frac{(x + 3)}{(x + 4)} \cdot \frac{(x - 4)}{(x - 4)} = \frac{(x + 4)(x + 2)}{(x + 4)(x - 4)} - \frac{(x + 3)(x - 4)}{(x + 4)(x - 4)}$$

Use the FOIL method to multiply each numerator.

$$= \frac{x^2 + 6x + 8}{(x + 4)(x - 4)} - \frac{x^2 - x - 12}{(x + 4)(x - 4)}$$

$$= \frac{x^2 + 6x + 8 - (x^2 - x - 12)}{(x + 4)(x - 4)}$$

$$= \frac{x^2 + 6x + 8 - x^2 + x + 12}{(x + 4)(x - 4)}$$

$$= \frac{7x + 20}{(x + 4)(x - 4)}$$ ■

Consider the problem

$$\frac{4}{x - 3} + \frac{x + 5}{3 - x}$$

How do we add these rational expressions? One method would be to write each fraction with the denominator $(x - 3)(3 - x)$. However, there is an easier way to do this problem. Study the following Helpful Hint.

| HELPFUL HINT | When adding or subtracting fractions whose denominators are opposites (and therefore differ only in signs), multiply both numerator *and* denominator of *either* of the fractions by -1. This will result in both fractions having a common denominator. |

$$\frac{x}{a-b} + \frac{y}{b-a} = \frac{x}{a-b} + \frac{-1}{-1} \cdot \frac{y}{(b-a)}$$

$$= \frac{x}{a-b} + \frac{-y}{a-b}$$

$$= \frac{x-y}{a-b}$$

EXAMPLE 6   Add $\dfrac{4}{x-3} + \dfrac{x+5}{3-x}$.

Solution: Since the denominators differ only in sign, we will multiply both the numerator and the denominator of a second fraction by -1. This will result in both fractions having a common denominator of $x-3$.

$$\frac{4}{x-3} + \frac{x+5}{3-x} = \frac{4}{x-3} + \frac{-1}{-1} \cdot \frac{(x+5)}{(3-x)}$$

$$= \frac{4}{x-3} + \frac{-x-5}{x-3}$$

$$= \frac{4-x-5}{x-3}$$

$$= \frac{-x-1}{x-3} \quad \text{or} \quad \frac{-(x+1)}{x-3} \quad \blacksquare$$

Recall from Section 6.1 that we can change any *two* signs of a fraction without changing its value. In Example 7 we will work a problem where we change two signs of the fraction to simplify the solution to a problem.

EXAMPLE 7   Subtract $\dfrac{x+2}{2x-5} - \dfrac{3x+5}{5-2x}$.

Solution: The denominators of the two fractions differ only in sign. If we change the signs of one of the denominators, we will have a common denominator. In this example we will change *two* of the fraction signs in the second fraction to obtain a common denominator.

$$\frac{x+2}{2x-5} - \frac{3x+5}{5-2x} = \frac{x+2}{2x-5} + \frac{3x+5}{-(5-2x)}$$

$$= \frac{x+2}{2x-5} + \frac{3x+5}{2x-5}$$

$$= \frac{x+2+3x+5}{2x-5}$$

$$= \frac{4x+7}{2x-5} \quad \blacksquare$$

In Example 7 we elected to change two signs of the second fraction. The same results could be obtained by multiplying the numerator and denominator of either the first or second fraction by -1 as was done in Example 6. Try this now.

EXAMPLE 8   Add $\dfrac{3}{x^2 + 5x + 6} + \dfrac{1}{3x^2 + 8x - 3}$.

Solution: $\dfrac{3}{(x + 2)(x + 3)} + \dfrac{1}{(3x - 1)(x + 3)}$

The LCD is $(x + 2)(x + 3)(3x - 1)$.

$$\frac{(3x - 1)}{(3x - 1)} \cdot \frac{3}{(x + 2)(x + 3)} + \frac{1}{(3x - 1)(x + 3)} \cdot \frac{(x + 2)}{(x + 2)}$$

$$= \frac{9x - 3}{(3x - 1)(x + 2)(x + 3)} + \frac{x + 2}{(3x - 1)(x + 2)(x + 3)}$$

$$= \frac{10x - 1}{(3x - 1)(x + 2)(x + 3)}$$

COMMON STUDENT ERROR

A common error in an addition or subtraction problem is to add or subtract the numerators and the denominators. Here is one such example.

Correct	*Wrong*

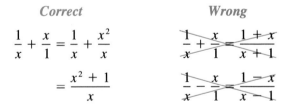

Remember that to add or subtract fractions you must first have a common denominator. Then you add or subtract the numerators while maintaining the common denominator.

Another common mistake is to treat an addition or subtraction problem as a multiplication problem. You can only divide out common factors when *multiplying* expressions, and not when adding or subtracting them.

Correct	*Wrong*

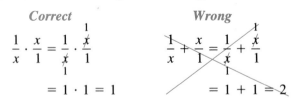

Exercise Set 6.5

Add or subtract as indicated.

1. $\dfrac{4}{x} + \dfrac{3}{2x}$

2. $\dfrac{1}{3x} + \dfrac{1}{2}$

3. $\dfrac{6}{x^2} + \dfrac{3}{2x}$

4. $3 + \dfrac{5}{x}$

5. $2 - \dfrac{1}{x^2}$

6. $\dfrac{5}{6y} + \dfrac{3}{4y^2}$

7. $\dfrac{1}{x^2} + \dfrac{3}{5x}$

8. $\dfrac{3x}{4y} + \dfrac{5}{6xy}$

9. $\dfrac{3}{4x^2y} + \dfrac{7}{5xy^2}$

10. $\dfrac{5}{12x^4y} - \dfrac{1}{5x^2y^3}$

11. $x + \dfrac{x}{y}$

12. $\dfrac{5}{x} + \dfrac{4}{x^2}$

13. $\dfrac{3x - 1}{x} + \dfrac{2}{3x}$

14. $\dfrac{3}{x} + 4$

15. $\dfrac{5x}{y} + \dfrac{y}{x}$

16. $\dfrac{3}{5p} - \dfrac{5}{2p^2}$

17. $\dfrac{4}{5x^2} - \dfrac{6}{y}$

18. $\dfrac{x - 3}{x} - \dfrac{x}{4x^2}$

19. $\dfrac{5}{x} + \dfrac{3}{x - 2}$

20. $6 - \dfrac{3}{x - 3}$

21. $\dfrac{9}{a + 3} + \dfrac{2}{a}$

22. $\dfrac{b}{a - b} + \dfrac{a + b}{b}$

23. $\dfrac{4}{3x} - \dfrac{2x}{3x + 6}$

24. $\dfrac{2}{x - 3} - \dfrac{4}{x - 1}$

25. $\dfrac{3}{x - 2} + \dfrac{1}{2 - x}$

26. $\dfrac{3}{x - 2} - \dfrac{1}{2 - x}$

27. $\dfrac{5}{x + 3} - \dfrac{4}{-x - 3}$

28. $\dfrac{5}{2x - 5} - \dfrac{3}{5 - 2x}$

29. $\dfrac{3}{x + 1} + \dfrac{4}{x - 1}$

30. $\dfrac{x}{2x - 4} + \dfrac{3}{x - 2}$

31. $\dfrac{x + 5}{x - 5} - \dfrac{x - 5}{x + 5}$

32. $\dfrac{x + 7}{x + 3} - \dfrac{x - 3}{x + 7}$

33. $\dfrac{x}{x^2 - 9} + \dfrac{4}{x + 3}$

34. $\dfrac{3}{(x + 5)^2} + \dfrac{2}{x + 5}$

35. $\dfrac{x + 2}{x^2 - 4} - \dfrac{2}{x + 2}$

36. $\dfrac{3}{(x - 2)(x + 3)} + \dfrac{5}{(x + 2)(x + 3)}$

37. $\dfrac{2x + 3}{x^2 - 7x + 12} - \dfrac{2}{x - 3}$

38. $\dfrac{x + 3}{x^2 - 3x - 10} - \dfrac{2}{x - 5}$

39. $\dfrac{x^2}{x^2 + 2x - 8} - \dfrac{x - 4}{x + 4}$

40. $\dfrac{x + 1}{x^2 - 2x + 1} - \dfrac{x + 1}{x - 1}$

41. $\dfrac{x - 1}{x^2 + 4x + 4} + \dfrac{x - 1}{x + 2}$

42. $\dfrac{y}{xy - x^2} - \dfrac{x}{y^2 - xy}$

43. $\dfrac{3}{x^2 + 2x - 8} + \dfrac{2}{x^2 - 3x + 2}$

44. $\dfrac{5}{x^2 - 9x + 8} - \dfrac{3}{x^2 - 6x - 16}$

45. $\dfrac{1}{x^2 - 4} + \dfrac{3}{x^2 + 5x + 6}$

46. $\dfrac{x}{2x^2 + 7x - 4} + \dfrac{2}{x^2 - x - 20}$

47. $\dfrac{x}{3x^2 + 5x - 2} - \dfrac{4}{2x^2 + 7x + 6}$

48. $\dfrac{x}{6x^2 + 7x + 2} + \dfrac{5}{2x^2 - 3x - 2}$

49. $\dfrac{x}{3x^2 + 5x - 2} - \dfrac{3}{2x^2 + 7x + 6}$

50. **(a)** Explain in your own words a step-by-step procedure to add or subtract two rational expressions that have unlike denominators.

(b) Using the procedure outlined in part (a), add $\dfrac{x}{x^2 - x - 6} + \dfrac{3}{x^2 - 4}$.

Cumulative Review Exercises

[2.6] **51.** A video cassette recorder counter will go from 0 to 18 in 2 minutes. There are two movies on a VCR tape. If Phong Nguyen wishes to watch the second movie and the first movie is $1\frac{1}{2}$ hours long, what will be the number on the counter at the end of the first movie, where the second movie starts? 810

[2.7] **52.** Solve the inequality $3(x - 2) + 2 < 4(x + 1)$ and graph the solution on the number line.

[4.6] **53.** Divide $(8x^2 + 6x - 13) \div (2x + 3)$.

[6.2] **54.** Multiply $\dfrac{x^2 + xy - 6y^2}{x^2 - xy - 2y^2} \cdot \dfrac{x^2 - y^2}{x^2 + 2xy - 3y^2}$.

JUST FOR FUN

Add or subtract as indicated.

1. $\dfrac{x}{x - 2} + \dfrac{3}{x + 2} + \dfrac{4}{x^2 - 4}$

2. $\dfrac{4}{x^2 + x - 6} + \dfrac{x}{x + 3} - \dfrac{5}{x - 2}$

3. $\dfrac{x + 6}{4 - x^2} - \dfrac{x + 3}{x + 2} + \dfrac{x - 3}{2 - x}$

4. $\dfrac{3x - 1}{x + 2} + \dfrac{x}{x - 3} - \dfrac{4}{2x + 3}$

6.6

Complex Fractions (Optional)

▶ **1**   Identify complex fractions.

▶ **2**   Simplify complex fractions using method 1.

▶ **3**   Simplify complex fractions using method 2.

▶ **1**   A **complex fraction** is one that has a fraction in its numerator or its denominator or in both its numerator and denominator. Examples of complex fractions are

$$\dfrac{\dfrac{2}{3}}{5} \qquad \dfrac{\dfrac{x + 1}{x}}{3x} \qquad \dfrac{\dfrac{x}{y}}{x + 1} \qquad \dfrac{\dfrac{a + b}{a}}{\dfrac{a - b}{b}}$$

Numerator of complex fraction $\left\{ \dfrac{a + b}{a} \right.$

$\longleftarrow$ Main fraction line

Denominator of complex fraction $\left\{ \dfrac{a - b}{b} \right.$

The expression above the main fraction line is the numerator, and the expression below the main fraction line is the denominator of the complex fraction. The numerator and denominator of a complex fraction are called **secondary fractions**.

▶ **2**   There are two methods to simplify complex fractions. Method 1 reinforces many of the concepts used in this chapter because we may need to add, subtract, multiply, and divide fractions. Many students prefer to use method 2, because the answer may be obtained more quickly. We will give three examples using method 1 and then work the same three examples using method 2.

Method 1

> **To Simplify a Complex Fraction**
>
> 1. Add or subtract the fractions in both the numerator and denominator of the complex fractions to obtain single fractions in both the numerator and the denominator.
> 2. Invert and multiply the denominator of the complex fraction by the numerator of the complex fraction.
> 3. Simplify when possible.

EXAMPLE 1   Simplify $\dfrac{\dfrac{2}{3} + \dfrac{3}{4}}{\dfrac{3}{4} - \dfrac{2}{3}}$.

Solution: First, simplify both the numerator and denominator of the complex fraction to obtain single fractions in both the numerator and denominator. The LCD of both the numerator and the denominator is 12.

$$\frac{\dfrac{2}{3} + \dfrac{3}{4}}{\dfrac{3}{4} - \dfrac{2}{3}} = \frac{\dfrac{4}{4} \cdot \dfrac{2}{3} + \dfrac{3}{4} \cdot \dfrac{3}{3}}{\dfrac{3}{3} \cdot \dfrac{3}{4} - \dfrac{2}{3} \cdot \dfrac{4}{4}} = \frac{\dfrac{8}{12} + \dfrac{9}{12}}{\dfrac{9}{12} - \dfrac{8}{12}} = \frac{\dfrac{17}{12}}{\dfrac{1}{12}}$$

Next, invert and multiply the denominator of the complex fraction.

$$\frac{\dfrac{17}{12}}{\dfrac{1}{12}} = \frac{17}{\cancel{12}} \cdot \frac{\cancel{12}^{1}}{1} = \frac{17}{1} = 17$$

EXAMPLE 2   Simplify $\dfrac{a + \dfrac{1}{x}}{x + \dfrac{1}{a}}$.

Solution: Express both the numerator and denominator of the complex fraction as single fractions; then invert and multiply.

$$\frac{\dfrac{x}{x} \cdot a + \dfrac{1}{x}}{\dfrac{a}{a} \cdot x + \dfrac{1}{a}} = \frac{\dfrac{ax}{x} + \dfrac{1}{x}}{\dfrac{ax}{a} + \dfrac{1}{a}}$$

$$= \frac{\dfrac{ax + 1}{x}}{\dfrac{ax + 1}{a}}$$

$$= \frac{\cancel{(ax + 1)}}{x} \cdot \frac{a}{\cancel{(ax + 1)}} = \frac{a}{x}$$

EXAMPLE 3   Simplify $\dfrac{a}{\dfrac{1}{a} + \dfrac{1}{b}}$.

Solution: $\dfrac{a}{\dfrac{1}{a} + \dfrac{1}{b}} = \dfrac{a}{\dfrac{b}{b} \cdot \dfrac{1}{a} + \dfrac{1}{b} \cdot \dfrac{a}{a}} = \dfrac{a}{\dfrac{b}{ab} + \dfrac{a}{ab}} = \dfrac{a}{\dfrac{b+a}{ab}}$

$= \dfrac{a}{1} \cdot \dfrac{ab}{b+a} = \dfrac{a^2 b}{b+a}$ or $\dfrac{a^2 b}{a+b}$ ∎

▶ **3**   Here is the second method that may be used to simplify a complex fraction.

Method 2

> **To Simplify a Complex Fraction**
>
> **1.** Find the least common denominator of *all* the denominators appearing in the complex fraction.
> **2.** Multiply both the numerator and denominator of the complex fraction by the LCD found in step 1.
> **3.** Simplify when possible.

We will now rework Examples 1, 2, and 3 using method 2.

EXAMPLE 4   Simplify $\dfrac{\dfrac{2}{3} + \dfrac{3}{4}}{\dfrac{3}{4} - \dfrac{2}{3}}$.

Solution: *Step 1:* The denominators in the complex fraction are 3 and 4. The LCD of 3 and 4 is 12. Thus 12 is the LCD of the complex fraction.

Step 2: Multiply both the numerator and denominator of the complex fraction by 12.

$$\dfrac{12}{12} \cdot \dfrac{\left(\dfrac{2}{3} + \dfrac{3}{4}\right)}{\left(\dfrac{3}{4} - \dfrac{2}{3}\right)} = \dfrac{12\left(\dfrac{2}{3}\right) + 12\left(\dfrac{3}{4}\right)}{12\left(\dfrac{3}{4}\right) - 12\left(\dfrac{2}{3}\right)}$$

Step 3: Simplify.

$$= \dfrac{12\left(\dfrac{2}{3}\right) + 12\left(\dfrac{3}{4}\right)}{12\left(\dfrac{3}{4}\right) - 12\left(\dfrac{2}{3}\right)}$$

$$= \dfrac{8 + 9}{9 - 8} = \dfrac{17}{1} = 17$$ ∎

Note that the answers to Examples 1 and 4 are the same.

EXAMPLE 5   Simplify $\dfrac{a + \dfrac{1}{x}}{x + \dfrac{1}{a}}$.

Solution: The denominators in the complex fraction are x and a. Therefore, the LCD of the complex fraction is ax. Multiply both the numerator and denominator of the complex fraction by ax.

$$\frac{ax}{ax} \cdot \frac{\left(a + \dfrac{1}{x}\right)}{\left(x + \dfrac{1}{a}\right)} = \frac{a^2 x + a}{ax^2 + x}$$

$$= \frac{a(ax + 1)}{x(ax + 1)} = \frac{a}{x} \qquad\blacksquare$$

Note that the answers to Examples 2 and 5 are the same.

EXAMPLE 6   Simplify $\dfrac{a}{\dfrac{1}{a} + \dfrac{1}{b}}$.

Solution: The denominators in the complex fraction are 1 (from a in the numerator), a, and b. Therefore, the LCD of the complex fraction is ab. Multiply both the numerator and denominator of the complex fraction by ab.

$$\frac{ab}{ab} \cdot \frac{a}{\left(\dfrac{1}{a} + \dfrac{1}{b}\right)} = \frac{a^2 b}{b + a} \qquad\blacksquare$$

Note that the answers to Examples 3 and 6 are the same. When asked to simplify a complex fraction, you may use either method unless you are told by your instructor to use a specific method.

Exercise Set 6.6

Simplify each expression.

1. $\dfrac{1 + \dfrac{3}{5}}{2 + \dfrac{1}{5}}$

2. $\dfrac{1 - \dfrac{9}{16}}{3 + \dfrac{4}{5}}$

3. $\dfrac{2 + \dfrac{3}{8}}{1 + \dfrac{1}{3}}$

4. $\dfrac{\dfrac{3}{5} + \dfrac{2}{7}}{\dfrac{1}{5} + \dfrac{5}{6}}$

5. $\dfrac{\dfrac{4}{9} - \dfrac{3}{8}}{4 - \dfrac{3}{5}}$

6. $\dfrac{1 - \dfrac{x}{y}}{x}$

7. $\dfrac{\dfrac{x^2 y}{4}}{\dfrac{2}{x}}$

8. $\dfrac{\dfrac{15a}{b^2}}{\dfrac{b^3}{5}}$

9. $\dfrac{\dfrac{8x^2 y}{3z^3}}{\dfrac{4xy}{9z^5}}$

10. $\dfrac{\dfrac{36x^4}{5y^4 z^5}}{\dfrac{9xy^2}{15z^5}}$

11. $\dfrac{x + \dfrac{1}{y}}{\dfrac{x}{y}}$

12. $\dfrac{x - \dfrac{x}{y}}{\dfrac{1 + x}{y}}$

13. $\dfrac{\dfrac{9}{x} + \dfrac{3}{x^2}}{3 + \dfrac{1}{x}}$

14. $\dfrac{\dfrac{2}{a} + \dfrac{1}{2a}}{a + \dfrac{a}{2}}$

15. $\dfrac{3 - \dfrac{1}{y}}{2 - \dfrac{1}{y}}$

16. $\dfrac{\dfrac{x}{x - y}}{\dfrac{x^2}{y}}$

17. $\dfrac{\dfrac{x}{y} - \dfrac{y}{x}}{\dfrac{x + y}{x}}$

18. $\dfrac{1}{\dfrac{1}{x} + y}$

19. $\dfrac{\dfrac{a^2}{b} - b}{\dfrac{b^2}{a} - a}$

20. $\dfrac{\dfrac{1}{x} + \dfrac{2}{x^2}}{2 + \dfrac{1}{x^2}}$

21. $\dfrac{\dfrac{a}{b} - 2}{\dfrac{-a}{b} + 2}$

22. $\dfrac{\dfrac{x^2 - y^2}{x}}{\dfrac{x + y}{x^3}}$

23. $\dfrac{\dfrac{4x + 8}{3x^2}}{\dfrac{4x}{6}}$

24. $\dfrac{\dfrac{1}{a} + \dfrac{1}{b}}{ab}$

25. $\dfrac{\dfrac{1}{a} + \dfrac{1}{b}}{\dfrac{1}{ab}}$

26. $\dfrac{\dfrac{1}{a} + 1}{\dfrac{1}{b} - 1}$

27. $\dfrac{\dfrac{a}{b} + \dfrac{1}{a}}{\dfrac{b}{a} + \dfrac{1}{a}}$

28. $\dfrac{\dfrac{1}{a} + \dfrac{1}{b}}{\dfrac{1}{a}}$

29. $\dfrac{\dfrac{1}{x} - \dfrac{1}{y}}{\dfrac{1}{x} + \dfrac{1}{y}}$

30. $\dfrac{\dfrac{1}{x^2} + \dfrac{1}{x}}{\dfrac{1}{x} + \dfrac{1}{x^2}}$

31. $\dfrac{\dfrac{1}{x^2} + \dfrac{1}{x}}{\dfrac{1}{y} + \dfrac{1}{y^2}}$

32. What is a complex fraction?

33. (a) Select the method you prefer to use to simplify complex fractions. Then write in your own words a step-by-step procedure for simplifying complex fractions using this method.

(b) Using the procedure given in part (a) simplify
$$\dfrac{\dfrac{2}{x} - \dfrac{3}{y}}{x + \dfrac{1}{y}}.$$

Cumulative Review Exercises

[2.5] **34.** Solve the equation $4x + 3(x - 2) = 7x - 6$.

[4.4] **35.** What is polynomial?

[5.1] **36.** Simplify $(4x^2y^3)^2(3xy^4)$.

[6.5] **37.** Subtract $\dfrac{x}{3x^2 + 17x - 6} - \dfrac{2}{x^2 + 3x - 18}$.

JUST FOR FUN

The efficiency of a jack, E, is expressed by the formula $E = \dfrac{\frac{1}{2}h}{h + \frac{1}{2}}$, where h is determined by the pitch of the jack's thread. Determine the efficiency of a jack whose value of h is **(a)** $\dfrac{2}{3}$ **(b)** $\dfrac{4}{5}$

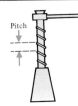

Pitch

6.7

Solving Equations Containing Rational Expressions

▶1 Solve equations with rational expressions.

▶1 In Sections 6.1 through 6.6, we focused on how to add, subtract, multiply, and divide rational expressions. In this section we are ready to solve equations containing fractions.

> **To Solve Equations Containing Fractions**
>
> 1. Determine the LCD of all fractions in the equation.
> 2. Multiply **both** sides of the equation by the LCD. **This will result in every term in the equation being multiplied by the LCD.**
> 3. Remove any parentheses and combine like terms on each side of the equation.
> 4. Solve the equation using the properties discussed in earlier sections.
> 5. Check your solution in the original equation.

The purpose of multiplying both sides of the equation by the LCD (step 2) is to eliminate all fractions from the equation. After both sides of the equation are multiplied by the LCD, the resulting equation should contain no fractions. We will omit some of the checks to save space.

EXAMPLE 1 Solve $\dfrac{x}{3} + 2x = 7$.

Solution:
$$3\left(\frac{x}{3} + 2x\right) = 7 \cdot 3 \qquad \text{Multiply both sides of the equation by the LCD, 3.}$$

$$3\left(\frac{x}{3}\right) + 3 \cdot 2x = 7 \cdot 3 \qquad \text{Distributive property.}$$

$$x + 6x = 21$$
$$7x = 21$$
$$x = 3$$

Check: $\dfrac{x}{3} + 2x = 7$

$$\frac{3}{3} + 2(3) = 7$$
$$1 + 6 = 7$$
$$7 = 7 \qquad \text{true}$$

EXAMPLE 2 Solve the equation $\dfrac{3}{4} + \dfrac{5x}{9} = \dfrac{x}{6}$.

Solution: Multiply both sides of the equation by the LCD, 36.

$$36\left(\frac{3}{4} + \frac{5x}{9}\right) = \frac{x}{6} \cdot 36$$

$$36\left(\frac{3}{4}\right) + 36\left(\frac{5x}{9}\right) = \frac{x}{6} \cdot 36$$

$$27 + 20x = 6x$$
$$27 = -14x$$
$$x = \frac{27}{-14} \quad \text{or} \quad \frac{-27}{14}$$

EXAMPLE 3   Solve the equation $\dfrac{x}{4} + 3 = 2(x - 2)$.

Solution: Multiply both sides of the equation by the LCD, 4.

$$\frac{x}{4} + 3 = 2(x - 2)$$

$$4\left(\frac{x}{4} + 3\right) = 4[2(x - 2)]$$

$$4\left(\frac{x}{4}\right) + 4(3) = 4[2(x - 2)]$$

$$4\left(\frac{x}{4}\right) + 4(3) = 8(x - 2)$$

$$x + 12 = 8(x - 2)$$

$$x + 12 = 8x - 16$$

$$12 = 7x - 16$$

$$28 = 7x$$

$$4 = x$$

∎

Warning: **Whenever a variable appears in any denominator, it is necessary to check your answer in the original equation. If the answer obtained makes any denominator equal to zero, that value is not a solution to the equation.** Such values are called **extraneous roots** or **extraneous solutions.**

EXAMPLE 4   Solve the equation $3 - \dfrac{4}{x} = \dfrac{5}{2}$.

Solution: Multiply both sides of the equation by the LCD, $2x$.

$$2x\left(3 - \frac{4}{x}\right) = \left(\frac{5}{2}\right) \cdot 2x$$

$$2x(3) - 2x\left(\frac{4}{x}\right) = \frac{5}{2} \cdot 2x$$

$$6x - 8 = 5x$$

$$x - 8 = 0$$

$$x = 8$$

Check: $3 - \dfrac{4}{x} = \dfrac{5}{2}$

$$3 - \frac{4}{8} = \frac{5}{2}$$

$$3 - \frac{1}{2} = \frac{5}{2}$$

$$\frac{5}{2} = \frac{5}{2} \qquad \text{true}$$

Since 8 does check, it is the solution to the equation.

∎

EXAMPLE 5   Solve the equation $\dfrac{x - 7}{x + 2} = \dfrac{1}{4}$.

Solution: The LCD is $4(x + 2)$. Multiply both sides of the equation by the LCD.

$$4(x + 2) \cdot \frac{(x - 7)}{(x + 2)} = \frac{1}{4} \cdot 4(x + 2)$$

$$4(x - 7) = 1(x + 2)$$

$$4x - 28 = x + 2$$

$$3x - 28 = 2$$

$$3x = 30$$

$$x = 10$$

A check will show that 10 is the solution.

In Section 2.6 we illustrated that proportions of the form

$$\frac{a}{b} = \frac{c}{d}$$

can be cross-multiplied to obtain $a \cdot d = b \cdot c$. Example 5 is a proportion and can also be solved by cross-multiplying.

EXAMPLE 6 Solve the following equation using cross-multiplication $\dfrac{3}{x + 4} = \dfrac{4}{x - 1}$.

Solution: $3(x - 1) = 4(x + 4)$

$3x - 3 = 4x + 16$

$-x - 3 = 16$

$-x = 19$

$x = -19$

A check will show that -19 is the solution to the equation.

Now let us examine some examples that involve quadratic equations. Recall from Section 5.6 that quadratic equations are of the form $ax^2 + bx + c = 0$, $a \neq 0$.

EXAMPLE 7 Solve the equation $x + \dfrac{12}{x} = -7$.

Solution: $x \cdot \left(x + \dfrac{12}{x}\right) = -7 \cdot x$ Multiply both sides of the equation by x.

$$x(x) + x\left(\frac{12}{x}\right) = -7x$$

$$x^2 + 12 = -7x$$

$$x^2 + 7x + 12 = 0$$

$$(x + 3)(x + 4) = 0$$

$$x + 3 = 0 \quad \text{or} \quad x + 4 = 0$$

$$x = -3 \quad \text{or} \qquad x = -4$$

Check:

$x = -3$	$x = -4$
$x + \dfrac{12}{x} = -7$	$x + \dfrac{12}{x} = -7$
$-3 + \dfrac{12}{-3} = -7$	$-4 + \dfrac{12}{-4} = -7$
$-3 + (-4) = -7$	$-4 + (-3) = -7$
$-7 = -7$ true	$-7 = -7$ true

EXAMPLE 8   Solve the equation $\dfrac{x^2}{x-4} = \dfrac{16}{x-4}$.

Solution: $(x-4) \cdot \dfrac{x^2}{(x-4)} = \dfrac{16}{(x-4)} \cdot (x-4)$

$$x^2 = 16$$

$$x^2 - 16 = 0 \qquad \text{This is a difference of two squares.}$$

$$(x+4)(x-4) = 0$$

$$x+4 = 0 \quad \text{or} \quad x-4 = 0$$

$$x = -4 \qquad\qquad x = 4$$

Check: $x = -4$ $\qquad\qquad\qquad\qquad x = 4$

$$\dfrac{x^2}{x-4} = \dfrac{16}{x-4} \qquad\qquad \dfrac{x^2}{x-4} = \dfrac{16}{x-4}$$

$$\dfrac{(-4)^2}{-4-4} = \dfrac{16}{-4-4} \qquad\qquad \dfrac{(4)^2}{4-4} = \dfrac{16}{4-4}$$

$$\dfrac{16}{-8} = \dfrac{16}{-8} \qquad\qquad\qquad \dfrac{16}{0} = \dfrac{16}{0} \qquad \text{not a solution}$$

$$-2 = -2 \qquad \text{true}$$

Since 4 results in a denominator of 0, $x = 4$ is *not* a solution to the equation. The 4 is an *extraneous root*. The only solution to the equation is $x = -4$.   ∎

Notice that the equation in Example 8 is a proportion. There are many proportions that are difficult to solve by cross-multiplication. What would happen if you attempted to solve the proportion in Example 8 by cross-multiplication? Try solving example 8 by cross-multiplying and see what happens.

EXAMPLE 9   Solve the equation $\dfrac{2x}{x^2-4} + \dfrac{1}{x-2} = \dfrac{2}{x+2}$.

Solution: $\dfrac{2x}{(x+2)(x-2)} + \dfrac{1}{x-2} = \dfrac{2}{x+2}$

Multiply both sides of the equation by the LCD, $(x+2)(x-2)$.

$$(x+2)(x-2) \cdot \left[\dfrac{2x}{(x+2)(x-2)} + \dfrac{1}{x-2} \right] = \dfrac{2}{x+2} \cdot (x+2)(x-2)$$

$$(x+2)(x-2) \cdot \dfrac{2x}{(x+2)(x-2)} + (x+2)(x-2) \cdot \dfrac{1}{(x-2)} = \dfrac{2}{(x+2)} \cdot (x+2)(x-2)$$

$$(x+2)(x-2) \cdot \dfrac{2x}{(x+2)(x-2)} + (x+2)(x-2) \cdot \dfrac{1}{(x-2)} = \dfrac{2}{(x+2)} \cdot (x+2)(x-2)$$

$$2x + (x+2) = 2(x-2)$$

$$2x + x + 2 = 2x - 4$$

$$3x + 2 = 2x - 4$$

$$x + 2 = -4$$

$$x = -6$$

A check will show that -6 is the solution to the equation.   ∎

HELPFUL HINT

At this point some students confuse adding and subtracting rational expressions with solving rational equations. When adding or subtracting rational expressions, we must rewrite each expression with a common denominator. When solving a rational equation, we multiply both sides of the equation by the LCD to eliminate fractions from the equation. Consider the following two problems. Note that the one on the right is an equation because it contains an equal sign. We will work both problems. The LCD for both problems is $x(x + 4)$.

Adding Rational Expressions

$$\frac{x + 2}{x + 4} + \frac{3}{x}$$

$$= \frac{x}{x} \cdot \frac{x + 2}{x + 4} + \frac{3}{x} \cdot \frac{x + 4}{x + 4}$$

$$= \frac{x(x + 2)}{x(x + 4)} + \frac{3(x + 4)}{x(x + 4)}$$

$$= \frac{x^2 + 2x}{x(x + 4)} + \frac{3x + 12}{x(x + 4)}$$

$$= \frac{x^2 + 2x + 3x + 12}{x(x + 4)}$$

$$= \frac{x^2 + 5x + 12}{x(x + 4)}$$

Solving Rational Equations

$$\frac{x + 2}{x + 4} = \frac{3}{x}$$

$$(x)(x + 4)\left(\frac{x + 2}{x + 4}\right) = \frac{3}{x}(x)(x + 4)$$

$$x(x + 2) = 3(x + 4)$$

$$x^2 + 2x = 3x + 12$$

$$x^2 - x - 12 = 0$$

$$(x - 4)(x + 3) = 0$$

$$x - 4 = 0 \quad \text{or} \quad x + 3 = 0$$

$$x = 4 \quad \text{or} \quad x = -3$$

The numbers 4 and -3 will both check and are thus solutions to the equation.

Note that when adding and subtracting rational expressions we usually end up with an algebraic expression. When solving rational equations, the solution will be a numerical value or values. The equation on the right could also be solved using cross-multiplication.

Exercise Set 6.7

Solve each equation, and check your solution.

1. $\dfrac{2}{5} = \dfrac{x}{10}$

2. $\dfrac{3}{k} = \dfrac{9}{6}$

3. $\dfrac{5}{12} = \dfrac{20}{x}$

4. $\dfrac{x}{8} = \dfrac{-15}{4}$

5. $\dfrac{a}{25} = \dfrac{12}{10}$

6. $\dfrac{9c}{10} = \dfrac{9}{5}$

7. $\dfrac{9}{3b} = \dfrac{-6}{2}$

8. $\dfrac{5}{8} = \dfrac{2b}{80}$

9. $\dfrac{x + 4}{9} = \dfrac{5}{9}$

10. $\dfrac{1}{4} = \dfrac{z + 1}{8}$

11. $\dfrac{4x + 5}{6} = \dfrac{7}{2}$

12. $\dfrac{a}{5} = \dfrac{a - 3}{2}$

13. $\dfrac{6x + 7}{10} = \dfrac{2x + 9}{6}$

14. $\dfrac{n}{10} = 9 - \dfrac{n}{5}$

15. $\dfrac{x}{3} - \dfrac{3x}{4} = \dfrac{1}{12}$

16. $\dfrac{2}{8} + \dfrac{3}{4} = \dfrac{w}{5}$

17. $\dfrac{3}{4} - x = 2x$

18. $\dfrac{2}{y} + \dfrac{1}{2} = \dfrac{5}{2y}$

19. $\dfrac{5}{3x} + \dfrac{3}{x} = 1$

20. $\dfrac{x}{4} - \dfrac{x}{6} = \dfrac{1}{4}$

21. $\dfrac{x - 1}{x - 5} = \dfrac{4}{x - 5}$

22. $\dfrac{2x + 3}{x + 1} = \dfrac{3}{2}$

23. $\dfrac{5y - 3}{7} = \dfrac{15y - 2}{28}$

24. $\dfrac{2}{x + 1} = \dfrac{1}{x - 2}$

25. $\dfrac{5}{-x-6} = \dfrac{2}{x}$

26. $\dfrac{4}{y-3} = \dfrac{6}{y+3}$

27. $\dfrac{2x-3}{x-4} = \dfrac{5}{x-4}$

28. $\dfrac{3}{x} + 4 = \dfrac{3}{x}$

29. $\dfrac{x-2}{x+4} = \dfrac{x+1}{x+10}$

30. $\dfrac{x-3}{x+1} = \dfrac{x-6}{x+5}$

31. $\dfrac{2x-1}{3} - \dfrac{3x}{4} = \dfrac{5}{6}$

32. $x + \dfrac{3}{x} = \dfrac{12}{x}$ 3,

33. $x + \dfrac{6}{x} = -5$

34. $\dfrac{15}{x} + \dfrac{9x-7}{x+2} = 9$

35. $\dfrac{3y-2}{y+1} = 4 - \dfrac{y+2}{y-1}$

36. $\dfrac{2b}{b+1} = 2 - \dfrac{5}{2b}$

37. $\dfrac{1}{x+3} + \dfrac{1}{x-3} = \dfrac{-5}{x^2-9}$

38. $c - \dfrac{c}{3} + \dfrac{c}{5} = 26$

39. $\dfrac{a}{a-3} + \dfrac{3}{2} = \dfrac{3}{a-3}$

40. $\dfrac{3x}{x^2-9} + \dfrac{1}{x-3} = \dfrac{3}{x+3}$

41. $\dfrac{2}{x-3} - \dfrac{4}{x+3} = \dfrac{8}{x^2-9}$

42. $\dfrac{x+1}{x+3} + \dfrac{x-3}{x-2} = \dfrac{2x^2-15}{x^2+x-6}$

43. $\dfrac{y}{2y+2} + \dfrac{2y-16}{4y+4} = \dfrac{y-3}{y+1}$

44. $\dfrac{3}{x+3} + \dfrac{5}{x+4} = \dfrac{12x+19}{x^2+7x+12}$

45. $\dfrac{1}{2} + \dfrac{1}{x-1} = \dfrac{2}{x^2-1}$

46. $\dfrac{2y}{y+2} = \dfrac{y}{y+3} - \dfrac{3}{y^2+5y+6}$

47. $\dfrac{x+2}{x^2-x} = \dfrac{6}{x^2-1}$

48. $\dfrac{2}{x-2} - \dfrac{1}{x+1} = \dfrac{5}{x^2-x-2}$

49. (a) Explain in your own words a step-by-step procedure to use to solve equations that contain rational expressions (or fractions).

  (b) Using the procedure given in part (a), solve the equation $\dfrac{1}{x-1} - \dfrac{1}{x+1} = \dfrac{3x}{x^2-1}$.

50. Consider the equation $\dfrac{x^2}{x+3} = \dfrac{25}{x+3}$.

  (a) Explain why you may have difficulty trying to solve this equation by cross multiplication.

  (b) Find the solution to the equation.

51. Consider the following two problems:

  Simplify: Solve:

  $\dfrac{x}{3} - \dfrac{x}{4} + \dfrac{1}{x-1}$, $\dfrac{x}{3} - \dfrac{x}{4} = \dfrac{1}{x-1}$

  (a) Explain the difference between the two types of problems.

  (b) Explain how you would work each problem to obtain the correct answer.

  (c) Find the correct answer to each problem.

Cumulative Review Exercises

[3.3] **52.** With a monthly bus pass that costs $18 per month, each bus ride costs 20 cents. Without the monthly bus pass, each bus ride costs 80 cents. How many bus rides would Steve have to take per month so that the total cost with the bus pass is the same as the total cost without the bus pass?

[3.4] **53.** Two angles are supplementary angles if their sum measures 180°. Find the two supplementary an-

gles if the larger angle is 30° greater than twice the smaller angle.

[4.7] **54.** How long will it take to fill a 600-gallon Jacuzzi if water is flowing into the Jacuzzi at a rate of 8 gallons a minute?

[2.2]
[5.6] **55.** Explain the difference between a linear equation and a quadratic equation, and give an example of each.

JUST FOR FUN

1. A formula frequently used in optics is

$$\frac{1}{p} + \frac{1}{q} = \frac{1}{f}$$

where p represents the distance of the object from a mirror (or lens), q represents the distance of the image from the mirror (or lens), and f represents the focal length of the mirror (or lens). If a mirror has a focal length of 10 centimeters, how far from the mirror will the image appear when the object is 30 centimeters from the mirror?

2. In electronics the total resistance, R_T, of resistors wired in a parallel circuit is determined by the formula

$$\frac{1}{R_T} = \frac{1}{R_1} + \frac{1}{R_2} + \frac{1}{R_3} + \cdots + \frac{1}{R_n}$$

where $R_1, R_2, R_3, \ldots, R_n$ are the resistances of the individual resistors (measured in ohms) in the circuit.

(a) Find the total resistance if two resistors, one of 200 ohms and the other of 300 ohms, are wired in a parallel circuit.

(b) If three identical resistors are to be wired in parallel, what should be the resistance of each resistor if the total resistance of the circuit is to be 300 ohms?

6.8

Applications of Rational Equations

▶ **1** Set up and solve problems containing rational expressions.

▶ **2** Set up and solve rate problems.

▶ **3** Set up and solve work problems.

▶ **1** Many applications of algebra involve rational equations. After we represent the application as an equation, we solve the rational equation as we did in Section 6.7.

EXAMPLE 1 The area of a triangle is 27 square feet. Find the base and height if its height is 3 feet less than twice its base.

Solution: Let $\quad x =$ base

Then $\quad 2x - 3 =$ height   See Fig. 6.1.

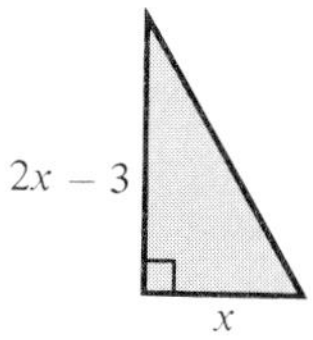

FIGURE 6.1

$$\text{area} = \frac{1}{2} \cdot \text{base} \cdot \text{height}$$

$$27 = \frac{1}{2}(x)(2x - 3) \qquad \text{Multiply both sides of the equation by 2.}$$

$$2(27) = 2\left[\frac{1}{2}(x)(2x - 3)\right]$$

$$54 = x(2x - 3)$$

$$54 = 2x^2 - 3x$$

$$0 = 2x^2 - 3x - 54$$

$$\text{or} \quad 2x^2 - 3x - 54 = 0$$

$$(2x + 9)(x - 6) = 0$$

$$2x + 9 = 0 \qquad \text{or} \quad x - 6 = 0$$

$$2x = -9 \quad \text{or} \qquad x = 6$$

$$x = -\frac{9}{2}$$

Since the dimensions of a geometric figure cannot be negative, we can eliminate $-\frac{9}{2}$ as an answer to our problem.

$$\text{Base} = x = 6 \text{ feet}$$

$$\text{Height} = 2x - 3 = 2(6) - 3 = 9 \text{ feet}$$

Check: $\quad a = \frac{1}{2}bh$

$$27 = \frac{1}{2}(6)(9)$$

$$27 = 27 \qquad \text{true}$$

EXAMPLE 2   One number is 3 times another number. The sum of their reciprocals is 4. Find the numbers.

Solution: Let x = smaller number

then $3x$ = larger number

The sum of their reciprocals is 4; thus

$$\frac{1}{x} + \frac{1}{3x} = 4$$

$$3x\left(\frac{1}{x} + \frac{1}{3x}\right) = 3x(4)$$

$$3x\left(\frac{1}{x}\right) + 3x\left(\frac{1}{3x}\right) = 12x$$

$$3 + 1 = 12x$$

$$4 = 12x$$

$$\frac{4}{12} = x$$

$$\frac{1}{3} = x$$

The smaller number is $\frac{1}{3}$; the larger number is $3x = 3\left(\frac{1}{3}\right) = 1$.

Check: $$\frac{1}{x} + \frac{1}{3x} = 4$$

$$\frac{1}{\frac{1}{3}} + \frac{1}{3\left(\frac{1}{3}\right)} = 4$$

$$3 + 1 = 4$$

$$4 = 4 \qquad \text{true}$$

■

Rate Problems   ▶ **2**   In Chapter 4 we discussed rate problems. Recall that

$$\text{distance} = \text{rate} \cdot \text{time}$$

If we solve this equation for time, we obtain

$$\text{time} = \frac{\text{distance}}{\text{rate}} \quad \text{or} \quad t = \frac{d}{r}$$

This equation is useful in solving rate problems when the total time of travel for two objects or the time of travel between two points is known.

EXAMPLE 3   A river has a current of 3 miles per hour. If it takes Jack's motorboat the same time to go 10 miles downstream as 6 miles upstream, find the speed of his boat in still water.

Solution: Let r = speed (or rate) of boat in still water

then $r + 3$ = speed of boat downstream (with current)

and $r - 3$ = speed of boat upstream (against current)

	Distance	Rate	Time
Downstream	10	$r + 3$	$\dfrac{10}{r + 3}$
Upstream	6	$r - 3$	$\dfrac{6}{r - 3}$

Since the time it takes to travel 10 miles downstream is the same as the time to travel 6 miles upstream, we set the times equal to each other and then solve the resulting equation.

$$\text{time downstream} = \text{time upstream}$$

$$\frac{10}{r + 3} = \frac{6}{r - 3} \qquad \text{Now cross-multiply.}$$

$$10(r - 3) = 6(r + 3)$$

$$10r - 30 = 6r + 18$$

$$4r - 30 = 18$$

$$4r = 48$$

$$r = 12$$

The speed of the boat in still water is 12 miles per hour. A check will show that 12 is the solution to the problem. ∎

EXAMPLE 4 Mr. Blake rides his bike every Saturday morning in the county park. During the first part of the ride he is peddling mostly uphill and travels at an average speed of 12 miles an hour. After a certain point, he is traveling mostly downhill and averages 18 miles per hour. If the total distance he travels is 30 miles and the total time he is riding is 2 hours, how long did he ride at each speed?

Solution: Let d = distance traveled at 12 miles per hour

then $30 - d$ = distance traveled at 18 miles per hour

	Distance	Rate	Time
Uphill	d	12	$\dfrac{d}{12}$
Downhill	$30 - d$	18	$\dfrac{30 - d}{18}$

Since the total time spent riding is 2 hours, we write

$$\text{time going uphill} + \text{time going downhill} = 2 \text{ hours}$$

$$\frac{d}{12} + \frac{30 - d}{18} = 2$$

$$36\left(\frac{d}{12} + \frac{30 - d}{18}\right) = 36 \cdot 2$$

$$\overset{3}{\cancel{36}}\left(\frac{d}{\cancel{12}}\right) + \overset{2}{\cancel{36}}\left(\frac{30 - d}{\cancel{18}}\right) = 72$$

$$3d + 60 - 2d = 72$$

$$d + 60 = 72$$

$$d = 12$$

The answer to the problem is not 12. Remember that the question asked us to *find the time spent* traveling at each speed. The variable d does not represent time, but represents the distance traveled at 12 miles per hour. To find the time traveled and to answer the question asked, we need to evaluate $\dfrac{d}{12}$ and $\dfrac{30 - d}{18}$ for $d = 12$.

Time at 12 mph   *Time at 18 mph*

$$\frac{d}{12} = \frac{12}{12} = 1 \qquad \frac{30 - d}{18} = \frac{30 - 12}{18} = \frac{18}{18} = 1$$

Thus 1 hour was spent traveling at each rate. ■

EXAMPLE 5   A car and train take parallel routes from Los Angeles to the California State Fair in Sacramento. The train averages 70 miles per hour and the car averages 50 miles per hour. If the train arrives at the fair 2.2 hours before the car, find the distance from Los Angeles to the fair.

Solution: Let d = distance from Los Angeles to the fair.

	Distance	Rate	Time
Train	d	70	$\dfrac{d}{70}$
Car	d	50	$\dfrac{d}{50}$

We are given that the car ride is 2.2 hours longer than the train ride. Therefore, to make the two times equal, we need to add 2.2 hours to the time of the train ride. Using this information, we set up the following equation.

time for car ride = time for train ride + 2.2 hours

$$\frac{d}{50} = \frac{d}{70} + 2.2$$

Now multiply both sides of the equation by the LCD, 350.

$$350\left(\frac{d}{50}\right) = 350\left(\frac{d}{70} + 2.2\right)$$

$$7d = 350\left(\frac{d}{70}\right) + 350(2.2)$$

$$7d = 5d + 770$$

$$2d = 770$$

$$d = 385$$

Therefore, the distance from Los Angeles to the State Fair in Sacramento is 385 miles. ■

Work Problems

▶ **3**   Problems where two or more machines or people work together to complete a certain task are sometimes referred to as **work problems.** Work problems often involve equations containing fractions. Generally, work problems are based on the fact that the fractional part of the work done by person 1 (or machine 1) plus the fractional part of the work done by person 2 (or machine 2) is equal to the total amount of work done by both people (or both machines). *We represent the total amount of* *work done by the number 1, which represents one whole job completed.*

Part of task done by first person or machine	+	Part of task done by second person or machine	=	1 (one whole task completed)

To determine the part of the total task completed by each person or machine, we use the formula

Part of task completed = rate · time

This formula is very similar to the formula

$$\text{Amount} = \text{rate} \cdot \text{time}$$

that was discussed in Section 4.7. To determine the part of the task completed, we need to determine the rate. Suppose Paul can do a particular task in 6 hours. Then he would complete $\frac{1}{6}$ of the task per hour. Thus, his rate is $\frac{1}{6}$ of the task per hour. If Audrey can do a particular task in 5 minutes, then her rate is $\frac{1}{5}$ of the task per minute. In general, if a person or machine can complete a task in x units of time, then the rate is $1/x$.

EXAMPLE 6 Mr. Donaldson can paint a house by himself in 20 hours. Mr. Cronkite can paint the same house by himself in 30 hours. How long will it take them to paint the house if they work together?

Solution: Let $t =$ the time, in hours, for both men to paint the house together. We will construct a table to help us in finding the part of the task completed by Mr. Donaldson and Mr. Cronkite in t hours.

	Rate of Work	Time Worked	Part of Task
Mr. Donaldson	$\frac{1}{20}$	t	$\frac{t}{20}$
Mr. Cronkite	$\frac{1}{30}$	t	$\frac{t}{30}$

$$\left(\begin{array}{c}\text{part of house painted}\\\text{by Mr. Donaldson in } t \text{ hours}\end{array}\right) + \left(\begin{array}{c}\text{part of house painted}\\\text{by Mr. Cronkite in } t \text{ hours}\end{array}\right) = 1 \text{ (whole house painted)}$$

$$\frac{t}{20} \qquad\qquad + \qquad\qquad \frac{t}{30} \qquad\qquad = 1$$

Now multiply both sides of the equation by the LCD, 60.

$$\frac{t}{20} + \frac{t}{30} = 1$$

$$60\left(\frac{t}{20} + \frac{t}{30}\right) = 60 \cdot 1$$

$$60\left(\frac{t}{20}\right) + 60\left(\frac{t}{30}\right) = 60$$

$$3t + 2t = 60$$

$$5t = 60$$

$$t = 12$$

Thus the two men working together can paint the house in 12 hours. ∎

EXAMPLE 7   A tank can be filled by one pipe in 4 hours and can be emptied by another pipe in 6 hours. If the valves to both pipes are open, how long will it take to fill the tank?

Solution: Let t = amount of time to fill the tank.

	Rate of Work	Time	Part of Task
Pipe filling tank	$\dfrac{1}{4}$	t	$\dfrac{t}{4}$
Pipe emptying tank	$\dfrac{1}{6}$	t	$\dfrac{t}{6}$

As one pipe is filling, the other is emptying the tank. Thus the pipes are working against each other. Therefore, instead of adding the parts of the task, as was done in Example 6, where the peopled worked together, we will subtract the parts of the task.

$$\left(\begin{array}{c} \text{part of tank} \\ \text{filled in } x \text{ hours} \end{array} \right) - \left(\begin{array}{c} \text{part of tank} \\ \text{emptied in } x \text{ hours} \end{array} \right) = 1 \text{ (total tank filled)}$$

$$\frac{t}{4} - \frac{t}{6} = 1$$

$$12\left(\frac{t}{4} - \frac{t}{6}\right) = 12 \cdot 1$$

$$12\left(\frac{t}{4}\right) - 12\left(\frac{t}{6}\right) = 12$$

$$3t - 2t = 12$$

$$t = 12$$

The tank will be filled in 12 hours.   ■

EXAMPLE 8   Dolores and Maryann are both auto mechanics at Simpson's garage. When Dolores removes and rebuilds a car's transmission by herself, it takes her 10 hours. When Dolores and Maryann work together to remove and rebuild the transmission, it takes them 6 hours. How long does it take Maryann by herself to remove and rebuild the transmission?

Solution: Let m = time for Maryann to remove and rebuild the transmission by herself.
    Let us make a table to help analyze the problem. In the table we make use of the fact that together they can remove and rebuild the transmission in 6 hours.

	Rate of Work	Time	Part of Task
Dolores	$\dfrac{1}{10}$	6	$\dfrac{6}{10}$ or $\dfrac{3}{5}$
Maryann	$\dfrac{1}{m}$	6	$\dfrac{6}{m}$

$$\begin{pmatrix} \text{part of task} \\ \text{completed by} \\ \text{Dolores} \end{pmatrix} + \begin{pmatrix} \text{part of task} \\ \text{completed by} \\ \text{Maryann} \end{pmatrix} = 1$$

$$\frac{3}{5} + \frac{6}{m} = 1$$

Now multiply both sides of the equation by the LCD, $5m$.

$$5m\left(\frac{3}{5} + \frac{6}{m}\right) = 5m \cdot 1$$

$$5m\left(\frac{3}{5}\right) + 5m\left(\frac{6}{m}\right) = 5m$$

$$3m + 30 = 5m$$

$$30 = 2m$$

$$15 = m$$

Thus it takes Maryann 15 hours by herself to remove and rebuild the transmission. ∎

EXAMPLE 9 Mr. and Mrs. O'Connor are handwriting thank-you notes to people who attended their wedding. Mrs. O'Connor by herself could write all the notes in 8 hours and Mr. O'Connor could write all the notes by himself in 7 hours. After Mrs. O'Connor has been writing thank-you notes for 5 hours by herself, she must leave town on business. Mr. O'Connor then continues the task of writing the thank-you notes. How long will it take Mr. O'Connor to finish writing the remaining notes?

Solution: Let t = time it will take Mr. O'Connor to finish writing the notes.

	Rate	Time	Part of task
Mrs. O'Connor	$\frac{1}{8}$	5	$\frac{5}{8}$
Mr. O'Connor	$\frac{1}{7}$	t	$\frac{t}{7}$

$$\begin{pmatrix} \text{part of cards} \\ \text{written by} \\ \text{Mrs. O'Connor} \end{pmatrix} + \begin{pmatrix} \text{part of cards} \\ \text{written by} \\ \text{Mr. O'Connor} \end{pmatrix} = 1$$

$$\frac{5}{8} + \frac{t}{7} = 1$$

$$56\left(\frac{5}{8} + \frac{t}{7}\right) = 56 \cdot 1$$

$$56\left(\frac{5}{8}\right) + 56\left(\frac{t}{7}\right) = 56$$

$$35 + 8t = 56$$

$$8t = 21$$

$$t = \frac{21}{8} \text{ or } 2\frac{5}{8}$$

Thus it will take Mr. O'Connor $2\frac{5}{8}$ hours to complete the cards. ∎

Exercise Set 6.8

For each problem (a) write an equation that can be used to solve the problem and (b) solve the problem.

1. The base of a triangle is 6 centimeters greater than its height. Find the base and height if the area is 80 square centimeters.

2. The height of a triangle is 1 centimeter less than twice its base. Find the base and height if the triangle's area is 33 square centimeters.

3. One number is three times as large as another. The sum of their reciprocals is $\frac{4}{3}$. Find the two numbers.

4. The numerator of the fraction $\frac{3}{4}$ is increased by an amount so that the value of the resulting fraction is $\frac{5}{2}$. Find the amount that the numerator was increased.

5. The reciprocal of 3 plus the reciprocal of 5 is the reciprocal of what number?

6. One number is 4 times as large as another. The sum of their reciprocals is $\frac{5}{8}$. Find the two numbers.

7. One positive number is 4 more than another. The sum of their reciprocals is $\frac{2}{3}$. Find the two numbers.

8. Jim can row 4 miles per hour in still water. It takes him as long to row 6 miles upstream as 10 miles downstream. How fast is the current?

9. In the Pixie River a boat travels 9 miles upstream in the same amount of time it travels 11 miles downstream. If the current of the river is 2 miles per hour, find the speed of the boat in still water.

10. Ms. Duncan took her two sons water skiing in still water. She drove the motor boat one way on the water pulling the younger son at 30 miles per hour. Then she turned around and pulled her older son in the opposite direction the same distance at 30 miles per hour. If the total time spent skiing was $\frac{1}{2}$ hour, how far did each son travel?

11. A business executive traveled 1800 miles by jet and then traveled an additional 300 miles on a private propeller plane. If the rate of the jet is four times the rate of the prop plane and the entire trip took 5 hours, find the speed of each plane.

12. One car travels 30 kilometers per hour faster than another. In the time it takes the slower car to travel 250 kilometers the faster car travels 400 kilometers. Find the speed of both cars.

13. Maria walked at a speed of 2 miles per hour and jogged at a speed of 4 miles per hour. If she jogged 3 miles farther than she walked and the total time for the trip was 3 hours, how far did she walk and how far did she jog?

14. On a treadmill, Mario walks a distance of 2 miles. He then doubles the speed of the treadmill and jogs for another 2 miles. If the total time he spent on the tread-

mill was 1 hour, find the speeds at which he walks and jogs.

15. In her daily morning activity, Dawn jogs a specific distance at 8 miles per hour and then walks another distance at 4 miles per hour. If the total distance she travels is 6 miles and the total time spent in her outing is 1.2 hours, find the distance she jogs and the distance she walks.

16. Apostolos jogs and then walks in alternating intervals. When he jogs he averages 5 miles per hour and when he walks he averages 2 miles per hour. If he walks and jogs a total of 3 miles in a total of 0.9 hour, what length of time is spent jogging and what length of time is spent walking?

17. A Boeing 747 flew from San Francisco to Honolulu, a distance of 2800 miles. Flying with the wind it averaged 600 miles per hour. When the wind changed from a tailwind to a headwind, the plane's speed dropped to 500 miles per hour. If the total time of the trip was 5 hours, determine the length of time it flew at each speed.

18. Sean and his father Scott begin skiing the same cross-country ski trail at the same time. If Sean, who averages 9 miles per hour, finishes the trail 0.5 hour quicker than his father, who averages 6 miles per hour, find the length of the trail.

19. Jan swims freestyle at an average speed of 40 meters per minute, and he swims using the breast stroke at an average speed of 30 meters per minute. Jan decides to swim freestyle across Echo Lake. After resting he swims back across the lake using the breast stroke. If his return trip took 20 minutes longer than his trip going, what is the width of the lake at the point where he crossed?

20. Marsha can construct a small retaining wall in 3 hours. Her apprentice can complete the same job in 6 hours. How long would it take them to complete the job working together?

21. Mr. Dell fertilizes the farm in 6 hours. Mrs. Dell fertilizes the farm in 7 hours. How long will it take them to fertilize their farm if they work together?

22. A $\frac{1}{2}$-inch-diameter hose can fill a swimming pool in 8 hours. A $\frac{4}{5}$-inch-diameter hose can fill the same pool in 5 hours. How long will it take to fill the pool when both hoses are used?

23. A conveyor belt operating at full speed can fill a tank with topsoil in 3 hours. When a valve at the bottom of the tank is opened, the tank will empty in 4 hours. If the conveyor belt is operating at full speed and the valve

at the bottom of the tank is open, how long will it take to fill the tank?

24. If the stopper is in the basin, the water from the rinse cycle of a washing machine will fill the basin in 4 minutes. If the stopper from the full basin is removed and no additional water is coming into the basin, the basin will empty in 5 minutes. If there is no stopper in the empty basin and water from the rinse cycle starts filling the basin, how long will it take for the basin to be filled?

25. One bottling machine can meet the company's daily production level of filled and capped bottles in 8 hours. When a second bottling machine is also running, the daily production of bottles can be completed in 3 hours. How long would it take the second bottling machine to meet the daily production level if it were working alone?

26. At the NCNB Savings Bank it takes a computer 4 hours to process and print payroll checks. When a second computer is used and the two computers work together, the checks can be processed and printed in 3 hours. How long would it take the second computer by itself to process and print the payroll checks?

27. The Wilsons own a large farm where they grow wheat. With the large tractor Mrs. Wilson can plow the entire farm in 6 days. With a smaller tractor Mr. Wilson can plow it in 10 days. Mrs. Wilson begins plowing the farm but after 4 days has problems with the large tractor and must stop. Mr. Wilson then begins plowing with the smaller tractor. How much longer will it take him to finish plowing the farm?

28. A construction company with two backhoes has contracted to dig a long trench for drainage pipes. The larger backhoe can dig the entire trench by itself in 12 days. The smaller backhoe can dig the entire trench by itself in 15 days. The large backhoe begins working on the trench by itself, but after 5 days it is transferred to a different location and the smaller backhoe begins working on the trench. How long will it take for the smaller backhoe to complete the job?

29. A boat designed to skim oil off the surface of the water has two skimmers. One skimmer can fill the boat's holding tank in 60 hours while the second skimmer can fill the boat's holding tank in 50 hours. There is also a valve in the holding tank that is used to transfer the oil to a larger vessel. If no new oil is coming into the holding tank, a full holding tank of skimmed oil can be transferred to a larger tank in 30 hours. If both skimmers begin skimming and the valve on the holding tank is opened, how long will it take for the empty holding tank on the skimmer to fill?

30. When only the cold-water valve is opened, a washtub will fill in 8 minutes. When only the hot-water valve is opened, the washtub will fill in 12 minutes. When the drain of the washtub is open, it will drain completely in 7 minutes. If both the hot- and cold-water valves are open and the drain is open, how long will it take for the washtub to fill?

31. Susan can knit an afghan in 20 hours, and Patty can knit the same afghan in 25 hours. After Susan has been knitting by herself for 11 hours, Patty decides to join her; how long will it take them working together to complete the afghan?

Cumulative Review Exercises

[1.9] **32.** Evaluate $6 - [(3 - 5^2) \div 11]^2 + 18 \div 3$.

[2.1] **33.** Simplify $\frac{1}{2}(x + 3) - (2x + 6)$.

[6.2] **34.** Divide $\dfrac{x^2 - 14x + 48}{x^2 - 5x - 24} \div \dfrac{2x^2 - 13x + 6}{2x^2 + 5x - 3}$.

[6.5] **35.** Add $\dfrac{x}{6x^2 - x - 15} - \dfrac{5}{9x^2 - 12x - 5}$.

JUST FOR FUN

1. The reciprocal of the difference of a certain number and 3 is twice the reciprocal of the difference of twice the number and 6. Find the number(s).

2. If three times a number is added to twice the reciprocal of the number, the answer is 5. Find the number(s).

3. Donald and Juniper McDonald, whose parents own a strawberry farm, have a responsibility to each pick the same number of buckets of strawberries a day. Donald, who is older, picks an average of 6 buckets of berries per hour, while Juniper picks an average of 3 buckets of berries per hour. If Donald and Juniper begin picking strawberries at the same time, and Donald finishes 1.5 hours before Juniper, how many buckets of strawberries must each pick?

SUMMARY

GLOSSARY

Algebraic fraction (or rational expression) *(245):*
An algebraic expression of the form $\frac{p}{q}$, where p and q are polynomials and $q \neq 0$.
Complex fraction *(270):* A fraction that has a fraction in its numerator or its denominator, or in both its numerator and denominator.
Extraneous root or extraneous solution *(276):* A

number obtained when solving an equation that is not a solution to the original equation.
Reduced to lowest terms *(247):* An algebraic fraction is reduced to its lowest terms when the numerator and denominator have no common factors other than 1.
Secondary fractions *(270):* The numerator and denominator of a complex fraction are secondary fractions.

IMPORTANT FACTS

For any fraction: $-\dfrac{a}{b} = \dfrac{-a}{b} = \dfrac{a}{-b}, b \neq 0$

To add fractions: $\dfrac{a}{c} + \dfrac{b}{c} = \dfrac{a + b}{c}, c \neq 0$

To subtract fractions: $\dfrac{a}{c} - \dfrac{b}{c} = \dfrac{a - b}{c}, c \neq 0$

To multiply fractions: $\dfrac{a}{b} \cdot \dfrac{c}{d} = \dfrac{ac}{bd}, b \neq 0, d \neq 0$

To divide fractions: $\dfrac{a}{b} \div \dfrac{c}{d} = \dfrac{a}{b} \cdot \dfrac{d}{c} = \dfrac{ad}{bc},$
$$b \neq 0, c \neq 0, d \neq 0$$

$$\text{Time} = \dfrac{\text{distance}}{\text{rate}}$$

Review Exercises

[6.1] *Determine the values of the variable for which the following expressions are defined.*

1. $\dfrac{6}{2x - 8}$

2. $\dfrac{5}{x^2 - 7x + 12}$

Reduce each expression to its lowest terms.

3. $\dfrac{x}{x - xy}$

4. $\dfrac{x^3 + 4x^2 + 12x}{x}$

5. $\dfrac{9x^2 + 6xy}{3x}$

6. $\dfrac{x^2 + x - 12}{x - 3}$

7. $\dfrac{x^2 - 4}{x - 2}$

8. $\dfrac{2x^2 - 7x + 3}{3 - x}$

9. $\dfrac{x^2 - 2x - 24}{x^2 + 6x + 8}$

10. $\dfrac{3x^2 - 8x - 16}{x^2 - 8x + 16}$

[6.2] *Multiply as indicated.*

11. $\dfrac{4y}{3x} \cdot \dfrac{4x^2 y}{2}$

12. $\dfrac{15x^2 y^3}{3z} \cdot \dfrac{6z^3}{5xy^3}$

13. $\dfrac{40a^3 b^4}{7c^3} \cdot \dfrac{14c^5}{5a^5 b}$

14. $\dfrac{1}{x - 2} \cdot \dfrac{2 - x}{2}$

15. $\dfrac{-x + 2}{3} \cdot \dfrac{6x}{x - 2}$

16. $\dfrac{4x + 4y}{x^2 y} \cdot \dfrac{y^3}{8x}$

17. $\dfrac{a - 2}{a + 3} \cdot \dfrac{a^2 + 4a + 3}{a^2 - a - 2}$

18. $\dfrac{x^2 - y^2}{x - y} \cdot \dfrac{x + y}{xy + x^2}$

Divide as indicated.

19. $\dfrac{6y^3}{x} \div \dfrac{y^3}{6x}$

20. $\dfrac{8xy^2}{z} \div \dfrac{x^4 y^2}{4z^2}$

21. $\dfrac{3x + 3y}{x^2} \div \dfrac{x^2 - y^2}{x^2}$

22. $\dfrac{1}{a^2 + 8a + 15} \div \dfrac{3}{a + 5}$

23. $\dfrac{4x}{a + 2} \div \dfrac{8x^2}{a - 2}$

24. $(x + 3) \div \dfrac{x^2 - 4x - 21}{x - 7}$

25. $\dfrac{x^2 - 3xy - 10y^2}{6x} \div \dfrac{x + 2y}{12x^2}$

26. $\dfrac{4x^2 - 16y^2}{9} \div \dfrac{(x + 2y)^2}{12}$

[6.3] *Add or subtract as indicated.*

27. $\dfrac{x}{x + 2} + \dfrac{2}{x + 2}$

28. $\dfrac{x}{x + 2} - \dfrac{2}{x + 2}$

29. $\dfrac{4x}{x + 2} + \dfrac{8}{x + 2}$

30. $\dfrac{6x}{3y} - \dfrac{8}{3y}$

31. $\dfrac{9x - 4}{x + 8} + \dfrac{76}{x + 8}$

32. $\dfrac{7x - 3}{x^2 + 7x - 30} - \dfrac{3x + 9}{x^2 + 7x - 30}$

33. $\dfrac{4x^2 - 11x + 4}{x - 3} - \dfrac{x^2 - 4x + 10}{x - 3}$

34. $\dfrac{6x^2 - 4x}{2x - 3} - \dfrac{(-3x + 12)}{2x - 3}$

[6.4] *Find the least common denominator.*

35. $\dfrac{x}{3} + \dfrac{5x}{8}$

36. $\dfrac{4}{3x} + \dfrac{8}{5x^2}$

37. $\dfrac{6}{x + 1} - \dfrac{3x}{x}$

38. $\dfrac{6x + 3}{x + 2} + \dfrac{4}{x - 3}$

39. $\dfrac{7x - 12}{x^2 + x} - \dfrac{4}{x + 1}$

40. $\dfrac{9x - 3}{x + y} - \dfrac{4x + 7}{x^2 - y^2}$

41. $\dfrac{4x^2}{x - 7} + 8x^2$

42. $\dfrac{19x - 5}{x^2 + 2x - 35} + \dfrac{3x - 2}{x^2 + 9x + 14}$

[6.5] *Add or subtract as indicated.*

43. $\dfrac{4}{2x} + \dfrac{x}{x^2}$

44. $\dfrac{1}{4x} + \dfrac{6x}{xy}$

45. $\dfrac{5x}{3xy} - \dfrac{4}{x^2}$

46. $6 + \dfrac{x}{x + 2}$

47. $5 - \dfrac{3}{x + 3}$

48. $\dfrac{a + c}{c} - \dfrac{a - c}{a}$

49. $\dfrac{3}{x + 3} + \dfrac{4}{x}$

50. $\dfrac{2}{3x} - \dfrac{3}{3x - 6}$

51. $\dfrac{x + 4}{x + 3} - \dfrac{x - 3}{x + 4}$

52. $\dfrac{4}{x + 5} + \dfrac{6}{(x + 5)^2}$

53. $\dfrac{x + 3}{x^2 - 9} + \dfrac{2}{x + 3}$

54. $\dfrac{4}{(x + 2)(x - 3)} - \dfrac{4}{(x - 2)(x + 2)}$

55. $\dfrac{x + 2}{x^2 - x - 6} + \dfrac{x - 3}{x^2 - 8x + 15}$

56. $\dfrac{x + 5}{x^2 - 15x + 50} - \dfrac{x - 2}{x^2 - 25}$

[6.6] *Simplify each complex fraction.*

57. $\dfrac{1 + \dfrac{5}{12}}{\dfrac{3}{8}}$

58. $\dfrac{4 - \dfrac{9}{16}}{1 + \dfrac{5}{8}}$

59. $\dfrac{\dfrac{15xy}{6z}}{\dfrac{3x}{z^2}}$

60. $\dfrac{\dfrac{36x^4y^2}{9xy^5}}{4z^2}$

61. $\dfrac{x + \dfrac{1}{y}}{y^2}$

62. $\dfrac{x - \dfrac{x}{y}}{\dfrac{1 + x}{y}}$

63. $\dfrac{\dfrac{4}{x} + \dfrac{2}{x^2}}{6 - \dfrac{1}{x}}$

64. $\dfrac{\dfrac{x}{x + y}}{\dfrac{x^2}{2x + 2y}}$

65. $\dfrac{\dfrac{1}{a}}{\dfrac{1}{a^2}}$

66. $\dfrac{\dfrac{1}{a} + 2}{\dfrac{1}{a} + \dfrac{1}{a}}$

67. $\dfrac{\dfrac{1}{x^2} + \dfrac{1}{x}}{\dfrac{1}{x^2} - \dfrac{1}{x}}$

68. $\dfrac{\dfrac{3x}{y} - x}{\dfrac{y}{x} - 1}$

[6.7] *Solve each equation.*

69. $\dfrac{3}{x} = \dfrac{8}{24}$

70. $\dfrac{4}{a} = \dfrac{16}{4}$

71. $\dfrac{x + 3}{5} = \dfrac{9}{5}$

72. $\dfrac{x}{6} = \dfrac{x - 4}{2}$

73. $\dfrac{3x + 4}{5} = \dfrac{2x - 8}{3}$

74. $\dfrac{x}{5} + \dfrac{x}{2} = -14$

75. $4 - \dfrac{5}{x + 5} = \dfrac{x}{x + 5}$

76. $\dfrac{4}{x} - \dfrac{1}{6} = \dfrac{1}{x}$

77. $\dfrac{1}{x - 2} + \dfrac{1}{x + 2} = \dfrac{1}{x^2 - 4}$

78. $\dfrac{x - 3}{x - 2} + \dfrac{x + 1}{x + 3} = \dfrac{2x^2 + x + 1}{x^2 + x - 6}$

79. $\dfrac{x}{x^2 - 9} + \dfrac{2}{x + 3} = \dfrac{4}{x - 3}$

[6.8] *Solve each problem.*

80. It takes Lee 5 hours to mow Mr. McKane's lawn. It takes Pat 4 hours to mow the same lawn. How long will it take them working together to mow Mr. McKane's lawn?

81. A $\frac{3}{4}$-inch-diameter hose can fill a swimming pool in 7 hours. A $\frac{5}{16}$-inch-diameter hose can siphon water out of a full pool in 12 hours. How long will it take to fill the pool if while one hose is filling the pool the other hose is siphoning water from the pool?

82. One number is four times as large as another. The sum of their reciprocals is $\frac{1}{2}$. Find the numbers.

83. A Greyhound bus can travel 400 kilometers in the same time that an Amtrak train can travel 600 kilometers. If the speed of the train is 40 kilometers per hour greater than that of the bus, find the speeds of the bus and the train.

Practice Test

Perform the operations indicated.

1. $\dfrac{3x^2y}{4z^2} \cdot \dfrac{8xz^3}{9y^4}$

2. $\dfrac{a^2 - 9a + 14}{a - 2} \cdot \dfrac{a^2 - 4a - 21}{(a - 7)^2}$

3. $\dfrac{x^2 - 9y^2}{3x + 6y} \div \dfrac{x + 3y}{x + 2y}$

4. $\dfrac{16}{y^2 + 2y - 15} \div \dfrac{4y}{y - 3}$

5. $\dfrac{6x + 3}{2y} + \dfrac{x - 5}{2y}$

6. $\dfrac{7x^2 - 4}{x + 3} - \dfrac{6x + 7}{x + 3}$

7. $\dfrac{5}{x} + \dfrac{3}{2x^2}$

8. $5 - \dfrac{6x}{x + 2}$

9. $\dfrac{x - 5}{x^2 - 16} - \dfrac{x - 2}{x^2 + 2x - 8}$

Simplify each expression.

10. $\dfrac{3 + \dfrac{5}{8}}{2 - \dfrac{3}{4}}$

11. $\dfrac{x + \dfrac{x}{y}}{\dfrac{1}{x}}$

Solve each equation.

12. $\dfrac{x}{3} - \dfrac{x}{4} = 5$

13. $\dfrac{x}{x - 8} + \dfrac{6}{x - 2} = \dfrac{x^2}{x^2 - 10x + 16}$

Solve the problem.

14. Mr. Johnson, on his tractor, can level a 1-acre field in 8 hours. Mr. Hackett, on his tractor, can level a 1-acre field in 5 hours. If they work together, how long will it take them to level a 1-acre field?

Cumulative Review Test

1. Evaluate $3x^2 - 2xy - 7$ when $x = -3$ and $y = 5$.

2. Evaluate $-4 - [2(-6 \div 3)^2] \div 2$.

3. Solve the equation $4y + 3 = -2(y + 6)$.

4. Simplify $\left(\dfrac{6x^2 y^3}{2x^5 y}\right)^3$.

5. Solve the formula $P = 2E + 3R$ for R.

6. Simplify $(6x^2 - 3x - 5) - (3x^2 + 8x - 9)$.

7. Multiply $(4x^2 - 6x + 3)(3x - 5)$.

8. Factor $6a^2 - 6a - 5a + 5$.

9. Factor $10x^2 - 5x + 5$.

10. Factor $x^2 - 10x + 24$.

11. Factor $6x^2 - 11x - 10$.

12. Solve $2x^2 = 11x - 12$.

13. Multiply $\dfrac{x^2 - 9}{x^2 - x - 6} \cdot \dfrac{x^2 - 2x - 8}{2x^2 - 7x - 4}$.

14. Subtract $\dfrac{x}{x + 4} - \dfrac{3}{x - 5}$.

15. Add $\dfrac{4}{x^2 - 3x - 10} + \dfrac{2}{x^2 + 5x + 6}$.

16. Solve the equation $\dfrac{x}{6} - \dfrac{x}{4} = \dfrac{1}{8}$.

17. Solve the equation $\dfrac{1}{x - 4} + \dfrac{2}{x - 3} = \dfrac{4}{x^2 - 7x + 12}$.

18. A school district allows its employees to choose from two medical plans. With plan 1, the employee pays 10% of all medical bills (the school district pays the balance to the doctor). With plan 2, the teacher pays the school district a one-time payment of $100, then the teacher pays 5% of all medical bills. What total medical bills would result in the teacher paying the same amount with the two plans?

19. A grocer wishes to mix 6 pounds of Chippy dog food worth $3 per pound with Hippy dog food, worth $4 per pound. How much of the Hippy dog food should be mixed if the dog food mixture is to sell for $3.20 per pound?

20. Chiquita rides her bike up a long hill at an average speed of 4 miles per hour. Then she rides down the hill and rides to her friend's house at an average speed of 12 miles per hour. If the total distance traveled is 6 miles and the total time spent riding is 1 hour, find the distance she rides at 4 miles per hour and the distance she rides at 12 miles per hour.

7.1 The Cartesian Coordinate System

7.2 Graphing Linear Equations

7.3 Slope of a Line

7.4 Slope—Intercept Form of a Linear Equation

7.5 Point—Slope Form of a Linear Equation (Optional)

7.6 Graphing Linear Inequalities

Summary

Review Exercises

Practice Test

See Section 7.3, Just for Fun 4.

7.1

The Cartesian Coordinate System

▶**1**   Use the Cartesian coordinate system.

▶**2**   Plot ordered pairs.

▶**1**   In this chapter we discuss several procedures that can be used to draw graphs. A **graph** shows the relationship between two variables in an equation. Many algebraic relationships are easier to understand if we can see a picture of them. We draw graphs using the **Cartesian (or rectangular) coordinate system.** The Cartesian coordinate system is named for its developer, the French mathematician and philosopher René Descartes (1596–1650).

Before you learn how to construct a graph, you must understand the Cartesian coordinate system. The Cartesian coordinate system consists of two axes (or number lines) drawn perpendicular to each other. The two intersecting axes form four **quadrants** (see Fig. 7.1).

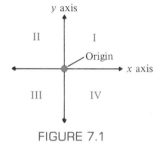

FIGURE 7.1

The horizontal axis is called the **x axis.** The vertical axis is called the **y axis.** The point of intersection of the two axes is called the **origin.** The origin has an *x* value of 0 and a *y* value of 0. Starting from the origin and moving to the right along the *x* axis, the numbers increase. Starting from the origin and moving to the left, the numbers decrease (see Fig. 7.2). Starting from the origin and moving up the *y* axis, the numbers increase. Starting from the origin and moving down, the numbers decrease.

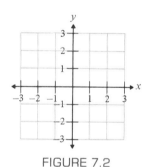

FIGURE 7.2

▶**2**   To locate a point, it is necessary to know both the *x* and *y* values, or **coordinates,** of the point. When the *x* and *y* coordinates of a point are placed in parentheses, with the *x* coordinate listed first, we have an **ordered pair.** In the ordered pair (3, 5) the *x* coordinate is 3 and the *y* coordinate is 5. The point representing the ordered pair (3, 5) is plotted in Fig. 7.3.

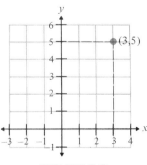

FIGURE 7.3

EXAMPLE 1 Plot each point on the same set of axes.

(a) $A(4, 2)$ (b) $B(2, 4)$ (c) $C(-3, 1)$ (d) $D(4, 0)$ (e) $E(-2, -5)$
(f) $F(0, -3)$ (g) $G(0, 3)$ (h) $H(6, -\frac{7}{2})$ (i) $I(-\frac{3}{2}, -\frac{5}{2})$

Solution: The first number in each ordered pair is the x coordinate and the second number is the y coordinate. The points are plotted in Fig. 7.4.

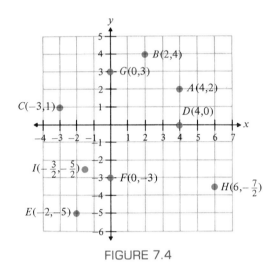

FIGURE 7.4

Note that when the x coordinate is 0, as in parts (f) and (g), the point is on the y axis. When the y coordinate is 0, as in part (d), the point is on the x axis.

EXAMPLE 2 List the ordered pairs for each point shown in Fig. 7.5.

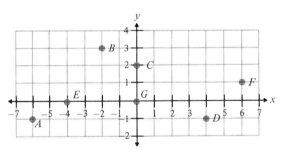

FIGURE 7.5

Solution: Remember to give the x value first in the ordered pair.

Point	Ordered Pair
A	$(-6, -1)$
B	$(-2, 3)$
C	$(0, 2)$
D	$(4, -1)$
E	$(-4, 0)$
F	$(6, 1)$
G	$(0, 0)$

Exercise Set 7.1

1. List the ordered pairs corresponding to each of the following points.

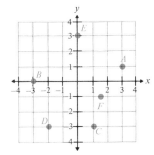

2. List the ordered pairs corresponding to each of the following points.

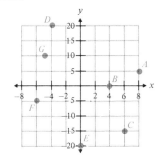

3. Plot each point on the same set of axes.
  (a) $(4, 2)$ **(b)** $(-3, 2)$ **(c)** $(0, -3)$
  (d) $(-2, 0)$ **(e)** $(-3, -4)$ **(f)** $(-4, -2)$

4. Plot each point on the same set of axes.
  (a) $(-3, -1)$ **(b)** $(2, 0)$ **(c)** $(-3, 2)$
  (d) $(\frac{1}{2}, -4)$ **(e)** $(-4, 2)$ **(f)** $(0, 4)$

5. Plot the following points. Then determine if they are all in a straight line.
  (a) $(1, -1)$ **(b)** $(5, 3)$ **(c)** $(-3, -5)$
  (d) $(0, -2)$ **(e)** $(2, 0)$

6. Plot the following points. Then determine if they are all in a straight line.
  (a) $(1, -2)$ **(b)** $(0, -5)$ **(c)** $(3, 1)$
  (d) $(-1, -8)$ **(e)** $(\frac{1}{2}, -\frac{7}{2})$

7. In an ordered pair which coordinate is always listed first?

8. Explain how to plot the point $(-3, 5)$ in the Cartesian coordinate system.

Cumulative Review Exercises

[4.4] **9.** Subtract $6x^2 - 4x + 5$ from $-2x^2 - 5x + 9$.

[4.5] **10.** Multiply $(3x^2 - 4x + 5)(2x - 3)$.

[5.2] **11.** Factor by grouping $x^2 - 2x + 3xy - 6y$.

[6.5] **12.** Subtract $\dfrac{3}{x + 2} - \dfrac{4}{x + 1}$

7.2

Graphing Linear Equations

▸ **1** Identify linear equations in two variables.

▸ **2** Know that linear equations in two variables have an infinite number of solutions.

▸ **3** Realize that the graph of a linear equation will be a straight line.

▸ **4** Graph linear equations by plotting points.

▸ **5** Graph linear equations using the x and y intercepts.

▸ **6** Identify horizontal and vertical lines in equation form.

▶ **1**   Most of the equations we have discussed thus far have contained only one variable. Exceptions to this include formulas used in application sections. In this chapter we consider linear equations in two variables.

A **linear equation in two variables** is an equation that can be put in the form

$$ax + by = c$$

where a, b, and c are real numbers.

Equations of the form $ax + by = c$ will be straight lines when graphed. For this reason such equations are called linear. Linear equations may be written in various forms, as we will show later. A linear equation in the form $ax + by = c$ is said to be in **standard form.**

Examples of Linear Equations
$$3x - 2y = 4$$
$$y = 5x + 3$$
$$x - 3y + 4 = 0$$

Note in the examples that only the equation $3x - 2y = 4$ is given in standard form. However, the bottom two equations can be written in standard form, as follows:

$$y = 5x + 3 \qquad x - 3y + 4 = 0$$
$$-5x + y = 3 \qquad\qquad x - 3y = -4$$

▶ **2**   Consider the linear equation in *one* variable, $2x + 3 = 5$. What is its solution?

$$2x + 3 = 5$$
$$2x = 2$$
$$x = 1$$

This equation has only one solution, 1.

Check: $2x + 3 = 5$
$$2(1) + 3 = 5$$
$$5 = 5 \qquad \text{true}$$

Now consider the linear equation in *two* variables, $y = x + 1$. What is the solution? Since the equation contains two variables, its solutions must contain two numbers, one for each variable. One set of numbers that satisfies this equation is $x = 1$ and $y = 2$. To see that this is true, we substitute both values into the equation and see that the equation checks.

$$y = x + 1$$
$$2 = 1 + 1$$
$$2 = 2 \qquad \text{true}$$

We write this answer as an ordered pair by writing the x and y values within parentheses separated by a comma. Remember the x value is always listed first since the form of an ordered pair is (x, y). Therefore, one possible solution to this equation is the ordered pair $(1, 2)$. The equation $y = x + 1$ has other possible solutions, as follows.

Solution	*Solution*	*Solution*
$x = 2, y = 3$	$x = -1, y = 0$	$x = -3, y = -2$
$y = x + 1$	$y = x + 1$	$y = x + 1$
$3 = 2 + 1$	$0 = -1 + 1$	$-2 = -3 + 1$
$3 = 3$ true	$0 = 0$ true	$-2 = -2$ true

Solution Written as an Ordered Pair

$(2, 3)$	$(-1, 0)$	$(-3, -2)$

How many possible solutions does the equation $y = x + 1$ have? The equation $y = x + 1$ has an unlimited or *infinite number* of possible solutions. Since it is not possible to list all the specific solutions to the equation, the solutions are illustrated with a graph.

EXAMPLE 1 Determine which of the following ordered pairs satisfy the equation $2x + 3y = 12$.
(a) $(2, 3)$ (b) $(3, 2)$ (c) $(8, -\frac{4}{3})$

Solution: To determine if the ordered pairs are solutions, we substitute them into the equation.

(a) $2x + 3y = 12$ (b) $2x + 3y = 12$ (c) $2x + 3y = 12$
    $2(2) + 3(3) = 12$ $2(3) + 3(2) = 12$ $2(8) + 3(-\frac{4}{3}) = 12$
       $4 + 9 = 12$ $6 + 6 = 12$ $16 - 4 = 12$
           $13 \neq 12$ $12 = 12$ $12 = 12$
$(2, 3)$ is not a solution. $(3, 2)$ is a solution. $(8, -\frac{4}{3})$ is a solution. ∎

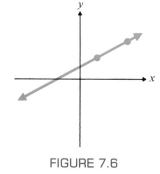

▶ 3 As mentioned earlier, **every linear equation of the form $ax + by = c$ will be a straight line when graphed.** Consider the straight line shown in Fig. 7.6. Note that only two points are needed to draw a straight line.

Since all linear equations are straight lines, only two ordered pairs that satisfy the equation are needed to graph the equation. However, it is always a good idea to use a third ordered pair as a check point. If the three points are not in a straight line, you have made a mistake. A set of points that are in a straight line are said to be **collinear.**

FIGURE 7.6

EXAMPLE 2 Determine if the three points given are collinear.
(a) $(2, 7)$, $(0, 3)$, and $(-2, -1)$
(b) $(0, 5)$, $(\frac{5}{2}, 0)$, and $(5, -5)$
(c) $(-2, -5)$, $(0, 1)$, and $(5, 8)$

Solution: We plot the points to determine if they are collinear. The solution is shown in Fig. 7.7.

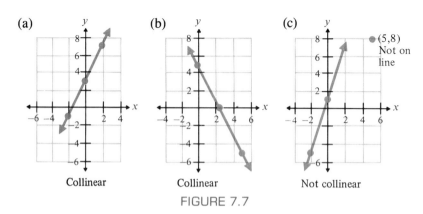

FIGURE 7.7

We will discuss three methods that can be used to graph linear equations. The methods are (1) graphing by plotting points, (2) graphing using the x and y intercepts, and (3) graphing using the slope and y intercept. In this section we discuss graphing by plotting points and by using the x and y intercepts. In the next section we discuss graphing using the slope and y intercept. We begin by discussing graphing by plotting points.

Graphing by Plotting Points

▶ **4**   Graphing by plotting points is the most versatile method of graphing because we can also use it to graph second- and higher-degree equations. We will graph quadratic equations, which are second-degree equations, by plotting points in Chapter 10.

Graphing Linear Equations by Plotting Points

---

1. Solve the linear equation for the variable y. That is, get the variable y by itself on the left side of the equal sign.
2. Select a value for the variable x. Substitute this value in the equation for x and find the corresponding value of y. Record the ordered pair (x, y).
3. Repeat step 2 with two different values of x. This will give you two additional ordered pairs.
4. Plot the three ordered pairs. The three points should be collinear. If they are not collinear, recheck your work for mistakes.
5. *With a straight-edge,* draw a straight line through the three points. Draw an arrow tip on each end of the line to show that the line continues indefinitely in both directions.

In step 1, you need to solve the equation for y. If you have forgotten how to do this, review Section 3.1. In steps 2 and 3, you need to select values for x. The values you choose to select are up to you. However, you should choose values small enough so that the ordered pairs obtained can be plotted on the axes. Since y is often easy to find when $x = 0$, 0 is always a good value to select for x.

EXAMPLE 3   Graph the equation $y = 3x + 6$.

Solution: First we determine that this is a linear equation. The graph must therefore be a straight line. The equation is already solved for y. Select three values for x, substitute them in the equation, and find the corresponding values for y. We will arbitrarily select the values 0, 2, and -3 for x. The calculations that follow show that when $x = 0$, $y = 6$, when $x = 2$, $y = 12$, and when $x = -3$, $y = -3$.

x	$y = 3x + 6$	Ordered Pair
0	$y = 3(0) + 6 = 6$	$(0, 6)$
2	$y = 3(2) + 6 = 12$	$(2, 12)$
-3	$y = 3(-3) + 6 = -3$	$(-3, -3)$

It is sometimes convenient to list the x and y values in tabular form. Then plot the three ordered pairs on the same set of axes (Fig. 7.8).

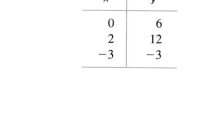

x	y
0	6
2	12
-3	-3

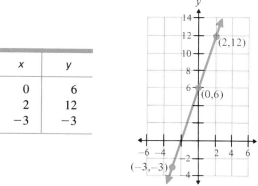

FIGURE 7.8

Since the three points are collinear, the graph appears correct. Connect the three points with a straight line. Place arrows at the ends of the line to show that the line continues infinitely in both directions. ∎

To graph the equation $y = 3x + 6$, we arbitrarily used the three values $x = 0$, $x = 2$, and $x = -3$. We could have selected three entirely different values and obtained exactly the same graph. When selecting values to substitute for x, use values that make the equation easy to evaluate.

The graph drawn in Example 3 represents the set of *all* ordered pairs that satisfy the equation $y = 3x + 6$. If we select any point on this line, the ordered pair represented by that point will be a solution to the equation $y = 3x + 6$. Similarly, any solution to the equation will be represented by a point on the line. Let us select some points on the graph and verify that they are solutions to the equation (see Fig. 7.9).

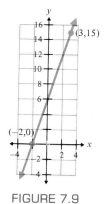

FIGURE 7.9

Points selected on line: $(3, 15), (-2, 0)$

Check $(3, 15)$: $y = 3x + 6$ $\qquad$ Check $(-2, 0)$: $y = 3x + 6$

$\qquad\qquad\quad 15 = 3(3) + 6$ $\qquad\qquad\qquad\qquad 0 = 3(-2) + 6$

$\qquad\qquad\quad 15 = 9 + 6$ $\qquad\qquad\qquad\qquad\quad 0 = -6 + 6$

$\qquad\qquad\quad 15 = 15 \qquad$ true $\qquad\qquad\qquad\quad 0 = 0 \qquad$ true

A graph of an equation is an illustration of the set of points whose coordinates satisfy the equation.

EXAMPLE 4 Graph the equation $2y = 4x - 12$.

Solution: We begin by solving the equation for y. This will make it easier to determine or-

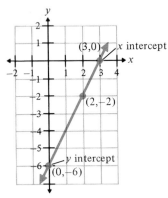

FIGURE 7.10

dered pairs that satisfy the equation. To solve the equation for y, divide both sides of the equation by 2.

$$2y = 4x - 12$$

$$y = \frac{4x - 12}{2} = \frac{4x}{2} - \frac{12}{2} = 2x - 6$$

Now select values for x and solve for y in the equation $y = 2x - 6$.

	$y = 2x - 6$	x	y
Let $x = 0$,	$y = 2(0) - 6 = -6$	0	-6
Let $x = 2$,	$y = 2(2) - 6 = -2$	2	-2
Let $x = 3$,	$y = 2(3) - 6 = 0$	3	0

Plot the points and draw the straight line (Fig. 7.10).   ■

Graphing Using the x and y Intercepts

▶ **5**   Now we will discuss graphing linear equations using the x and y intercepts. Let us examine two points on the graph in Fig. 7.10. Note that the graph crosses the x axis at the point (3, 0). Therefore, 3 is called the **x intercept.** Note that the x intercept has a y coordinate of 0. The graph crosses the y axis at the point (0, −6). Therefore, −6 is called the **y intercept.** Note that the y intercept has an x coordinate of 0. It is often convenient to graph linear equations by finding their x and y intercepts. To graph an equation using the x and y intercepts, use the procedure that follows.

Graphing Linear Equations Using the x and y Intercepts

1. **Find the y intercept by setting x equal to 0 and solving the resulting equation for y.**
2. **Find the x intercept by setting y equal to 0 and solving the resulting equation for x.**
3. **Determine a check point by selecting a nonzero value for x and finding the corresponding value for y.**
4. **Plot the y intercept (where the graph crosses the y axis), the x intercept (where the graph crosses the x axis), and the check point. The three points should be collinear. If not, recheck your work.**
5. *Using a straight-edge,* draw a straight line through the three points. Draw an arrow tip at both ends of the line to show that the line continues indefinitely in both directions.

Note that since only two points are needed to determine a straight line it is not absolutely necessary to determine and plot the check point in step 3. However, if you use only the x and y intercepts to draw your graph and one of those points is wrong, your graph will be incorrect and you will not know it. It is always a good idea to use three points when graphing a linear equation.

EXAMPLE 5　Graph the equation $3y = 6x + 12$ by plotting the x and y intercepts.

Solution:　To find the y intercept (where the graph crosses the y axis), set $x = 0$ and solve for y.

$$3y = 6x + 12$$
$$3y = 6(0) + 12$$
$$3y = 0 + 12$$
$$3y = 12$$
$$y = \frac{12}{3} = 4$$

The graph crosses the y axis at 4. The ordered pair representing the y intercept is $(0, 4)$.

Check:　$3y = 6x + 12$
$3(4) = 6(0) + 12$
$12 = 12$　true

To find the x intercept (where the graph crosses the x axis), set $y = 0$ and solve for x.

$$3y = 6x + 12$$
$$3(0) = 6x + 12$$
$$0 = 6x + 12$$
$$-12 = 6x$$
$$\frac{-12}{6} = x$$
$$-2 = x$$

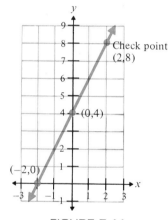

FIGURE 7.11

The graph crosses the x axis at -2. The ordered pair representing the x intercept is $(-2, 0)$.

Check:　$3y = 6x + 12$
$3(0) = 6(-2) + 12$
$0 = 0$　true

Now plot the intercepts (Fig. 7.11).

Before we graph the equation, we will arbitrarily select a nonzero value for x, find the corresponding value of y, and make sure that it is collinear with the x and y intercepts. This third point is our check point.

$$\text{Let } x = 2$$
$$3y = 6x + 12$$
$$3y = 6(2) + 12$$
$$3y = 12 + 12$$
$$3y = 24$$
$$y = \frac{24}{3} = 8$$

Plot the check point $(2, 8)$. Since the three points are collinear, draw the straight line through all three points.　∎

EXAMPLE 6   Graph the equation $2x + 3y = 9$ by finding the x and y intercepts.

Solution:

Find y Intercept	Find x Intercept	Check Point
Let $x = 0$	Let $y = 0$	Let $x = 2$
$2x + 3y = 9$	$2x + 3y = 9$	$2x + 3y = 9$
$2(0) + 3y = 9$	$2x + 3(0) = 9$	$2(2) + 3y = 9$
$0 + 3y = 9$	$2x + 0 = 9$	$4 + 3y = 9$
$3y = 9$	$2x = 9$	$3y = 5$
$y = 3$	$x = \dfrac{9}{2}$	$y = \dfrac{5}{3}$

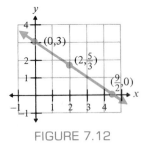

The three ordered pairs are $(0, 3)$, $(\frac{9}{2}, 0)$, and $(2, \frac{5}{3})$.

The three points appear to be collinear. Draw the straight line through the three points (Fig. 7.12).

FIGURE 7.12

EXAMPLE 7   Graph the equation $y = 20x + 60$.

Solution:

Find y Intercept	Find x Intercept	Check Point
Let $x = 0$	Let $y = 0$	Let $x = 3$
$y = 20x + 60$	$y = 20x + 60$	$y = 20x + 60$
$y = 20(0) + 60$	$0 = 20x + 60$	$y = 20(3) + 60$
$y = 60$	$-60 = 20x$	$y = 60 + 60$
	$-3 = x$	$y = 120$

Since the values of y to be plotted are large, we let each interval on the y axis be 15 units rather than 1 (Fig. 7.13). In addition, the length of the intervals on the y axis will be made smaller than those on the x axis. Occasionally, you will have to use different scales on the x and y axes, as illustrated, to accommodate the graph.

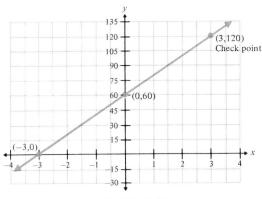

FIGURE 7.13

When selecting the scales to be used on your axes, you should realize that different scales will result in the same equation having a different appearance. Consider the graphs shown in Fig. 7.14. Both graphs represent the same equation, $y = x$. In Fig. 7.14a, both the x and y axes have the same scale. In Fig. 7.14b, the x and y axes do not have the same scale. Both graphs are correct in that each represents the graph of $y = x$. The difference in appearance is due solely to the difference in scales on the x axis. When possible, keep the scales on the x and y axis the same, as in Fig. 7.14a.

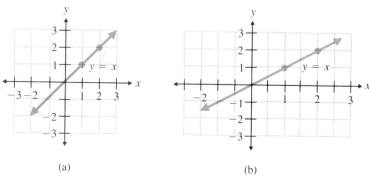

(a) (b)

FIGURE 7.14

EXAMPLE 8 Graph the equation $3y + 2x = -4$.

(a) By solving the equation for y, selecting three arbitrary values of x, and finding the corresponding values of y.

(b) By using the x and y intercepts.

Solution: (a) $3y + 2x = -4$

$$3y = -2x - 4$$

$$y = \frac{-2x - 4}{3} = -\frac{2}{3}x - \frac{4}{3}$$

When selecting values for x, we will select those that are multiples of 3 so that the arithmetic will be easier. Let us select 0, 3, and -3.

$$y = -\frac{2}{3}x - \frac{4}{3}$$

x	y
0	$-\frac{4}{3}$
3	$-\frac{10}{3}$
-3	$\frac{2}{3}$

Let $x = 0$, $y = -\dfrac{2}{3}(0) - \dfrac{4}{3} = 0 - \dfrac{4}{3} = -\dfrac{4}{3}$

Let $x = 3$, $y = -\dfrac{2}{3}(3) - \dfrac{4}{3} = -2 - \dfrac{4}{3} = -\dfrac{6}{3} - \dfrac{4}{3} = -\dfrac{10}{3}$

Let $x = -3$, $y = -\dfrac{2}{3}(-3) - \dfrac{4}{3} = 2 - \dfrac{4}{3} = \dfrac{6}{3} - \dfrac{4}{3} = \dfrac{2}{3}$

The graph is shown in Fig. 7.15. Note that the points do not always come out to be integral values.

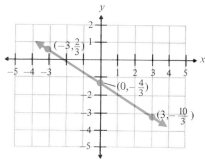

FIGURE 7.15

(b) $3y + 2x = -4$

Find y Intercept	*Find x Intercept*	*Check Point*
Let $x = 0$	Let $y = 0$	Let $x = 3$, *then*
$3y + 2x = -4$	$3y + 2x = -4$	$y = -\dfrac{10}{3}$
$3y + 2(0) = -4$	$3(0) + 2x = -4$	from part (a)
$3y = -4$	$2x = -4$	$\left(3, -\dfrac{10}{3}\right)$
$y = -\dfrac{4}{3}$	$x = -2$	

The graph is shown in Fig. 7.16. Note that the graphs in Figs. 7.15 and 7.16 are the same.

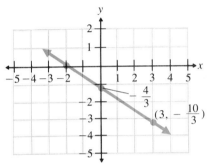

FIGURE 7.16

▶ **6**   When a linear equation contains only one variable, its graph will be either a horizontal or a vertical line, as is explained in Examples 9 and 10.

EXAMPLE 9   Graph the equation $y = 3$.

Solution: This equation can be written as $y = 3 + 0x$. Thus, for any value of x selected, y will be 3. The graph of $y = 3$ is illustrated in Fig. 7.17.

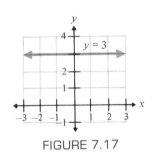

FIGURE 7.17

The graph of any equation of the form $y = a$ will always be a horizontal line with a y intercept of a.

EXAMPLE 10   Graph the equation $x = -2$.

Solution: This equation can be written as $x = -2 + 0y$. Thus, for any value of y selected, x will have a value of -2. The graph of $x = -2$ is illustrated in Fig. 7.18.

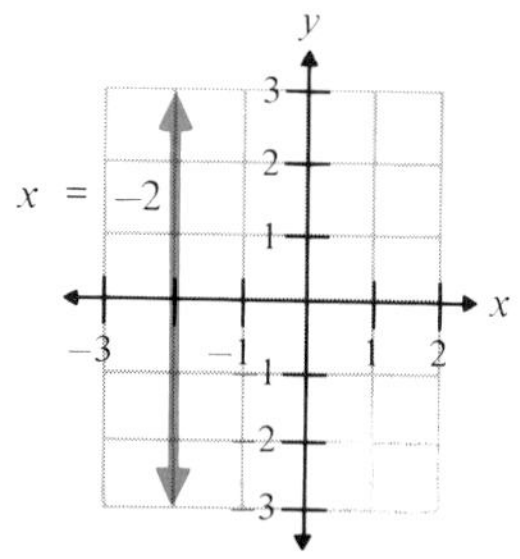

FIGURE 7.18

The graph of any equation of the form $x = a$ will always be a vertical line with an x intercept of a.

Exercise Set 7.2

1. Determine which of the following ordered pairs satisfy the equation $y = 2x - 1$.

(a) $(0, -1)$ (b) $(-1, 0)$ (c) $(5, 3)$

(d) $(6, 11)$ (e) $(\frac{1}{2}, 0)$

2. Determine which of the following ordered pairs satisfy the equation $2x + y = -4$.

(a) $(2, 0)$ (b) $(5, -14)$ (c) $(-2, 0)$

(d) $(0, -4)$ (e) $(\frac{5}{8}, -\frac{3}{4})$

3. Determine which of the following ordered pairs satisfy the equation $5x - 6 = 2y$.

(a) $(9, 20)$ (b) $(-2, -8)$ (c) $(0, -3)$

(d) $(\frac{6}{5}, 0)$ (e) $(-\frac{3}{8}, 6)$

4. Determine which of the following ordered pairs satisfy the equation $-3x + 8y = 12$.

(a) $(4, 3)$ (b) $(-3, \frac{21}{8})$ (c) $(0, \frac{3}{2})$

(d) $(\frac{1}{3}, \frac{35}{24})$ (e) $(-4, 0)$

5. Find the missing coordinate in the following solutions for $2x + y = 6$.

(a) $(2, ?)$ (b) $(-1, ?)$ (c) $(?, -5)$

(d) $(?, -3)$ (e) $(?, 0)$ (f) $(\frac{1}{2}, ?)$

6. Find the missing coordinate in the following solutions for $3x - 2y = 8$.

(a) $(2, ?)$ (b) $(0, ?)$ (c) $(?, 0)$

(d) $(?, -\frac{1}{2})$ (e) $(-3, ?)$ (f) $(?, -5)$

Graph each equation.

7. $y = 6$ **8.** $x = -2$ **9.** $x = 3$ **10.** $y = 5$

Graph each equation by plotting points. Plot at least three points for each graph.

11. $y = 4x - 2$ **12.** $y = -x + 3$ **13.** $y = 6x + 2$

14. $y = x - 4$ **15.** $y = -\frac{1}{2}x + 3$ **16.** $2y = 2x + 4$

17. $6x - 2y = 4$ **18.** $4x - y = 5$ **19.** $5x - 2y = 8$

20. $-2x + 4y = 8$ **21.** $6x + 5y = 30$ **22.** $-2x - 3y = 6$

23. $-4x - y = -2$ **24.** $8y - 16x = 24$ **25.** $y = 20x + 40$

26. $2y - 50 = 100x$ **27.** $y = \dfrac{2}{3}x$ **28.** $y = -\dfrac{3}{5}x$

29. $y = \dfrac{1}{2}x + 4$ **30.** $y = -\dfrac{2}{5}x + 2$ **31.** $2y = 3x + 6$

32. $4x - 6y = 10$

Graph each equation using the x and y intercepts.

33. $y = 2x + 4$ **34.** $y = -2x + 6$ **35.** $y = 4x - 3$

36. $y = -3x + 8$ **37.** $y = -6x + 5$ **38.** $y = 4x + 16$

39. $2y + 3x = 12$ **40.** $-2x + 3y = 10$ **41.** $4x = 3y - 9$

42. $7x + 14y = 21$

43. $\frac{1}{2}x + y = 4$

44. $30x + 25y = 50$

45. $6x - 12y = 24$

46. $25x + 50y = 100$

47. $8y = 6x - 12$

48. $-3y - 2x = -6$

49. $30y + 10x = 45$

50. $120x - 360y = 720$

51. $40x + 6y = 40$

52. $20x - 240 = -60y$

53. $\frac{1}{3}x + \frac{1}{4}y = 12$

54. $\frac{1}{5}x - \frac{2}{3}y = 60$

55. $\frac{1}{2}x = \frac{2}{5}y - 80$

56. $\frac{2}{3}y = \frac{5}{4}x + 120$

Write the equation represented by the given graph. See Examples 9 and 10.

57.
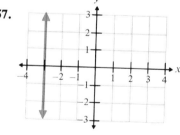

58.
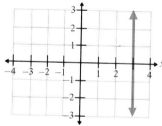

59.
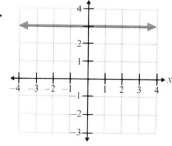

60.
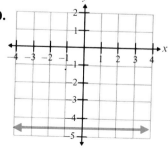

61. What does the graph of a linear equation illustrate?

62. Explain how to find the x and y intercepts of a line.

63. How many points are needed to graph a straight line? How many points should be used?

64. What will the graph of $y = a$ look like for any real number a?

65. What will the graph of $x = a$ look like for any real number a?

Consider the equations that follow. Determine the coefficients to be placed in the shaded areas so that the equation, when graphed, will be a line with the x and y intercepts specified. Explain how you determined your answer.

66. ▓x + ▓$y = 20$; x intercept of 4, y intercept of 5

67. ▓x + ▓$y = 18$; x intercept of -3, y intercept of 6

68. ▓x − ▓$y = -12$; x intercept of -2, y intercept of 3

69. ▓x − ▓$y = 30$; x intercept of -5, y intercept of -15

Cumulative Review Exercises

[2.5] **70.** Solve the equation $4(x - 2) - (3 - x) = 2x + 4$. 5

[4.7] **71.** Two people on bicycles are 18 miles apart headed toward each other. One rider is traveling at a speed of 3 miles per hour faster than the other rider. If the two riders meet in 1.5 hours, find the speed of each rider.

[5.6] **72.** Solve the equation $2x^2 = -23x + 12$.

[6.7] **73.** Solve the equation $x - 14 = \dfrac{-48}{x}$.

7.3

▶**1** Learn the meaning of slope.

▶**2** Find the slope of a line.

▶**3** Examine the slopes of horizontal and vertical lines.

▶**1** The slope of a line is an important concept in many areas of mathematics. A knowledge of slope is helpful in understanding linear equations.

The slope of a line is a measure of the *steepness* of the line. The **slope of a line** is a ratio of the vertical change to the horizontal change between any two selected points on the line. As an example, consider the two points (3, 6) and (1, 2), see Fig. 7.19a.

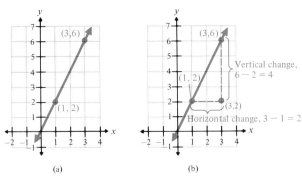

(a)   (b)

FIGURE 7.19

If we draw a line parallel to the x axis through the point (1, 2) and a line parallel to the y axis through the point (3, 6), the two lines intersect at (3, 2), see Fig. 7.19b. From Fig. 7.19b, we can determine the slope of the line. The vertical

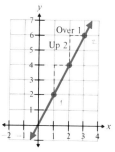

FIGURE 7.20

change (along the y axis) is $6 - 2$ or 4 units. The horizontal change (along the x axis) is $3 - 1$ or 2 units.

$$\text{Slope} = \frac{\text{vertical change}}{\text{horizontal change}} = \frac{4}{2} = 2$$

Thus the slope of the line through these two points is 2. By examining the line connecting these two points, we can see that as the graph moves up two units on the y axis it moves to the right one unit on the x axis (see Fig. 7.20).

▶**2** Let us now determine the procedure to find the slope of a line between any two points (x_1, y_1) and (x_2, y_2). Consider Fig. 7.21.

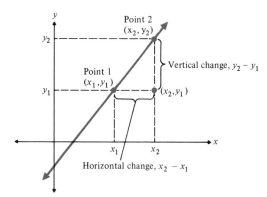

FIGURE 7.21

The vertical change can be found by subtracting y_1 from y_2. The horizontal change can be found by subtracting x_1 from x_2.

Slope of a Line through the Points (x_1, y_1) and (x_2, y_2)

$$\text{Slope} = \frac{\text{change in } y \text{ (vertical change)}}{\text{change in } x \text{ (horizontal change)}} = \frac{y_2 - y_1}{x_2 - x_1}$$

It makes no difference which two points are selected when finding the slope of a line. It also makes no difference which point you label (x_1, y_1) or (x_2, y_2). The Greek capital letter delta, Δ, is often used to represent the words "the change in." Thus the slope, which is symbolized by the letter m, is sometimes indicated as

$$m = \frac{\Delta y}{\Delta x} = \frac{y_2 - y_1}{x_2 - x_1}$$

EXAMPLE 1 Find the slope of the line through the points $(-6, -1)$ and $(3, 5)$.

Solution: We will designate the ordered pair $(-6, -1)$ as (x_1, y_1) and the ordered pair $(3, 5)$ as (x_2, y_2).

$$m = \frac{y_2 - y_1}{x_2 - x_1} = \frac{5 - (-1)}{3 - (-6)} = \frac{5 + 1}{3 + 6} = \frac{6}{9} = \frac{2}{3}$$

Thus the slope is $\frac{2}{3}$.

If we had designated $(3, 5)$ as (x_1, y_1) and $(-6, -1)$ as (x_2, y_2), we would have obtained the same results.

$$m = \frac{y_2 - y_1}{x_2 - x_1} = \frac{-1 - 5}{-6 - 3} = \frac{-6}{-9} = \frac{2}{3}$$ ■

A straight line for which the value of y increases as x increases has a **positive slope** (see Fig. 7.22a). A line with a positive slope rises as it moves from left to right. A straight line for which the value of y decreases as x increases has a **negative slope** (see Fig. 7.22b). A line with a negative slope falls as it moves from left to right.

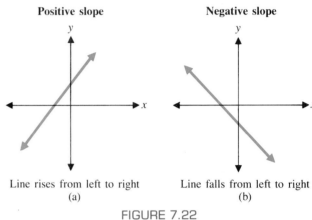

Line rises from left to right
(a)

Line falls from left to right
(b)

FIGURE 7.22

EXAMPLE 2   Consider the line illustrated in Fig. 7.23.

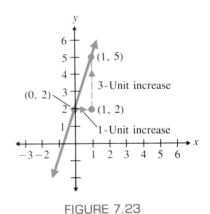

FIGURE 7.23

(a) Determine its slope by observing the vertical change and horizontal change between the two given points.
(b) Calculate the slope of the line using the two given points.

Solution: (a) The first thing you should notice is that the slope is positive since the line rises from left to right. Now determine the vertical change between the two points. The vertical change is $+3$ units. Next determine the horizontal change between the two points. The horizontal change is $+1$ unit. Since the slope is the ratio of the vertical change to the horizontal change between any two points, and since the slope is positive, the slope of this line is $\frac{3}{1}$ or 3.

(b) We can use any two points on the line to determine its slope. Since we are given the ordered pairs (1, 5) and (0, 2) we will use them.

$$\text{Let } (x_2, y_2) \text{ be } (1, 5) \qquad \text{Let } (x_1, y_1) \text{ be } (0, 2)$$

$$m = \frac{y_2 - y_1}{x_2 - x_1} = \frac{5 - 2}{1 - 0} = \frac{3}{1} = 3$$

Note that the answer obtained in part (b) agrees with the answer obtained in part (a). If we had designated the point (1, 5) to be (x_1, y_1) and the point (0, 2) to be (x_2, y_2), the slope would not have changed. Try reversing (x_1, y_1) and (x_2, y_2) and see that you will still obtain a slope of 3.   ■

EXAMPLE 3   Find the slope of the line in Fig. 7.24 by observing the vertical change and horizontal change between the two given points.

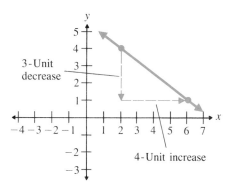

FIGURE 7.24

Solution: Since the graph falls from left to right we should realize that the line has a negative slope. The vertical change between the two given points is -3 units since it is decreasing. The horizontal change between the two given points is 4 units since it is increasing. Since the ratio of the vertical change to the horizontal change is -3 units to 4 units, the slope of this line is $\frac{-3}{4}$ or $-\frac{3}{4}$.   ■

Using the two points indicated in Fig. 7.24 and the definition of slope, calculate the slope of the line. You should obtain the same answer as we did in Example 3.

▶ **3**   Now we consider the slope of horizontal and vertical lines.

Slope of Horizontal Lines

Consider the graph of $y = 3$ (see Fig. 7.25). What is its slope?

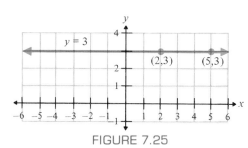

FIGURE 7.25

The graph is parallel to the x axis and goes through the points $(2, 3)$ and $(5, 3)$. Let the point $(5, 3)$ represent (x_2, y_2) and let $(2, 3)$ represent (x_1, y_1). Then the slope of the line is

$$m = \frac{y_2 - y_1}{x_2 - x_1} = \frac{3 - 3}{5 - 2} = \frac{0}{3} = 0$$

Since there is no change in y, this graph has a slope of 0. **Every horizontal line has a slope of 0.**

Slope of Vertical Lines

Consider the graph of $x = 3$ (Fig. 7.26). What is its slope?

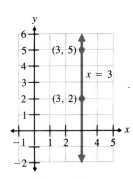

FIGURE 7.26

The graph is parallel to the y axis and goes through the points $(3, 2)$ and $(3, 5)$. Let the point $(3, 5)$ represent (x_2, y_2) and let $(3, 2)$ represent (x_1, y_1). Then the slope of the line is

$$m = \frac{y_2 - y_1}{x_2 - x_1} = \frac{5 - 2}{3 - 3} = \frac{3}{0}$$

We learned in Section 1.7 that $\frac{3}{0}$ is undefined. Thus, we say that the slope of this line is undefined. **The slope of any vertical line is undefined.**

Exercise Set 7.3

Find the slope of the line through the given points.

1. $(4, 1)$ and $(5, 6)$

2. $(8, -2)$ and $(6, -4)$

3. $(9, 0)$ and $(5, -2)$

4. $(5, -6)$ and $(6, -5)$

5. $(3, 8)$ and $(-3, 8)$

6. $(-4, 2)$ and $(6, 5)$

7. $(-4, 6)$ and $(-2, 6)$

8. $(9, 3)$ and $(5, -6)$

9. $(3, 4)$ and $(3, -2)$

10. $(-7, 5)$ and $(3, -4)$

11. $(-4, 2)$ and $(5, -3)$

12. $(-9, -6)$ and $(-3, -1)$

13. $(-1, 7)$ and $(4, -3)$

14. $(0, 4)$ and $(6, -2)$

By observing the vertical and horizontal change of the line between the two points indicated, determine the slope of the line.

15.

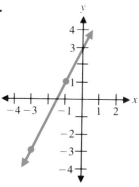

16.

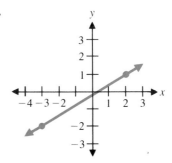

17.

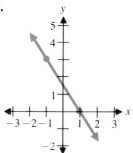

18.

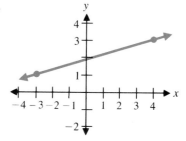

19.

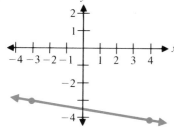

20.

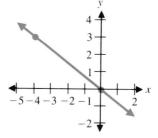

21.

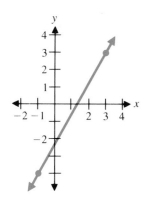

22.

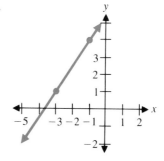

23.

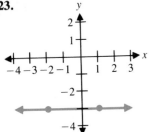

24.

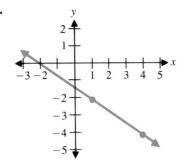

25.

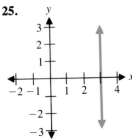

26.

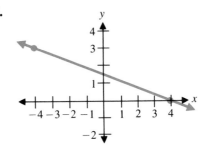

✏ **27.** Explain what is meant by the slope of a line.

✏ **28.** Explain how to find the slope of a line.

✏ **29.** Explain how to tell by observation if the line has a positive slope or negative slope.

✏ **30.** What is the slope of any horizontal line? Explain your answer.

✏ **31.** Do vertical lines have a slope? Explain.

Cumulative Review Exercises

We have spent a great deal of time discussing and solving equations. A list of the various types of equations we have discussed follows. For each type of equation (a) give a general description of the equation, and (b) give a specific example of the equation.

[2.2] **32.** A linear equation in one variable.

[5.6] **33.** A quadratic equation in one variable.

[6.7] **34.** A rational equation in one variable.

[7.2] **35.** A linear equation in two variables.

JUST FOR FUN

1. Find the slope of the line through the points $(\frac{1}{2}, -\frac{3}{8})$ and $(-\frac{4}{9}, -\frac{7}{2})$.

2. If one point on a line is $(6, -4)$ and the slope of the line is $-\frac{5}{3}$, identify another point on the line.

3. One point on a line is $(-5, 2)$ and the slope of the line is $-\frac{3}{4}$. A second point on the line has a y coordinate of -7. Find the x coordinate of the point.

✏ **4.** The slope of a hill and the slope of a line both measure steepness. However, there are several important differences.

 (a) Explain how you think the slope of a hill is determined.

 (b) Is the slope of a line, graphed in the Cartesian coordinate system, measured in any specific unit? Is the slope of a hill measured in any specific unit?

7.4

Slope–Intercept Form of a Linear Equation

▶ **1** Write an equation in slope–intercept form.

▶ **2** Graph a linear equation using the slope and y intercept.

▶ **3** Determine the equation of a line in slope–intercept form.

▶ **4** Determine if two lines are parallel.

▶ **5** Compare the three methods of graphing linear equations.

▶ **1** A very important form of a linear equation is called the **slope–intercept form, $y = mx + b$.** The graph of an equation of the form $y = mx + b$ will always be a straight line with a **slope of m** and a **y intercept of b.** For example, the graph of the equation $y = 3x - 4$ will be a straight line with a slope of 3 and a y intercept of -4. The graph of $y = -2x + 5$ will be a straight line with a slope of -2 and a y intercept of 5.

Slope–Intercept Form of a Linear Equation

$$y = mx + b$$

where m is the slope, and b is the y intercept of the line.

---

slope y intercept

$$y = mx + b$$

Equations in Slope–Intercept Form

Equation	Slope	y Intercept
$y = 3x - 6$	3	-6
$y = \dfrac{1}{2}x + \dfrac{3}{2}$	$\dfrac{1}{2}$	$\dfrac{3}{2}$
$y = -5x + 3$	-5	3
$y = -\dfrac{2}{3}x - \dfrac{3}{5}$	$-\dfrac{2}{3}$	$-\dfrac{3}{5}$

To write a linear equation in slope–intercept form, solve the equation for y.

Once the equation is solved for y, the numerical coefficient of the x term will be the slope, and the constant term will be the y intercept.

EXAMPLE 1   Write the equation $-3x + 4y = 8$ in slope–intercept form. State the slope and y intercept.

Solution: To write this equation in slope–intercept form, we solve the equation for y.

$$-3x + 4y = 8$$
$$4y = 3x + 8$$
$$y = \frac{3x + 8}{4}$$
$$y = \frac{3}{4}x + \frac{8}{4}$$
$$y = \frac{3}{4}x + 2$$

The slope is $\frac{3}{4}$, and the y intercept is 2.   ■

Graphing Linear Equations Using the Slope and y Intercept

▸**2**   In the last section we discussed two methods of graphing a linear equation. They were (1) by plotting points and (2) by using the x and y intercepts. Now we will illustrate a third method to graph linear equations. This method makes use of the slope and the y intercept of a line. Remember that when we solve an equation for y we put the equation in slope–intercept form. Once in this form we can determine the slope and the y intercept of the graph by observation. The procedure to use to graph by this method follows.

To Graph Linear Equations Using the Slope and y Intercept

1. Solve the linear equation for y. That is, get the equation in slope–intercept form, $y = mx + b$.
2. Determine the slope, m, and y intercept, b.
3. Plot the y intercept on the y axis.
4. Use the slope to determine a second point.
   (a) If the slope is **positive,** a second point can be determined by moving **up and to the right.** Thus, if the slope is of the form $\dfrac{p}{q}$, we can obtain a second point by moving up p units and to the *right q* units.
   (b) If the slope is **negative,** a second point can be determined by moving **down and to the right** (or **up and to the left**). Thus if the slope is of the form $-\dfrac{p}{q}$ $\left(\text{or } \dfrac{-p}{q} \text{ or } \dfrac{p}{-q} \right)$, we can obtain a second point by moving *down p* units and to the *right q* units (or *up p* units and to the *left q* units).
5. With a straight-edge, draw a straight line through the two points. Draw arrow tips at the ends of the line to show that the line continues indefinitely in both directions.

EXAMPLE 2   Write the equation $-3x + 4y = 8$ in slope–intercept form; then use the slope and y intercept to graph $-3x + 4y = 8$.

Solution: In Example 1 we solved $-3x + 4y = 8$ for y. We found that

$$y = \frac{3}{4}x + 2$$

The slope of the line is $\frac{3}{4}$ and the y intercept is 2. Begin by marking 2 at the y intercept (Fig. 7.27). Now we determine a second point by making use of the slope $\frac{3}{4}$. Since the slope is positive, we will move up 3 units and to the right 4 units to find the second point. A second point will be at (4, 5). We can continue this process to obtain a third point at (8, 8). Now draw a straight line through the three points. Notice that the line has a positive slope, which is what we expected.

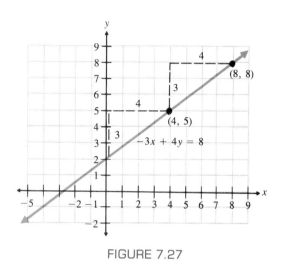

FIGURE 7.27

EXAMPLE 3 Graph the equation $5x + 3y = 12$ using the slope and y intercept.

Solution: Solve the equation for y.

$$5x + 3y = 12$$
$$3y = -5x + 12$$
$$y = \frac{-5x + 12}{3}$$
$$y = -\frac{5}{3}x + 4$$

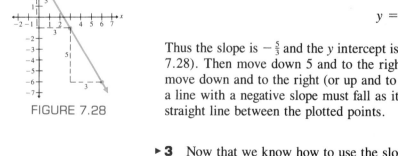

FIGURE 7.28

Thus the slope is $-\frac{5}{3}$ and the y intercept is 4. Begin by marking 4 on the y axis (Fig. 7.28). Then move down 5 and to the right 3 units to determine the next point. We move down and to the right (or up and to the left) because the slope is negative and a line with a negative slope must fall as it goes from left to right. Finally, draw the straight line between the plotted points.

▶ **3** Now that we know how to use the slope–intercept form of a line, we can use it to write the equation of a given line. To do so, we need to determine the slope, m, and y intercept, b, of the line. Once we determine these values we can write the equation in slope–intercept form, $y = mx + b$. For example, if we determine the slope of the given line to be -4 and the y intercept to be 6, then the equation of the line is $y = -4x + 6$.

EXAMPLE 4   Determine the equation of the line shown in Fig. 7.29.

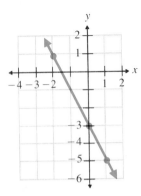

FIGURE 7.29

Solution: The graph shows that the y intercept is -3. Now we need to determine the slope of the line. Since the graph falls from left to right, it has a negative slope. We can see that the vertical change is 2 units for each horizontal change of 1 unit. Thus the slope of the line is -2. The slope can also be determined by selecting any two points on the line and calculating the slope. Let us use the point $(-2, 1)$ to represent (x_2, y_2) and the point $(0, -3)$ to represent (x_1, y_1).

$$m = \frac{\Delta y}{\Delta x} = \frac{y_2 - y_1}{x_2 - x_1} = \frac{1 - (-3)}{-2 - 0} = \frac{1 + 3}{-2} = \frac{4}{-2} = -2$$

Again we obtain a slope of -2. The slope–intercept form of a line is $y = mx + b$, where m is the slope and b is the y intercept. Substituting -2 for m and -3 for b gives the equation $y = -2x - 3$. Thus the equation of the line in Fig. 7.29 is $y = -2x - 3$.   ∎

Parallel Lines

▶ **4**   We will discuss the meaning of parallel lines shortly, but before we do we will work Example 5.

EXAMPLE 5   Determine if both equations represent lines that have the same slope.

$$6x + 3y = 8$$
$$-4x - 2y = -3$$

Solution: Solve each equation for y to get the equations in slope–intercept form.

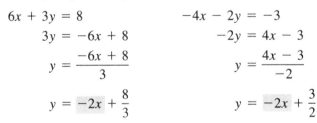

Both lines have the same slope of -2. Notice, however, that their y intercepts are different.   ∎

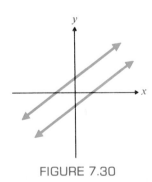

FIGURE 7.30

Two lines are **parallel** when they do not intersect no matter how far they are extended. Figure 7.30 illustrates two parallel lines. **Lines with the same slope will**

be parallel (or identical) **lines when graphed.** The graphs of the equations in Example 5 will be parallel lines since they both have a slope of -2. Note that the two equations represent different lines since their y intercepts are different.

To Determine If Two Lines Are Parallel

Write both equations in slope–intercept form and compare the slopes of the two lines. If both lines have the same slope, but different y intercepts, then the lines are parallel. If the slopes are not the same, the lines are not parallel. Note that if both equations have the same slope and the same y intercept then both equations represent the same line.

EXAMPLE 6 (a) Determine whether or not the following equations represent parallel lines.
(b) Graph both equations on the same set of axes.

$$y = 2x + 4$$

$$-4x + 2y = -2$$

Solution: (a) Write each equation in slope–intercept form and compare their slopes. The equation $y = 2x + 4$ is already in slope–intercept form.

$$-4x + 2y = -2$$
$$2y = 4x - 2$$
$$y = \frac{4x - 2}{2} = 2x - 1$$

The two equations we now need to consider are

$$y = 2x + 4$$
$$y = 2x - 1$$

Since both equations have the same slope, 2, but different y intercepts, the equations represent parallel lines.

(b) We now graph $y = 2x + 4$ and $y = 2x - 1$ on the same set of axes (Fig. 7.31). Remember that $y = 2x - 1$ is the equation $-4x + 2y = -2$ in slope–intercept form.

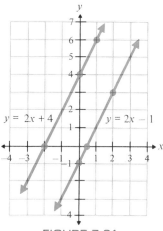

FIGURE 7.31

Summary of Three Methods of Graphing Linear Equations

▶ **5** We have discussed three methods to graph a linear equation: (1) plotting points, (2) using the x and y intercepts, and (3) using the slope and y intercept. In Example 7 we graph an equation using all three methods. No one method is always the easiest to use. If the equation is given in slope–intercept form, $y = mx + b$, then perhaps graphing by plotting points or by using the slope and y intercept might be easier. If the equation is given in standard form, $ax + by = c$, then perhaps graphing using the intercepts might be easier. Unless your teacher specifies that you should graph by a specific method, you can use the method that you feel most comfortable with. Graphing by plotting points is the most versatile method since it can also be used to graph equations that are not straight lines.

EXAMPLE 7   Graph $3x - 2y = 8$ (a) by plotting points; (b) using the x and y intercepts; and (c) using the slope and y intercept.

Solution: For parts (a) and (c) we must write the equation in slope–intercept form.

$$3x - 2y = 8$$
$$-2y = -3x + 8$$
$$y = \frac{-3x + 8}{-2} = \frac{3}{2}x - 4$$

(a) *Plotting points:* Substitute values for x and solve for y. Then plot the ordered pairs and draw the graph. See Fig. 7.32(a).

$$y = \frac{3}{2}x - 4$$

x	y
0	−4
2	−1
4	2

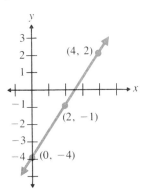

FIGURE 7.32(a)

(b) *Intercepts:* Find the x and y intercepts and a check point. Then plot the points and draw the graph. See Fig. 7.32(b)

$$3x - 2y = 8$$

x Intercept	*y Intercept*	*Check Point*
Let $y = 0$	Let $x = 0$	Let $x = 2$
$3x - 2(0) = 8$	$3(0) - 2y = 8$	$3(2) - 2y = 8$
$3x = 8$	$-2y = 8$	$6 - 2y = 8$
$x = \frac{8}{3}$	$y = -4$	$-2y = 2$
		$y = -1$

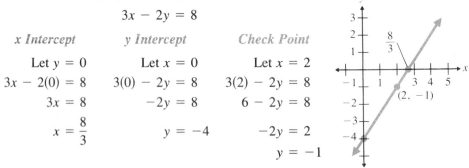

FIGURE 7.32(b)

(c) *Slope and y intercept:* Start by plotting the y intercept -4. Since the slope is $\frac{3}{2}$, we obtain a second point by moving up 3 units and to the right 2 units. The graph is illustrated in Fig. 7.32(c)

$$y = \frac{3}{2}x - 4$$

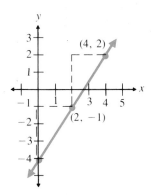

FIGURE 7.32(c)

Exercise Set 7.4

Determine the slope and y intercept of each equation. Graph the equation using the slope and y intercept.

1. $y = 2x - 1$

2. $y = 3x + 2$

3. $y = -x + 5$

4. $y = 2x$

5. $y = -4x$

6. $2x + y = 5$

7. $-2x + y = -3$

8. $3x - y = -2$

9. $3x + 3y = 9$

10. $5x - 2y = 10$

11. $-x + 2y = 8$

12. $5x + 10y = 15$

13. $4x = 6y + 9$

14. $4y = 5x - 12$

15. $-6x = -2y + 8$

16. $6y = 5x - 9$

17. $-3x + 8y = -8$

18. $16y = 8x + 32$

19. $3x = 2y - 4$

20. $15x + 20y = 30$

21. $20x = 80y + 40$

Determine the equation of each line.

22.

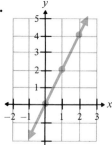

23.

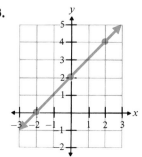

24.

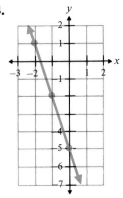

25.

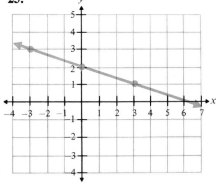

26.

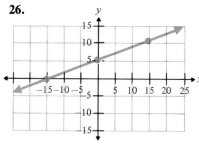

27.

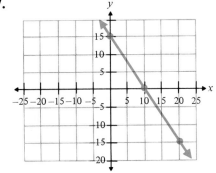

28.

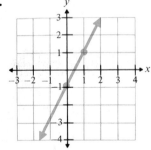

29.

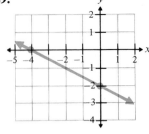

Determine if the two given lines are parallel.

30. $y = 2x - 4$
$y = 2x + 3$

31. $2x + 3y = 8$
$y = \dfrac{-2}{3}x + 5$

32. $4x + 2y = 9$
$8x = 4 - 4y$

33. $3x - 5y = 7$
$5y + 3x = 2$

34. $2x + 5y = 9$
$-x + 3y = 9$

35. $6x + 2y = 8$
$4x - 9 = -y$

36. $y = \dfrac{1}{2}x - 6$
$3y = 6x + 9$

37. $2y - 6 = -5x$
$y = -\dfrac{5}{2}x - 2$

*Graph the equation by (a) plotting points, (b) using the x and y intercepts, and (c) using the slope and y intercept.
See Example 7.*

38. $3x - 2y = 4$

39. $4x + 3y = 6$

40. What does it mean when a line has a positive slope?

41. What does it mean when a line has a negative slope?

42. When you are given an equation in a form other than slope–intercept form, how can you change it to slope–intercept form?

43. Explain how you can determine if two lines are parallel without actually graphing them.

44. (a) Explain in your own words, in a step-by-step manner, how to graph a linear equation using each of the three methods discussed: plotting points, using the x and y intercepts, and using the slope and y intercept.

(b) Graph $-3x + 2y = 4$ using each of the three methods.

Cumulative Review Exercises

[2.6] **45.** If the tire on a car spins 250 revolutions per minute when the car is traveling at 20 miles per hour, how many revolutions will it make when the car is traveling at 30 miles per hour?

[4.7] **46.** Consuella has 1 liter of a 5% saltwater solution. How much pure water, without salt, will Consuella need to add to the 5% solution to obtain a 2% saltwater solution?

[5.5] **47.** Factor $4x^2 - 16y^4$.

[6.5] **48.** Add $\dfrac{3}{x - 2} + \dfrac{5}{x + 2}$.

See Exercise 45.

7.5

Point–Slope Form of a Linear Equation (Optional)

▶ **1** Determine the equation of a line, given a point on the line and the slope of the line.

Thus far we have discussed the standard form of a line, $ax + by = c$, and the slope–intercept form of a line, $y = mx + b$. Now we will discuss another form of the line called the *point–slope form*.

▶ **1** When the slope of a line and a point on the line are known, we can use the point–slope form to determine the equation of the line. The point–slope form can be developed by beginning with the slope between any arbitrary point (x, y) and a fixed point (x_1, y_1) on a line.

$$m = \frac{y - y_1}{x - x_1} \quad \text{or} \quad \frac{m}{1} = \frac{y - y_1}{x - x_1}$$

Now cross-multiply to obtain

$$m(x - x_1) = y - y_1 \quad \text{or} \quad y - y_1 = m(x - x_1)$$

Point–Slope Form of a Linear Equation

$$y - y_1 = m(x - x_1)$$

where m is the slope of the line and (x_1, y_1) is a point on the line.

EXAMPLE 1 Write an equation of the line that goes through the point $(2, 3)$ and has a slope of 4.

Solution: The slope m is 4. The point on the line is $(2, 3)$; call this point (x_1, y_1). Substitute 4 for m, 2 for x_1, and 3 for y_1 in the point–slope form of a line.

$$y - y_1 = m(x - x_1)$$
$$y - 3 = 4(x - 2)$$
$$y - 3 = 4x - 8$$
$$y = 4x - 5$$

The graph of $y = 4x - 5$ has a slope of 4 and passes through the point $(2, 3)$. ∎

The answer to Example 1 was given in slope–intercept form. The answer could have also been given in standard form. Therefore, two other acceptable answers are $-4x + y = -5$ and $4x - y = 5$. Your instructor may specify the form in which the answer is to be given.

EXAMPLE 2 Find an equation of the line through the points $(-1, -3)$ and $(4, 2)$.

Solution: To use the point–slope form, we must first find the slope of the line through the two points. To determine the slope, let us designate $(-1, -3)$ as (x_1, y_1) and $(4, 2)$ as (x_2, y_2).

$$m = \frac{y_2 - y_1}{x_2 - x_1} = \frac{2 - (-3)}{4 - (-1)} = \frac{2 + 3}{4 + 1} = \frac{5}{5} = 1$$

We can use either point (one at a time) in determining the equation of the line. This example will be worked out using both points to show that the solutions obtained are identical.

Using the point $(-1, -3)$ as (x_1, y_1),

$$y - y_1 = m(x - x_1)$$
$$y - (-3) = 1[x - (-1)]$$
$$y + 3 = x + 1$$
$$y = x - 2$$

Using the point $(4, 2)$ as (x_1, y_1),

$$y - y_1 = m(x - x_1)$$
$$y - 2 = 1(x - 4)$$
$$y - 2 = x - 4$$
$$y = x - 2$$

The solutions are identical.   ■

EXAMPLE 3   Write the equation $3x = 4y + 6$ in (a) standard form, (b) slope–intercept form, and (c) point–slope form.

Solution: (a) Standard form is $ax + by = c$.

$$3x = 4y + 6$$
$$3x - 4y = 4y - 4y + 6$$
$$3x - 4y = 6$$

Another acceptable answer is $-3x + 4y = -6$.

(b) Slope–intercept form is $y = mx + b$. To write the equation in slope–intercept form, we solve the equation for y.

$$3x = 4y + 6$$
$$3x - 6 = 4y + 6 - 6$$
$$3x - 6 = 4y$$
$$\frac{3x - 6}{4} = \frac{4y}{4}$$
$$\frac{3}{4}x - \frac{6}{4} = y$$

or $\qquad y = \frac{3}{4}x - \frac{3}{2}$

The slope of the line is $\frac{3}{4}$ and the y intercept is $-\frac{3}{2}$.

(c) Point–slope form is $y - y_1 = m(x - x_1)$. We can begin by starting with the equation in slope–intercept form and rewriting the equation with only the x term on the right.

$$y = \frac{3}{4}x - \frac{3}{2}$$

$$y + \frac{3}{2} = \frac{3}{4}x - \frac{3}{2} + \frac{3}{2}$$

$$y + \frac{3}{2} = \frac{3}{4}x$$

or $\qquad y + \frac{3}{2} = \frac{3}{4}(x - 0)$

Note that $\frac{3}{4}(x - 0)$ is the same as $\frac{3}{4}x$ when the distributive property is used. The slope of the line is $\frac{3}{4}$ and a point on the line is $\left(0, -\frac{3}{2}\right)$. Other procedures which result in equivalent answers can be used to write the equation in point–slope form; see Just for Fun, Exercise 3. ∎

Exercise Set 7.5

Use the point–slope form to determine an equation of a line with the properties given. Write the equation in slope–intercept form.

1. Slope = 5, through $(0, 4)$

2. Slope = 4, through $(2, 3)$

3. Slope = -2, through $(-4, 5)$

4. Slope = -1, through $(6, 0)$

5. Slope = $\frac{1}{2}$, through $(-1, -5)$

6. Slope = $-\frac{2}{3}$, through $(-1, -2)$

7. Slope = $\frac{3}{5}$, through $(4, -2)$

8. Through $(4, 6)$ and $(-1, 1)$

9. Through $(-4, -2)$ and $(-2, 4)$

10. Through $(6, 3)$ and $(5, 2)$

11. Through $(-4, 6)$ and $(4, -6)$

12. Through $(1, 0)$ and $(-2, 4)$

13. Through $(10, 3)$ and $(0, -2)$

14. Through $(-6, -2)$ and $(5, -3)$

15. What are three different forms of a linear equation? Write the equation $4 - 2y = 6x$ in each of the three forms.

16. Write the equation $4y = 6x - 8$ in standard form, slope–intercept form, and point–slope form.

17. Write the equation $5x = 3y + 8$ in standard form, slope–intercept form, and point–slope form.

Cumulative Review Exercises

[4.6] **18.** Divide $\dfrac{9x^3 - 3x^2 - 9x + 4}{3x + 2}$.

[6.2] **19.** Divide $\dfrac{x^2 + 2x - 8}{x^2 - 16} \div \dfrac{2x^2 - 5x - 3}{x^2 - 7x + 12}$.

[6.7] **20.** Solve the equation $x + \dfrac{30}{x} = 11$.

[6.8] **21.** The area of a triangle is 36 square feet. Find the base and height if its height is 7 feet less than twice its base.

JUST FOR FUN

1. Write an equation of the line parallel to $3x - 4y = 6$ that passes through the point $(-4, -1)$.

2. Two lines are **perpendicular** and cross at right angles when their slopes are negative reciprocals of each other. The negative reciprocal of any number a is $-1/a$. Write an equation of the line perpendicular to $-5x + 2y = -4$ that passes through the point $(2, \frac{1}{2})$.

3. In Example 3(c), we changed a linear equation from slope–intercept form to an equation in point–slope form. A second method to convert a linear equation in slope–intercept form to point–slope form follows. Explain in a step-by-step manner the procedure used to make this change.

$$y = \frac{3}{4}x - \frac{3}{2}$$

$$y = \frac{3}{4}(x - 2)$$

$$y - 0 = \frac{3}{4}(x - 2)$$

7.6

Graphing Linear Inequalities

▶ **1** Graph linear inequalities in two variables.

▶ **1**   A linear inequality results when the equal sign in a linear equation is replaced with an inequality sign. Examples of linear inequalities in two variables are

$$3x + 2y > 4 \qquad -x + 3y < -2$$
$$-x + 4y \geq 3 \qquad 4x - y \leq 4$$

To Graph a Linear Inequality

1. Replace the inequality symbol with an equal sign.
2. Draw the graph of the equation in step 1. If the original inequality contained a $\geq$ or $\leq$ symbol, draw the graph using a solid line. If the original inequality contained a $>$ or $<$ symbol, draw the graph using a dashed line.
3. Select any point not on the line and determine if this point is a solution to the original inequality. If the selected point is a solution, shade the region on the side of the line containing this point. If the selected point does not satisfy the inequality, shade the region on the side of the line not containing this point.

EXAMPLE 1   Graph the inequality $y < 2x - 4$.

Solution: Graph the equation $y = 2x - 4$ (Fig. 7.33). Since the original inequality contains a less than sign, $<$, use a dashed line when drawing the graph. The dashed line indicates that the points on this line are not solutions to the inequality $y < 2x - 4$.

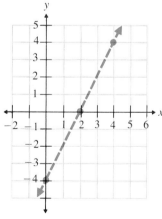

FIGURE 7.33

Next select a point not on the line and determine if this point satisfies the inequality. Often the easiest point to use is the origin, $(0, 0)$.

$$y < 2x - 4$$
$$0 < 2(0) - 4$$
$$0 < 0 - 4$$
$$0 < -4 \qquad \text{false}$$

Since 0 is not less than -4, the point $(0, 0)$ does not satisfy the inequality. The solution will therefore be all the points on the opposite side of the line from the point $(0, 0)$. Shade in this region (Fig. 7.34).

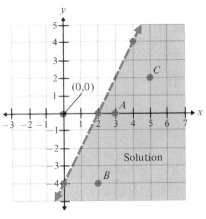

FIGURE 7.34

Every point in the shaded area satisfies the given inequality. Let us check a few selected points A, B, and C.

Point A	*Point B*	*Point C*
$(3, 0)$	$(2, -4)$	$(5, 2)$
$y < 2x - 4$	$y < 2x - 4$	$y < 2x - 4$
$0 < 2(3) - 4$	$-4 < 2(2) - 4$	$2 < 2(5) - 4$
$0 < 2$ true	$-4 < 0$ true	$2 < 6$ true ∎

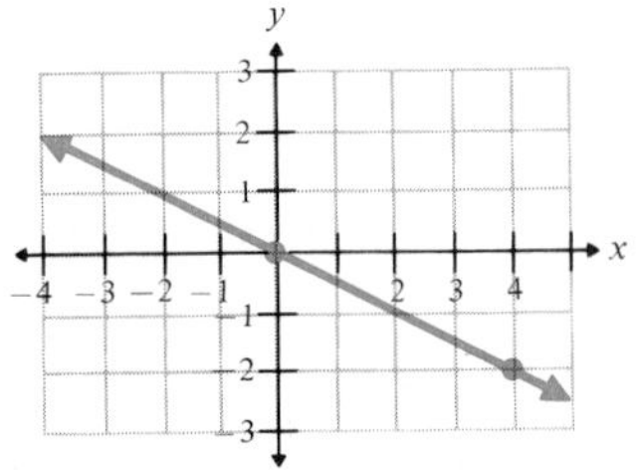

The shaded area in Figure 7.34 satisfies the inequality $y < 2x - 4$. The unshaded area to the left of the dashed line would satisfy the inequality $y > 2x - 4$.

FIGURE 7.35

EXAMPLE 2   Graph the inequality $y \geq -\frac{1}{2}x$.

Solution: Graph the equation $y = -\frac{1}{2}x$. Since the inequality is $\geq$, we use a solid line to indicate that the points on the line are solutions to the inequality (Fig. 7.35). Since the point $(0, 0)$ is on the line, we cannot select that point to find the solution. Let us arbitrarily select the point $(3, 1)$.

$$y \geq -\frac{1}{2}x$$

$$1 \geq -\frac{1}{2}(3)$$

$$1 \geq -\frac{3}{2} \quad \text{true}$$

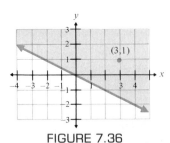

FIGURE 7.36

Since the point $(3, 1)$ satisfies the inequality, every point on the same side of the line as $(3, 1)$ will also satisfy the inequality $y \geq -\frac{1}{2}x$. Shade this region as indicated (Fig. 7.36). Every point in the shaded region as well as every point on the line satisfies the inequality. ∎

Exercise Set 7.6

Graph each inequality.

1. $x > 3$

2. $y < -2$

3. $x \geq \dfrac{5}{2}$

4. $y < x$

5. $y \geq 2x$

6. $y > -2x$

7. $y < 2x + 1$

8. $y \geq 3x -$

9. $y < -3x + 4$

10. $y \geq 2x + 4$

11. $y \geq \dfrac{1}{2}x - 4$

12. $y < 3x + 5$

13. $y \leq \dfrac{1}{3}x + 6$

14. $y > 6x + 1$

15. $y \leq -3x + 5$

16. $y \leq \dfrac{2}{3}x + 3$

17. $y > 5x - 4$

18. $y > \dfrac{2}{3}x - 1$

19. $y \leq -x + 4$

20. $y \geq 2x - 3$

21. $y > 3x - 2$

22. $y \leq -2x + 3$

23. $y \geq -4x + 3$

24. $y > -\dfrac{x}{2} + 4$

25. $y < -\dfrac{x}{3} - 2$

26. When graphing inequalities that contain either a ≤ or ≥, explain why the points on the line will be solutions to the inequality.

27. When graphing inequalities that contain either a < or >, explain why the points on the line will not be solutions to the inequality.

Cumulative Review Exercises

[2.7] **28.** Solve the inequality $2(x - 3) < 4(x - 2) - 4$ and graph the solution on the number line.

[3.1] **29.** Use the simple interest formula $i = prt$ to find the principal if the simple interest Manuel gained over a 3-year period at a rate of 8% is $300.

[6.7] **30.** Solve the formula $C = \dfrac{5}{9}(F - 32)$ for F.

[7.4] **31.** The equation of a line is $6x - 5y = 9$. Find the slope and y intercept of the line.

SUMMARY

GLOSSARY

Cartesian (or rectangular) coordinate system *(295):* Two axes intersecting at right angles that are used when drawing graphs.

Collinear: *(299):* A set of points in a straight line is collinear.

Graph: *(301):* An illustration of the set of points whose coordinates satisfy an equation or an inequality.

Linear equation in two variables: *(298):* An equation of the form $ax + by = c$.

Negative slope: *(312):* A line has a negative slope when the values of y decrease as the values of x increase.

Ordered pair: *(295):* The x and y coordinates of a point listed within parentheses, x coordinate first: (x, y).

Origin: *(295):* The point of intersection of the x and y axes.

Parallel lines: *(320):* Lines that never intersect.

Positive slope: *(312):* A line has a positive slope when the values of y increase as the values of x increase.

Slope of a line: *(310):* The ratio of the vertical change to the horizontal change between any two selected points on the line.

x axis: *(295):* The horizontal axis in the Cartesian coordinate system.

x intercept: *(302):* The value of x at the point where the graph crosses the x axis.

y axis: *(295):* The vertical axis in the Cartesian coordinate system.

y intercept: *(302):* The value of y at the point where the graph crosses the y axis.

IMPORTANT FACTS

To find the x intercept: Set $y = 0$ and solve the equation for x.

To find the y intercept: Set $x = 0$ and solve the equation for y.

Slope of line, m, through points (x_1, y_1) and (x_2, y_2):

$$m = \frac{y_2 - y_1}{x_2 - x_1}$$

Methods of Graphing

$$y = 3x - 4$$

By Plotting Points   *Using Intercepts*   *Using Slope and y Intercept*

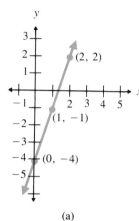

(a)

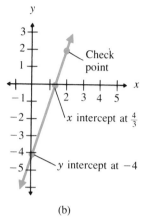

(b)

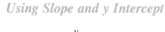

(c)

Standard form of a linear equation: $ax + by = c$.

Slope–intercept form of a linear equation: $y = mx + b$.

Point–slope form of a linear equation: $y - y_1 = m(x - x_1)$.

Review of slope

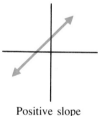

Positive slope
(rises to right)

Negative slope
(falls to right)

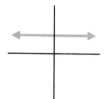

Slope is 0
(horizontal line)

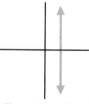

Slope is undefined
(vertical line)

Review Exercises

[7.1] **1.** Plot each ordered pair on the same set of axes.
  (a) $A(5, 3)$
  (b) $B(0, 6)$
  (c) $C(5, \frac{1}{2})$
  (d) $D(-4, 3)$
  (e) $E(-6, -1)$
  (f) $F(-2, 0)$

2. Determine if the points given are collinear:
  $(0, -4), (6, 8), (-2, 0), (4, 4)$

[7.3] **3.** Determine which of the following ordered pairs satisfy the equation $2x + 3y = 9$.
  (a) $(4, 3)$ **(b)** $(0, 3)$
  (c) $(-1, 4)$ **(d)** $(2, \frac{5}{3})$

4. Find the missing coordinate in the following solutions: $3x - 2y = 8$ has solutions (a) $(2, ?)$; (b) $(0, ?)$; (c) $(?, 4)$; (d) $(?, 0)$.

Graph each equation using the method of your choice.

5. $y = 6$

6. $x = -3$

7. $y = 3x$

8. $y = 2x - 1$

9. $y = -3x + 4$

10. $y = -\dfrac{1}{2}x + 4$

11. $2x + 3y = 6$

12. $3x - 2y = 12$

13. $2y = 3x - 6$

14. $4x - y = 8$

15. $-5x - 2y = 10$

16. $3x = 6y + 9$

17. $25x + 50y = 100$

18. $3x - 2y = 270$

19. $\dfrac{2}{3}x = \dfrac{1}{4}y + 20$

[7.3] *In Exercises 20–22, find the slope of the line through the given points.*

20. $(3, -7)$ and $(-2, 5)$

21. $(-4, -2)$ and $(8, -3)$

22. $(-2, -1)$ and $(-4, 3)$

23. What is the slope of a horizontal line?

24. What is the slope of a vertical line?

Find the slope of the lines indicated by observing the horizontal change and the vertical change between the two given points.

25.

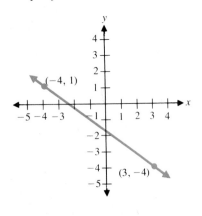

26.

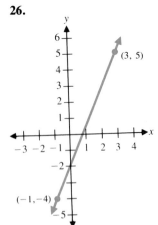

27.

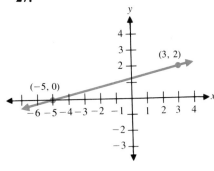

[7.4] *Determine the slope and y intercept of each equation.*

28. $y = -x + 4$

29. $y = 3x + 5$

30. $y = -4x + \dfrac{1}{2}$

31. $2x + 3y = 8$

32. $3x + 6y = 9$

33. $4y = 6x + 12$

34. $3x + 5y = 12$

35. $9x + 7y = 15$

36. $36x - 72y = 144$

37. $4x - 8 = 0$

38. $3y + 9 = 0$

Write the equation of each line.

39.

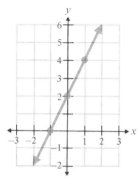

40.

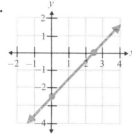

41.

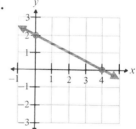

Determine if the two lines are parallel.

42. $y = 3x - 6$
    $6y = 18x + 6$

43. $2x - 3y = 9$
    $3x - 2y = 6$

44. $y = \dfrac{4}{9}x + 5$
    $4x = 9y + 2$

45. $4x = 6y + 3$
    $-2x = -3y + 10$

[7.5] *Find the equation of the line with the properties given.*

46. Slope $= 2$, through $(3, 4)$

47. Slope $= -3$, through $(-1, 5)$

48. Slope $= -\frac{2}{3}$, through $(3, 2)$

49. Slope $= 0$, through $(4, 2)$

50. Slope is undefined, through $(3, 5)$

51. Through $(4, 3)$ and $(2, 1)$

52. Through $(-2, 3)$ and $(0, -4)$

53. Through $(-4, -2)$ and $(-4, 3)$

[7.6] *Graph each inequality.*

54. $y \geq -3$

55. $x < 4$

56. $y < 3x$

57. $y > 2x + 1$

58. $y \leq 4x - 3$

59. $y \geq 6x + 5$

60. $y < -x + 4$

61. $y \leq \dfrac{1}{3}x - 2$

Practice Test

1. Determine which of the following ordered pairs satisfy the equation $3y = 5x - 9$.
   (a) $(3, 2)$ **(b)** $(\frac{9}{5}, 0)$
   (c) $(-2, -6)$ **(d)** $(0, 3)$

2. Find the slope of the line through the points $(-4, 3)$ and $(2, -5)$.

3. Find the slope and y intercept of $4x - 9y = 15$.

4. Write an equation of the graph in the accompanying figure.

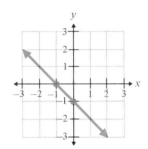

*5. Write, in slope–intercept form, an equation of the line with a slope of 3 passing through the point $(1, 3)$.

*6. Write, in slope–intercept form, an equation of the line passing through the points $(3, -1)$ and $(-4, 2)$.

7. Determine if the following equations represent parallel lines. Explain how you determined your answer.
   $$2y = 3x - 6 \text{ and } y - \frac{3}{2}x = -5.$$

8. Graph $x = -5$.

9. Graph $y = 3x - 2$.

10. Graph $3x + 5y = 15$.

11. Graph $4x = -y + 10$.

12. Graph $3x - 2y = 8$.

13. Graph $y \geq -3x + 5$.

14. Graph $y < 4x - 2$.

* From optional Section 7.5

Systems of Linear Equations

8.1 Introduction

8.2 Solving Systems of Equations Graphically

8.3 Solving Systems of Equations by Substitution

8.4 Solving Systems of Equations by the Addition Method

8.5 Applications of Systems of Equations

8.6 Systems of Linear Inequalities (Optional)

Summary

Review Exercises

Practice Test

Cumulative Review Test

See Section 8.5, Exercise 8.

8.1

Introduction

▶**1** Identify a system of equations.

▶**2** Determine whether or not an ordered pair is a solution to a system of equations.

▶**1** When we must find the common solution to two or more linear equations, we refer to the equations in this type of problem as **simultaneous linear equations** or as a **system of linear equations.**

$$\left.\begin{array}{ll} (1) & y = x + 5 \\ (2) & y = 2x + 4 \end{array}\right\} \quad \text{system of linear equations}$$

▶**2** The **solution to a system of equations** is the ordered pair or pairs that satisfy all equations. The solution to the system above is (1, 6).

Check: ***In Equation (1)***   ***In Equation (2)***

(1, 6)	(1, 6)
$y = x + 5$	$y = 2x + 4$
$6 = 1 + 5$	$6 = 2(1) + 4$
$6 = 6$ true	$6 = 6$ true

The ordered pair (1, 6) satisfies *both* equations and is a solution to the system of equations. Notice that the ordered pair (2, 7) satisfies the first equation but does not satisfy the second equation.

Check: ***In Equation (1)***   ***In Equation (2)***

(2, 7)	(2, 7)
$y = x + 5$	$y = 2x + 4$
$7 = 2 + 5$	$7 = 2(2) + 4$
$7 = 7$ true	$7 = 8$ false

Since the ordered pair (2, 7) does not satisfy both equations, it is *not* a solution to the system of equations.

EXAMPLE 1   Determine which of the following ordered pairs satisfy the system of equations.

$$y = 2x - 8$$
$$2x + y = 4$$

(a) (2, −4)   (b) (4, −4)   (c) (3, −2)

Solution: (a) Substitute 2 for x and −4 for y in each equation.

$y = 2x - 8$	$2x + y = 4$
$-4 = 2(2) - 8$	$2(2) + (-4) = 4$
$-4 = 4 - 8$	$4 - 4 = 4$
$-4 = -4$ true	$0 = 4$ false

Since $(2, -4)$ does not satisfy both equations, it is not a solution to the system of equations.

(b)

$$y = 2x - 8 \qquad\qquad 2x + y = 4$$
$$-4 = 2(4) - 8 \qquad 2(4) + (-4) = 4$$
$$-4 = 8 - 8 \qquad\qquad 8 - 4 = 4$$
$$-4 = 0 \quad \text{false} \qquad\qquad 4 = 4 \quad \text{true}$$

Since $(4, -4)$ does not satisfy both equations, it is not a solution to the system of equations.

(c)

$$y = 2x - 8 \qquad\qquad 2x + y = 4$$
$$-2 = 2(3) - 8 \qquad 2(3) + (-2) = 4$$
$$-2 = 6 - 8 \qquad\qquad 6 - 2 = 4$$
$$-2 = -2 \quad \text{true} \qquad\qquad 4 = 4 \quad \text{true}$$

Since $(3, -2)$ satisfies both equations, it is a solution to the system of linear equations.

In this chapter we discuss three different methods for finding the solution to a system of equations: the graphical method, the substitution method, and the addition method.

Exercise Set 8.1

Determine which, if any, of the following ordered pairs satisfy each system of linear equations.

1. $y = 3x - 4$
$y = -x + 4$
(a) $(-2, 2)$ (b) $(-4, -8)$ (c) $(2, 2)$

2. $y = -4x$
$y = -2x + 8$
(a) $(0, 0)$ (b) $(-4, 16)$ (c) $(2, -8)$

3. $y = 2x - 3$
$y = x + 5$
(a) $(8, 13)$ (b) $(4, 5)$ (c) $(4, 9)$

4. $x + 2y = 4$
$y = 3x + 3$
(a) $(0, 2)$ (b) $(-2, 3)$ (c) $(4, 15)$

5. $3x - y = 6$
$2x + y = 9$
(a) $(3, 3)$ (b) $(4, -2)$ (c) $(-6, 3)$

6. $y = 2x + 4$
$y = 2x - 1$
(a) $(0, 4)$ (b) $(3, 10)$ (c) $(-2, 0)$

7. $2x - 3y = 6$
$y = \dfrac{2}{3}x - 2$
(a) $(3, 0)$ (b) $(3, -2)$ (c) $\left(1, -\dfrac{4}{3}\right)$

8. $y = -x + 4$
$2y = -2x + 8$
(a) $(2, 5)$ (b) $(0, 4)$ (c) $(5, -1)$

9. $3x - 4y = 8$
$2y = \dfrac{3}{2}x - 4$
(a) $(0, -2)$ (b) $(1, -6)$ (c) $\left(-\dfrac{1}{3}, -\dfrac{9}{4}\right)$

10. $2x + 3y = 6$
$-2x + 5 = y$
(a) $\left(\dfrac{1}{2}, \dfrac{5}{3}\right)$ (b) $(2, 1)$ (c) $\left(\dfrac{9}{4}, \dfrac{1}{2}\right)$

11. $y = 2x - 3$
$2x - 3y = 4$
(a) $\left(\dfrac{1}{2}, -2\right)$ (b) $\left(\dfrac{5}{4}, -\dfrac{1}{2}\right)$ (c) $\left(\dfrac{1}{5}, -\dfrac{10}{3}\right)$

12. What does the solution to a system of equations represent?

Cumulative Review Exercises

[1.3] **13.** Consider this set of numbers

$$\{6, -4, 0, \sqrt{3}, 2\tfrac{1}{2}, -\tfrac{9}{5}, 4.22, -\sqrt{7}\}$$

List those that are:

(a) Natural numbers   **(b)** Whole numbers   **(c)** Integers

(d) Rational numbers   **(e)** Irrational numbers   **(f)** Real numbers

[1.4] *Insert either* $>$, $<$, *or* $=$ *in the shaded area to make each statement true.*

14. $|-4|$ ▨ $-|4|$ **15.** $|-6|$ ▨ $|-2|$ **16.** $-|-3|$ ▨ -3

[1.8] *Evaluate each expression.*

17. -3^3 **18.** $(-3)^3$ **19.** -3^4 **20.** $(-3)^4$

[4.1] **21.** Simplify $\left(\dfrac{3x^2y^4}{x^3y^2}\right)^2$.

8.2

Solving Systems of Equations Graphically

▶ **1** Recognize the three possible types of systems of equations.

▶ **2** Solve a system of equations graphically.

▶ **1** The **solution to a system of linear equations** is the ordered pair (or pairs) common to all lines in the system when the lines are graphed. When two lines are graphed, three situations are possible, as illustrated in Fig. 8.1.

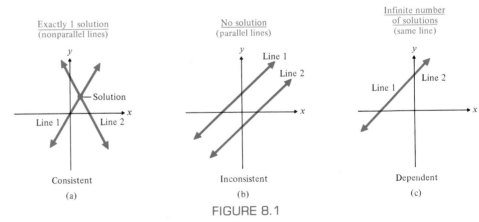

FIGURE 8.1

In Fig. 8.1a, lines 1 and 2 are nonparallel lines and intersect at exactly one point. This system of equations has *exactly one solution*. This is an example of a **consistent system of equations.** A consistent system of equations is a system of equations that has a solution.

Lines 1 and 2 of Fig. 8.1b are two different parallel lines. The lines do not intersect, and this system of equations has *no solution*. This is an example of an **inconsistent system of equations.** An inconsistent system of equations is a system of equations that has no solution.

In Fig. 8.1c, lines 1 and 2 are actually the same line. In this case, every point on the line satisfies both equations and is a solution to the system of equations. This system has *an infinite number of solutions*. This is an example of a **dependent system of equations.** A dependent system of linear equations is a system of equations that has an infinite number of solutions. If a system of two linear equations is dependent, then both equations represent the same line. *Note that a dependent system is also a consistent system since it has a solution.*

We can determine if a system of linear equations is consistent, inconsistent, or dependent by writing each equation in slope–intercept form and comparing the slopes and y intercepts. Note that if the slopes of the lines are different, Fig. 8.1a, the system is consistent. If the slopes are the same but the y intercepts are different, Fig. 8.1b, the system is inconsistent, and if both the slopes and the y intercepts are the same, Fig. 8.1c, the system is dependent.

EXAMPLE 1 Determine whether the system is consistent, inconsistent, or dependent.

$$3x + 4y = 8$$
$$6x + 8y = 4$$

Solution: Write each equation in slope–intercept form and then compare the slopes and y intercepts.

$$
\begin{array}{ll}
3x + 4y = 8 & 6x + 8y = 4 \\
4y = -3x + 8 & 8y = -6x + 4 \\
y = \dfrac{-3x + 8}{4} & y = \dfrac{-6x + 4}{8} \\
y = \dfrac{-3}{4}x + 2 & y = \dfrac{-6}{8}x + \dfrac{4}{8} \\
& y = \dfrac{-3}{4}x + \dfrac{1}{2}
\end{array}
$$

Since the equations have the same slope, $-\frac{3}{4}$, and different y intercepts, the lines are parallel. This system of equations is therefore inconsistent and has no solution.   ∎

▶ **2**

> **To obtain the solution to a system of equations graphically,** graph each equation and determine the point or points of intersection.

EXAMPLE 2 Solve the following system of equations graphically.

$$2x + y = 11$$
$$x + 3y = 18$$

Solution: Find the x and y intercepts of each graph; then draw the graphs.

$2x + y = 11$	*Ordered Pair*	$x + 3y = 18$	*Ordered Pair*
Let $x = 0$; then $y = 11$	$(0, 11)$	Let $x = 0$; then $y = 6$	$(0, 6)$
Let $y = 0$; then $x = \dfrac{11}{2}$	$\left(\dfrac{11}{2}, 0\right)$	Let $y = 0$; then $x = 18$	$(18, 0)$

The two graphs (Fig. 8.2) appear to intersect at the point (3, 5). The point (3, 5) is the possible solution to the system of equations. To be sure, however, we must check to see that (3, 5) satisfies *both* equations.

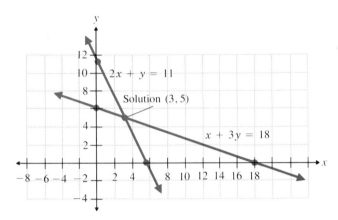

FIGURE 8.2

Check: $2x + y = 11$ $x + 3y = 18$
           $2(3) + 5 = 11$ $3 + 3(5) = 18$
           $11 = 11$ true $18 = 18$ true

Since the ordered pair (3, 5) checks in both equations, it is the solution to the system of equations. This is a consistent system of equations. ■

EXAMPLE 3   Solve the following system of equations graphically.

$$2x + y = 3$$
$$4x + 2y = 12$$

Solution:

$2x + y = 3$	*Ordered Pair*	$4x + 2y = 12$	*Ordered Pair*
Let $x = 0$; then $y = 3$	(0, 3)	Let $x = 0$; then $y = 6$	(0, 6)
Let $y = 0$; then $x = \dfrac{3}{2}$	$\left(\dfrac{3}{2}, 0\right)$	Let $y = 0$; then $x = 3$	(3, 0)

The two lines (Fig. 8.3) appear to be parallel.

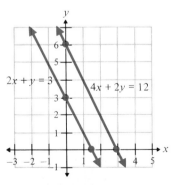

FIGURE 8.3

To show that the two lines are indeed parallel, write each equation in slope–intercept form.

$$2x + y = 3 \qquad 4x + 2y = 12$$
$$y = -2x + 3 \qquad 2y = -4x + 12$$
$$y = -2x + 6$$

Both equations have the same slope, -2, and different y intercepts; thus the lines must be parallel. Since parallel lines do not intersect, this system of equations has no solution. This is an inconsistent system of equations. ■

EXAMPLE 4 Solve the following system of equations graphically.

$$x - \frac{1}{2}y = 2$$
$$y = 2x - 4$$

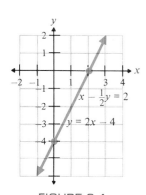

FIGURE 8.4

Solution:

$x - \frac{1}{2}y = 2$	*Ordered Pair*	$y = 2x - 4$	*Ordered Pair*
Let $x = 0$; then $y = -4$	$(0, -4)$	Let $x = 0$; then $y = -4$	$(0, -4)$
Let $y = 0$; then $x = 2$	$(2, 0)$	Let $y = 0$; then $x = 2$	$(2, 0)$

Both equations represent the same line (Fig. 8.4). When the equations are changed to slope–intercept form, it becomes clear that the equations are identical and the system is dependent.

$$x - \frac{1}{2}y = 2 \qquad\qquad y = 2x - 4$$
$$2\left(x - \frac{1}{2}y\right) = 2(2)$$
$$2x - y = 4$$
$$-y = -2x + 4$$
$$y = 2x - 4$$

The solution to this system of equations is all the points on the line. ■

When graphing a system of equations, the intersection of the lines is not always easy to read on the graph. For example, the true solution to a system of equations may be the ordered pair $(\frac{5}{9}, -\frac{4}{11})$. In cases like this, it is not easy to find the exact value of the solution by observation, but you should be able to give an approximate answer. An approximate answer to this system might be $(\frac{1}{2}, -\frac{1}{3})$ or $(0.6, -0.3)$. The accuracy of your answer will depend on how carefully you draw the graphs and on the scale of the graph paper used.

In Example 8 on page 132 we solved a problem involving computers using only one variable. In the example that follows, we will work that same problem using two variables and illustrate the solution in the form of a graph. Although an answer may sometimes be easier to obtain using only one variable, a graph of the situation may help you to visualize the total picture better. Before you read Example 5, you may wish to review the solution to Example 8 on page 132.

EXAMPLE 5 A major university is doing research to determine the most cost-efficient method to set up a number of computer terminals throughout the university. The university is considering two options. Option 1 is an \$80,000 minicomputer whose terminals cost

$1000 each. Option 2 is a $20,000 network system whose terminals cost $2500 each. How many terminals would the university have to install to make the total cost of the network system equal to the total cost of the minicomputer?

Solution: We need to determine the number of terminals where both systems will have the same total cost.

$$\text{Let } c = \text{total cost of the system}$$
$$n = \text{number of terminals}$$

Now write an equation to represent the cost of each option using the two variables c and n.

Option 1:   total cost = cost of minicomputer + cost of n terminals
$$c = 80{,}000 + 1000n$$

Option 2:   total cost = cost of network + cost of n terminals
$$c = 20{,}000 + 2500n$$

$$\text{System of equations} \begin{cases} c = 80{,}000 + 1000n \\ c = 20{,}000 + 2500n \end{cases}$$

Now graph each equation.

$$c = 80{,}000 + 1000n$$

	n	c
Let $n = 0$, $c = 80{,}000 + 1000(0) = 80{,}000$	0	80,000
Let $n = 30$, $c = 80{,}000 + 1000(30) = 110{,}000$	30	110,000
Let $n = 50$, $c = 80{,}000 + 1000(50) = 130{,}000$	50	130,000

$$c = 20{,}000 + 2500n$$

	n	c
Let $n = 0$, $c = 20{,}000 + 2500(0) = 20{,}000$	0	20,000
Let $n = 30$, $c = 20{,}000 + 2500(30) = 95{,}000$	30	95,000
Let $n = 50$, $c = 20{,}000 + 2500(50) = 145{,}000$	50	145,000

The graph (Fig. 8.5) shows that the total cost of the network system equals the cost of the minicomputer when 40 terminals are used. This is the same answer obtained in Example 8 on page 132. If the university plans to have fewer than 40 terminals, the network system (option 2) would be less expensive.

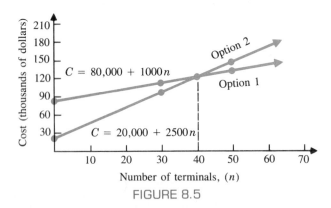

FIGURE 8.5

Exercise Set 8.2

Identify each system of linear equations (labeled 1 and 2) as consistent, inconsistent, or dependent. State whether the system has exactly one solution, no solution, or an infinite number of solutions.

1.
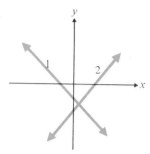

2.
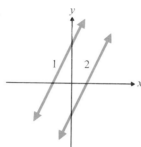

3.
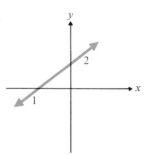

4.
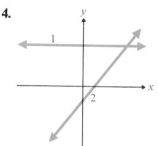

5.
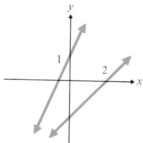

6.
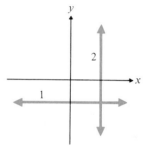

7.
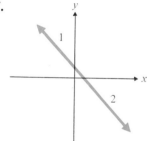

8.
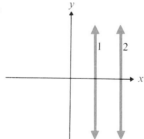

Express each equation in slope–intercept form. Without graphing the equations, state whether the system of equations has exactly one solution, no solution, or an infinite number of solutions.

9. $y = 3x - 2$
   $2y = 4x - 6$

10. $x + y = 6$
   $x - y = 6$

11. $3y = 2x + 3$
   $y = \dfrac{2}{3}x - 2$

12. $y = \dfrac{1}{2}x + 4$
   $2y = x + 8$

13. $4x = y - 6$
   $3x = 4y + 5$

14. $x + 2y = 6$
   $2x + y = 4$

15. $2x = 3y + 4$
   $6x - 9y = 12$

16. $x - y = 2$
   $2x - 2y = -2$

17. $y = \dfrac{3}{2}x + \dfrac{1}{2}$
   $3x - 2y = -\dfrac{1}{2}$

18. $x - y = 3$
   $\dfrac{1}{2}x - 2y = -6$

Determine the solution to each system of equations graphically. If the system is dependent or inconsistent, so state.

19. $y = x + 2$
$y = -x + 2$

20. $y = 2x + 4$
$y = -3x - 6$

21. $y = 3x - 6$
$y = -x + 6$

22. $y = 2x - 1$
$2y = 4x + 6$

23. $2x = 4$
$y = -3$

24. $2x + 3y = 6$
$4x = -6y + 12$

25. $y = x + 2$
$x + y = 4$

26. $2x + y = 6$
$2x - y = -2$

27. $y = -\dfrac{1}{2}x + 4$
$x + 2y = 6$

28. $x + 2y = -4$
$2x - y = -3$

29. $x + 2y = 8$
$2x - 3y = 2$

30. $4x - y = 5$
$2y = 8x - 10$

31. $x + y = 5$
$2y = x - 2$

32. $2x + 3y = 6$
$2x + y = -2$

33. $y = 3$
$y = 2x - 3$

34. $x = 3$
$y = 2x - 2$

35. $x - 2y = 4$
$2x - 4y = 8$

36. $3x + y = -6$
$2x = 1 + y$

37. $2x + y = -2$
$6x + 3y = 6$

38. $y = 2x - 3$
$y = -x$

39. $4x - 3y = 6$
$2x + 4y = 14$

40. $2x + 6y = 6$
$y = -\dfrac{1}{3}x + 1$

41. $2x - 3y = 0$
$x + 2y = 0$

42. $2x = 4y - 12$
$-4x + 8y = 8$

43. $6x + 8y = 36$
$-3x - 4y = -9$

44. $x + 4y = -8$
$3x + 2y = 6$

45. If the minicomputer system in Example 5 costs $60,000 plus $1500 per terminal, and the network system costs $20,000 plus $3500 per terminal, how many terminals would have to be ordered for the cost of the minicomputer system to equal the cost of the network system? Find the solution by graphing the system of equations.

46. Given the system of equations $5x - 4y = 10$ and $12y = 15x - 20$, determine without graphing if the two equations will be parallel lines when graphed. Explain how you determined your answer.

47. (a) What is a consistent system of equations? (b) What is an inconsistent system of equations? (c) What is a dependent system of equations?

48. Explain how to determine if a system of linear equations has exactly one solution, no solution, or an infinite number of solutions without graphing the equations.

49. When a dependent system of two linear equations is graphed, what will be the results?

Cumulative Review Exercises

[5.2] **50.** Factor $xy - 4x + 3y - 12$.

[5.4] **51.** Factor $8x^2 + 2x - 15$.

[6.5] **52.** Subtract $\dfrac{x}{x + 3} - 2$.

[6.7] **53.** Solve $\dfrac{x}{x + 3} - 2 = 0$.

8.3

Solving Systems of Equations by Substitution

▶ **1**   Solve a system of equations by substitution.

Often a graphic solution to a system of equations may be inaccurate since you must estimate the answer by observation. For example, can you determine the answer to the system of equations shown in Fig. 8.6? You may estimate the solution to be $\left(\frac{7}{10}, \frac{3}{2}\right)$ when it may actually be $\left(\frac{4}{5}, \frac{8}{5}\right)$. When an exact answer is necessary, the system should be solved algebraically, either by substitution or by addition of equations.

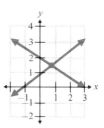

FIGURE 8.6

▶ **1**   The procedure used to solve a system of equations by substitution is illustrated in Example 1. The procedure used to solve by addition is presented in the next section. Regardless of which of the two algebraic techniques is used to solve a system of equations, our immediate goal remains the same, that is, to obtain one equation containing only one unknown.

EXAMPLE 1   Solve the following system of equations by substitution.

$$2x + y = 11$$
$$x + 3y = 18$$

Solution: Begin by solving for one of the variables in either of the equations. You may solve for any of the variables; however, if you solve for a variable with a numerical coefficient of 1, you may avoid working with fractions. In this system the y term in $2x + y = 11$ and the x term in $x + 3y = 18$ both have a numerical coefficient of 1.
    Let us elect to solve for y in $2x + y = 11$.

$$2x + y = 11$$
$$y = -2x + 11$$

Next, substitute $-2x + 11$ for y in the *other equation, $x + 3y = 18$,* and solve for the remaining variable, x.

$$x + 3y = 18$$
$$x + 3(-2x + 11) = 18$$
$$x - 6x + 33 = 18$$
$$-5x + 33 = 18$$
$$-5x = -15$$
$$x = 3$$

Finally, substitute $x = 3$ in the equation solved for y and find the value of y.

$$y = -2x + 11$$
$$y = -2(3) + 11$$
$$y = -6 + 11$$
$$y = 5$$

The solution is the ordered pair $(3, 5)$.   ■

Note that this solution is identical to the graphic solution obtained in Example 2 of Section 8.2.

> ### To Solve a System of Equations by Substitution
>
> 1. Solve for a variable in either equation. (If possible, solve for a variable with a numerical coefficient of 1 to avoid working with fractions.)
> 2. Substitute the expression found for the variable in step 1 into the other equation.
> 3. Solve the equation determined in step 2 to find the value of one variable.
> 4. Substitute the value found in step 3 into the equation obtained in step 1 to find the remaining variable.

EXAMPLE 2   Solve the following system of equations by substitution.

$$2x + y = 3$$
$$4x + 2y = 12$$

Solution: Solve for y in $2x + y = 3$.

$$2x + y = 3$$
$$y = -2x + 3$$

Now substitute the expression $-2x + 3$ for y in the *other equation,* $4x + 2y = 12$, and solve for x.

$$4x + 2y = 12$$
$$4x + 2(-2x + 3) = 12$$
$$4x - 4x + 6 = 12$$
$$6 = 12 \qquad \text{false}$$

Since the statement 6 = 12 is false, the system has no solution. (Therefore, the lines will be parallel and the system is inconsistent.) ∎

Note that the solution in Example 2 is identical to the graphic solution obtained in Example 3 of Section 8.2.

EXAMPLE 3   Solve the following system of equations by substitution.

$$x - \frac{1}{2}y = 2$$
$$y = 2x - 4$$

Solution: The equation $y = 2x - 4$ is already solved for y. Substitute $2x - 4$ for y in the other equation, $x - \frac{1}{2}y = 2$, and solve for the remaining variable, x.

$$x - \frac{1}{2}y = 2$$

$$x - \frac{1}{2}(2x - 4) = 2$$

$$x - x + 2 = 2$$

$$2 = 2 \qquad \text{true}$$

Since the statement 2 = 2 is true, this system has an infinite number of solutions. Therefore, the lines will be the same when graphed and the system is dependent. ∎

Note that the solution in Example 3 is identical to the solution obtained graphically in Example 4 of Section 8.2.

EXAMPLE 4   Solve the following system of equations by substitution.

$$2x + 4y = 6$$
$$4x - 2y = -8$$

Solution: None of the variables in either equation has a numerical coefficient of 1. However, since the numbers 4 and 6 are both divisible by 2, if we solve the first equation for x, we will avoid having to work with fractions.

$$2x + 4y = 6$$
$$2x = -4y + 6$$
$$\frac{2x}{2} = \frac{-4y + 6}{2}$$
$$x = -2y + 3$$

Now substitute $-2y + 3$ for x in the other equation, $4x - 2y = -8$, and solve for the remaining variable y.

$$4x - 2y = -8$$
$$4(-2y + 3) - 2y = -8$$
$$-8y + 12 - 2y = -8$$
$$-10y + 12 = -8$$
$$-10y = -20$$
$$y = 2$$

Finally, solve for x by substituting $y = 2$ in the equation previously solved for x, $x = -2y + 3$.

$$x = -2y + 3$$
$$x = -2(2) + 3 = -4 + 3 = -1$$

The solution is $(-1, 2)$.   ■

COMMON STUDENT ERROR

Remember that a solution to a system of linear equations must contain *both* an x and a y value. Don't solve the system for one of the variables and forget to solve for the other.

EXAMPLE 5   Solve the following system of equations by substitution.

$$4x + 4y = 3$$
$$2x = 2y + 5$$

Solution: We will elect to solve for x in the second equation.

$$2x = 2y + 5$$
$$x = \frac{2y + 5}{2}$$
$$x = y + \frac{5}{2}$$

Now substitute $y + \frac{5}{2}$ for x in the other equation.

$$4x + 4y = 3$$

$$4\left(y + \frac{5}{2}\right) + 4y = 3$$

$$4y + 10 + 4y = 3$$

$$8y + 10 = 3$$

$$8y = -7$$

$$y = -\frac{7}{8}$$

Finally, find the value of x.

$$x = y + \frac{5}{2}$$

$$x = -\frac{7}{8} + \frac{5}{2} = -\frac{7}{8} + \frac{20}{8} = \frac{13}{8}$$

The solution is $\left(\frac{13}{8}, -\frac{7}{8}\right)$. ■

Exercise Set 8.3

Find the solution to each system of equations using substitution.

1. $x + 2y = 4$
$2x - 3y = 1$

2. $y = x + 3$
$y = -x - 5$

3. $x + y = -2$
$x - y = 0$

4. $2x + y = 3$
$2y = 6 - 4x$

5. $2x + y = 3$
$2x + y + 5 = 0$

6. $y = 2x + 4$
$y = -2$

7. $x = 4$
$x + y + 5 = 0$

8. $y = \frac{1}{3}x - 2$
$x - 3y = 6$

9. $x - \frac{1}{2}y = 2$
$y = 2x - 4$

10. $2x + 3y = 7$
$6x - y = 1$

11. $3x + y = -1$
$y = 3x + 5$

12. $y = -2x + 5$
$x + 3y = 0$

13. $y = 2x - 13$
$-4x - 7 = 9y$

14. $x = y + 4$
$3x + 7y = -18$

15. $2x + 3y = 7$
$6x - 2y = 10$

16. $4x - 3y = 6$
$2x + 4y = 5$

17. $3x - y = 14$
$6x - 2y = 10$

18. $5x - 2y = -7$
$5 = y - 3x$

19. $2x - 7y = 6$
$5x - 8y = -4$

20. $4x - 5y = -4$
$3x = 2y - 3$

21. $3x + 4y = 10$
$4x + 5y = 14$

22. $5x + 4y = -7$
$x - \frac{5}{3}y = -2$

23. When solving the system of equations

$$3x + 6y = 9$$
$$4x + 3y = 5$$

by substitution, which variable, in which equation, would you choose to solve for to make the solution easier? Explain your answer.

24. When solving a system of linear equations by substitution, how will you know if the system is inconsistent?

25. When solving a system of linear equations by substitution, how will you know if the system is dependent?

Cumulative Review Exercises

[2.6] **26.** If the directions on a box of Quik-Bake Brownie Mix state that 2 eggs must be added to every 3 cups of brownie mix, how many eggs must be added to 9 cups of brownie mix?

27. If triangles ABC and A′B′C′ at right are similar triangles, find the length of side x.

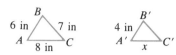

[4.1] **28.** Simplify $(4x^2y^3)^3 (3x^4y^5)^2$.

[6.7] **29.** Solve the equation
$$\frac{3}{x - 12} + \frac{5}{x - 5} = \frac{5}{x^2 - 17x + 60}.$$

8.4

Solving Systems of Equations by the Addition Method

▶ **1** Solve a system of equations by addition.

▶ **1** A third, and often the easiest, method of solving a system of equations is by the addition (or elimination) method. The object of this process is to obtain two equations whose sum will be an equation containing only one variable. Always keep in mind that our immediate goal is to obtain one equation containing only one unknown.

EXAMPLE 1 Solve the following system of equations using the addition method.

$$x + y = 6$$
$$2x - y = 3$$

Solution: Note that one equation contains $+y$ and the other contains $-y$. By adding the equations, we can eliminate the variable y and obtain one equation containing only one unknown.

$$\begin{array}{r} x + y = 6 \\ 2x - y = 3 \\ \hline 3x \quad\quad = 9 \end{array}$$

Now solve for the remaining variable, x.

$$\frac{3x}{3} = \frac{9}{3}$$
$$x = 3$$

Finally, solve for y by inserting $x = 3$ in either of the original equations.

$$x + y = 6$$
$$3 + y = 6$$
$$y = 3$$

The solution is (3, 3).
Check the answer in *both* equations.

$$x + y = 6 \qquad\qquad 2x - y = 3$$
$$3 + 3 = 6 \qquad\qquad 2(3) - 3 = 3$$
$$6 = 6 \quad \text{true} \qquad\qquad 6 - 3 = 3$$
$$3 = 3 \quad \text{true}$$

EXAMPLE 2   Solve the following system of equations using the addition method.

$$-x + 3y = 8$$
$$x + 2y = -13$$

Solution: By adding the equations we can eliminate the variable x.

$$
\begin{array}{rcr}
-x + 3y = & & 8 \\
x + 2y = & & -13 \\
\hline
5y = & & -5 \\
\end{array}
$$

$$\frac{5y}{5} = \frac{-5}{5}$$

$$y = -1$$

Now solve for x by substituting $y = -1$ in either of the original equations.

$$x + 2y = -13$$
$$x + 2(-1) = -13$$
$$x - 2 = -13$$
$$x = -11$$

The solution is $(-11, -1)$.   ∎

To Solve a System of Equations by the Addition (or Elimination) Method

1. If necessary, rewrite each equation so that the terms containing variables appear on the left side of the equal sign and any constants appear on the right side of the equal sign.
2. If necessary, multiply one or both equations by a constant(s) so that when the equations are added the resulting sum will contain only one variable.
3. Add the equations. This will result in a single equation containing only one variable.
4. Solve for the variable in the equation in step 3.
5. Substitute the value found in step 4 into either of the original equations. Solve that equation to find the value of the remaining variable.

In step 2 we indicate it may be necessary to multiply one or both equations by a constant. In this text we will use brackets, [ ], to indicate that both sides of the equation within the brackets are to be multiplied by some constant. Thus, for example, $2[x + y = 1]$ means that both sides of the equation $x + y = 1$ are to be multiplied by 2. We write

$$2[x + y = 1] \qquad \text{gives} \qquad 2x + 2y = 2$$

Similarly, $-3[4x - 2y = 5]$ means both sides of the equation $4x - 2y = 5$ are to be multiplied by -3. We write

$$-3[4x - 2y = 5] \qquad \text{gives} \qquad -12x + 6y = -15$$

The use of this notation may make it easier for you to follow the procedure used to solve the problem.

EXAMPLE 3　Solve the following system of equations using the addition method.

$$2x + y = 6$$
$$3x + y = 5$$

Solution:　The object of the addition process is to obtain two equations whose sum will be an equation containing only one variable. If we add these two equations, none of the variables will be eliminated. However, if we multiply either equation by -1 and then add, we will accomplish our goal. We will multiply the top equation by -1.

$$-1[2x + y = 6] \qquad \text{gives} \qquad -2x - y = -6$$
$$3x + y = 5 \qquad\qquad\qquad\qquad 3x + y = 5$$

Remember that both sides of the equation must be multiplied by the -1. This process has the effect of changing the sign of each term in the equation being multiplied without changing the solution to the system of equations. Now add the two equations on the right.

$$-2x - y = -6$$
$$\underline{3x + y = 5}$$
$$x = -1$$

Solve for y in either of the original equations.

$$2x + y = 6$$
$$2(-1) + y = 6$$
$$-2 + y = 6$$
$$y = 8$$

The solution is $(-1, 8)$.　　　　　　■

EXAMPLE 4　Solve the following system of equations using the addition method.

$$2x + y = 11$$
$$x + 3y = 18$$

Solution:　To eliminate the variable x, we multiply the second equation by -2 and add the two equations.

$$2x + y = 11 \qquad \text{gives} \qquad 2x + y = 11$$
$$-2[x + 3y = 18] \qquad\qquad\qquad -2x - 6y = -36$$

Now add:

$$2x + y = 11$$
$$\underline{-2x - 6y = -36}$$
$$-5y = -25$$
$$y = 5$$

Solve for x.

$$2x + y = 11$$
$$2x + 5 = 11$$
$$2x = 6$$
$$x = 3$$

The solution $(3, 5)$ is identical to the solution obtained graphically in Example 2 of Section 8.2 and by substitution in Example 1 of Section 8.3.　　　■

In Example 4, we could have multiplied the first equation by -3 to eliminate the variable y.

$$-3[2x + y = 11] \qquad \text{gives} \qquad -6x - 3y = -33$$
$$x + 3y = 18 \qquad\qquad\qquad x + 3y = 18$$

Now add:

$$\begin{array}{r} -6x - 3y = -33 \\ x + 3y = 18 \\ \hline -5x = -15 \\ x = 3 \end{array}$$

Solve for y.

$$2x + y = 11$$
$$2(3) + y = 11$$
$$6 + y = 11$$
$$y = 5$$

The solution remains the same, $(3, 5)$.

EXAMPLE 5   Solve the following system of equations using the addition method.

$$4x + 2y = -18$$
$$-2x - 5y = 10$$

Solution: To eliminate the variable x, we can multiply the second equation by 2 and then add.

$$4x + 2y = -18 \qquad \text{gives} \qquad 4x + 2y = -18$$
$$2[-2x - 5y = 10] \qquad\qquad\qquad -4x - 10y = 20$$

$$\begin{array}{r} 4x + 2y = -18 \\ -4x - 10y = 20 \\ \hline -8y = 2 \\ y = -\dfrac{1}{4} \end{array}$$

Solve for x:

$$4x + 2y = -18$$
$$4x + 2\left(-\frac{1}{4}\right) = -18$$
$$4x - \frac{1}{2} = -18$$

$$2\left(4x - \frac{1}{2}\right) = 2(-18) \qquad \text{Multiply both sides of the equation by 2 to remove fractions.}$$
$$8x - 1 = -36$$
$$8x = -35$$
$$x = -\frac{35}{8}$$

The solution is $\left(-\frac{35}{8}, -\frac{1}{4}\right)$.
Check the solution $\left(-\frac{35}{8}, -\frac{1}{4}\right)$ in both equations.

Check:

$$4x + 2y = -18 \qquad\qquad -2x - 5y = 10$$

$$4\left(-\frac{35}{8}\right) + 2\left(-\frac{1}{4}\right) = -18 \qquad -2\left(-\frac{35}{8}\right) - 5\left(-\frac{1}{4}\right) = 10$$

$$-\frac{35}{2} - \frac{1}{2} = -18 \qquad\qquad \frac{35}{4} + \frac{5}{4} = 10$$

$$-\frac{36}{2} = -18 \qquad\qquad \frac{40}{4} = 10$$

$$-18 = -18 \quad \text{true} \qquad\qquad 10 = 10 \quad \text{true} \quad \blacksquare$$

Note that the solution to Example 5 contains fractions. You should not expect to always get integers as answers.

EXAMPLE 6 Solve the following system of equations using the addition method.

$$2x + 3y = 6$$
$$5x - 4y = -8$$

Solution: The variable x can be eliminated by multiplying the first equation by -5 and the second by 2 and then adding the equations.

$$
\begin{array}{lll}
-5[2x + 3y = 6] & \text{gives} & -10x - 15y = -30 \\
2[5x - 4y = -8] & & 10x - 8y = -16
\end{array}
$$

$$
\begin{array}{r}
-10x - 15y = -30 \\
10x - 8y = -16 \\
\hline
-23y = -46 \\
y = 2
\end{array}
$$

The same value could be obtained for y by multiplying the first equation by 5 and the second by -2 and then adding. Try it now and see.

Solve for x.

$$2x + 3y = 6$$
$$2x + 3(2) = 6$$
$$2x + 6 = 6$$
$$2x = 0$$
$$x = 0$$

The solution to this system is $(0, 2)$. $\qquad\qquad\blacksquare$

EXAMPLE 7 Solve the following system of equations using the addition method.

$$2x + y = 3$$
$$4x + 2y = 12$$

Solution: The variable y can be eliminated by multiplying the first equation by -2 and then adding the two equations.

$$
\begin{array}{lll}
-2[2x + y = 3] & \text{gives} & -4x - 2y = -6 \\
4x + 2y = 12 & & 4x + 2y = 12
\end{array}
$$

$$
\begin{array}{r}
-4x - 2y = -6 \\
4x + 2y = 12 \\
\hline
0 = 6 \quad \text{false}
\end{array}
$$

Since $0 = 6$ is a false statement, this system has no solution. The system is inconsistent and the lines will be parallel when graphed.

This solution is identical to the solutions obtained by graphing in Example 3 of Section 8.2 and by substitution in Example 2 of Section 8.3. ∎

EXAMPLE 8   Solve the following system of equations using the addition method.

$$x - \frac{1}{2}y = 2$$
$$y = 2x - 4$$

Solution: First align the x and y terms on the left side of the equation.

$$x - \frac{1}{2}y = 2$$
$$-2x + y = -4$$

Now proceed as in the previous examples.

$$2\left[x - \frac{1}{2}y = 2\right] \qquad \text{gives} \qquad 2x - y = 4$$
$$-2x + y = -4 \qquad\qquad\qquad -2x + y = -4$$

$$\begin{array}{r} 2x - y = 4 \\ -2x + y = -4 \\ \hline 0 = 0 \qquad \text{true} \end{array}$$

Since $0 = 0$ is a true statement, the system is dependent and has an infinite number of solutions. When graphed, both equations will be the same line. This solution is identical to the solutions obtained by graphing in Example 4 of Section 8.2 and by substitution in Example 3 of Section 8.3. ∎

EXAMPLE 9   Solve the following system of linear equations using the addition method.

$$2x + 3y = 7$$
$$5x - 7y = -3$$

Solution: We can eliminate the variable x by multiplying the first equation by -5 and the second by 2.

$$-5[2x + 3y = 7] \qquad \text{gives} \qquad -10x - 15y = -35$$
$$2[5x - 7y = -3] \qquad\qquad\qquad 10x - 14y = -6$$

$$\begin{array}{r} -10x - 15y = -35 \\ 10x - 14y = -6 \\ \hline -29y = -41 \\ y = \dfrac{41}{29} \end{array}$$

We can now find x by substituting $y = \frac{41}{29}$ into one of the original equations and solving for x. If you try this, you will see that although it can be done, it gets pretty messy. An easier method that can be used to solve for x is to go back to the original equations and eliminate the variable y.

$$7[2x + 3y = 7] \qquad \text{gives} \qquad 14x + 21y = 49$$
$$3[5x - 7y = -3] \qquad\qquad 15x - 21y = -9$$

$$14x + 21y = 49$$
$$\underline{15x - 21y = -9}$$
$$29x = 40$$
$$x = \frac{40}{29}$$

Thus the solution is $\left(\frac{40}{29}, \frac{41}{29}\right)$.

Check by substituting $\left(\frac{40}{29}, \frac{41}{29}\right)$ in both equations.

$$2x + 3y = 7 \qquad\qquad\qquad 5x - 7y = -3$$
$$2\left(\frac{40}{29}\right) + 3\left(\frac{41}{29}\right) = 7 \qquad\qquad 5\left(\frac{40}{29}\right) - 7\left(\frac{41}{29}\right) = -3$$
$$\frac{80}{29} + \frac{123}{29} = 7 \qquad\qquad\qquad \frac{200}{29} - \frac{287}{29} = -3$$
$$\frac{203}{29} = 7 \qquad\qquad\qquad\qquad \frac{-87}{29} = -3$$
$$7 = 7 \quad \text{true} \qquad\qquad\qquad\qquad -3 = -3 \quad \text{true} \quad \blacksquare$$

HELPFUL HINT	We have illustrated three possible methods for solving a system of linear equations: graphing, substitution, and addition. When you are given a system of equations, which method should you use to solve the system? When you need an exact solution, graphing should not be used. Of the two algebraic methods, the addition method may be easier to use if there are no numerical coefficients of 1 in the system. If one or more of the equations has a coefficient of 1, you may wish to use either method.

Exercise Set 8.4

Solve each system of equations using the addition method.

1. $x + y = 8$
   $x - y = 4$

2. $x - y = 6$
   $x + y = 4$

3. $-x + y = 5$
   $x + y = 1$

4. $x + y = 10$
   $-x + y = -2$

5. $x + 2y = 15$
   $x - 2y = -7$

6. $3x + y = 10$
   $4x - y = 4$

7. $4x + y = 6$
   $-8x - 2y = 20$

8. $5x + 3y = 30$
   $3x + 3y = 18$

9. $-5x + y = 14$
   $-3x + y = -2$

10. $2x - y = 7$
   $3x + 2y = 0$

11. $3x + y = 10$
   $3x - 2y = 16$

12. $-4x + 3y = 0$
   $5x - 6y = 9$

13. $4x - 3y = 8$
   $2x + y = 14$

14. $2x - 3y = 4$
   $2x + y = -4$

15. $5x + 3y = 6$
   $2x - 4y = 5$

16. $6x - 4y = 9$
   $2x - 8y = 3$

17. $4x - 2y = 6$
   $y = 2x - 3$

18. $5x - 2y = -4$
   $-3x - 4y = -34$

19. $3x - 2y = -2$
   $3y = 2x + 4$

20. $5x + 4y = 10$
   $-3x - 5y = 7$

21. $5x - 4y = 20$
   $-3x + 2y = -15$

22. $5x = 2y - 4$
   $3x - 5y = 6$

23. $6x + 2y = 5$
   $3y = 5x - 8$

24. $4x - 3y = -4$
   $3x - 5y = 10$

25. $4x + 5y = 0$
   $3x = 6y + 4$

26. $4x - 3y = 8$
   $-3x + 4y = 9$

27. $x - \frac{1}{2}y = 4$
   $3x + y = 6$

28. $2x - \frac{1}{3}y = 6$
   $5x - y = 4$

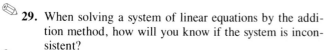

29. When solving a system of linear equations by the addition method, how will you know if the system is inconsistent?

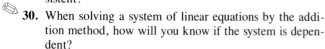

30. When solving a system of linear equations by the addition method, how will you know if the system is dependent?

31. **(a)** In your own words explain the procedure to follow to solve a system of linear equations by the addition method.

(b) Solve the system that follows by following the procedure given in part (a).

$$3x - 2y = 10$$
$$2x + 5y = 13$$

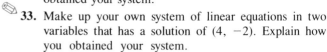

32. Make up your own system of linear equations in two variables that has a solution of (2, 3). Explain how you obtained your system.

33. Make up your own system of linear equations in two variables that has a solution of (4, −2). Explain how you obtained your system.

Cumulative Review Exercises

[4.1] **34.** Simplify $\left(\dfrac{16x^4 y^6 z^3}{4x^6 y^5 z^7} \right)^3$.

[4.7] **35.** A highway crew is paving a new highway. On the average, they pave 110 feet of highway each day. How long will it take the crew to pave a distance of 2420 feet?

[6.2] **36.** Simplify $\dfrac{2x^2 - x - 6}{x^2 - 7x + 10} \div \dfrac{2x^2 - 7x - 4}{x^2 - 9x + 20}$.

[6.5] **37.** Simplify $\dfrac{x}{x^2 - 1} - \dfrac{3}{x^2 - 16x + 15}$.

JUST FOR FUN

Solve each system of equations using the addition method. (*Hint:* First remove all fractions.)

1. $\dfrac{x + 2}{2} - \dfrac{y + 4}{3} = 4.$

$\dfrac{x + y}{2} = \dfrac{1}{2} + \dfrac{x - y}{3}$

2. $\dfrac{5x}{2} + 3y = \dfrac{9}{2} + y$

$\dfrac{1}{4}x - \dfrac{1}{2}y = 6x + 12$

8.5

Applications of Systems of Equations

▶ **1**   Use systems of equations to solve practical problems.

▶ **1**   The method you use to solve a system of equations may depend on whether you wish to see "the entire picture" or are interested in finding the exact answer. If you are interested in the trend as the variable changes, you might decide to graph the equations. If you just want the answer, that is, the ordered pair common to both equations, you might use one of the two algebraic methods to find the common solution.

Many of the applications solved in earlier chapters using only one variable can now be solved using two variables. The following example illustrates how Example 3 of Section 3.3 can be solved using two variables.

EXAMPLE 1   The sum of two numbers is 17. Find the two numbers if one number is 5 more than twice the other number.

Solution: Let x = one number

y = second number

Statement	*Equation*
The sum of two numbers is 17	$x + y = 17$
One number is 5 more than twice the other number	$y = 2x + 5$

$$\text{System of equations} \begin{cases} x + y = 17 \\ y = 2x + 5 \end{cases}$$

Substitute $2x + 5$ for y in first equation.

$$x + y = 17$$
$$x + (2x + 5) = 17$$
$$3x + 5 = 17$$
$$3x = 12$$
$$x = 4$$

Find second number: $y = 2x + 5$
$$y = 2(4) + 5 = 13$$

Thus the two numbers are 4 and 13.   ■

EXAMPLE 2   Diedre has a total of 18 coins in her purse. If the coins consist of only dimes and quarters and the total value of the coins is $3.60, find the number of dimes and the number of quarters she has.

Solution: Let d = number of dimes
q = number of quarters

We must write all money amounts in the same units, either all amounts in dollars or all amounts in cents. Rather than writing $3.60 as 360 cents, we will choose to use dollars as our units. Thus we will write dimes and quarters in terms of dollars.
   Since a dime is $0.10, the value of d dimes is $0.10d$. Similarly, the value of q quarters is $0.25q$.

Statement	*Equation*
The total number of coins is 18	$d + q = 18$
Value of dimes + value of quarters is 3.60	$0.10d + 0.25q = 3.60$

$$\text{System of equations} \begin{cases} d + q = 18 \\ 0.10d + 0.25q = 3.60 \end{cases}$$

Solve $d + q = 18$ for d.

$$d + q = 18$$
$$d = 18 - q$$

Now substitute $18 - q$ for d in $0.10d + 0.25q = 3.60$.

$$0.10d + 0.25q = 3.60$$
$$0.10(18 - q) + 0.25q = 3.60$$
$$1.80 - 0.10q + 0.25q = 3.60$$
$$1.80 + 0.15q = 3.60$$
$$0.15q = 1.80$$
$$q = 12$$

Therefore, there are 12 quarters. The number of dimes is $d = 18 - q = 18 - 12 = 6$. Thus Diedre has 12 quarters and 6 dimes.   ■

EXAMPLE 3   A plane can travel 600 miles per hour with the wind and 450 miles per hour against the wind. Find the speed of the wind and the speed of the plane in still air.

Solution: Let p = speed of plane in still air

w = speed of wind

If p equals the speed of the plane in still air and w equals the speed of the wind, then $p + w$ equals the speed of the plane going with the wind, and $p - w$ equals the speed of the plane going against the wind. We make use of this information when writing the system of equations.

$$\left.\begin{array}{ll}\text{Speed of plane going with wind:} & p + w = 600 \\ \text{Speed of plane going against wind:} & p - w = 450\end{array}\right\} \textit{system of equations}$$

$$\begin{array}{rl} p + w = & 600 \\ \underline{p - w = } & \underline{450} \\ 2p \quad\;\; = & 1050 \\ p \quad\;\; = & 525 \end{array}$$

The plane's speed is 525 miles per hour in still air.

$$p + w = 600$$
$$525 + w = 600$$
$$w = 75$$

The wind's speed is 75 miles per hour.   ■

Distance Problems   We introduced the distance formula, distance = rate · time or $d = rt$, in Section 4.7 and worked additional distance problems in Section 6.8. Now we will introduce a method, using two variables and a system of equations, to solve distance problems that involve two rates. When working problems using this method, we use the information given to write two equations in two variables. Then we proceed to solve the system of equations using one of the three methods introduced in this chapter. Often when working distance problems with two different rates it is helpful to construct a table indicating the information given. We will do this in Examples 4 and 5.

EXAMPLE 4   It takes Malcolm 3 hours in his motorboat to make a 48-mile trip downstream with the current. The return trip against the current takes him 4 hours. Find (a) the speed of the motorboat in still water and (b) the speed of the current.

Solution: (a) Let us make a sketch of the situation; see Fig. 8.7.

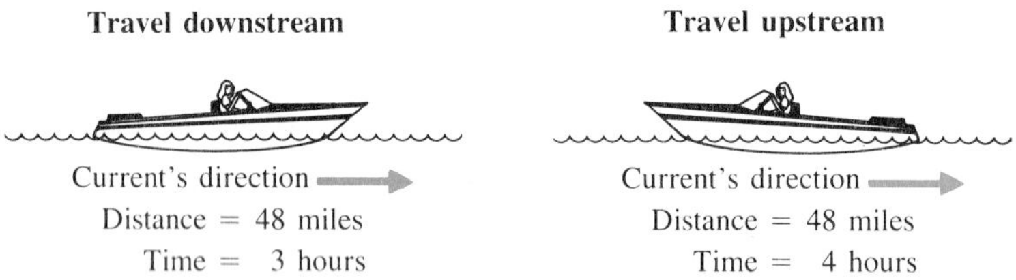

Travel downstream	**Travel upstream**
Current's direction ⟶	Current's direction ⟶
Distance = 48 miles	Distance = 48 miles
Time = 3 hours	Time = 4 hours

FIGURE 8.7

Let m = speed of the motorboat in still water

c = speed of the current

Boat's Direction	Rate	Time	Distance
With current	$m + c$	3	$3(m + c)$
Against current	$m - c$	4	$4(m - c)$

Since the distances traveled downstream and upstream are both 48 miles, our system of equations is

$$3(m + c) = 48$$
$$4(m - c) = 48$$

If we divide both sides of the top equation by 3 and both sides of the bottom equation by 4, we obtain a simplified system of equations.

$$\frac{3(m + c)}{3} = \frac{48}{3} \qquad\qquad \frac{4(m - c)}{4} = \frac{48}{4}$$
$$m + c = 16 \qquad\qquad\qquad m - c = 12$$

Now we solve the simplified system of equations.

$$
\begin{aligned}
m + c &= 16 \\
\underline{m - c} &= \underline{12} \\
2m &= 28 \\
m &= 14
\end{aligned}
$$

Therefore, the speed of the boat in still water is 14 miles per hour.

(b) The speed of the current may be found by substituting 14 for x in either of the simplified equations. We will use $m + c = 16$.

$$
\begin{aligned}
m + c &= 16 \\
14 + c &= 16 \\
c &= 2
\end{aligned}
$$

Thus the current is 2 miles per hour. ∎

EXAMPLE 5 Dawn and Chris go hiking down the Grand Canyon. Chris begins hiking 0.5 hour before Dawn. Chris travels at 2 miles per hour and Dawn travels at 1.5 miles per hour. How long will it take, after Chris begins hiking, for Chris and Dawn to be 2 miles apart?

Solution: We need to find the time it takes for Chris and Dawn to become separated by 2 miles.

Let x = time Chris is hiking

y = time Dawn is hiking

Hiker	Rate	Time	Distance
Chris	2	x	$2x$
Dawn	1.5	y	$1.5y$

Since Chris begins hiking 0.5 hour before Dawn, our first equation is

$$x = y + 0.5$$

Note that if we add 0.5 hour to Dawn's time we get the time Chris has been hiking.

Our second equation is obtained from the fact that the distance between the two hikers must be 2 miles. Since the hikers are traveling in the same direction, we must subtract their distances to obtain a difference of 2 miles. Since Chris is traveling at a faster speed and started first, we subtract Dawn's distance from Chris's distance.

$$\text{Chris's distance} - \text{Dawn's distance} = 2 \text{ miles}$$
$$2x - 1.5y = 2$$

The system of equations is

$$x = y + 0.5$$
$$2x - 1.5y = 2$$

Substituting $y + 0.5$ for x in the second equation, we get

$$2(y + 0.5) - 1.5y = 2$$
$$2y + 1 - 1.5y = 2$$
$$0.5y + 1 = 2$$
$$0.5y = 1$$
$$y = \frac{1}{0.5} = 2$$

Thus the time Dawn has been hiking is 2 hours. Since Chris has been hiking 0.5 hour longer than Dawn, Chris has been hiking for 2.5 hours when Chris and Dawn become separated by 2 miles. ∎

Mixture Problems

Mixture problems were solved with one variable in Section 4.7. Now we will solve mixture problems using two variables and systems of equations. Recall that any problem in which two or more quantities are combined to produce a different quantity, or a single quantity is separated into two or more different quantities, may be considered a mixture problem. Example 6 uses the **simple interest formula** that was presented in Section 3.1. Recall that simple interest = principal · rate · time.

EXAMPLE 6   Emil has invested a total of $12,000 in two savings accounts. One account gives 5% simple interest and the other gives 8% simple interest. Find the amount invested in each account if he receives a total of $840 interest after 1 year.

Solution: Let x = principal invested at 5%
y = principal invested at 8%

	Principal	Rate	Time	Interest
5% account	x	0.05	1	$0.05x$
8% account	y	0.08	1	$0.08y$

Since the total interest is $840, one of our equations is

$$0.05x + 0.08y = 840$$

The second equation comes from our knowledge that the total principal invested is $12,000. Thus our second equation is

$$x + y = 12{,}000$$

$$\left.\begin{array}{r} 0.05x + 0.08y = 840 \\ x + y = 12{,}000 \end{array}\right\} \text{ system of equations}$$

We will multiply our first equation by 100 to eliminate the decimal numbers. This gives the system

$$5x + 8y = 84{,}000$$
$$x + y = 12{,}000$$

$$\begin{array}{l} 5x + 8y = 84{,}000 \\ -5[x + y = 12{,}000] \end{array} \quad \text{gives} \quad \begin{array}{r} 5x + 8y = 84{,}000 \\ -5x - 5y = -60{,}000 \\ \hline 3y = 24{,}000 \\ y = 8000 \end{array}$$

Now solve for x.

$$x + y = 12{,}000$$
$$x + 8000 = 12{,}000$$
$$x = 4000$$

Thus $4000 is invested at 5% and $8000 is invested at 8%. ■

EXAMPLE 7 Deborah, who owns a coffee shop that specializes in selling special blend coffees, wishes to mix Amaretto coffee that sells for $6 per pound with 12 pounds of Kona coffee that sells for $7.50 per pound.

(a) How many pounds of Amaretto coffee must be mixed with the 12 pounds of Kona coffee to obtain a mixture worth $6.50 per pound?
(b) How much of the mixture will be produced?

Solution: (a) We are asked to find the number of pounds of Amaretto coffee.

Let x = number of pounds of Amaretto coffee
y = number of pounds of mixture

Often it is helpful to make a sketch of the situation. After we draw a sketch we will construct a table. In our sketch we will use a cup to put the coffee in.

	Amaretto coffee		Kona coffee		Mixture
Number of pounds	x	+	12	=	y
Price per pound	$6		$7.50		$6.50

The value of the coffee is found by multiplying the number of pounds by the price per pound.

Coffee	Price	Number of Pounds	Value of Coffee
Amaretto	6	x	$6x$
Kona	7.50	12	7.50(12)
Mixture	6.50	y	$6.50y$

Our two equations come from the following information:

$$\begin{pmatrix} \text{number of pounds} \\ \text{of Amaretto coffee} \end{pmatrix} + \begin{pmatrix} \text{number of pounds} \\ \text{of Kona coffee} \end{pmatrix} = \begin{pmatrix} \text{number of pounds} \\ \text{in mixture} \end{pmatrix}$$

$$x + 12 = y$$

Value of Amaretto coffee + value of Kona coffee = value of mixture

$$6x + 7.50(12) = 6.50y$$

$$\left. \begin{array}{l} x + 12 = y \\ 6x + 7.50(12) = 6.50y \end{array} \right\} \textit{system of equations}$$

Now substitute $x + 12$ for y in the bottom equation and solve for x.

$$6x + 7.50(12) = 6.50y$$
$$6x + 7.50(12) = 6.50(x + 12)$$
$$6x + 90 = 6.50x + 78$$
$$90 = 0.50x + 78$$
$$12 = 0.50x$$
$$24 = x$$

Thus 24 pounds of the Amaretto coffee must be mixed with the 12 pounds of Kona coffee.

(b) The total mixture will be 24 + 12 or 36 pounds.   ∎

EXAMPLE 8   A 50% sulfuric acid solution is to be mixed with a 75% sulfuric acid solution to get 60 liters of a 60% sulfuric acid solution. How many liters of the 50% solution and the 75% solution should be mixed?

Solution: Let x = number of liters of 50% solution

y = number of liters of 75% solution

	50% Acid solution		75% Acid solution		Mixture
Number of liters	x	+	y	=	60
Percent strength	50%		75%		60%

The acid content of a solution is found by multiplying the amount of the solution by the strength of the solution.

	Number of Liters	Percent Acid	Acid Content
50% Solution	x	0.50	$0.50x$
75% Solution	y	0.75	$0.75y$
Mixture	60	0.60	$0.60(60)$

Statement	*Equation*
Total volume of combination is 60 liters	$x + y = 60$

$$\left(\begin{array}{c}\text{Acid content of}\\\text{50\% solution}\end{array}\right) + \left(\begin{array}{c}\text{Acid content of}\\\text{75\% solution}\end{array}\right) = \left(\begin{array}{c}\text{Acid content}\\\text{of mixture}\end{array}\right) \qquad 0.5x + 0.75y = 0.6(60)$$

System of equations $\begin{cases} x + y = 60 \\ 0.5x + 0.75y = 0.6(60) \end{cases}$

Solve for y in the first equation.

$$x + y = 60$$
$$y = 60 - x$$

Substitute $60 - x$ for y in the second equation.

$$0.5x + 0.75y = 0.6(60)$$
$$0.5x + 0.75(60 - x) = 36$$
$$0.5x + 45 - 0.75x = 36$$
$$-0.25x + 45 = 36$$
$$-0.25x = -9$$
$$x = \frac{-9}{-0.25} = 36$$

Now solve for y.

$$y = 60 - x$$
$$y = 60 - 36$$
$$y = 24$$

Thus 36 liters of a 50% acid solution should be mixed with 24 liters of a 75% acid solution to obtain 60 liters of a 60% acid solution. ∎

Exercise Set 8.5

Express Exercises 1 through 28 as a system of linear equations and then find the solution.

1. The sum of two integers is 37. Find the numbers if one number is 1 greater than twice the other.

2. The sum of two consecutive even integers is 54. Find the two numbers.

3. The sum of two consecutive odd integers is 76. Find the two numbers.

4. The difference of two integers is 25. Find the two numbers if the larger is 1 less than three times the smaller.

5. The difference of two integers is 28. Find the two numbers if the larger is three times the smaller.

6. Eunice has 20 coins worth a total of $1.40. If the coins are all nickels or dimes, how many of each coin does she have?

7. Paul has 25 rare currency bills in his safe deposit box. If their face values total $101 and Paul has only $2 and $5 bills, find the number of each type of currency in his collection.

8. Robin collects 1-ounce gold dollar coins. Her collection consists of 14 coins, which are either gold United States Eagles or gold Canadian Maple Leafs. The total value of her collection is $6560. If the Eagles have a value of $480 each and the Maple Leafs have a value of $460 each, find the number of Eagles and Maple Leafs that Robin owns.

9. A plane can travel 540 miles per hour with the wind and 490 against the wind. Find the speed of the plane in still air and the speed of the wind.

10. Carlos can paddle a kayak 4.5 miles per hour with the current and 3.2 miles per hour against the current. Find the speed of the kayak in still water and the speed of the current.

See Exercise 10.

11. During a race, Teresa's speed boat travels at a speed of 4 miles per hour faster than Jill's speed boat. If Teresa's boat finishes the race in 3 hours and Jill finishes the race in 3.2 hours, find the speed of each boat.

12. Bob drives from Columbus, Ohio, toward Chicago, Illinois, a distance of 903 miles. At the same time, Mickey starts driving to Columbus, Ohio, from Chicago, Illinois. If the two meet after 7 hours, and Mickey's speed averages 15 miles per hour greater than Bob's speed, find the speed of each car.

13. A United Airlines jet leaves New York's Kennedy Airport headed for Los Angeles International Airport, a distance of 2700 miles. At the same time, a Delta Airlines jet leaves the Los Angeles Airport headed for Kennedy Airport. Due to headwinds and tailwinds, the United jet's average speed is 100 miles per hour greater than that of the Delta jet. If the two jets pass one another after 3 hours, find the speed of each plane.

14. Two cars start at the same time 240 miles apart and travel toward each other. One car travels 6 miles per hour faster than the other. If they meet after 2 hours, what was the average speed of each car?

15. Micki and Petra go jogging along the same trail. Micki starts 0.3 hour before Petra. If Micki jogs at a rate of 5 miles per hour and Petra jogs at 8 miles per hour, how long after Petra starts would it take for Petra to catch up to Micki?

16. Simon trots his horse Chipmunk east at 8 miles per hour. One half-hour later, Michelle starts at the same point and canters her horse Buttermilk west at 16 miles per hour. How long will it take after Michelle starts riding for Michelle and Simon to be separated by 10 miles?

17. Mr. and Mrs. McAdams invest a total of $8000 in two savings accounts. One account gives 10% simple interest and the other 8% simple interest. Find the amount placed in each account if they receive a total of $750 in interest after 1 year.

18. The Webers wish to invest a total of $12,500 in two savings accounts. One account is giving 10% simple interest and the other $5\frac{1}{4}$% simple interest. The Webers wish their interest from the two accounts to be at least $1200 at the end of the year. Find the minimum amount that can be placed in the account giving 10% interest.

19. Marie, a chemist, has a solution of hydrochloric acid with a 25% acid concentration and a second hydrochloric acid solution with a 50% acid concentration. How many liters of each should she mix to get 10 liters of a hydrochloric acid solution with a 40% acid concentration?

20. Moura Williams, a druggist, needs 1000 milliliters of a 10% phenobarbital solution. She has only 5% and 25% phenobarbital solutions available. How many milliliters of each solution should she mix to obtain the desired solution?

21. The total cost of printing an insurance flyer consists of a fixed charge for typesetting and an additional charge for each flyer. If the total cost for 1000 flyers is $550 and the total cost of 2000 flyers is $800, find the fixed charge and the charge for each flyer.

22. Janet wishes to mix 30 pounds of coffee to sell for a total cost of $160. To obtain the mixture, she will mix coffee that sells for $5 per pound with coffee that sells

for $7 per pound. How many pounds of each type coffee should she use?

23. Jason has milk that is 5% butterfat and skim milk without butterfat. How much 5% milk and how much skim milk should he mix to make 100 gallons of milk that is 3.5% butterfat?

24. Max Cisneros plants corn and wheat on his 62-acre farm in Albuquerque. He estimates that his income before deducting expenses is $3000 per acre of corn and $2200 per acre of wheat. Find the number of acres of corn and wheat planted if his total income before expenses is $158,800.

25. Karla bought five times as many shares of Avon stock as she did of Coca-Cola stock. The Avon stock cost $37 a share and the Coca-Cola stock cost $75 a share. If her total cost for all the stock was $7,800, how many shares of each did she purchase?

26. The All Natural Juice Company sells apple juice for 12 cents an ounce and apple drink for 6 cents an ounce. They wish to market and sell for 10 cents an ounce cans of juice drink that are part juice and part drink. How many ounces of each will be used if the juice drink is to be sold in 8-ounce cans?

27. Pierre's recipe for Quiche Lorraine calls for 16 ounces (or 2 cups) of light cream, which is 20% milk fat. It is often difficult to find light cream with 20% milk fat at the supermarket. What is commonly found is heavy cream which is 36% milk fat and half and half which is 10.5% milk fat. How many ounces of the heavy cream and how much of the half and half should be mixed to obtain the 16 ounces of light cream that is 20% milkfat?

28. The perimeter of a rectangular piece of land is 800 feet. If the length is 100 feet greater than the width, find the dimensions of the piece of land.

Cumulative Review Exercises

[1.9] **29.** Evaluate $3(4x - 3)^2 - 2y^2 - 1$ when $x = 3$ and $y = -2$.

[4.2] **30.** Simplify $(3x^4)^{-2}$.

[4.4] **31.** Indicate whether or not each of the following is a polynomial. If it is not a polynomial explain why. Give the degree of each polynomial.

(a) $3x^3 + 2x - 6$ (b) $\frac{1}{2}x^4 - 3x^2 - 2$

(c) $x^3 - 2x^2 - x^{-1}$

[4.6] **32.** Divide $(2x^2 + 5x - 10) \div (x + 4)$.

JUST FOR FUN

1. Two brothers jog to school daily. The older jogs at 9 miles per hour, the younger at 5 miles per hour. When the older brother reaches the school, the younger brother is $\frac{1}{2}$ mile from the school. How far is the school from the boys' house?

2. By weight, an alloy of brass is 70% copper and 30% zinc. Another alloy of brass is 40% copper and 60% zinc. How many grams of each of these alloys must be melted and combined to obtain 300 grams of a brass alloy that is 60% copper and 40% zinc?

See Just for Fun Exercise 1.

8.6

Systems of Linear Inequalities (Optional)

▶ **1** Solve systems of linear inequalities graphically.

▶ **1** In Section 7.6, we learned how to graph linear inequalities in two variables. In Section 8.2, we learned how to solve systems of equations graphically. In this section, we discuss how to solve systems of linear inequalities graphically. The **solution to a system of linear inequalities** is the set of points that satisfies all inequalities in the system. Although a system of linear inequalities may contain more than two inequalities, in this book we will consider only systems with two inequalities.

To Solve a System of Linear Inequalities

Graph each inequality on the same set of axes. The solution is the set of points that satisfies all the inequalities in the system.

EXAMPLE 1   Determine the solution to the system of inequalities.

$$x + 2y \leq 6$$
$$y > 2x - 4$$

Solution: First graph the inequality $x + 2y \leq 6$ (see Fig. 8.8).

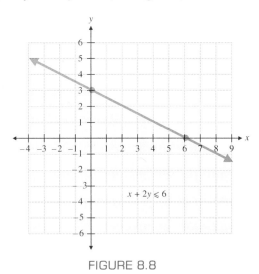

FIGURE 8.8

Now, on the same set of axes, graph the inequality $y > 2x - 4$ (see Fig. 8.9).

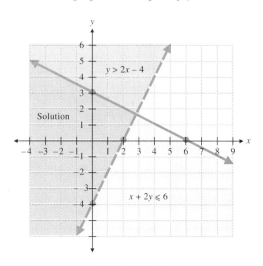

FIGURE 8.9

The solution is the set of points common to both inequalities. It is the part of the graph that contains both shadings. The dashed line is not part of the solution. However, the part of the solid line that satisfies both inequalities is part of the solution.

EXAMPLE 2   Determine the solution to the system of inequalities.

$$2x + 3y \geq 4$$
$$2x - y > -6$$

Solution: Graph $2x + 3y \geq 4$ (see Fig. 8.10). Graph $2x - y > -6$ on the same set of axes (Fig. 8.11). The solution is the part of the graph with both shadings and the part of the solid line that satisfies both inequalities.

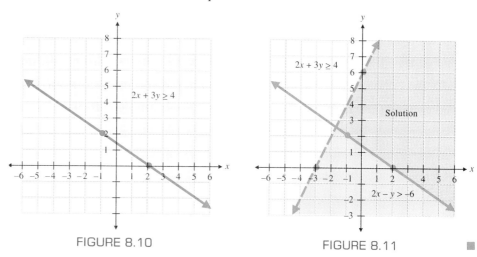

FIGURE 8.10   FIGURE 8.11

EXAMPLE 3   Determine the solution to the system of inequalities.

$$y < 2$$
$$x > -3$$

Solution: Graph both inequalities on the same set of axes (Fig. 8.12).

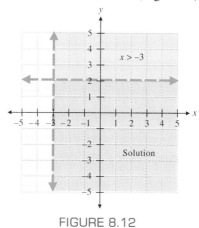

FIGURE 8.12

Exercise Set 8.6

Determine the solution to each system of inequalities.

1. $x + y > 2$
$x - y < 2$

2. $y \leq 3x - 2$
$y > -4x$

3. $y \leq x$
$y < -2x + 4$

4. $2x + 3y < 6$
$4x - 2y \geq 8$

5. $y > x + 1$
$y \geq 3x + 2$

6. $x + 3y \geq 6$
$2x - y > 4$

7. $x - 2y < 6$
   $y \leq -x + 4$

8. $y \leq 3x + 4$
   $y < 2$

9. $4x + 5y < 20$
   $x \geq -3$

10. $3x - 4y \leq 12$
    $y > -x + 4$

11. $x \leq 4$
    $y \geq -2$

12. $x \geq 0$
    $y \leq 0$

13. $x > -3$
    $y > 1$

14. $4x + 2y > 8$
    $y \leq 2$

15. $-2x + 3y \geq 6$
    $x + 4y \geq 4$

 16. Is it possible that a system of linear inequalities can have no solution? Explain your answer with the use of your own specific example.

Cumulative Review Exercises

[4.5] **17.** Multiply $(x^2 - 2x + 7)(3x - 4)$.

[5.2] **18.** Factor $xy + x - 3y - 3$.

[5.3] **19.** Factor $x^2 - 13x + 42$.

[5.4] **20.** Factor $6x^2 - x - 2$.

JUST FOR FUN

Determine the solution to the following system of inequalities.

$$x + 2y \leq 6$$
$$2x - y < 2$$
$$x > 2$$

SUMMARY

GLOSSARY

Consistent system of linear equations *(340):* A system of linear equations that has a solution.
Dependent system of linear equations *(341):* A system of linear equations that has an infinite number of solutions.
Inconsistent system of linear equations *(340):* A system of linear equations that has no solution.
Simultaneous linear equations or a system of linear

equations *(338):* Two or more linear equations considered together.
Solution to a system of equations *(338):* The ordered pair or pairs that satisfy all equations in the system.
Solution to a system of inequalities *(367):* The set of ordered pairs that satisfies all inequalities in the system.

IMPORTANT FACTS

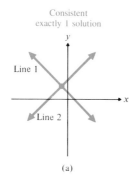

(a)

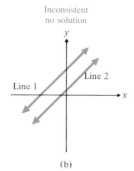

(b)

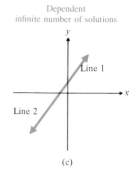

(c)

Three methods that can be used to solve a system of linear equations are the (1) graphical method, (2) substitution method, and (3) addition (or elimination) method.

Review Exercises

[8.1] *Determine which, if any, of the ordered pairs satisfy each system of equations.*

1. $y = 3x - 2$
   $2x + 3y = 5$
   (a) $(0, -2)$ **(b)** $(2, 4)$ **(c)** $(1, 1)$

2. $y = -x + 4$
   $3x + 5y = 15$
   (a) $\left(\dfrac{5}{2}, \dfrac{3}{2}\right)$ **(b)** $(0, 4)$ **(c)** $\left(\dfrac{1}{2}, \dfrac{3}{5}\right)$

[8.2] *Identify each system of linear equations as consistent, inconsistent, or dependent. State whether the system has exactly one solution, no solution, or an infinite number of solutions.*

3.

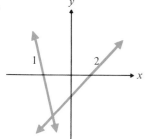

4.

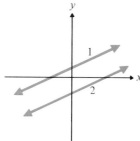

5.

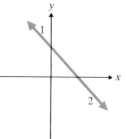

6.

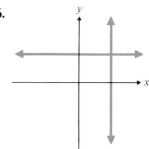

Express each equation in slope–intercept form. Without graphing or solving the system of equations, state whether the system of linear equations has exactly one solution, no solution, or an infinite number of solutions.

7. $x + 2y = 8$
$3x + 6y = 12$

8. $y = -3x - 6$
$2x + 3y = 8$

9. $y = \dfrac{1}{2}x - 4$
$x - 2y = 8$

10. $6x = 4y - 8$
$4x = 6y + 8$

Determine the solution to each system of equations graphically.

11. $y = x + 3$
$y = 2x + 5$

12. $x = -2$
$y = 3$

13. $y = 3$
$y = -2x + 5$

14. $x + 3y = 6$
$y = 2$

15. $x + 2y = 8$
$2x - y = -4$

16. $y = x - 3$
$2x - 2y = 6$

17. $2x + y = 0$
$4x - 3y = 10$

18. $x + 2y = 4$
$\dfrac{1}{2}x + y = -2$

[8.3] *Find the solution to each system of equations using substitution.*

19. $y = 2x - 8$
$2x - 5y = 0$

20. $x = 3y - 9$
$x + 2y = 1$

21. $2x + y = 5$
$3x + 2y = 8$

22. $2x - y = 6$
$x + 2y = 13$

23. $3x + y = 17$
$2x - 3y = 4$

24. $x = -3y$
$x + 4y = 6$

25. $4x - 2y = 10$
$y = 2x + 3$

26. $2x + 4y = 8$
$4x + 8y = 16$

27. $2x - 3y = 8$
$6x + 5y = 10$

28. $4x - y = 6$
$x + 2y = 8$

[8.4] *Find the solution to each system of equations using the addition method.*

29. $x + y = 6$
$x - y = 10$

30. $x + 2y = -3$
$2x - 2y = 6$

31. $2x + 3y = 4$
$x + 2y = -6$

32. $x + y = 12$
$2x + y = 5$

33. $4x - 3y = 8$
$2x + 5y = 8$

34. $-2x + 3y = 15$
$3x + 3y = 10$

35. $2x + y = 9$
$-4x - 2y = 4$

36. $2x + 2y = 8$
$y = 4x - 3$

37. $3x + 4y = 10$
$-6x - 8y = -20$

38. $2x - 5y = 12$
$3x - 4y = -6$

[8.5] *Express the problem as a system of linear equations, and then find the solution.*

39. The sum of two integers is 48. Find the two numbers if the larger is 3 less than twice the smaller.

40. Donald has 15 coins in his pocket that have a total value of $2.10. If he has only dimes and quarters, how many of each does he have?

41. Green Turf's grass seed costs 60 cents a pound and Agway's grass seed costs 45 cents a pound. How many pounds of each were used to make a 40-pound mixture of the two grass seeds that cost $20.25?

42. A chemist has a 30% acid solution and a 50% acid solution. How much of each must be mixed to get 6 liters of a 40% acid solution?

43. A plane can travel 600 miles per hour with the wind and 530 miles per hour against the wind. Find the speed of the wind and the speed of the plane in still air.

[8.6] *Determine the solution to each system of inequalities.*

44. $x + y > 2$
$2x - y \le 4$

45. $2x - 3y \le 6$
$x + 4y > 4$

46. $2x - 6y > 6$
$x > -2$

47. $x < 2$
$y \ge -1$

Practice Test

1. Determine which, if any, of the ordered pairs satisfy the system of equations.

   (a) $(0, -6)$ **(b)** $\left(-3, -\dfrac{3}{2}\right)$ **(c)** $(2, -4)$

   $$x + 2y = -6$$
   $$3x + 2y = -12$$

Identify each system as consistent, inconsistent, or dependent. State whether the system has exactly one solution, no solution, or an infinite number of solutions.

2.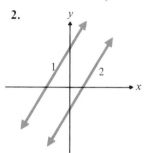

3.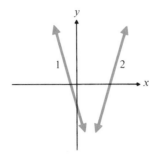

4.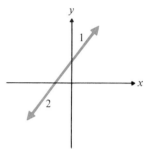

Express each equation in slope–intercept form. Then determine, without solving the system, whether the system of equations has exactly one solution, no solution, or an infinite number of solutions.

5. $3y = 6x - 9$
   $2x - y = 6$

6. $3x + 2y = 10$
   $3x - 2y = 10$

Solve each system of equations graphically.

7. $y = 3x - 2$
   $y = -2x + 8$

8. $3x - 2y = -3$
   $3x + y = 6$

Solve each system of equations using substitution.

9. $3x + y = 8$
   $x - y = 6$

10. $4x - 3y = 9$
    $2x + 4y = 10$

Solve each system of equations using the addition method.

11. $2x + y = 5$
    $x + 3y = -10$

12. $3x + 2y = 12$
    $-2x + 5y = 8$

Express the problem as a system of linear equations, and then find the solution.

13. Budget Rent a Car Agency charges $40 per day plus 8 cents per mile to rent a certain car. Hertz charges $45 per day plus 3 cents per mile to rent the same car. How many miles will have to be driven for the cost of Budget's car to equal the cost of Hertz's car?

14. The Modern Grocery has cashews that sell for $6.00 a pound and peanuts that sell for $4.50 a pound. How much of each must Albert, the grocer, mix to get 20 pounds of a mixture that he can sell for $5.00 per pound?

*15. Determine the solution to the system of inequalities.

$$2x + 4y < 8$$
$$x - 3y \geq 6$$

* From optional Section 8.6

Cumulative Review Test

1. Evaluate $\dfrac{|-4| + |-16| \div 2^2}{3 - [2 - (4 \div 2)]}$.

2. Solve the equation $4(x - 2) + 6(x - 3) = 2 - 4x$.

3. Solve the equation $3x^2 - 13x + 12 = 0$.

4. Solve the equation $\dfrac{1}{3}(x + 2) + \dfrac{1}{4} = 8$.

5. Solve the equation $\dfrac{1}{x - 3} + \dfrac{1}{x + 3} = \dfrac{1}{x^2 - 9}$.

6. Find the length of side x if the two figures are similar.

7. Simplify $(x^5 y^3)^4 (2x^3 y^5)$.

8. Factor $6x^2 - 11x + 4$.

9. Add $\dfrac{4}{x^2 - 9} - \dfrac{3}{x^2 - 9x + 18}$.

10. Divide $\dfrac{x^2 - 7x + 12}{2x^2 - 11x + 12} \div \dfrac{x^2 - 9}{x^2 - 16}$.

11. Graph the equation $2x - 3y = 6$ by plotting points.

12. Graph the equation $3x + 2y = 9$ using the x and y intercepts.

13. Graph the inequality $2x - y < 6$.

14. Without graphing the equation, determine if the following system of equations has exactly one solution, no solutions, or an infinite number of solutions. Explain how you determine your answer.

$$3x = 2y + 8$$
$$-4y = -6x + 12$$

15. Solve the system of equations graphically.

$$x + 2y = 2$$
$$2x - 3y = -3$$

16. Solve the system of equations using the addition method.

$$2x + 3y = 4$$
$$x - 4y = 6$$

17. A factory worker can inspect 40 units in 15 minutes. How long will it take her to inspect 160 units?

18. **(a)** An author is trying to decide between two publishing contract offers. The PCR Publishing Company offers an initial grant of $20,000 plus a 10% royalty rate on all dollar sales. The ARA Publishing Company offers an initial grant of $10,000 plus a 12% royalty rate on all dollar sales. How many dollars of sales are needed for the author's total income to be the same from both companies?

(b) If the author expects total dollar sales of $200,000, which company would result in the higher income?

19. How many liters of a 20% hydrochloric acid solution and how many liters of a 35% hydrochloric acid solution should be mixed to get 10 liters of a 25% hydrochloric acid solution?

20. Mr. and Mrs. Pontilo own a pizza shop. Mr. Pontilo can clean the pizza shop by himself in 50 minutes. Mrs. Pontilo can clean the pizza shop by herself in 60 minutes. How long will it take them to clean the pizza shop if they work together?

9.1 Introduction

9.2 Multiplying and Simplifying Square Roots

9.3 Dividing and Simplifying Square Roots

9.4 Addition and Subtraction of Square Roots

9.5 Solving Radical Equations

9.6 Applications of Radicals

9.7 Higher Roots and Fractional Exponents (Optional)

Summary

Review Exercises

Practice Test

See Section 9.6, Exercise 29.

9.1

Introduction

► **1** Learn the terminology associated with radical expressions.

► **2** Determine principal or positive square roots.

► **3** Recognize that not all radical expressions represent real numbers.

► **4** Recognize which radical expressions are perfect squares and which are rational or irrational numbers.

► **5** Write square roots as exponential expressions.

► **1** In this section we will introduce a number of important concepts related to radicals. We will first discuss some terminology associated with radicals. Let us begin with square roots. Square roots are one type of radical expression that you will use in both mathematics and science.

$$\sqrt{x} \text{ is read "the square root of } x."$$

The $\sqrt{}$ is called the **radical sign.** The number or expression inside the radical sign is called the **radicand.**

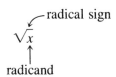

The entire expression, including the radical sign and radicand, is called the **radical expression.**

Another part of a radical expression is its **index.** The index tells the "root" of the expression. Square roots have an index of 2. The index of square roots is generally not written.

$$\sqrt{x} \qquad \text{means} \qquad \overset{\displaystyle \text{index}}{\underset{}{\sqrt[2]{x}}}$$

Other types of radical expressions have different indexes. For example, $\sqrt[3]{x}$ is the third or cube root of x. The index of cube roots is 3. Cube roots are discussed in optional Section 9.7.

Examples of Square Roots	*How Read*	*Radicand*
$\sqrt{8}$	the square root of 8	8
$\sqrt{5x}$	the square root of $5x$	$5x$
$\sqrt{\dfrac{x}{2y}}$	the square root of x over $2y$	$\dfrac{x}{2y}$

► **2** Every positive number has two square roots, a principal or positive square root and a negative square root.

The **principal or positive square root** of a positive real number x, written $\sqrt{x}$, is that *positive* number whose square equals x.

Examples

$$\sqrt{25} = 5 \qquad \text{since } 5^2 = 5 \cdot 5 = 25$$
$$\sqrt{36} = 6 \qquad \text{since } 6^2 = 6 \cdot 6 = 36$$
$$\sqrt{\frac{1}{4}} = \frac{1}{2} \qquad \text{since } \left(\frac{1}{2}\right)^2 = \left(\frac{1}{2}\right)\left(\frac{1}{2}\right) = \frac{1}{4}$$
$$\sqrt{\frac{4}{9}} = \frac{2}{3} \qquad \text{since } \left(\frac{2}{3}\right)^2 = \left(\frac{2}{3}\right)\left(\frac{2}{3}\right) = \frac{4}{9}$$

The **negative square root** of a positive real number x, written $-\sqrt{x}$, is the additive inverse or opposite of the principal square root. For example, $-\sqrt{25} = -5$ and $-\sqrt{36} = -6$. **Whenever we use the term square root in this book, we will be referring to the principal or positive square root.**

EXAMPLE 1 Evaluate: (a) $\sqrt{64}$ (b) $\sqrt{100}$

Solution: (a) $\sqrt{64} = 8$ since $8^2 = (8)(8) = 64$.
(b) $\sqrt{100} = 10$ since $(10)^2 = 100$. ■

EXAMPLE 2 Evaluate: (a) $-\sqrt{64}$ (b) $-\sqrt{100}$

Solution: (a) $-\sqrt{64} = -8$.
(b) $-\sqrt{100} = -10$. ■

▶ **3** An important point you must understand is that **square roots of negative numbers are not real numbers.** Consider $\sqrt{-4}$; to what is $\sqrt{-4}$ equal? To evaluate $\sqrt{-4}$, we must find some number whose square equals -4. But we know that the square of any nonzero number must be a positive number. Therefore, no number squared gives -4, and $\sqrt{-4}$ has no real value. Numbers like $\sqrt{-4}$, or the square root of any negative number, are called **imaginary numbers.** The study of imaginary numbers is beyond the scope of this book.

EXAMPLE 3 Indicate whether the radical expression is a real or imaginary number.
(a) $-\sqrt{9}$ (b) $\sqrt{-9}$ (c) $\sqrt{-37}$ (d) $-\sqrt{37}$

Solution: (a) Real (equal to -3);   (b) imaginary;   (c) imaginary;   (d) real ■

Suppose we have an expression like $\sqrt{x}$, where x represents some number. To make sure that the radical $\sqrt{x}$ is a real number, and not imaginary, we will always assume that x is a nonnegative number.

In this chapter, unless stated otherwise, we will assume that all variables that appear in a radicand represent nonnegative numbers.

▶ **4** To help in our discussion of rational and irrational numbers, we will define perfect square numbers. The numbers 1, 4, 9, 16, 25, 36, 49, . . . are called **perfect square numbers** because each number is *the square of a natural number*. When a perfect square number is a factor of a radicand, we may refer to it as a **perfect square factor.**

1, 2, 3, 4, 5, 6, 7, . . .	natural numbers
$1^2, 2^2, 3^2, 4^2, 5^2, 6^2, 7^2, \ldots$	the squares of the natural numbers
1, 4, 9, 16, 25, 36, 49, . . .	perfect square number

What are the next two perfect square numbers? Note that the square root of a perfect square number is an integer. That is, $\sqrt{1} = 1$, $\sqrt{4} = 2$, $\sqrt{9} = 3$, $\sqrt{16} = 4$, and so on.

A **rational number** is one that can be written in the form $\dfrac{a}{b}$, where a and b are integers, $b \neq 0$. Examples of rational numbers are $\frac{1}{2}, \frac{3}{5}, -\frac{9}{2}, 4$, and 0. All integers are rational numbers since they can be expressed with a denominator of 1. For example, $4 = \frac{4}{1}$ and $0 = \frac{0}{1}$. The square roots of perfect square numbers are also rational numbers since each is an integer. When a rational number is written as a decimal, it will be either a terminating or repeating decimal.

Terminating Decimals	*Repeating Decimals*
$\dfrac{1}{2} = 0.5$	$\dfrac{1}{3} = 0.333\ldots$
$\dfrac{1}{4} = 0.25$	$\dfrac{2}{3} = 0.666\ldots$
$\sqrt{4} = 2.0$	$\dfrac{1}{6} = 0.1666\ldots$

Real numbers that are not rational numbers are called **irrational numbers.** Irrational numbers when written as decimals are nonterminating, nonrepeating decimals. The square root of every positive integer that is not a perfect square is an irrational number. For example, $\sqrt{2}$ and $\sqrt{3}$ are irrational numbers.

Appendix D gives the square roots of integers from 1 to 100 rounded to the nearest thousandth. The symbol $\approx$ means "is approximately equal to." From Appendix D we can determine that $\sqrt{2} \approx 1.414$. The actual value of the square root of 2, and every other irrational number, when written as a decimal, will continue indefinitely in a nonrepeating pattern. Thus, whenever we give a value of a square root for an irrational number, it is only an approximation.

The table in Appendix D indicates the positive square root of integers between 1 and 100. Of these, only 10 are rational numbers. Table 9.1 illustrates the 20 smallest perfect square numbers. You may wish to refer to this table from time to time.

TABLE 9.1 The 20 Smallest Perfect Square Numbers

Perfect Square Number	Square Root of Perfect Square Number		Value	Perfect Square Number	Square Root of Perfect Square Number		Value
1	$\sqrt{1}$	$=$	1	121	$\sqrt{121}$	$=$	11
4	$\sqrt{4}$	$=$	2	144	$\sqrt{144}$	$=$	12
9	$\sqrt{9}$	$=$	3	169	$\sqrt{169}$	$=$	13
16	$\sqrt{16}$	$=$	4	196	$\sqrt{196}$	$=$	14
25	$\sqrt{25}$	$=$	5	225	$\sqrt{225}$	$=$	15
36	$\sqrt{36}$	$=$	6	256	$\sqrt{256}$	$=$	16
49	$\sqrt{49}$	$=$	7	289	$\sqrt{289}$	$=$	17
64	$\sqrt{64}$	$=$	8	324	$\sqrt{324}$	$=$	18
81	$\sqrt{81}$	$=$	9	361	$\sqrt{361}$	$=$	19
100	$\sqrt{100}$	$=$	10	400	$\sqrt{400}$	$=$	20

The 20 square roots listed in Table 9.1 are rational numbers. All other square roots of integers between 1 and 400 will be irrational numbers. For example, if we look at

$\sqrt{230}$ we can see that, since it is not in Table 9.1, it is an irrational number. Furthermore, since $\sqrt{230}$ is between $\sqrt{225}$ and $\sqrt{256}$, the value of $\sqrt{230}$ will be between the values of 15 and 16.

EXAMPLE 4 Refer to Table 9.1. Then indicate whether the following square roots are rational or irrational numbers.

(a) $\sqrt{118}$ (b) $\sqrt{169}$ (c) $\sqrt{64}$ (d) $\sqrt{200}$

Solution: (a) Irrational; (b) rational, equal to 13; (c) rational, equal to 8; (d) irrational.

Calculator Corner

Finding Square Roots on the Calculator

The square root key, $\boxed{\sqrt{x}}$, on calculators can be used to find square roots of nonnegative numbers. For example, to find the square root of 4, we press

$$\boxed{c}\ 4\ \boxed{\sqrt{x}}\ 2$$

The calculator would display the answer 2. What would the calculator display if we evaluate $\sqrt{7}$?

$$\boxed{c}\ 7\ \boxed{\sqrt{x}}\ 2.6457513$$

Note that $\sqrt{7}$ is an irrational number since 7 is not a perfect square. $\sqrt{7}$ is therefore a nonrepeating, nonterminating decimal. The exact decimal value of $\sqrt{7}$, or any other irrational number, can never be given exactly. The answers given on a calculator for irrational numbers are only close approximations of their value.

Suppose that we tried to evaluate $\sqrt{-4}$ on a calculator. What would the calculator give as an answer?

$$\boxed{c}\ 4\ \boxed{^{+}/_{-}}\ \boxed{\sqrt{x}}\ \text{Error}$$

The calculator would give some type of error message since the square root of -4, or the square root of any other negative number, is not a real number.

Changing Square Roots to Exponential Form

▶ **5** Radical expressions can be written in exponential form. Since we are discussing square roots, we will show how to write square roots in exponential form. Writing radicals in exponential form will be discussed in more detail in optional Section 9.7. We introduce this information here because your instructor may wish to use exponential form to help explain certain concepts.

Recall that the index of square roots is 2. For example,

$$\sqrt{x} = \sqrt[2]{x}$$

We use the index, 2, when writing square roots in exponential form. To change from a square root to an expression in exponential form, simply write the radicand of the square root to the $\frac{1}{2}$ power, as follows:

Changing Square Root to Exponential Form

$$\sqrt{} \quad = \quad \frac{1}{2} \leftarrow \text{index of square root}$$

$\uparrow$ radicand $\uparrow$

For example, $\sqrt{8}$ in exponential form is $8^{1/2}$, and $\sqrt{2xy} = (2xy)^{1/2}$.

Examples

Square Root Form	Exponential Form
$\sqrt{16}$	$(16)^{1/2}$
$\sqrt{3x}$	$(3x)^{1/2}$
$\sqrt{12x^2y}$	$(12x^2y)^{1/2}$

EXAMPLE 5   Write each radical expression in exponential form.

(a) $\sqrt{6}$ (b) $\sqrt{8x}$

Solution: (a) $6^{1/2}$ (b) $(8x)^{1/2}$ ∎

We can also convert an expression from exponential form to radical form. To do so, we reverse the process. For example, $(5x)^{1/2}$ can be written $\sqrt{5x}$ and $(20x^4)^{1/2}$ can be written $\sqrt{20x^4}$.

The rules of exponents presented in Sections 4.1 and 4.2 are still true when the exponents are fractions. For example,

$$(x^2)^{1/2} = x^{2 \cdot 1/2} = x^1 = x$$
$$(xy)^{1/2} = x^{1/2}y^{1/2}$$

and

$$x^{1/2} \cdot x^{3/2} = x^{(1/2)+(3/2)} = x^{4/2} = x^2$$

Exercise Set 9.1

Evaluate each expression.

1. $\sqrt{1}$
2. $\sqrt{4}$
3. $\sqrt{0}$
4. $\sqrt{64}$
5. $-\sqrt{81}$
6. $\sqrt{9}$
7. $\sqrt{121}$
8. $\sqrt{100}$
9. $-\sqrt{16}$
10. $-\sqrt{36}$
11. $\sqrt{144}$
12. $\sqrt{49}$
13. $\sqrt{169}$
14. $\sqrt{225}$
15. $-\sqrt{1}$
16. $-\sqrt{100}$
17. $\sqrt{81}$
18. $-\sqrt{25}$
19. $-\sqrt{121}$
20. $-\sqrt{169}$
21. $\sqrt{\dfrac{1}{4}}$
22. $\sqrt{\dfrac{4}{9}}$
23. $\sqrt{\dfrac{9}{16}}$
24. $\sqrt{\dfrac{25}{64}}$
25. $-\sqrt{\dfrac{4}{25}}$
26. $-\sqrt{\dfrac{100}{144}}$

Use your calculator or Appendix D to evaluate each expression. Round the answer to the nearest thousandth.

27. $\sqrt{8}$
28. $\sqrt{2}$
29. $\sqrt{15}$
30. $\sqrt{30}$
31. $\sqrt{80}$
32. $\sqrt{79}$
33. $\sqrt{81}$
34. $\sqrt{52}$
35. $\sqrt{97}$
36. $\sqrt{5}$
37. $\sqrt{3}$
38. $\sqrt{40}$

Answer true or false.

39. $\sqrt{25}$ is a rational number.
40. $\sqrt{-4}$ is not a real number.
41. $\sqrt{-5}$ is not a real number.
42. $\sqrt{5}$ is an irrational number.
43. $\sqrt{9}$ is an irrational number.
44. $\sqrt{\frac{1}{4}}$ is a rational number.
45. $\sqrt{\frac{4}{9}}$ is a rational number.
46. $\sqrt{231}$ is a rational number.
47. $\sqrt{125}$ is a rational number.
48. $\sqrt{27}$ is an irrational number.
49. $\sqrt{(18)^2}$ is an integer.
50. $\sqrt{(12)^2}$ is an integer.

Write each of the following in exponential form.

51. $\sqrt{7}$ **52.** $\sqrt{24}$ **53.** $\sqrt{17}$ **54.** $\sqrt{16}$

55. $\sqrt{5x}$ **56.** $\sqrt{6y}$ **57.** $\sqrt{12x^2}$ **58.** $\sqrt{25x^2y}$

59. $\sqrt{19xy^2}$ **60.** $\sqrt{34x^3y}$ **61.** $\sqrt{40x^3}$ **62.** $\sqrt{36x^3y^3}$

63. Whenever we see a variable in a square root, what assumption do we make about the variable? Why do we make this assumption?

64. In your own words, explain how you would determine if the square root of a given positive integer less than 400 is a rational or irrational number (a) by using a calculator, and (b) without the use of a calculator.

65. In your own words, explain why the square root of a negative number is not a real number.

Cumulative Review Exercises

[6.7] *Solve the equations.*

66. $\dfrac{2x}{x^2 - 4} + \dfrac{1}{x - 2} = \dfrac{2}{x + 2}.$

67. $\dfrac{4x}{x^2 + 6x + 9} - \dfrac{2x}{x + 3} = \dfrac{x + 1}{x + 3}.$

[6.8] **68.** A conveyor belt operating at full speed can fill a large silo with corn in 6 hours. A second conveyor belt operating at full speed can fill the same silo in 5 hours. How long will it take both conveyor belts operating together to fill the silo?

[7.3] **69.** Find the slope of the line through the points $(-5, 3)$ and $(6, 7)$.

9.2

Multiplying and Simplifying Square Roots

▶ **1** Use the product rule to simplify radicals containing numbers.

▶ **2** Use the product rule to simplify radicals containing variables.

▶ **1** To simplify square roots in this section we will make use of the **product rule for radicals**.

Product Rule for Radicals

$$\sqrt{a} \cdot \sqrt{b} = \sqrt{a \cdot b}, \qquad a \geq 0, b \geq 0 \qquad \text{Rule 1}$$

The product rule says that the product of two square roots is equal to the square root of the product. The product rule applies only when both a and b are nonnegative, since the square roots of negative numbers are not real numbers.

Example of the Product Rule

$$\left. \begin{array}{c} \sqrt{1} \cdot \sqrt{60} \\ \sqrt{2} \cdot \sqrt{30} \\ \sqrt{3} \cdot \sqrt{20} \\ \sqrt{4} \cdot \sqrt{15} \\ \sqrt{6} \cdot \sqrt{10} \end{array} \right\} = \sqrt{60}$$

Note that $\sqrt{60}$ can be factored into any of these forms.

> **To Simplify a Square Root Containing Only a Numerical Value**
>
> 1. Write the numerical value as a product of the largest perfect square factor and another factor.
> 2. Use the product rule to write the expression as a product of square roots, with each square root containing one of the factors.
> 3. Find the square root of the perfect square factor.

EXAMPLE 1   Simplify $\sqrt{60}$.

Solution: The only perfect square factor of 60 is 4.

$$\begin{aligned} \sqrt{60} &= \sqrt{4 \cdot 15} \\ &= \sqrt{4} \cdot \sqrt{15} \\ &= 2\sqrt{15} \end{aligned}$$

$2\sqrt{15}$ is read "two times the square root of fifteen" or "two radical fifteen."   ∎

EXAMPLE 2   Simplify $\sqrt{75}$.

Solution: $$\begin{aligned} \sqrt{75} &= \sqrt{25 \cdot 3} \\ &= \sqrt{25} \cdot \sqrt{3} \\ &= 5\sqrt{3} \end{aligned}$$ ∎

EXAMPLE 3   Simplify $\sqrt{80}$.

Solution: $$\begin{aligned} \sqrt{80} &= \sqrt{16 \cdot 5} \\ &= \sqrt{16} \cdot \sqrt{5} \\ &= 4\sqrt{5} \end{aligned}$$ ∎

HELPFUL HINT

When simplifying a square root, it is not uncommon for students to use a perfect square factor that is not the largest perfect square factor of the radicand. Let us consider Example 3 again. Four is also a perfect square factor of 80.

$$\sqrt{80} = \sqrt{4 \cdot 20} = \sqrt{4} \cdot \sqrt{20} = 2\sqrt{20}$$

Since 20 itself contains a perfect square factor of 4, the problem is not complete. Rather than starting the entire problem again, you can continue the simplification process as follows:

$$\sqrt{80} = 2\sqrt{20} = 2\sqrt{4 \cdot 5} = 2\sqrt{4} \cdot \sqrt{5} = 2 \cdot 2\sqrt{5} = 4\sqrt{5}$$

Now the result checks with the answer in Example 3.

EXAMPLE 4   Simplify $\sqrt{180}$.

Solution: $$\begin{aligned} \sqrt{180} &= \sqrt{36 \cdot 5} \\ &= \sqrt{36} \cdot \sqrt{5} \\ &= 6\sqrt{5} \end{aligned}$$ ∎

EXAMPLE 5   Simplify $\sqrt{156}$.

Solution: $\sqrt{156} = \sqrt{4 \cdot 39}$
$= \sqrt{4} \cdot \sqrt{39}$
$= 2\sqrt{39}$

Although 39 can be factored into $3 \cdot 13$, neither of these factors is a perfect square. Thus the answer can be simplified no further. ■

▶ **2**   Now we will simplify radicals that contain variables in the radicand.

In Section 9.1 we noted that certain numbers in a radicand were **perfect square factors.** When a radical contains a variable (or number) raised to an **even exponent,** that variable (or number) and exponent together also form a perfect square factor. For example, in the expression $\sqrt{x^4}$, the x^4 is a perfect square factor since the exponent, 4, is even. In the expression $\sqrt{x^5}$, the x^5 is not a perfect square factor since the exponent is odd.

To evaluate square roots when the radicand is a perfect square factor, we use the following rule.

$$\sqrt{a^{2 \cdot n}} = a^n \qquad a \geq 0 \qquad \text{Rule 2}$$

This rule states that **the square root of a variable raised to an even power equals the variable raised to one-half that power.**

Examples
$$\sqrt{x^2} = x$$
$$\sqrt{y^4} = y^2$$
$$\sqrt{x^{12}} = x^6$$
$$\sqrt{x^{20}} = x^{10}$$

A special case of rule 2 is
$$\sqrt{a^2} = a, \qquad a \geq 0$$

EXAMPLE 6   Simplify: (a) $\sqrt{x^{54}}$ (b) $\sqrt{x^4 y^6}$ (c) $\sqrt{x^8 y^2}$ (d) $\sqrt{y^8 z^{12}}$

Solution: (a) $\sqrt{x^{54}} = x^{27}$
(b) $\sqrt{x^4 y^6} = \sqrt{x^4} \sqrt{y^6} = x^2 y^3$
(c) $\sqrt{x^8 y^2} = x^4 y$
(d) $\sqrt{y^8 z^{12}} = y^4 z^6$ ■

To Evaluate the Square Root of a Radicand Containing a Variable Raised to an Odd Power

---

1. Express the variable as the product of two factors, one of which has an exponent of 1 (the other will therefore be a perfect square factor).
2. Use the product rule to simplify.

Examples 7 and 8 illustrate this procedure.

EXAMPLE 7   Simplify each expression.

(a) $\sqrt{x^3}$ (b) $\sqrt{y^{11}}$ (c) $\sqrt{x^{99}}$

Solution: (a) $\sqrt{x^3} = \sqrt{x^2 \cdot x} = \sqrt{x^2} \cdot \sqrt{x}$

$$= x \cdot \sqrt{x} \quad \text{or} \quad x\sqrt{x}$$

(Remember that x is the same as x^1.)

(b) $\sqrt{y^{11}} = \sqrt{y^{10} \cdot y} = \sqrt{y^{10}} \cdot \sqrt{y}$

$$= y^5\sqrt{y}$$

(c) $\sqrt{x^{99}} = \sqrt{x^{98} \cdot x} = \sqrt{x^{98}} \cdot \sqrt{x}$

$$= x^{49}\sqrt{x}$$ ■

More complex radicals can be simplified using the product rule for radicals and the principles discussed in this section.

EXAMPLE 8   Simplify each expression.

(a) $\sqrt{16x^3}$ (b) $\sqrt{32x^2}$ (c) $\sqrt{32x^3}$

Solution: Write each expression as the product of square roots, one of which has a perfect square radicand.

(a) $\sqrt{16x^3} = \sqrt{16x^2} \cdot \sqrt{x}$

$$= 4x\sqrt{x}$$

(b) $\sqrt{32x^2} = \sqrt{16x^2} \cdot \sqrt{2}$

$$= 4x\sqrt{2}$$

(c) $\sqrt{32x^3} = \sqrt{16x^2} \cdot \sqrt{2x}$

$$= 4x\sqrt{2x}$$ ■

EXAMPLE 9   Simplify each expression.

(a) $\sqrt{50x^2y}$ (b) $\sqrt{48x^3y^2}$ (c) $\sqrt{98x^9y^7}$

Solution: (a) $\sqrt{50x^2y} = \sqrt{25x^2} \cdot \sqrt{2y}$

$$= 5x\sqrt{2y}$$

(b) $\sqrt{48x^3y^2} = \sqrt{16x^2y^2} \cdot \sqrt{3x}$

$$= 4xy\sqrt{3x}$$

(c) $\sqrt{98x^9y^7} = \sqrt{49x^8y^6} \cdot \sqrt{2xy}$

$$= 7x^4y^3\sqrt{2xy}$$ ■

The radicand of your simplified answer should not contain any perfect square factors or any variables with an exponent greater than 1.

Now let us look at an example where we use the product rule to multiply two radicals together before simplifying.

EXAMPLE 10   Multiply and then simplify.
(a) $\sqrt{2} \cdot \sqrt{8}$ (b) $\sqrt{2x} \cdot \sqrt{8}$ (c) $\sqrt{2x} \cdot \sqrt{8x}$

Solution: (a) $\sqrt{2} \cdot \sqrt{8} = \sqrt{2 \cdot 8} = \sqrt{16} = 4$
(b) $\sqrt{2x} \cdot \sqrt{8} = \sqrt{16x} = \sqrt{16} \cdot \sqrt{x} = 4\sqrt{x}$
(c) $\sqrt{2x} \cdot \sqrt{8x} = \sqrt{16x^2} = 4x$

EXAMPLE 11   Multiply and then simplify.
(a) $\sqrt{8x^3y} \cdot \sqrt{4xy^5}$ (b) $\sqrt{5xy^6} \sqrt{6x^3y}$

Solution: (a) $\sqrt{8x^3y} \cdot \sqrt{4xy^5} = \sqrt{32x^4y^6} = \sqrt{16x^4y^6} \cdot \sqrt{2}$
$= 4x^2y^3\sqrt{2}$
(b) $\sqrt{5xy^6} \cdot \sqrt{6x^3y} = \sqrt{30x^4y^7} = \sqrt{x^4y^6} \cdot \sqrt{30y}$
$= x^2y^3\sqrt{30y}$

Note: In part (b), 30 can be factored in many ways. However, none of the factors are perfect squares, so we leave the answer as given.

Exercise Set 9.2

Simplify each expression.

1. $\sqrt{16}$	**2.** $\sqrt{64}$	**3.** $\sqrt{8}$	**4.** $\sqrt{75}$
5. $\sqrt{96}$	**6.** $\sqrt{125}$	**7.** $\sqrt{32}$	**8.** $\sqrt{52}$
9. $\sqrt{160}$	**10.** $\sqrt{28}$	**11.** $\sqrt{48}$	**12.** $\sqrt{27}$
13. $\sqrt{108}$	**14.** $\sqrt{128}$	**15.** $\sqrt{156}$	**16.** $\sqrt{180}$
17. $\sqrt{256}$	**18.** $\sqrt{212}$	**19.** $\sqrt{900}$	**20.** $\sqrt{x^4}$
21. $\sqrt{y^6}$	**22.** $\sqrt{x^9}$	**23.** $\sqrt{x^2y^4}$	**24.** $\sqrt{x^2y}$
25. $\sqrt{x^9y^{12}}$	**26.** $\sqrt{x^4y^5z^6}$	**27.** $\sqrt{a^2b^4c}$	**28.** $\sqrt{a^3b^9c^{11}}$

29. $\sqrt{3x^3}$	**30.** $\sqrt{12x^4y^2}$	**31.** $\sqrt{50x^2y^3}$
32. $\sqrt{125x^3y^5}$	**33.** $\sqrt{200y^5z^{12}}$	**34.** $\sqrt{64xyz^5}$
35. $\sqrt{243q^2b^3c}$	**36.** $\sqrt{500ab^4c^3}$	**37.** $\sqrt{128x^3yz^5}$
38. $\sqrt{112x^6y^8}$	**39.** $\sqrt{250x^4yz}$	**40.** $\sqrt{98x^4y^4z}$

Simplify each expression.

41. $\sqrt{8} \cdot \sqrt{3}$	**42.** $\sqrt{5} \cdot \sqrt{5}$	**43.** $\sqrt{18} \cdot \sqrt{3}$
44. $\sqrt{60} \cdot \sqrt{5}$	**45.** $\sqrt{75} \cdot \sqrt{6}$	**46.** $\sqrt{30} \cdot \sqrt{5}$
47. $\sqrt{3x}\sqrt{5x}$	**48.** $\sqrt{4x^3}\sqrt{4x}$	**49.** $\sqrt{5x^2}\sqrt{8x^3}$
50. $\sqrt{15x^2}\sqrt{6x^5}$	**51.** $\sqrt{12x^2y}\sqrt{6xy^3}$	**52.** $\sqrt{20xy^4}\sqrt{6x^5}$
53. $\sqrt{18xy^4}\sqrt{3x^2y}$	**54.** $\sqrt{40x^2y^4}\sqrt{6x^3y^5}$	**55.** $\sqrt{15xy^6}\sqrt{6xyz}$
56. $\sqrt{14xyz^5}\sqrt{3xy^2z^6}$	**57.** $\sqrt{9x^4y^6}\sqrt{4x^2y^4}$	**58.** $\sqrt{3x^3yz^6}\sqrt{6x^4y^5z^6}$
59. $(\sqrt{4x})^2$	**60.** $(\sqrt{6x^2})^2$	**61.** $(\sqrt{13x^4y^6})^2$

62. (a) Explain why $\sqrt{32x^3}$ is not a simplified expression.
(b) Simplify $\sqrt{32x^3}$.

Consider the radical expressions that follow. Indicate the coefficients and exponents that should be placed in the shaded areas to make a true statement. Explain how you obtained your answer.

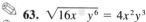

63. $\sqrt{16x\ \ y^6} = 4x^2y^3$

64. $\sqrt{\ \ x^4y\ } = 4x^2y^4$

65. $\sqrt{4x\ \ y\ } = 2x^3y^2\sqrt{y}$

66. $\sqrt{3x^4y\ } \cdot \sqrt{3x\ \ y^5} = 3x^5y^7\sqrt{xy}$

67. $\sqrt{2x\ \ y^5} \cdot \sqrt{\ \ x^3y\ } = 4x^7y^6\sqrt{x}$

68. $\sqrt{32x^4z\ } \cdot \sqrt{\ \ x\ \ z^{12}} = 8x^5z^9\sqrt{z}$

Cumulative Review Exercises

[6.2] **69.** Divide $\dfrac{3x^2 - 16x - 12}{3x^2 - 10x - 8} \div \dfrac{x^2 - 7x + 6}{3x^2 - 11x - 4}$.

[7.4] **70.** Write the equation $3x + 6y = 9$ in slope–intercept form and indicate the slope and the y-intercept.

[7.6] **71.** Graph $6x - 5y \geq 30$.

[8.4] **72.** Solve the system of equations
$$3x - 4y = 6,$$
$$5x - 3y = 5.$$

9.3

Dividing and Simplifying Square Roots

▶ **1** Understand what it means for a radical to be simplified.

▶ **2** Use the quotient rule to simplify radicals.

▶ **3** Rationalize denominators.

▶ **1** In this section we will use a new rule, the quotient rule, to simplify radicals containing fractions. However, before we do that, we need to discuss what it means for a square root to be simplified.

We Will Consider a Square Root Simplified When
1. There are no perfect square factors in any radicand.
2. No radicand contains a fraction.
3. There are no square roots in any denominator.

All three criteria must be met for an expression to be simplified. Let us look at some radical expressions that *are not simplified.*

Radical	*Reason Not Simplified*
$\sqrt{8}$	(1) Contains perfect square factor, 4. (note $\sqrt{8} = \sqrt{4} \cdot \sqrt{2} = 2\sqrt{2}$)
$\sqrt{x^3}$	(1) Contains perfect square factor, x^2. (note $\sqrt{x^3} = \sqrt{x^2} \cdot \sqrt{x} = x\sqrt{x}$)
$\sqrt{\dfrac{1}{2}}$	(2) Radicand contains a fraction.
$\dfrac{1}{\sqrt{2}}$	(3) Square root in the denominator.

▶ **2**   The quotient rule for radicals states that the quotient of the square root of a divided by the square root of b is equal to the square root of the quotient of a divided by b.

Quotient Rule for Radicals
___

$$\frac{\sqrt{a}}{\sqrt{b}} = \sqrt{\frac{a}{b}}, \qquad a \geq 0, b > 0 \qquad \text{Rule 3}$$

Examples 1 through 4 illustrate how the quotient rule is used to simplify square roots.

EXAMPLE 1   Simplify each expression.

(a) $\sqrt{\dfrac{8}{2}}$ (b) $\sqrt{\dfrac{25}{5}}$ (c) $\sqrt{\dfrac{9}{4}}$

Solution: When the square root contains a fraction, divide out any factor common to both the numerator and denominator. If the square root still contains a fraction, use the quotient rule for radicals to simplify.

(a) $\sqrt{\dfrac{8}{2}} = \sqrt{4} = 2$

(b) $\sqrt{\dfrac{25}{5}} = \sqrt{5}$

(c) $\sqrt{\dfrac{9}{4}} = \dfrac{\sqrt{9}}{\sqrt{4}} = \dfrac{3}{2}$ ∎

EXAMPLE 2   Simplify each expression.

(a) $\sqrt{\dfrac{16x^2}{8}}$ (b) $\sqrt{\dfrac{64x^4y}{2x^2y}}$ (c) $\sqrt{\dfrac{3x^2y^4}{27x^4}}$ (d) $\sqrt{\dfrac{15xy^5z^2}{3x^5yz}}$

Solution: First divide out any common factors to both the numerator and denominator; then use the quotient rule for radicals to simplify.

(a) $\sqrt{\dfrac{16x^2}{8}} = \sqrt{2x^2} = \sqrt{2}\sqrt{x^2} = \sqrt{2}x \text{ or } x\sqrt{2}$

(b) $\sqrt{\dfrac{64x^4y}{2x^2y}} = \sqrt{32x^2} = \sqrt{16x^2}\sqrt{2} = 4x\sqrt{2}$

(c) $\sqrt{\dfrac{3x^2y^4}{27x^4}} = \sqrt{\dfrac{y^4}{9x^2}} = \dfrac{\sqrt{y^4}}{\sqrt{9x^2}} = \dfrac{y^2}{3x}$

(d) $\sqrt{\dfrac{15xy^5z^2}{3x^5yz}} = \sqrt{\dfrac{5y^4z}{x^4}} = \dfrac{\sqrt{5y^4z}}{\sqrt{x^4}} = \dfrac{\sqrt{y^4}\sqrt{5z}}{\sqrt{x^4}} = \dfrac{y^2\sqrt{5z}}{x^2}$ ∎

When given a fraction containing radical expressions in both the numerator and denominator, we use the quotient rule to simplify, as illustrated below in Examples 3 and 4.

EXAMPLE 3 Simplify each expression.

(a) $\dfrac{\sqrt{2}}{\sqrt{8}}$ (b) $\dfrac{\sqrt{75}}{\sqrt{3}}$

Solution: (a) $\dfrac{\sqrt{2}}{\sqrt{8}} = \sqrt{\dfrac{2}{8}} = \sqrt{\dfrac{1}{4}} = \dfrac{\sqrt{1}}{\sqrt{4}} = \dfrac{1}{2}$

(b) $\dfrac{\sqrt{75}}{\sqrt{3}} = \sqrt{\dfrac{75}{3}} = \sqrt{25} = 5$ ■

EXAMPLE 4 Simplify each expression.

(a) $\dfrac{\sqrt{32x^4y^3}}{\sqrt{8xy}}$ (b) $\dfrac{\sqrt{75x^8y^4}}{\sqrt{3x^5y^8}}$

Solution: (a) $\dfrac{\sqrt{32x^4y^3}}{\sqrt{8xy}} = \sqrt{\dfrac{32x^4y^3}{8xy}} = \sqrt{4x^3y^2} = \sqrt{4x^2y^2} \cdot \sqrt{x} = 2xy\sqrt{x}$

(b) $\dfrac{\sqrt{75x^8y^4}}{\sqrt{3x^5y^8}} = \sqrt{\dfrac{75x^8y^4}{3x^5y^8}} = \sqrt{\dfrac{25x^3}{y^4}} = \dfrac{\sqrt{25x^3}}{\sqrt{y^4}} = \dfrac{\sqrt{25x^2}\sqrt{x}}{\sqrt{y^4}} = \dfrac{5x\sqrt{x}}{y^2}$ ■

▶ **3** When the denominator of a fraction contains the square root of a nonperfect square number, we generally simplify the expression by **rationalizing the denominator. To rationalize a denominator means to remove all radicals from the denominator.** We rationalize the denominator because it is easier (without a calculator) to obtain the approximate value of a number like $\sqrt{2}/2$ than a number like $1/\sqrt{2}$.

> **To rationalize a denominator,** multiply *both* the numerator and the denominator of the fraction by the square root that appears in the denominator or by the square root of a number that makes the denominator a perfect square.

EXAMPLE 5 Simplify $\dfrac{1}{\sqrt{2}}$.

Solution: Since $\sqrt{2} \cdot \sqrt{2} = \sqrt{4} = 2$, we multiply both numerator and denominator by $\sqrt{2}$.

$$\dfrac{1}{\sqrt{2}} = \dfrac{1}{\sqrt{2}} \cdot \dfrac{\sqrt{2}}{\sqrt{2}} = \dfrac{\sqrt{2}}{\sqrt{4}} = \dfrac{\sqrt{2}}{2}$$

The answer $\dfrac{\sqrt{2}}{2}$ is simplified since it satisfies the three requirements stated earlier.   ■

In Example 5, multiplying both the numerator and denominator by $\sqrt{2}$ is equivalent to multiplying the fraction by 1, which does not change the value of the original fraction.

EXAMPLE 6 Simplify each expression.

(a) $\sqrt{\dfrac{2}{3}}$ (b) $\sqrt{\dfrac{x^2}{18}}$

Solution: (a) $\sqrt{\dfrac{2}{3}} = \dfrac{\sqrt{2}}{\sqrt{3}} = \dfrac{\sqrt{2}}{\sqrt{3}} \cdot \dfrac{\sqrt{3}}{\sqrt{3}} = \dfrac{\sqrt{6}}{3}$

(b) $\sqrt{\dfrac{x^2}{18}} = \dfrac{\sqrt{x^2}}{\sqrt{18}} = \dfrac{x}{\sqrt{9} \cdot \sqrt{2}} = \dfrac{x}{3\sqrt{2}}$

Now rationalize the denominator.

$$\dfrac{x}{3\sqrt{2}} \cdot \dfrac{\sqrt{2}}{\sqrt{2}} = \dfrac{x\sqrt{2}}{3\sqrt{4}} = \dfrac{x\sqrt{2}}{3 \cdot 2} = \dfrac{x\sqrt{2}}{6}$$

Part (b) can also be rationalized as follows:

$$\sqrt{\dfrac{x^2}{18}} = \dfrac{\sqrt{x^2}}{\sqrt{18}} = \dfrac{x}{\sqrt{18}} \cdot \dfrac{\sqrt{2}}{\sqrt{2}} = \dfrac{x\sqrt{2}}{\sqrt{36}} = \dfrac{x\sqrt{2}}{6}$$

Note that $\dfrac{x}{\sqrt{18}} \cdot \dfrac{\sqrt{18}}{\sqrt{18}}$ will also give us the same result when simplified. ∎

COMMON STUDENT ERROR

A number within a square root *cannot* be divided by a number not within the square root.

Correct	*Wrong*
$\dfrac{\sqrt{2}}{2}$ cannot be simplified any further.	$\dfrac{\sqrt{2}^{\,1}}{\underset{1}{2}} = \sqrt{1} = 1$
$\dfrac{\sqrt{6}}{3}$ cannot be simplified any further.	$\dfrac{\sqrt{6}^{\,2}}{\underset{1}{3}} = \sqrt{2}$
$\dfrac{\sqrt{x^3}}{x} = \dfrac{\sqrt{x^2}\sqrt{x}}{x} = \dfrac{x\sqrt{x}}{x} = \sqrt{x}$	$\dfrac{\sqrt{x^3}^{\,2}}{\underset{1}{x}} = \sqrt{x^2} = x$

Each of the following simplifications is correct because the numbers divided out are not within square roots.

Correct	*Correct*
$\dfrac{\overset{2}{6}\sqrt{2}}{\underset{1}{3}} = 2\sqrt{2}$	$\dfrac{x\sqrt{2}}{x} = \sqrt{2}$
$\dfrac{\overset{1}{4}\sqrt{3}}{\underset{2}{8}} = \dfrac{\sqrt{3}}{2}$	$\dfrac{3x^2\sqrt{5}}{x} = 3x\sqrt{5}$

Exercise Set 9.3

Simplify each expression.

1. $\sqrt{\dfrac{12}{3}}$　　　**2.** $\sqrt{\dfrac{8}{2}}$　　　**3.** $\sqrt{\dfrac{27}{3}}$　　　**4.** $\sqrt{\dfrac{16}{4}}$

5. $\dfrac{\sqrt{18}}{\sqrt{2}}$　　　**6.** $\dfrac{\sqrt{3}}{\sqrt{27}}$　　　**7.** $\sqrt{\dfrac{1}{25}}$　　　**8.** $\sqrt{\dfrac{16}{25}}$

9. $\sqrt{\dfrac{9}{49}}$　　　**10.** $\sqrt{\dfrac{4}{81}}$　　　**11.** $\dfrac{\sqrt{10}}{\sqrt{490}}$　　　**12.** $\sqrt{\dfrac{16x^3}{4x}}$

13. $\sqrt{\dfrac{40x^3}{2x}}$　　　**14.** $\sqrt{\dfrac{45x^2}{16x^2y^4}}$　　　**15.** $\sqrt{\dfrac{9xy^4}{3y^3}}$　　　**16.** $\sqrt{\dfrac{50x^3y^6}{10x^3y^8}}$

17. $\sqrt{\dfrac{25x^6y}{45x^6y^3}}$　　　**18.** $\sqrt{\dfrac{14xyz^5}{56x^3y^3z^4}}$　　　**19.** $\sqrt{\dfrac{72xy}{72x^3y^5}}$　　　**20.** $\dfrac{\sqrt{16x^4}}{\sqrt{8x}}$

21. $\dfrac{\sqrt{32x^5}}{\sqrt{8x}}$　　　**22.** $\dfrac{\sqrt{60x^2y^2}}{\sqrt{6x^2y^4}}$　　　**23.** $\dfrac{\sqrt{16x^4y}}{\sqrt{25x^6y^3}}$　　　**24.** $\dfrac{\sqrt{72}}{\sqrt{36x^2y^6}}$

25. $\dfrac{\sqrt{45xy^6}}{\sqrt{9xy^4z^2}}$　　　**26.** $\dfrac{\sqrt{24x^2y^6}}{\sqrt{8x^4z^4}}$

Simplify each expression.

27. $\dfrac{3}{\sqrt{2}}$　　　**28.** $\dfrac{2}{\sqrt{3}}$　　　**29.** $\dfrac{4}{\sqrt{8}}$　　　**30.** $\dfrac{6}{\sqrt{6}}$

31. $\dfrac{5}{\sqrt{10}}$　　　**32.** $\dfrac{9}{\sqrt{50}}$　　　**33.** $\sqrt{\dfrac{2}{5}}$　　　**34.** $\sqrt{\dfrac{7}{12}}$

35. $\sqrt{\dfrac{3}{15}}$　　　**36.** $\sqrt{\dfrac{3}{10}}$　　　**37.** $\sqrt{\dfrac{x^2}{2}}$　　　**38.** $\sqrt{\dfrac{x^2}{7}}$

39. $\sqrt{\dfrac{x^2}{8}}$　　　**40.** $\sqrt{\dfrac{x^3}{18}}$　　　**41.** $\sqrt{\dfrac{x^4}{5}}$　　　**42.** $\sqrt{\dfrac{x^3}{11}}$

43. $\sqrt{\dfrac{x^6}{15y}}$　　　**44.** $\sqrt{\dfrac{x^5y}{12y^2}}$　　　**45.** $\sqrt{\dfrac{8x^4y^2}{32x^2y^3}}$　　　**46.** $\sqrt{\dfrac{27xz^4}{6y^4}}$

47. $\sqrt{\dfrac{18yz}{75x^4y^5z^3}}$　　　**48.** $\dfrac{\sqrt{25x^5}}{\sqrt{100xy^5}}$　　　**49.** $\dfrac{\sqrt{90x^4y}}{\sqrt{2x^5y^5}}$　　　**50.** $\dfrac{\sqrt{120xyz^2}}{\sqrt{9xy^2}}$

51. What are the three conditions necessary for a square root to be considered simplified?

52. (a) Explain, in your own words, how to rationalize the denominator of a fraction of the form $\dfrac{a}{\sqrt{b}}$. (b) Rationalize $\dfrac{a}{\sqrt{b}}$

Cumulative Review Exercises

[5.4] **53.** Factor $3x^2 - 12x - 96$.

[6.1] **54.** Reduce $\dfrac{x - 1}{x^2 - 1}$ to lowest terms.

[6.7] **55.** Solve the equation $x + \dfrac{24}{x} = 10$.

[8.2] **56.** Solve the system of equations graphically.

$$y = 2x - 2$$
$$2x + 3y = 10$$

9.4

Addition and Subtraction of Square Roots

▸**1** Add and subtract like and unlike square roots.

▸**2** Rationalize a binomial denominator.

▸**1** **Like square roots** are square roots having the same radicands. Like square roots are added in much the same manner that like terms are added. **To add like square roots,** add their coefficients and then multiply that sum by the like square root.

Examples of Adding Like Terms

$$2x + 3x = (2 + 3)x = 5x$$
$$4x + x = 4x + 1x = (4 + 1)x = 5x$$

Examples of Adding Like Square Roots

$$2\sqrt{7} + 3\sqrt{7} = (2 + 3)\sqrt{7} = 5\sqrt{7}$$
$$4\sqrt{x} + \sqrt{x} = 4\sqrt{x} + 1\sqrt{x} = (4 + 1)\sqrt{x} = 5\sqrt{x}$$

Note that adding like square roots is an application of the distributive property.

$$2\sqrt{7} + 3\sqrt{7} = (2 + 3)\sqrt{7}$$
$$= 5\sqrt{7}$$

Other Examples of Adding Like Square Roots

$$2\sqrt{5} - 3\sqrt{5} = (2 - 3)\sqrt{5} = -1\sqrt{5} = -\sqrt{5}$$

$$\sqrt{x} + \sqrt{x} = 1\sqrt{x} + 1\sqrt{x} = (1 + 1)\sqrt{x} = 2\sqrt{x}$$

$$6\sqrt{2} + 3\sqrt{2} - \sqrt{2} = (6 + 3 - 1)\sqrt{2} = 8\sqrt{2}$$

$$\frac{2\sqrt{3}}{5} + \frac{1\sqrt{3}}{5} = \left(\frac{2}{5} + \frac{1}{5}\right)\sqrt{3} = \frac{3}{5}\sqrt{3} \quad \text{or} \quad \frac{3\sqrt{3}}{5}$$

EXAMPLE 1   Simplify each expression, if possible.
(a) $4\sqrt{3} + 2\sqrt{3} - 2$ (b) $\sqrt{5} - 4\sqrt{5} + 5$
(c) $5 + 3\sqrt{2} - \sqrt{2} + 3$ (d) $2\sqrt{3} + 5\sqrt{2}$

Solution: (a) $4\sqrt{3} + 2\sqrt{3} - 2 = 6\sqrt{3} - 2$
(b) $\sqrt{5} - 4\sqrt{5} + 5 = -3\sqrt{5} + 5$
(c) $5 + 3\sqrt{2} - \sqrt{2} + 3 = 8 + 2\sqrt{2}$
(d) Cannot be simplified since the radicands are different. ■

EXAMPLE 2   Simplify each expression.
(a) $2\sqrt{x} - 3\sqrt{x} + 4\sqrt{x}$ (b) $3\sqrt{x} + x + 4\sqrt{x}$
(c) $x + \sqrt{x} + 2\sqrt{x} + 3$ (d) $x\sqrt{x} + 3\sqrt{x} + x$
(e) $\sqrt{xy} + 2\sqrt{xy} - \sqrt{x}$

Solution: (a) $2\sqrt{x} - 3\sqrt{x} + 4\sqrt{x} = 3\sqrt{x}$
(b) $3\sqrt{x} + x + 4\sqrt{x} = x + 7\sqrt{x}$ Only $3\sqrt{x}$ and $4\sqrt{x}$ can be combined.
(c) $x + \sqrt{x} + 2\sqrt{x} + 3 = x + 3\sqrt{x} + 3$ Only the $\sqrt{x}$ and $2\sqrt{x}$ can be combined.
(d) $x\sqrt{x} + 3\sqrt{x} + x = (x + 3)\sqrt{x} + x$ Only the $x\sqrt{x}$ and $3\sqrt{x}$ can be combined.
(e) $\sqrt{xy} + 2\sqrt{xy} - \sqrt{x} = 3\sqrt{xy} - \sqrt{x}$ Only the $\sqrt{xy}$ and $2\sqrt{xy}$ can be combined.
■

Unlike square roots are square roots having different radicands. It is sometimes possible to change unlike square roots into like square roots, as illustrated in Examples 3, 4, and 5.

EXAMPLE 3   Simplify $\sqrt{2} + \sqrt{18}$.

Solution: Since 18 has a perfect square factor, 9, we write it as a product of a perfect square factor and another factor.

$$\sqrt{2} + \sqrt{18} = \sqrt{2} + \sqrt{9 \cdot 2}$$
$$= \sqrt{2} + \sqrt{9}\sqrt{2}$$
$$= \sqrt{2} + 3\sqrt{2}$$
$$= 4\sqrt{2}$$
■

EXAMPLE 4   Simplify $\sqrt{24} - \sqrt{54}$.

Solution: Write each radicand as a product of a perfect square factor and another factor.
$$\sqrt{24} - \sqrt{54} = \sqrt{4 \cdot 6} - \sqrt{9 \cdot 6}$$
$$= \sqrt{4}\sqrt{6} - \sqrt{9}\sqrt{6}$$
$$= 2\sqrt{6} - 3\sqrt{6} = -\sqrt{6}$$
■

EXAMPLE 5   Simplify each expression.
(a) $2\sqrt{8} - \sqrt{32}$ (b) $3\sqrt{12} + 5\sqrt{27} + 2$ (c) $\sqrt{120} - \sqrt{75}$

Solution: (a) $2\sqrt{8} - \sqrt{32} = 2\sqrt{4 \cdot 2} - \sqrt{16 \cdot 2}$

$\qquad\qquad\qquad = 2\sqrt{4}\sqrt{2} - \sqrt{16}\sqrt{2}$

$\qquad\qquad\qquad = 2 \cdot 2\sqrt{2} - 4\sqrt{2}$

$\qquad\qquad\qquad = 4\sqrt{2} - 4\sqrt{2}$

$\qquad\qquad\qquad = 0$

(b) $3\sqrt{12} + 5\sqrt{27} + 2 = 3\sqrt{4 \cdot 3} + 5\sqrt{9 \cdot 3} + 2$

$\qquad\qquad\qquad\qquad\quad = 3\sqrt{4}\sqrt{3} + 5\sqrt{9}\sqrt{3} + 2$

$\qquad\qquad\qquad\qquad\quad = 3 \cdot 2\sqrt{3} + 5 \cdot 3\sqrt{3} + 2$

$\qquad\qquad\qquad\qquad\quad = 6\sqrt{3} + 15\sqrt{3} + 2$

$\qquad\qquad\qquad\qquad\quad = 21\sqrt{3} + 2$

(c) $\sqrt{120} - \sqrt{75} = \sqrt{4 \cdot 30} - \sqrt{25 \cdot 3}$

$\qquad\qquad\qquad\quad = \sqrt{4}\sqrt{30} - \sqrt{25}\sqrt{3}$

$\qquad\qquad\qquad\quad = 2\sqrt{30} - 5\sqrt{3}$

Since 30 has no perfect square factors and since the radicands are different, the expression $2\sqrt{30} - 5\sqrt{3}$ cannot be simplified any further. ■

COMMON STUDENT ERROR

The product rule presented in Section 9.2 was $\sqrt{a} \cdot \sqrt{b} = \sqrt{a \cdot b}$. The same principle **does not apply to** addition.

Wrong

$$\sqrt{a} + \sqrt{b} = \sqrt{a + b}$$

For example, to evaluate $\sqrt{9} + \sqrt{16}$,

Correct	*Wrong*
$\sqrt{9} + \sqrt{16} = 3 + 4$	$\sqrt{9} + \sqrt{16} = \sqrt{9 + 16}$
$\qquad\qquad\quad = 7$	$\qquad\qquad\quad = \sqrt{25}$
	$\qquad\qquad\quad = 5$

▶ **2** When the denominator of a rational expression is a binomial with a square root term, we again **rationalize the denominator.** We do this by multiplying both the numerator and the denominator of the fraction by the **conjugate** of the denominator. The conjugate of a binomial is a binomial having the same two terms with the sign of the second term changed.

Binomial	*Its Conjugate*
$3 + \sqrt{2}$	$3 - \sqrt{2}$
$\sqrt{5} - 3$	$\sqrt{5} + 3$
$2\sqrt{3} - \sqrt{5}$	$2\sqrt{3} + \sqrt{5}$
$x + \sqrt{3}$	$x - \sqrt{3}$

When a binomial is multiplied by its conjugate using the FOIL method, the outer and inner terms will add to zero.

EXAMPLE 6   Multiply $(2 + \sqrt{3})(2 - \sqrt{3})$ using the FOIL method.

Solution:

$(2 + \sqrt{3})(2 - \sqrt{3})$

$$
\begin{array}{cccc}
\text{F} & \text{O} & \text{I} & \text{L} \\
2(2) + & 2(-\sqrt{3}) + & 2(\sqrt{3}) + & \sqrt{3}(-\sqrt{3}) \\
= \quad 4 & - \quad 2\sqrt{3} & + \quad 2\sqrt{3} & - \quad \sqrt{9}
\end{array}
$$
$$= 4 - \sqrt{9}$$
$$= 4 - 3 = 1$$

EXAMPLE 7   Multiply $(\sqrt{3} - \sqrt{5})(\sqrt{3} + \sqrt{5})$ using the FOIL method.

Solution:

$(\sqrt{3} - \sqrt{5})(\sqrt{3} + \sqrt{5})$

$$
\begin{array}{cccc}
\text{F} & \text{O} & \text{I} & \text{L} \\
\sqrt{3} \cdot \sqrt{3} + & \sqrt{3} \cdot \sqrt{5} + & (-\sqrt{5})(\sqrt{3}) + & (-\sqrt{5})(\sqrt{5}) \\
= \quad \sqrt{9} & + \quad \sqrt{15} & - \quad \sqrt{15} & - \quad \sqrt{25}
\end{array}
$$
$$= \sqrt{9} - \sqrt{25}$$
$$= 3 - 5 = -2$$

Now let's try some examples where we rationalize the denominator when the denominator is a binomial with one or more terms a radical.

EXAMPLE 8   Simplify $\dfrac{5}{2 + \sqrt{3}}$.

Solution: To rationalize the denominator, multiply both the numerator and the denominator by $2 - \sqrt{3}$, which is the conjugate of $2 + \sqrt{3}$.

$$\frac{5}{2 + \sqrt{3}} \cdot \frac{2 - \sqrt{3}}{2 - \sqrt{3}} = \frac{5(2 - \sqrt{3})}{(2 + \sqrt{3})(2 - \sqrt{3})}$$
$$= \frac{5(2 - \sqrt{3})}{4 - 3}$$
$$= \frac{5(2 - \sqrt{3})}{1}$$
$$= 5(2 - \sqrt{3}) \quad \text{or} \quad 10 - 5\sqrt{3}$$

Note that $-5\sqrt{3} + 10$ is also an acceptable answer.

EXAMPLE 9   Simplify $\dfrac{6}{\sqrt{5} - \sqrt{2}}$.

Solution: Multiply both the numerator and the denominator of the fraction by $\sqrt{5} + \sqrt{2}$, the conjugate of $\sqrt{5} - \sqrt{2}$.

$$\frac{6}{\sqrt{5} - \sqrt{2}} \cdot \frac{\sqrt{5} + \sqrt{2}}{\sqrt{5} + \sqrt{2}} = \frac{6(\sqrt{5} + \sqrt{2})}{5 - 2}$$

$$= \frac{\overset{2}{\cancel{6}}(\sqrt{5} + \sqrt{2})}{\underset{1}{\cancel{3}}}$$

$$= 2(\sqrt{5} + \sqrt{2}) \quad \text{or} \quad 2\sqrt{5} + 2\sqrt{2} \qquad \blacksquare$$

EXAMPLE 10 Simplify $\dfrac{\sqrt{3}}{2 - \sqrt{6}}$.

Solution: Multiply both the numerator and the denominator of the fraction by $2 + \sqrt{6}$, the conjugate of $2 - \sqrt{6}$.

$$\frac{\sqrt{3}}{2 - \sqrt{6}} \cdot \frac{2 + \sqrt{6}}{2 + \sqrt{6}} = \frac{\sqrt{3}(2 + \sqrt{6})}{4 - 6}$$

$$= \frac{2\sqrt{3} + \sqrt{3} \cdot \sqrt{6}}{-2}$$

$$= \frac{2\sqrt{3} + \sqrt{18}}{-2}$$

$$= \frac{2\sqrt{3} + \sqrt{9} \cdot \sqrt{2}}{-2}$$

$$= \frac{2\sqrt{3} + 3\sqrt{2}}{-2}$$

$$= \frac{-2\sqrt{3} - 3\sqrt{2}}{2} \qquad \blacksquare$$

EXAMPLE 11 Simplify $\dfrac{x}{x + \sqrt{y}}$.

Solution: Multiply both the numerator and the denominator of the fraction by the conjugate of the denominator, $x - \sqrt{y}$.

$$\frac{x}{x + \sqrt{y}} \cdot \frac{x - \sqrt{y}}{x - \sqrt{y}} = \frac{x(x - \sqrt{y})}{x^2 - y} = \frac{x^2 - x\sqrt{y}}{x^2 - y}$$

Remember that you cannot divide out the x^2 terms because they are not factors. $\blacksquare$

Exercise Set 9.4

Similify each expression.

1. $4\sqrt{3} - 2\sqrt{3}$ **2.** $\sqrt{5} + 2\sqrt{5}$ **3.** $6\sqrt{7} - 8\sqrt{7}$

4. $4\sqrt{10} + 6\sqrt{10} - \sqrt{10} + 2$ **5.** $2\sqrt{3} - 2\sqrt{3} - 4\sqrt{3} + 5$ **6.** $12\sqrt{15} + 5\sqrt{15} - 8\sqrt{15}$

7. $4\sqrt{x} + \sqrt{x}$ **8.** $-2\sqrt{x} - 3\sqrt{x}$ **9.** $-\sqrt{x} + 6\sqrt{x} - 2\sqrt{x}$

10. $3\sqrt{y} - 6\sqrt{y}$ **11.** $3\sqrt{y} - \sqrt{y} + 3$ **12.** $3\sqrt{5} - \sqrt{x} + 4\sqrt{5} + 3\sqrt{x}$

13. $\sqrt{x} + \sqrt{y} + x + 3\sqrt{y}$ **14.** $2 + 3\sqrt{y} - 6\sqrt{y} + 5$ **15.** $3 + 4\sqrt{x} - 6\sqrt{x}$

16. $4\sqrt{x} + 6\sqrt{x} - 3\sqrt{x} + 2x$

Simplify each expression.

17. $\sqrt{8} - \sqrt{12}$ **18.** $\sqrt{27} + \sqrt{45}$ **19.** $\sqrt{200} - \sqrt{72}$

20. $\sqrt{75} + \sqrt{108}$ **21.** $\sqrt{125} + \sqrt{20}$ **22.** $\sqrt{60} - \sqrt{135}$

23. $4\sqrt{50} - \sqrt{72} + \sqrt{8}$ **24.** $-4\sqrt{90} + 3\sqrt{40} + 2\sqrt{10}$ **25.** $-6\sqrt{75} + 4\sqrt{125}$

26. $4\sqrt{80} - \sqrt{75}$ **27.** $5\sqrt{250} - 9\sqrt{80}$ **28.** $7\sqrt{108} - 6\sqrt{180}$

29. $8\sqrt{64} - \sqrt{96}$ **30.** $3\sqrt{250} + 5\sqrt{160}$

Multiply as indicated.

31. $(3 + \sqrt{2})(3 - \sqrt{2})$ **32.** $(\sqrt{6} + 3)(\sqrt{6} - 3)$ **33.** $(6 - \sqrt{5})(6 + \sqrt{5})$

34. $(\sqrt{8} - 3)(\sqrt{8} + 3)$ **35.** $(\sqrt{x} + 3)(\sqrt{x} - 3)$ **36.** $(\sqrt{x} + 5)(\sqrt{x} - 5)$

37. $(\sqrt{6} + x)(\sqrt{6} - x)$ **38.** $(\sqrt{y} - 3)(\sqrt{y} + 3)$ **39.** $(\sqrt{x} + y)(\sqrt{x} - y)$

40. $(\sqrt{5x} + \sqrt{y})(\sqrt{5x} - \sqrt{y})$ **41.** $(2\sqrt{x} + 3\sqrt{y})(2\sqrt{x} - 3\sqrt{y})$ **42.** $(4\sqrt{2x} + \sqrt{3y})(4\sqrt{2x} - \sqrt{3y})$

Simplify each expression.

43. $\dfrac{4}{2 + \sqrt{3}}$ **44.** $\dfrac{3}{\sqrt{6} - 5}$ **45.** $\dfrac{3}{\sqrt{2} + 5}$

46. $\dfrac{4}{\sqrt{2} - 7}$ **47.** $\dfrac{2}{\sqrt{2} + \sqrt{3}}$ **48.** $\dfrac{5}{\sqrt{5} - \sqrt{6}}$

49. $\dfrac{8}{\sqrt{5} - \sqrt{8}}$ **50.** $\dfrac{1}{\sqrt{17} - \sqrt{8}}$ **51.** $\dfrac{2}{6 + \sqrt{x}}$

52. $\dfrac{5}{\sqrt{x} - 3}$ **53.** $\dfrac{6}{4 - \sqrt{y}}$ **54.** $\dfrac{5}{3 + \sqrt{x}}$

55. $\dfrac{4}{\sqrt{x} - y}$ **56.** $\dfrac{9}{x + \sqrt{y}}$ **57.** $\dfrac{x}{\sqrt{x} + \sqrt{y}}$

58. $\dfrac{\sqrt{3}}{\sqrt{x} - \sqrt{3}}$ **59.** $\dfrac{\sqrt{x}}{\sqrt{5} + \sqrt{x}}$ **60.** $\dfrac{x}{\sqrt{x} - y}$

61. (a) Explain in your own words how to rationalize the denominator in a fraction of the form $\dfrac{a}{b + \sqrt{c}}$.

     (b) Rationalize $\dfrac{a}{b + \sqrt{c}}$.

Cumulative Review Exercises

[2.5] **62.** Solve the equation $3(2x - 6) = 4(x - 9) + 3x$.

[5.6] **63.** Solve the equation $2x^2 - x - 36 = 0$.

[6.5] **64.** Subtract $\dfrac{1}{x^2 - 4} - \dfrac{2}{x - 2}$.

[6.8] **65.** Mr. Moreno can stack a pile of wood in 20 minutes. With his wife's help, together they can stack the wood in 12 minutes. How long would it take Mrs. Moreno to stack the wood by herself?

9.5

Solving Radical Equations

▶**1** Solve radical equations with only one square root term.

▶**2** Solve radical equations with two square root terms.

▶**1** A **radical equation** is an equation that contains a variable in a radicand. Some examples of radical equations are

$$\sqrt{x} = 3, \qquad \sqrt{x + 4} = 6, \qquad \sqrt{x - 2} = x - 6$$

To Solve a Radical Equation Containing Only One Square Root Term

1. Use the appropriate properties to rewrite the equation with the square root term by itself on one side of the equation. We call this *isolating* the radical.
2. Combine like terms.
3. Square both sides of the equation to remove the square root.
4. Solve the equation for the variable.
5. Check the solution in the original equation for extraneous roots.

The following examples illustrate this procedure.

EXAMPLE 1 Solve the equation $\sqrt{x} = 6$.

Solution: The square root containing the variable is already by itself on one side of the equation. Square both sides of the equation.

$$\sqrt{x} = 6$$
$$(\sqrt{x})^2 = (6)^2$$
$$x = 36$$

Check:
$$\sqrt{x} = 6$$
$$\sqrt{36} = 6$$
$$6 = 6 \qquad \text{true}$$

EXAMPLE 2 Solve the equation $\sqrt{x + 4} = 6$.

Solution: The square root containing the variable is already by itself on one side of the equation. Square both sides of the equation.

$$\sqrt{x + 4} = 6$$
$$(\sqrt{x + 4})^2 = 6^2$$
$$x + 4 = 36$$
$$x + 4 - 4 = 36 - 4$$
$$x = 32$$

Check: $\sqrt{x + 4} = 6$
$$\sqrt{32 + 4} = 6$$
$$\sqrt{36} = 6$$
$$6 = 6 \quad \text{true}$$

EXAMPLE 3   Solve the equation $\sqrt{x} + 4 = 6$.

Solution: Since the 4 is outside the square root sign, we first subtract 4 from both sides of the equation to isolate the square root term.

$$\sqrt{x} + 4 = 6$$
$$\sqrt{x} + 4 - 4 = 6 - 4$$
$$\sqrt{x} = 2$$

Now square both sides of the equation.

$$(\sqrt{x})^2 = 2^2$$
$$x = 4$$

A check will show that 4 is the solution.

HELPFUL HINT   When you square both sides of an equation, you may introduce extraneous roots. An **extraneous root** is a number obtained when solving an equation that is not a solution to the original equation. Therefore, equations that are squared in the process of finding their solutions should always be checked for extraneous roots by substituting the numbers found back in the **original** equation.

Consider the equation

$$x = 5$$

Now square both sides.

$$x^2 = 25$$

Note that the equation $x = 5$ is only true when x is 5. However the equation $x^2 = 25$ is true for both 5 and -5. When we squared $x = 5$, we introduced the extraneous root -5.

EXAMPLE 4   Solve the equation $\sqrt{x} = -5$.

Solution: $\sqrt{x} = -5$ *Check:* $\sqrt{x} = -5$
$$(\sqrt{x})^2 = (-5)^2 \qquad\qquad \sqrt{25} = -5$$
$$x = 25 \qquad\qquad\qquad 5 = -5 \quad \textbf{false}$$

Since the check results in a false statement, the number 25 is an extraneous root and is not a solution to the given equation. Thus the equation $\sqrt{x} = -5$ has no real solutions.

In Example 4, some of you might have realized without working the problem that there is no solution. In the original equation, the left side is positive and the right side is negative; thus they cannot possibly be equal.

EXAMPLE 5   Solve the equation $\sqrt{2x - 3} = x - 3$.

Solution: Square both sides of the equation.

$$(\sqrt{2x - 3})^2 = (x - 3)^2$$
$$2x - 3 = (x - 3)(x - 3)$$
$$2x - 3 = x^2 - 6x + 9$$

Now solve the quadratic equation as explained earlier. Move the $2x$ and -3 to the right side of the equation to obtain

$$0 = x^2 - 8x + 12 \quad \text{or} \quad x^2 - 8x + 12 = 0$$

Now factor.

$$x^2 - 8x + 12 = 0$$
$$(x - 6)(x - 2) = 0$$
$$x - 6 = 0 \quad \text{or} \quad x - 2 = 0$$
$$x = 6 \qquad\qquad x = 2$$

Check:

$x = 6$	$x = 2$
$\sqrt{2x - 3} = x - 3$	$\sqrt{2x - 3} = x - 3$
$\sqrt{2(6) - 3} = 6 - 3$	$\sqrt{2(2) - 3} = 2 - 3$
$\sqrt{9} = 3?$	$\sqrt{1} = -1?$
$3 = 3$ true	$1 = -1$ **false**

The 6 is a solution, but 2 is not a solution to the equation.

EXAMPLE 6   Solve the equation $2x - 5\sqrt{x} - 3 = 0$.

Solution: First rewrite the equation so that the square root containing the variable is by itself on one side of the equation.

$$2x - 5\sqrt{x} - 3 = 0$$
$$-5\sqrt{x} = -2x + 3$$
$$\text{or} \quad 5\sqrt{x} = 2x - 3$$

Now square both sides of the equation.

$$(5\sqrt{x})^2 = (2x - 3)^2$$
$$25x = (2x - 3)(2x - 3)$$
$$25x = 4x^2 - 12x + 9$$
$$0 = 4x^2 - 37x + 9$$
$$0 = (4x - 1)(x - 9)$$
$$4x - 1 = 0 \quad \text{or} \quad x - 9 = 0$$
$$4x = 1 \qquad\qquad x = 9$$
$$x = \frac{1}{4}$$

Check: $x = \dfrac{1}{4}$

$2x - 5\sqrt{x} - 3 = 0$

$2\left(\dfrac{1}{4}\right) - 5\sqrt{\dfrac{1}{4}} - 3 = 0$

$\dfrac{1}{2} - 5\left(\dfrac{1}{2}\right) - 3 = 0$

$\dfrac{1}{2} - \dfrac{5}{2} - 3 = 0$

$\dfrac{1}{2} - \dfrac{5}{2} - \dfrac{6}{2} = 0$

$-\dfrac{10}{2} = 0$

$-5 = 0$ **false**

$x = 9$

$2x - 5\sqrt{x} - 3 = 0$

$2(9) - 5\sqrt{9} - 3 = 0$

$18 - 5(3) - 3 = 0$

$18 - 15 - 3 = 0$

$0 = 0$ true

The solution is 9; $\frac{1}{4}$ is not a solution. ∎

▸**2** Consider the radical equations

$$\sqrt{x + 1} = \sqrt{x - 3}, \qquad \sqrt{x + 5} - \sqrt{2x + 4} = 0$$

These equations are different from those previously discussed because they have two square root terms containing the variable x. To solve equations of this type, rewrite the equation, when necessary, so that there is only one square root term on each side of the equation. Then square both sides of the equation. Examples 7 and 8 illustrate this procedure.

EXAMPLE 7 Solve the equation $\sqrt{2x + 2} = \sqrt{3x - 5}$.

Solution: Since each side of the equation already contains one square root term, it is not necessary to rewrite the equation. Square both sides of the equation, then solve for x.

$$(\sqrt{2x + 2})^2 = (\sqrt{3x - 5})^2$$
$$2x + 2 = 3x - 5$$
$$2 = x - 5$$
$$7 = x$$

Check: $\sqrt{2x + 2} = \sqrt{3x - 5}$

$\sqrt{2(7) + 2} = \sqrt{3(7) - 5}$

$\sqrt{16} = \sqrt{16}$

$4 = 4$ true

The solution is 7.

EXAMPLE 8 Solve the equation $3\sqrt{x-2} - \sqrt{7x+4} = 0$.

Solution: Add $\sqrt{7x+4}$ to both sides of the equation to get one square root term on each side of the equation. Then square both sides of the equation.

$$3\sqrt{x-2} - \sqrt{7x+4} + \sqrt{7x+4} = 0 + \sqrt{7x+4}$$
$$3\sqrt{x-2} = \sqrt{7x+4}$$
$$(3\sqrt{x-2})^2 = (\sqrt{7x+4})^2$$
$$9(x-2) = 7x+4$$
$$9x - 18 = 7x + 4$$
$$2x - 18 = 4$$
$$2x = 22$$
$$x = 11$$

Check:
$$3\sqrt{x-2} - \sqrt{7x+4} = 0$$
$$3\sqrt{11-2} - \sqrt{7(11)+4} = 0$$
$$3\sqrt{9} - \sqrt{77+4} = 0$$
$$3(3) - \sqrt{81} = 0$$
$$9 - 9 = 0 \qquad \text{true}$$

Exercise Set 9.5

Solve each equation. If the equation has no real solution, so state.

1. $\sqrt{x} = 8$

2. $\sqrt{x} = 5$

3. $\sqrt{x} = -3$

4. $\sqrt{x-3} = 6$

5. $\sqrt{x+5} = 3$

6. $\sqrt{2x-4} = 2$

7. $\sqrt{2x+4} = -6$

8. $\sqrt{x-4} = 8$

9. $\sqrt{x} + 3 = 5$

10. $4 + \sqrt{x} = 9$

11. $6 = 4 + \sqrt{x}$

12. $2 = 8 - \sqrt{x}$

13. $4 + \sqrt{x} = 2$

14. $\sqrt{3x+4} = x - 2$

15. $\sqrt{2x-5} = x - 4$

16. $\sqrt{x^2+8} = x + 2$

17. $\sqrt{2x-6} = \sqrt{5x-27}$

18. $2\sqrt{x+3} = 10$

19. $\sqrt{3x+3} = \sqrt{5x-1}$

20. $\sqrt{2x-5} = \sqrt{x+2}$

21. $\sqrt{3x+9} = 2\sqrt{x}$

22. $x - 6 = \sqrt{3x}$

23. $\sqrt{4x-5} = \sqrt{x+9}$

24. $\sqrt{x^2+3} = x + 1$

25. $3\sqrt{x} = \sqrt{x+8}$

26. $x - 5 = \sqrt{x^2-35}$

27. $4\sqrt{x} = x + 3$

28. $2x - 1 = -\sqrt{x}$

29. $\sqrt{2x-3} = 2\sqrt{3x-2}$

30. $6 - 2\sqrt{3x} = 0$

31. $\sqrt{x^2+5} = x + 5$

32. $2\sqrt{4x-3} = 10$

33. $x - 4\sqrt{x} + 3 = 0$

34. $\sqrt{4x+5} + 5 = 2x$

35. Why is it necessary to always check solutions to radical equations?

36. (a) Write in your own words a step-by-step procedure for solving equations containing a single square root term. (b) Solve the equation $\sqrt{x+1} - 1 = 1$ using the procedure outlined in part (a).

Cumulative Review Exercises

[8.2] **37.** Solve the system of equations graphically.

$$3x - 2y = 6$$
$$y = 2x - 4$$

[8.3] **38.** Solve the system of equations by substitution.

$$3x - 2y = 6$$
$$y = 2x - 4$$

[8.4] **39.** Solve the system of equations by the addition method.

$$3x - 2y = 6$$
$$y = 2x - 4$$

[8.5] **40.** A boat can travel at a speed of 12 miles per hour with the current and 4 miles per hour against the current. Find the speed of the boat in still water and the speed of the current.

See Exercise 40.

JUST FOR FUN

Solve each equation. (*Hint:* It will be necessary to square both sides of the equation twice—good luck.)

1. $\sqrt{x} + 2 = \sqrt{x + 16}$
2. $\sqrt{x + 1} = 2 - \sqrt{x}$.
3. $\sqrt{x + 7} = 5 - \sqrt{x - 8}$.

9.6

Applications of Radicals

▶ **1**   Use the Pythagorean theorem.

▶ **2**   Use the distance formula.

▶ **3**   Learn some scientific applications of radicals.

In this section we will focus on some of the many important applications of radicals. We will discuss the Pythagorean theorem and the distance formula, and then give a few additional applications of radicals.

Pythagorean Theorem

▶ **1**   A **right triangle** is a triangle that contains a right, or 90°, angle. A right triangle is illustrated in Fig. 9.1.

The two smaller sides of a right triangle are called the **legs** and the side opposite the right angle is called the **hypotenuse**. The Pythagorean theorem expresses the relationship between the legs of a right triangle and its hypotenuse.

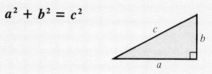

FIGURE 9.1

Pythagorean Theorem

The square of the hypotenuse of a right triangle is equal to the sum of the squares of the two legs.

If a and b represent the legs, and c represents the hypotenuse, then

$$a^2 + b^2 = c^2$$

Jn Section 9.5, when we solved radical equations containing square roots, we raised both sides of the equation to the second power to eliminate square roots. When we solve problems using the Pythagorean theorem, we will raise both sides of the equation to the $\frac{1}{2}$ power to remove the square on one of the variables. We can do this because the rules of exponents presented in Sections 4.1 and 4.2 also apply to fractional exponents. Since lengths are positive, we also make the assumption that the a, b, and c in the Pythagorean theorem represent positive values.

EXAMPLE 1 Find the hypotenuse of the right triangle whose legs are 3 feet and 4 feet.

Solution: Draw a picture of the problem (Fig. 9.2). When drawing the picture, it makes no difference which leg is called a and which leg is called b.

$$a^2 + b^2 = c^2$$
$$3^2 + 4^2 = c^2$$
$$9 + 16 = c^2$$
$$25 = c^2$$
$$(25)^{1/2} = (c^2)^{1/2} \qquad \text{Raise both sides of the}$$
$$\sqrt{25} = c \qquad\qquad \text{equation to the } \tfrac{1}{2} \text{ power.}$$
$$5 = c$$

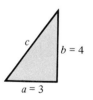

FIGURE 9.2

The hypotenuse is 5 feet.

Check: $a^2 + b^2 = c^2$
$$3^2 + 4^2 = 5^2$$
$$9 + 16 = 25$$
$$25 = 25 \qquad \text{true}$$

In Example 1, where we have $25 = c^2$, we can also solve this equation for c by taking the square root of both sides of the equation. We will discuss the square root property that allows us to do this in Section 10.1.

EXAMPLE 2   The hypotenuse of a right triangle is 12 inches. Find the second leg if one leg is 8 inches.

Solution: First, draw a sketch of the triangle (see Fig. 9.3).

$c = 12$ $b = ?$

$a = 8$

FIGURE 9.3

$$a^2 + b^2 = c^2$$
$$8^2 + b^2 = (12)^2$$
$$64 + b^2 = 144$$
$$b^2 = 80$$
$$(b^2)^{1/2} = (80)^{1/2}$$
$$b = \sqrt{80}, \quad \text{or} \quad \text{approximately 8.94 inches}$$

EXAMPLE 3   A regulation baseball diamond is a square with 90 feet between bases. How far is second base from home plate?

Solution: Draw the baseball diamond (see Fig. 9.4). We are asked to find the distance from second base to home plate. This distance is the hypotenuse of the triangle, c, shown in Fig. 9.5.

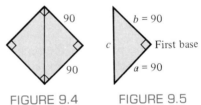

90

$b = 90$

c First base

90

$a = 90$

FIGURE 9.4   FIGURE 9.5

$$a^2 + b^2 = c^2$$
$$(90)^2 + (90)^2 = c^2$$
$$8100 + 8100 = c^2$$
$$16{,}200 = c^2$$
$$c = \sqrt{16{,}200}, \quad \text{or} \quad \text{approximately 127.28 feet}$$

The $\sqrt{16{,}200}$ was evaluated on a calculator using the $\boxed{\sqrt{}}$ key.

Distance Formula   ▶**2**   The distance formula can be used to find the distance between two points, (x_1, y_1) and (x_2, y_2), in the Cartesian coordinate system.

Distance Formula
_____

$$d = \sqrt{(x_2 - x_1)^2 + (y_2 - y_1)^2}$$

EXAMPLE 4 Find the length of the straight line between the points $(-1, -4)$ and $(5, -2)$.

Solution: The two points are illustrated in Fig. 9.6. It makes no difference which points are labeled (x_1, y_1), and (x_2, y_2). Let $(5, -2)$ be (x_2, y_2) and $(-1, -4)$ be (x_1, y_1). Thus $x_2 = 5$, $y_2 = -2$ and $x_1 = -1$, $y_1 = -4$.

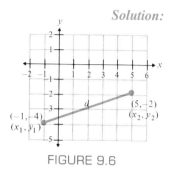

FIGURE 9.6

$$d = \sqrt{(x_2 - x_1)^2 + (y_2 - y_1)^2}$$
$$= \sqrt{[5 - (-1)]^2 + [-2 - (-4)]^2}$$
$$= \sqrt{(5 + 1)^2 + (-2 + 4)^2}$$
$$= \sqrt{6^2 + 2^2}$$
$$= \sqrt{36 + 4} = \sqrt{40}, \quad \text{or} \quad \text{approximately } 6.32$$

Thus the distance between $(-1, -4)$ and $(5, -2)$ is approximately 6.32 units. ■

Other Applications

▶ **3** Radicals are often used in science and mathematics courses. Examples 5 through 7 illustrate some scientific applications of radicals.

EXAMPLE 5 During the sixteenth and seventeenth centuries, Galileo Galilei performed many experiments with objects falling freely under the influence of gravity. He showed, for example, that a rock dropped from, say, 10 feet, hit the ground with a higher velocity than did the same rock dropped from 5 feet. A formula that can be used to determine the velocity of an object (neglecting air resistance) after it has fallen a certain height is

$$v = \sqrt{2gh}$$

where g is the acceleration of gravity and h is the height the object has fallen. On Earth the acceleration of gravity, g, is approximately 32 feet per second squared.
(a) Find the velocity of a rock after it has fallen 5 feet.
(b) Find the velocity of a coffee mug after it has fallen 10 feet.
(c) Find the velocity of a piano after it has fallen 100 feet.

Solution: (a) Begin by substituting 32 for g in the given equation.

$$v = \sqrt{2gh} = \sqrt{2(32)h} = \sqrt{64h}$$

At $h = 5$ feet,

$$v = \sqrt{64(5)} = \sqrt{320} \approx 17.9 \text{ feet per second}$$

After a rock has fallen 5 feet, its velocity is approximately 17.9 feet per second.
(b) After falling 10 feet,

$$v = \sqrt{64(10)} = \sqrt{640} \approx 25.3 \quad \text{feet per second}$$

The velocity of a coffee mug after falling 10 feet is approximately 25.3 feet per second.
(c) After falling 100 feet,

$$v = \sqrt{64(100)} = \sqrt{6400} = 80 \text{ feet per second}$$

The velocity of a piano after falling 100 feet is 80 feet per second. ■

EXAMPLE 6   The formula for the period in seconds, T, of a pendulum (the time required for the pendulum to make one complete swing both back and forth) is $T = 2\pi \sqrt{L/32}$, where L is the length of the pendulum in feet (see Fig. 9.7). Find the period of the pendulum if its length is 8 feet. Use 3.14 as an approximation for π.

Solution:

$$T = 2(3.14)\sqrt{\dfrac{8}{32}}$$

$$= 6.28\sqrt{\dfrac{1}{4}}$$

$$= 6.28\left(\dfrac{1}{2}\right) = 3.14 \text{ seconds}$$

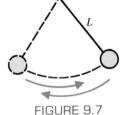

FIGURE 9.7

A pendulum 8 feet long takes 3.14 seconds to make one complete swing.   ■

EXAMPLE 7   For any planet, its "year" is the time it takes for the planet to revolve once around the sun. The number of "Earth days" in a given planet's year, N, is approximated by the formula

$$N = 0.2(\sqrt{R})^3$$

where R is the mean distance from the sun in millions of kilometers. Find the number of Earth days in the year of each of the planets illustrated in Fig. 9.8.

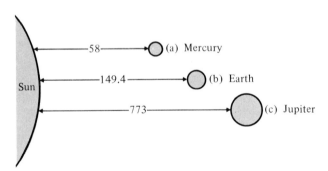

FIGURE 9.8

Solution: (a) Mercury:

$$N = 0.2(\sqrt{58})^3$$
$$\approx 0.2(7.6)^3$$
$$\approx 0.2(441.7)$$
$$\approx 88.3$$

It takes Mercury about 88 Earth days to revolve once around the sun.

(b) Earth:

$$N = 0.2(\sqrt{149.4})^3$$
$$\approx 0.2(12.2)^3$$
$$\approx 0.2(1826.1)$$
$$\approx 365.2$$

It takes Earth about 365 Earth days to revolve once around the sun (an answer that should not be surprising).

(c) Jupiter:

$$N \approx 0.2(\sqrt{773})^3$$
$$\approx 0.2(27.8)^3$$
$$\approx 0.2(21,484.9)$$
$$\approx 4297$$

It takes Jupiter about 4297 Earth days (or about 11.8 of Earth's years) to revolve once about the sun. ■

Exercise Set 9.6

Use the Pythagorean theorem to find the quantity indicated. You may leave your answer in square root form if a calculator with a square root key is not available. If a calculator is available, round answers to the nearest hundredth.

1.

2.

3.

4.

5.

6.

7.

8.

9.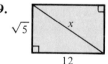

10.

11.

12.

13. A football field is 120 yards long from end zone to end zone. Find the length of the diagonal from one end zone to the other if the width of the field is 53.3 yards.

14. A boxing ring is a square 16 feet by 16 feet (actual ring size will vary with country and state). Find the distance from one boxer's corner to the other boxer's corner.

15. How long a length of wire is needed to reach from the top of a 4-meter telephone pole to a point 1.5 meters from the base of the pole?

16. An 8-meter extension ladder is placed against a house. The base of the ladder is 2 meters from the house. How high is the top of the ladder?

See Exercise 13.

Use the distance formula to answer questions 17 through 20. You may leave your answer in square root form if a calculator with a square root key is not available.

17. Find the length of the straight line between the points $(-4, 3)$ and $(-1, 4)$ (see Example 4).

18. Find the length of the straight line between the points $(4, -3)$ and $(6, 2)$.

19. Find the length of the straight line between the points $(-8, 4)$ and $(4, -8)$.

20. Find the length of the straight line between the points $(0, 5)$ and $(-6, -4)$.

Solve each problem. You may leave your answer in square root form if a calculator with a square root key is not available.

21. Find the side of a square that has an area of 144 square inches. Use $A = s^2$.

22. A formula for the area of a circle is $A = \pi r^2$, where π is approximately 3.14 and r is the radius of the circle. Find the radius of a circle of area 20 square inches.

23. Find the radius of a circle with an area of 80 square feet.

24. Find the length of the diagonal of a rectangle with a length of 12 inches and width of 5 inches.

25. Find the length of the diagonal of a rectangle with a length of 9 inches and width of 12 inches.

26. Find the length of the diagonal of a rectangle with a length of 25 inches and width of 10 inches.

27. Find the velocity of a lamp after it has fallen 80 feet (use $v = \sqrt{2gh}$, refer to Example 5).

28. Find the velocity of a plate after it has fallen 1000 feet.

29. Find the velocity with which an olive dropped from the Empire State Building, height 1250 feet, will strike the ground.

30. With what velocity will a suitcase dropped from a plane 1 mile (5280 feet) high strike the ground?

31. Find the period of a 40-foot pendulum (use $T = 2\pi\sqrt{L/32}$, refer to Example 6.)

32. Find the period of a 60-foot pendulum.

33. Find the period of a 10-foot pendulum.

34. Find the number of Earth days in the year of planet Mars, whose mean distance from the sun is 227 million kilometers [use $N = 0.2(\sqrt{R})^3$, and refer to Example 7].

35. Find the number of Earth days in the year of the planet Saturn, whose mean distance from the sun is 1418 million kilometers.

36. When two forces, F_1 and F_2, pull at right angles to each other as illustrated, the resultant, or the effective force, R, can be found by the formula

$$R = \sqrt{F_1^2 + F_2^2}$$

Two cars at a 90° angle to each other are trying to pull a third out of the mud, as shown. If car A is exerting a force of 600 pounds and car B is exerting a force of 800 pounds, find the resulting force on the car stuck in the mud.

37. The escape velocity in meters per second, or the velocity needed for a spacecraft to escape a planet's gravitational field, is found by the formula

$$v_e = \sqrt{2gR}$$

where g is the force of gravity of the planet and R is the radius of the planet in meters. Find the escape velocity for Earth where $g = 9.75$ meters per second squared and $R = 6,370,000$ meters.

Cumulative Review Exercises

[2.7] **38.** Solve the inequality $2(x + 3) < 4x - 6$.

[4.2] **39.** Simplify $(4x^{-4}y^3)^{-1}$.

[4.4] **40.** Simplify
$$x^3 - 2x^2 - 6x + 4 - (3x^3 - 6x^2 + 8).$$

[4.6] **41.** Divide $\dfrac{5x^4 - 9x^3 + 6x^2 - 4x - 3}{3x^2}$.

JUST FOR FUN

1. The length of a rectangle is 3 inches more than its width. If the length of the diagonal is 15 inches, find the dimensions of the rectangle.
2. The force of gravity on the moon is $\frac{1}{6}$ of that on Earth. If a camera falls from a rocket 100 feet above the surface of the moon, <u>with</u> what velocity will it strike the moon? Use $v = \sqrt{2gh}$; see Example 5.
3. Find the length of a <u>pendulum</u> if the period is to be 2 seconds (use $T = 2\pi \sqrt{L/32}$; refer to Example 6.)
4. The length of the diagonal of a rectangular solid (see the accompanying figure) is given by

$$d = \sqrt{a^2 + b^2 + c^2}$$

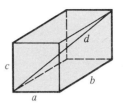

Find the length of the diagonal of a suitcase of length 37 inches, width 15 inches, and depth 9 inches.

9.7

Higher Roots and Fractional Exponents (Optional)

▸**1** Evaluate cube and fourth roots.

▸**2** Simplify cube and fourth roots.

▸**3** Write radical expressions in exponential form.

▸**1** In this section we will use the same basic concepts used in Sections 9.1 through 9.4 to work with roots that contain indexes of 3 and 4. In this section we introduce cube and fourth roots.

$$\sqrt[3]{a} \text{ is read "the cube root of } a\text{"}$$
$$\sqrt[4]{a} \text{ is read "the fourth root of } a\text{"}$$

Note that

$$\sqrt[3]{a} = b \quad \text{if} \quad b^3 = a$$

and

$$\sqrt[4]{a} = b \quad \text{if} \quad b^4 = a, \qquad b > 0$$

Examples

$$\sqrt[3]{8} = 2 \qquad \text{since } 2^3 = 2 \cdot 2 \cdot 2 = 8$$
$$\sqrt[3]{-8} = -2 \qquad \text{since } (-2)^3 = (-2)(-2)(-2) = -8$$
$$\sqrt[3]{27} = 3 \qquad \text{since } 3^3 = 3 \cdot 3 \cdot 3 = 27$$
$$\sqrt[4]{16} = 2 \qquad \text{since } 2^4 = 2 \cdot 2 \cdot 2 \cdot 2 = 16$$
$$\sqrt[4]{81} = 3 \qquad \text{since } 3^4 = 3 \cdot 3 \cdot 3 \cdot 3 = 81$$

EXAMPLE 1 Evaluate: (a) $\sqrt[3]{-27}$ (b) $\sqrt[3]{125}$

Solution: (a) To find $\sqrt[3]{-27}$, we must find the number that when cubed gives -27.

$$\sqrt[3]{-27} = -3 \qquad \text{since } (-3)^3 = -27$$

(b) To find $\sqrt[3]{125}$, we must find the number that when cubed gives 125.

$$\sqrt[3]{125} = 5 \qquad \text{since } 5^3 = 125$$

Note that the cube root of a positive number is a positive number and the cube root of a negative number is a negative number. The radicand of a fourth root (or any even root) must be a __nonnegative number__ for the expression to be a real number. For example, $\sqrt[4]{-16}$ is not a real number because no real number raised to the fourth power can be a negative number.

 Calculator Corner

Finding Cube and Higher Roots on a Calculator

Some calculators contain a key that finds cube roots and higher roots. If your calculator contains a $\boxed{x^y}$ or $\boxed{y^x}$ key, then you can find these roots. To find cube and higher roots, you need to use both the inverse key, $\boxed{\text{inv}}$, and either the $\boxed{x^y}$ or $\boxed{y^x}$ key. (Some calculators use a $\boxed{\text{2nd}}$ key in place of the $\boxed{\text{inv}}$ key.)

To find the value of $\sqrt[3]{216}$ using a calculator, press the following keys:

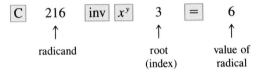

Thus $\sqrt[3]{216} = 6$. If your calculator has a $\boxed{y^x}$ key, use that key instead of the $\boxed{x^y}$ key in the preceding sequence. To find the value of $\sqrt[4]{618}$ using a calculator, press

$\boxed{\text{C}}$ 618 $\boxed{\text{inv}}$ $\boxed{x^y}$ 4 $\boxed{=}$ 4.98594

Thus $\sqrt[4]{618} \approx 4.98594$.

To find an odd root of a negative number, find the odd root of that positive number and then place a negative sign before the answer. For example, to find $\sqrt[3]{-64}$, find $\sqrt[3]{64}$, which is 4; then place a negative sign before the answer to get -4. Thus $\sqrt[3]{-64} = -4$.

It will be helpful in the explanations that follow if we define perfect cube numbers. A **perfect cube number** is a number that is the cube of a natural number.

$$1, \quad 2, \quad 3, \quad 4, \quad 5, \quad 6, \quad 7, \quad 8, \quad 9, \quad 10, \ldots \quad \text{Natural numbers}$$
$$1^3, 2^3, 3^3, 4^3, \quad 5^3, \quad 6^3, \quad 7^3, \quad 8^3, \quad 9^3, \quad 10^3, \ldots$$
$$1, \ 8, \ 27, \ 64, \quad 125, \ 216, \ 343, \ 512, \ 729, \ 1000, \ldots \quad \text{Perfect cube numbers}$$

Note that $\sqrt[3]{1} = 1$, $\sqrt[3]{8} = 2$, $\sqrt[3]{27} = 3$, $\sqrt[3]{64} = 4$, and so on.

Perfect fourth power numbers can be expressed in a similar manner.

$$1, \quad 2, \quad 3, \quad 4, \quad 5, \quad 6, \ldots \quad \text{Natural numbers}$$
$$1^4, 2^4, \ 3^4, \ 4^4, \quad 5^4, \quad 6^4, \ldots$$
$$1, \ 16, \ 81, \ 256, \ 625, \ 1296, \ldots \quad \text{Perfect fourth power numbers}$$

Note that $\sqrt[4]{1} = 1$, $\sqrt[4]{16} = 2$, $\sqrt[4]{81} = 3$, $\sqrt[4]{256} = 4$, and so on.

You may wish to refer back to these numbers when evaluating cube and fourth roots.

▶ **2**   The product rule used in simplifying square roots can be expanded for indexes greater than 2. We will again use the product rule to simplify radicals.

> **Product Rule for Radicals**
>
> $$\sqrt[n]{a}\,\sqrt[n]{b} = \sqrt[n]{ab}, \qquad \text{for } a \geq 0,\ b \geq 0$$

To simplify a cube root whose radicand is a number, write the radicand as the product of a perfect cube number and another number. Then simplify, using the product rule.

EXAMPLE 2   Simplify: (a) $\sqrt[3]{32}$ (b) $\sqrt[3]{54}$ (c) $\sqrt[4]{32}$

Solution: (a) $\sqrt[3]{32} = \sqrt[3]{8 \cdot 4} = \sqrt[3]{8}\sqrt[3]{4} = 2\sqrt[3]{4}$
(b) $\sqrt[3]{54} = \sqrt[3]{27 \cdot 2} = \sqrt[3]{27}\sqrt[3]{2} = 3\sqrt[3]{2}$
(c) Write $\sqrt[4]{32}$ as a product of a perfect fourth power number and another number, then simplify.

$$\sqrt[4]{32} = \sqrt[4]{16 \cdot 2} = \sqrt[4]{16}\sqrt[4]{2} = 2\sqrt[4]{2}$$ ∎

▶ **3**   A radical expression can be written in **exponential form** by using the following rule.

> $$\sqrt[n]{a} = a^{1/n}, \qquad a \geq 0 \qquad \text{Rule 4}$$

Examples

$$\sqrt{8} = 8^{1/2} \qquad\qquad \sqrt{x} = x^{1/2}$$
$$\sqrt[3]{4} = 4^{1/3} \qquad\qquad \sqrt[4]{9} = 9^{1/4}$$
$$\sqrt[3]{x} = x^{1/3} \qquad\qquad \sqrt[4]{y} = y^{1/4}$$
$$\sqrt[3]{5x^2} = (5x^2)^{1/3} \qquad \sqrt[4]{3y^2} = (3y^2)^{1/4}$$

Notice $\sqrt{8} = 8^{1/2}$ and $\sqrt{x} = x^{1/2}$, which is consistent with what we learned in Section 9.1. This concept just given can be expanded as follows.

> Power ↘   ↙ Index
> $$\sqrt[n]{a^m} = (\sqrt[n]{a})^m = a^{m/n}, \text{ for } a \geq 0 \text{ and } m \text{ and } n \text{ integers} \qquad \text{Rule 5}$$

As long as the radicand is nonnegative, we can change from one form to another.

Examples

$$\sqrt[3]{8^4} = (\sqrt[3]{8})^4 = 2^4 = 16 \qquad \sqrt[3]{x^3} = x^{3/3} = x^1 = x$$
$$27^{2/3} = (\sqrt[3]{27})^2 = 3^2 = 9 \qquad \sqrt[4]{y^{12}} = y^{12/4} = y^3$$

EXAMPLE 3   Simplify: (a) $\sqrt[3]{y^{15}}$ (b) $\sqrt[4]{x^{24}}$

Solution: (a) $\sqrt[3]{y^{15}} = y^{15/3} = y^5$
(b) $\sqrt[4]{x^{24}} = x^{24/4} = x^6$ ∎

EXAMPLE 4   Evaluate: (a) $8^{5/3}$ (b) $16^{5/4}$ (c) $8^{-2/3}$

Solution: (a) $8^{5/3} = (\sqrt[3]{8})^5 = 2^5 = 32$

(b) $16^{5/4} = (\sqrt[4]{16})^5 = 2^5 = 32$

(c) Recall from Section 5.2 that $x^{-m} = \dfrac{1}{x^m}$. Thus

$$8^{-2/3} = \frac{1}{8^{2/3}} = \frac{1}{(\sqrt[3]{8})^2} = \frac{1}{2^2} = \frac{1}{4}$$

COMMON STUDENT ERROR

Students often make mistakes simplifying expressions that contain negative exponents. Be careful when working such problems. The following is a common error.

Correct	Wrong
$27^{-2/3} = \dfrac{1}{27^{2/3}}$	$\cancel{27^{-2/3} = -27^{2/3}}$

The expression $27^{-2/3}$ simplifies to $\dfrac{1}{9}$. Can you show this?

EXAMPLE 5   Write each of the following radicals in exponential form.

(a) $\sqrt[3]{x^5}$ (b) $\sqrt[4]{y^7}$ (c) $\sqrt[4]{z^{15}}$

Solution: (a) $\sqrt[3]{x^5} = x^{5/3}$

(b) $\sqrt[4]{y^7} = y^{7/4}$

(c) $\sqrt[4]{z^{15}} = z^{15/4}$

EXAMPLE 6   Simplify: (a) $\sqrt{x} \cdot \sqrt[4]{x}$ (b) $(\sqrt[4]{x^2})^8$

Solution: (a) $\sqrt{x} \cdot \sqrt[4]{x} = x^{1/2} \cdot x^{1/4}$ (b) $(\sqrt[4]{x^2})^8 = (x^{2/4})^8$

$\qquad\qquad\qquad = x^{(1/2)+(1/4)}$ $\qquad = (x^{1/2})^8$

$\qquad\qquad\qquad = x^{(2/4)+(1/4)}$ $\qquad = x^4$

$\qquad\qquad\qquad = x^{3/4}$

$\qquad\qquad\qquad = \sqrt[4]{x^3}$

This section was meant to give you a brief introduction to roots other than square roots. If you take a course in intermediate algebra, you may study these concepts in more depth.

Exercise Set 9.7

Evaluate each of the following.

1. $\sqrt[3]{8}$ 2. $\sqrt[3]{27}$ 3. $\sqrt[3]{-8}$ 4. $\sqrt[3]{-27}$

5. $\sqrt[4]{16}$ 6. $\sqrt[3]{125}$ 7. $\sqrt[4]{81}$ 8. $\sqrt[4]{1}$

9. $\sqrt[3]{-1}$ 10. $\sqrt[3]{-125}$ 11. $\sqrt[3]{64}$ 12. $\sqrt[3]{-64}$

Simplify each of the following.

13. $\sqrt[3]{54}$ **14.** $\sqrt[3]{32}$ **15.** $\sqrt[3]{16}$ **16.** $\sqrt[3]{24}$

17. $\sqrt[3]{81}$ **18.** $\sqrt[3]{128}$ **19.** $\sqrt[4]{32}$ **20.** $\sqrt[3]{250}$

21. $\sqrt[3]{40}$ **22.** $\sqrt[4]{48}$

Simplify each of the following.

23. $\sqrt[3]{x^3}$ **24.** $\sqrt[3]{y^6}$ **25.** $\sqrt[4]{y^{12}}$ **26.** $\sqrt[4]{y^{16}}$

27. $\sqrt[3]{x^{12}}$ **28.** $\sqrt[3]{x^9}$ **29.** $\sqrt[3]{y^4}$ **30.** $\sqrt[4]{y^{24}}$

31. $\sqrt[3]{x^{15}}$ **32.** $\sqrt[3]{x^{18}}$

Evaluate each of the following.

33. $8^{4/3}$ **34.** $27^{4/3}$ **35.** $16^{3/4}$ **36.** $81^{3/4}$

37. $1^{5/3}$ **38.** $16^{5/2}$ **39.** $9^{3/2}$ **40.** $64^{2/3}$

41. $16^{3/4}$ **42.** $25^{3/2}$ **43.** $125^{4/3}$ **44.** $8^{-1/3}$

45. $27^{-2/3}$ **46.** $16^{-3/4}$ **47.** $8^{-5/3}$ **48.** $64^{-2/3}$

Write each radical in exponential form.

49. $\sqrt[3]{x^7}$ **50.** $\sqrt[3]{x^6}$ **51.** $\sqrt[3]{x^4}$ **52.** $\sqrt[4]{x^7}$

53. $\sqrt[4]{y^{15}}$ **54.** $\sqrt[4]{x^9}$ **55.** $\sqrt[4]{y^{21}}$ **56.** $\sqrt[4]{x^5}$

Simplify each of the following and write the answer in exponential form.

57. $\sqrt[3]{x} \cdot \sqrt[3]{x}$ **58.** $\sqrt[4]{x} \cdot \sqrt[4]{x}$ **59.** $\sqrt[4]{x^2} \cdot \sqrt[4]{x^2}$ **60.** $\sqrt[3]{x} \cdot \sqrt[3]{x^5}$

61. $(\sqrt[3]{x^2})^6$ **62.** $(\sqrt[4]{x^3})^4$ **63.** $(\sqrt[4]{x^2})^4$ **64.** $(\sqrt[3]{x^6})^2$

65. Show that for $x = 8$, $(\sqrt[3]{x})^2 = \sqrt[3]{x^2}$

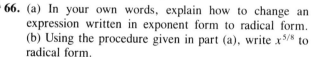 **66.** (a) In your own words, explain how to change an expression written in exponent form to radical form.
(b) Using the procedure given in part (a), write $x^{5/8}$ to radical form.

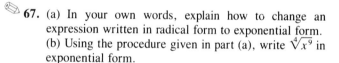 **67.** (a) In your own words, explain how to change an expression written in radical form to exponential form.
(b) Using the procedure given in part (a), write $\sqrt[4]{x^9}$ in exponential form.

Cumulative Review Exercises

[1.9] **68.** Evaluate $-x^2 + 4xy - 6$ when $x = 2$ and $y = -4$.

[5.4] **69.** Factor $3x^2 - 28x + 32$.

[7.2] **70.** Graph $y = \dfrac{2}{3}x - 4$.

[9.3] **71.** Simplify $\sqrt{\dfrac{64x^3y^7}{2x^4}}$.

JUST FOR FUN

Simplify each of the following:

1. $\sqrt[3]{xy} \cdot \sqrt[3]{x^2y^2}$ **2.** $\sqrt[4]{3x^2y} \cdot \sqrt[4]{27x^6y^3}$

3. $\sqrt[4]{32} - \sqrt[4]{2}$ **4.** $\sqrt[3]{3x^3y} + \sqrt[3]{24x^3y}$

5. $\dfrac{1}{\sqrt[3]{2}}$ **6.** $\dfrac{1}{\sqrt[3]{x}}$

SUMMARY

GLOSSARY

Conjugate *(393):* The conjugate of $a + b$ is $a - b$.

Hypotenuse *(402):* The side opposite the right angle in a right triangle.

Index of a radical *(376):* The root of a radical expression.

Irrational numbers *(378):* Real numbers that are not rational numbers.

Legs of a right triangle *(402):* The two smaller sides of the right triangle.

Like square roots *(391):* Square roots having the same radicand.

Perfect cube numbers *(410):* 1, 8, 27, 64, 125, 216, 343, 512, . . .

Perfect square factor *(377):* When a radicand contains a variable (or number) raised to an even exponent, that variable (or number) and exponent together form a perfect square factor.

Perfect square numbers *(377):* 1, 4, 9, 16, 25, 36, 49, 64, . . .

Principal or positive square root *(376):* The principal or positive square root of a positive real number x, written $\sqrt{x}$, is that positive number whose square equals x.

Radical equation *(397):* An equation that contains a variable in a radicand.

Radical expression *(376):* A mathematical expression containing a radical.

Radical sign *(376):* $\sqrt{}$.

Radicand *(376):* The expression within the radical sign.

Rational number *(378):* A number that can be written in the form $\dfrac{a}{b}$, where a and b are both integers, $b \neq 0$.

Rationalize the denominator *(388):* Removing radical expressions from the denominator of a fraction.

Right triangle *(402):* A triangle with a 90° angle.

Unlike square roots *(392):* Square roots having different radicands.

IMPORTANT FACTS

Product rule for radicals: $\sqrt{a} \cdot \sqrt{b} = \sqrt{ab}, \quad a \geq 0, b \geq 0$

$\sqrt{a^{2 \cdot n}} = a^n, \quad a \geq 0$

$\sqrt{a^2} = a, \quad a \geq 0$

Quotient rule for radicals: $\dfrac{\sqrt{a}}{\sqrt{b}} = \sqrt{\dfrac{a}{b}}, \quad a \geq 0, b > 0$

$\sqrt[n]{a} = a^{1/n}, \quad a \geq 0$

$\sqrt[n]{a^m} = \left(\sqrt[n]{a}\right)^m = a^{m/n}, \quad a \geq 0$

Pythagorean theorem: $a^2 + b^2 = c^2$

Distance formula: $d = \sqrt{(x_2 - x_1)^2 + (y_2 - y_1)^2}$

Review Exercises

[9.1] *Evaluate each expression.*

1. $\sqrt{25}$

2. $\sqrt{36}$

3. $-\sqrt{81}$

Write each expression in exponential form.

4. $\sqrt{8}$

5. $\sqrt{26x}$

6. $\sqrt{20xy^2}$

[9.2] *Simplify each expression.*

7. $\sqrt{32}$

8. $\sqrt{44}$

9. $\sqrt{45x^5y^4}$

10. $\sqrt{125x^4y^6}$

11. $\sqrt{15x^5yz^3}$

12. $\sqrt{48ab^4c^5}$

Simplify each expression.

13. $\sqrt{8} \cdot \sqrt{12}$

14. $\sqrt{5x} \cdot \sqrt{5x}$

15. $\sqrt{18x} \cdot \sqrt{2xy}$

16. $\sqrt{25x^2y} \cdot \sqrt{3y}$

17. $\sqrt{20xy^4} \cdot \sqrt{5xy^3}$

18. $\sqrt{8x^3y} \cdot \sqrt{3y^4}$

[9.3] *Simplify each expression.*

19. $\dfrac{\sqrt{32}}{\sqrt{2}}$

20. $\sqrt{\dfrac{10}{250}}$

21. $\sqrt{\dfrac{7}{28}}$

22. $\dfrac{3}{\sqrt{5}}$

23. $\sqrt{\dfrac{5x}{12}}$

24. $\sqrt{\dfrac{x}{6}}$

25. $\sqrt{\dfrac{x^2}{2}}$

26. $\sqrt{\dfrac{x^5}{8}}$

27. $\sqrt{\dfrac{60xy^5}{4x^5y^3}}$

28. $\sqrt{\dfrac{30x^4y}{15x^2y^4}}$

29. $\dfrac{\sqrt{90}}{\sqrt{8x^3y^2}}$

30. $\dfrac{\sqrt{2x^4yz^4}}{\sqrt{7x^5yz^2}}$

31. $\dfrac{3}{1+\sqrt{2}}$

32. $\dfrac{5}{3-\sqrt{6}}$

33. $\dfrac{\sqrt{3}}{2+\sqrt{x}}$

34. $\dfrac{2}{\sqrt{x}-5}$

35. $\dfrac{\sqrt{5}}{\sqrt{x}+\sqrt{3}}$

[9.4] *Simplify each expression.*

36. $6\sqrt{3}-2\sqrt{3}$

37. $6\sqrt{2}-8\sqrt{2}+\sqrt{2}$

38. $3\sqrt{x}-5\sqrt{x}$

39. $\sqrt{x}+3\sqrt{x}-4\sqrt{x}$

40. $\sqrt{8}-\sqrt{2}$

41. $7\sqrt{40}-2\sqrt{10}$

42. $2\sqrt{98}-4\sqrt{72}$

43. $3\sqrt{18}+5\sqrt{50}-2\sqrt{32}$

44. $4\sqrt{27}+5\sqrt{80}+2\sqrt{12}$

[9.5] *Solve each equation.*

45. $\sqrt{x}=9$

46. $\sqrt{x}=-2$

47. $\sqrt{x-3}=6$

48. $\sqrt{3x+1}=5$

49. $\sqrt{2x+4}=\sqrt{3x-5}$

50. $4\sqrt{x}-x=4$

51. $\sqrt{x^2+4}=x+2$

52. $\sqrt{3x+5}-\sqrt{5x-9}=0$

53. $3\sqrt{2x+3}=9$

[9.6] *Find the quantity indicated. You may leave your answer in square root form if the answer is not a perfect square.*

54.

55.

56.

57.

58. Jason leans a 12-foot ladder against a house. If the base of the ladder is 3 feet from the house, how high is the ladder on the house?

59. Find the diagonal of a rectangle with a length of 15 inches and a width of 6 inches.

60. Find the straight-line distance between the points $(-5, 4)$ and $(3, -2)$.

61. Find the straight-line distance between the points $(6, 5)$ and $(-6, 8)$.

62. Find the side of a square that has an area of 121 square feet.

[9.7] *Evaluate.*

63. $\sqrt[3]{8}$

64. $\sqrt[3]{-27}$

65. $\sqrt[4]{16}$

Simplify each of the following.

66. $\sqrt[3]{16}$

67. $\sqrt[3]{24}$

68. $\sqrt[4]{32}$

69. $\sqrt[3]{48}$

70. $\sqrt[3]{54}$

71. $\sqrt[4]{96}$

72. $\sqrt[3]{x^{15}}$

73. $\sqrt[3]{x^{12}}$

74. $\sqrt[4]{y^{16}}$

75. $\sqrt[4]{y^{20}}$

Evaluate each of the following.

76. $8^{2/3}$

77. $16^{1/2}$

78. $27^{-2/3}$

79. $64^{2/3}$

80. $16^{-3/4}$

81. $25^{3/2}$

Write each radical in exponential form.

82. $\sqrt[3]{x^5}$

83. $\sqrt[3]{x^{10}}$

84. $\sqrt[4]{y^9}$

85. $\sqrt{x^5}$

86. $\sqrt{y^{11}}$

87. $\sqrt[4]{x^7}$

Simplify each of the following.

88. $\sqrt{x} \cdot \sqrt{x}$

89. $\sqrt[3]{x} \cdot \sqrt[3]{x}$

90. $\sqrt[3]{x^2} \cdot$

91. $\sqrt[4]{x^2} \cdot \sqrt[4]{x^6}$

92. $(\sqrt[3]{x^3})^2$

93. $(\sqrt[3]{x^2})^3$

94. $(\sqrt[4]{x^2})^6$

95. $(\sqrt[4]{x^3})^8$

Practice Test

1. Write $\sqrt{3xy}$ in exponential form.

Simplify each expression.

2. $\sqrt{(x + 3)^2}$

3. $\sqrt{96}$

4. $\sqrt{12x^2}$

5. $\sqrt{32x^4y^5}$

6. $\sqrt{8x^2y} \cdot \sqrt{6xy}$

7. $\sqrt{15xy^2} \cdot \sqrt{5x^3y^3}$

8. $\sqrt{\dfrac{5}{125}}$

9. $\dfrac{\sqrt{3xy^2}}{\sqrt{48x^3}}$

10. $\dfrac{1}{\sqrt{2}}$

11. $\sqrt{\dfrac{4x}{5}}$

12. $\sqrt{\dfrac{40x^2y^5}{6x^3y^7}}$

13. $\dfrac{3}{2 + \sqrt{5}}$

14. $\dfrac{6}{\sqrt{x} - 3}$

15. $\sqrt{48} + \sqrt{75} + 2\sqrt{3}$

16. $4\sqrt{x} - 6\sqrt{x} - \sqrt{x}$

Solve each equation.

17. $\sqrt{x + 5} = 9$

18. $2\sqrt{x - 4} + 4 = x$

Solve each problem.

19. Find the value of x in the triangle shown.

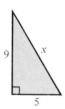

20. Find the length of the straight line between the points $(-2, -4)$ and $(5, 1)$.

*21. Evaluate $27^{-4/3}$

*22. Simplify $\sqrt[3]{x^4} \cdot \sqrt[3]{x^{11}}$.

10

Quadratic Equations

10.1 The Square Root Property

10.2 Solving Quadratic Equations by Completing the Square

10.3 Solving Quadratic Equations by the Quadratic Formula

10.4 Graphing Quadratic Equations

Summary

Review Exercises

Practice Test

Cumulative Review Test

See Section 10.1, Example 6.

10.1

The Square Root Property

▶**1** Know that every positive real number has two square roots.

▶**2** Solve quadratic equations using the square root property.

In Section 5.6 we solved quadratic equations by factoring. Recall that **quadratic equations** are equations of the form

$$ax^2 + bx + c = 0,$$

where a, b, and c are real numbers, $a \neq 0$. A quadratic equation in this form is said to be in **standard form.**

In this chapter we will give two techniques, completing the square and the quadratic formula, for solving quadratic equations that cannot be solved by factoring. In this section we introduce the square root property, which will be used in the completing the square procedure in the next section.

▶**1** In Section 9.1 we stated that every positive number has two square roots. Thus far we have been using only the positive or principal square root. In this section we use both the positive and negative square roots of a number.

The positive square root of 25 is 5.

$$\sqrt{25} = 5$$

The negative square root of 25 is -5.

$$-\sqrt{25} = -5$$

Notice that $5 \cdot 5 = 25$ and $(-5)(-5) = 25$. The two square roots of 25 are $+5$ and -5. A convenient way to indicate the two square roots of a number is to use the plus or minus symbol, $\pm$. For example, the square roots of 25 can be indicated ± 5, read "plus or minus 5."

Number	*Both Square Roots*
36	± 6
100	± 10
7	$\pm \sqrt{7}$

The value of a number like $-\sqrt{5}$ can be found by finding the value of $\sqrt{5}$ on your calculator, or in Appendix D, and then taking its opposite or negative value.

$$\sqrt{5} = 2.24 \qquad \text{(rounded to two decimal places)}$$
$$-\sqrt{5} = -2.24$$

Consider the equation

$$x^2 = 25$$

We can see by substitution that this equation has two solutions, 5 and -5.

Check:
$$
\begin{array}{cc}
x = 5 & x = -5 \\
x^2 = 25 & x^2 = 25 \\
5^2 = 25 & (-5)^2 = 25 \\
25 = 25 \quad \text{true} & 25 = 25 \quad \text{true}
\end{array}
$$

Therefore, the solutions to the equation $x^2 = 25$ are 5 or -5 (or ± 5).

▶ **2**   In general, for any quadratic equation of the form $x^2 = a$, we can use the square root property to obtain the solution.

Square Root Property

$$\text{If} \quad x^2 = a$$
$$\text{then} \quad x = \sqrt{a} \quad \text{or} \quad x = -\sqrt{a} \ (\text{or} \quad x = \pm\sqrt{a})$$

For example, if $x^2 = 7$, then by the square root property, $x = \sqrt{7}$ or $x = -\sqrt{7}$ (or $x = \pm\sqrt{7}$).

EXAMPLE 1   Solve the equation $x^2 - 9 = 0$.

Solution: Add 9 to both sides of the equation to get the variable all by itself on one side of the equation.

$$x^2 = 9$$

Now use the square root property.

$$x = \pm\sqrt{9}$$
$$x = \pm 3$$

Check in the original equation.

$$
\begin{array}{ll}
x = 3 & x = -3 \\
x^2 - 9 = 0 & x^2 - 9 = 0 \\
3^2 - 9 = 0 & (-3)^2 - 9 = 0 \\
9 - 9 = 0 & 9 - 9 = 0 \\
0 = 0 \quad \text{true} & 0 = 0 \quad \text{true}
\end{array}
$$

EXAMPLE 2   Solve the equation $x^2 + 7 = 71$.

Solution: Begin by subtracting 7 from both sides of the equation.

$$x^2 + 7 = 71$$
$$x^2 = 64$$
$$x = \pm\sqrt{64}$$
$$x = \pm 8$$

EXAMPLE 3   Solve the equation $x^2 - 7 = 0$.

Solution:
$$x^2 - 7 = 0$$
$$x^2 = 7$$
$$x = \pm\sqrt{7}$$

EXAMPLE 4   Solve the equation $(x - 3)^2 = 4$.

Solution: Begin by using the square root property.

$$(x - 3)^2 = 4$$
$$x - 3 = \pm\sqrt{4}$$
$$x - 3 = \pm 2$$
$$x - 3 + 3 = 3 \pm 2 \qquad \text{Add 3 to both sides of the equation.}$$
$$x = 3 \pm 2$$
$$x = 3 + 2 \qquad \text{or} \qquad x = 3 - 2$$
$$x = 5 \qquad \text{or} \qquad x = 1$$

The solutions are 1 and 5.   ■

EXAMPLE 5   Solve the equation $(3x + 4)^2 = 32$.

Solution:
$$(3x + 4)^2 = 32$$
$$3x + 4 = \pm\sqrt{32}$$
$$3x + 4 = \pm 4\sqrt{2}$$
$$3x = -4 \pm 4\sqrt{2}$$
$$x = \frac{-4 \pm 4\sqrt{2}}{3}$$

Thus the solutions are $\dfrac{-4 + 4\sqrt{2}}{3}$ and $\dfrac{-4 - 4\sqrt{2}}{3}$.   ■

EXAMPLE 6   Mrs. Albert wants a rectangular garden. To make her garden look most appealing, the length of the rectangle is to be 1.62 times its width (a rectangle with this length-to-width ratio is referred to as the *golden rectangle*); see Fig. 10.1. Find the dimensions of the rectangle if the rectangle is to have an area of 6000 square feet.

Solution: Let x = width of rectangle; then $1.62x$ = length of rectangle.

$$\text{Area} = \text{length} \cdot \text{width}$$
$$6000 = (1.62x)x$$
$$6000 = 1.62x^2$$
$$x^2 = \frac{6000}{1.62} \approx 3703.7$$
$$x \approx \pm\sqrt{3703.7} \approx \pm 60.86 \text{ feet}$$

x

$1.62x$

FIGURE 10.1

Since the distance cannot be negative, the width, x, is 60.86 feet. The length is $1.62(60.86) = 98.60$ feet.

Check: area = length · width
$$6000 = (98.60)(60.86)$$
$$6000 \approx 6000.79 \qquad \text{true} \quad \text{(There is a slight round-off error due to rounding off decimal answers.)}$$ ■

Note that the answer did not come out to be a whole number. In many real-life situations this is the case. You should not feel uncomfortable when this occurs. If you do not have a calculator with a square root $\boxed{\sqrt{}}$ key, leave your answers in terms of square roots.

Exercise Set 10.1

Solve each equation.

1. $x^2 = 16$ **2.** $x^2 = 25$ **3.** $x^2 = 100$

4. $x^2 = 49$ **5.** $y^2 = 36$ **6.** $z^2 = 9$

7. $x^2 = 10$ **8.** $a^2 = 15$ **9.** $x^2 = 8$

10. $w^2 = 24$ **11.** $3x^2 = 12$ **12.** $5y^2 = 45$

13. $2w^2 = 34$ **14.** $5x^2 = 90$ **15.** $2x^2 + 1 = 19$

16. $3x^2 - 4 = 8$ **17.** $4w^2 - 3 = 12$ **18.** $3y^2 + 8 = 36$

19. $5x^2 - 9 = 30$ **20.** $2x^2 + 3 = 51$

Solve the equation.

21. $(x + 1)^2 = 4$ **22.** $(x - 2)^2 = 9$ **23.** $(x - 3)^2 = 16$

24. $(x + 5)^2 = 25$ **25.** $(x + 4)^2 = 36$ **26.** $(x - 4)^2 = 100$

27. $(x - 1)^2 = 12$ **28.** $(x + 3)^2 = 18$ **29.** $(x + 6)^2 = 20$

30. $(x - 4)^2 = 32$ **31.** $(x + 2)^2 = 25$ **32.** $(x + 6)^2 = 75$

33. $(x - 9)^2 = 100$ **34.** $(x - 3)^2 = 15$ **35.** $(2x + 3)^2 = 18$

36. $(3x - 2)^2 = 30$ **37.** $(4x + 1)^2 = 20$ **38.** $(5x - 6)^2 = 100$

39. $(2x - 6)^2 = 18$ **40.** $(3x - 5)^2 = 90$

41. The length of a rectangle is twice its width. Find the length and width if the area is 80 square feet.

42. The length of a rectangle is 3 times its width. Find the length and width if its area is 96 square feet.

43. Write an equation that has solutions of 6 and −6.

44. Write an equation that has solutions of $\sqrt{7}$ and $-\sqrt{7}$.

45. Fill in the shaded area to make a true statement. The equation $x^2 - \blacksquare = 27$ has solutions of 6 and −6. Explain how you determined your answer.

46. Fill in the shaded area to make a true statement. The equation $x^2 + \blacksquare = 45$ has solutions of 8 and −8. Explain how you determined your answer.

Cumulative Review Exercises

[5.4] **47.** Factor $4x^2 - 10x - 24$.

[6.7] **48.** Solve $\dfrac{3x - 7}{x - 4} = \dfrac{5}{x - 4}$.

[6.8] **49.** Collette invests $10,000 in two saving accounts. Part of the money is put into an account paying 6% simple interest. The rest is put into a savings account paying 8% simple interest. If the interest for the year from both accounts totals $760, how much money was in each account?

[7.4] **50.** Determine the equation of the line.

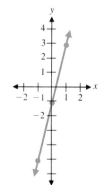

10.2

Solving Quadratic Equations by Completing the Square

▶ **1**   Write perfect square trinomials.

▶ **2**   Solve quadratic equations by completing the square.

Quadratic equations that cannot be solved using factoring can be solved by completing the square or by the quadratic formula. In this section we focus on the technique of completing the square. In Section 10.3 we will use the quadratic formula.

▶ **1**   A **perfect square trinomial** is a trinomial that can be expressed as the square of a binomial. Some examples follow.

Perfect Square Trinomials	Factors	Square of a Binomial
$x^2 + 6x + 9$	$= (x + 3)(x + 3)$	$= (x + 3)^2$
$x^2 - 6x + 9$	$= (x - 3)(x - 3)$	$= (x - 3)^2$
$x^2 + 10x + 25$	$= (x + 5)(x + 5)$	$= (x + 5)^2$
$x^2 - 10x + 25$	$= (x - 5)(x - 5)$	$= (x - 5)^2$

Notice that each of the squared terms in the preceding perfect square trinomials has a numerical coefficient of 1. When the coefficient of the squared term is 1, there is an important relationship between the coefficient of the x term and the constant. In every perfect square trinomial of this type, *the constant term is the square of one-half the coefficient of the x term.*

Consider the perfect square trinomial $x^2 + 6x + 9$. The coefficient of the x term is 6 and the constant is 9. Note that

$$\left[\frac{1}{2}(6) \right]^2 = 3^2 = 9$$

Consider the perfect square trinomial $x^2 - 10x + 25$. The coefficient of the x term is -10 and the constant is 25. Note that

$$\left[\frac{1}{2}(-10) \right]^2 = (-5)^2 = 25$$

Consider the expression $x^2 + 8x + \ \blacksquare$. Can you determine what number must be placed in the colored box to make the trinomial a perfect square trinomial? If you answered 16, you answered correctly. Note that

$$\left[\frac{1}{2}(8) \right]^2 = 4^2 = 16$$
$$x^2 + 8x + 16 = (x + 4)^2$$

Let us examine perfect square trinomials a little further.

Perfect Square Trinomial   *Square of a Binomial*

$$x^2 + 6x + 9 \quad = \quad (x + 3)^2$$

$$\frac{1}{2}(6) = 3$$

$$x^2 - 10x + 25 \quad = \quad (x - 5)^2$$

$$\frac{1}{2}(-10) = -5$$

Note that when a perfect square trinomial is written as the square of a binomial *the constant in the binomial is one-half the value of the coefficient of the x term in the perfect square trinomial.*

▶ **2**   The procedure for solving a quadratic equation by completing the square is illustrated in the following example.

EXAMPLE 1   Solve the equation $x^2 + 6x + 5 = 0$ by completing the square.

Solution: First we make sure that the squared term has a coefficient of 1. (We will explain what to do if the coefficient is not 1 in Example 5.) Next we wish to get the squared and x terms by themselves on the left side of the equation. Therefore, we subtract 5 from both sides of the equation.

$$x^2 + 6x + 5 = 0$$
$$x^2 + 6x = -5$$

Determine one-half the numerical coefficient of the x term. In this example the x term is $6x$.

$$\frac{1}{2}(6) = 3$$

Square this number.

$$(3)^2 = (3)(3) = 9$$

Add this product to both sides of the equation.

$$x^2 + 6x + 9 = -5 + 9$$

or

$$x^2 + 6x + 9 = 4$$

By following this procedure, we produce a perfect square trinomial on the left side of the equation. The expression $x^2 + 6x + 9$ is a perfect square trinomial that can be expressed as $(x + 3)^2$. Therefore,

$$x^2 + 6x + 9 = 4$$

can be written $(x + 3)^2 = 4$

Now take the square root of both sides of the equation.

$$x + 3 = \pm\sqrt{4}$$
$$x + 3 = \pm 2$$

Finally, solve for x by subtracting 3 from both sides of the equation.

$$x + 3 - 3 = -3 \pm 2$$
$$x = -3 \pm 2$$
$$x = -3 + 2 \quad \text{or} \quad x = -3 - 2$$
$$x = -1 \qquad\qquad x = -5$$

Thus the solutions are -1 and -5. Check both solutions in the original equation.

$$x = -1$$
$$x^2 + 6x + 5 = 0$$
$$(-1)^2 + 6(-1) + 5 = 0$$
$$1 - 6 + 5 = 0$$
$$0 = 0 \quad \text{true}$$

$$x = -5$$
$$x^2 + 6x + 5 = 0$$
$$(-5)^2 + 6(-5) + 5 = 0$$
$$25 - 30 + 5 = 0$$
$$0 = 0 \quad \text{true} \quad \blacksquare$$

To Solve a Quadratic Equation by Completing the Square

1. Use the multiplication (or division) property if necessary to make the numerical coefficient of the squared term equal to 1.
2. Rewrite the equation with the constant by itself on the right side of the equation.
3. Take one-half the numerical coefficient of the first-powered term, square it, and add this quantity to both sides of the equation.
4. Replace the trinomial with its equivalent squared binomial.
5. Take the square root of both sides of the equation.
6. Solve for the variable.
7. Check your answers in the original equation.

EXAMPLE 2   Solve the equation $x^2 - 8x + 15 = 0$ by completing the square.

Solution: $x^2 - 8x + 15 = 0$
$$x^2 - 8x = -15$$

Take half the numerical coefficient of the x term, square it, and add this product to both sides of the equation.

$$\frac{1}{2}(-8) = -4, \qquad (-4)^2 = 16$$

Now add 16 to both sides of the equation

$$x^2 - 8x + 16 = -15 + 16$$
$$x^2 - 8x + 16 = 1$$
$$\text{or} \qquad (x - 4)^2 = 1$$
$$x - 4 = \pm\sqrt{1}$$
$$x - 4 = \pm 1$$
$$x = 4 \pm 1$$
$$x = 4 + 1 \quad \text{or} \quad x = 4 - 1$$
$$x = 5 \qquad\qquad x = 3$$

A check will show that the solutions are 5 and 3. $\blacksquare$

EXAMPLE 3 Solve the equation $x^2 = 3x + 18$ by completing the square.

Solution: Place all terms except the constant on the left side of the equation.

$$x^2 = 3x + 18$$
$$x^2 - 3x = 18$$

Take half the numerical coefficient of the x term, square it, and add this product to both sides of the equation.

$$\frac{1}{2}(-3) = -\frac{3}{2}, \qquad \left(-\frac{3}{2}\right)^2 = \frac{9}{4}$$

$$x^2 - 3x + \frac{9}{4} = 18 + \frac{9}{4}$$

$$\left(x - \frac{3}{2}\right)^2 = 18 + \frac{9}{4}$$

$$\left(x - \frac{3}{2}\right)^2 = \frac{72}{4} + \frac{9}{4}$$

$$\left(x - \frac{3}{2}\right)^2 = \frac{81}{4}$$

$$x - \frac{3}{2} = \pm\sqrt{\frac{81}{4}}$$

$$x - \frac{3}{2} = \pm\frac{9}{2}$$

$$x = \frac{3}{2} \pm \frac{9}{2}$$

$$x = \frac{3}{2} + \frac{9}{2} \quad \text{or} \quad x = \frac{3}{2} - \frac{9}{2}$$

$$x = \frac{12}{2} = 6 \qquad x = -\frac{6}{2} = -3$$

The solutions are 6 and -3. ■

In the following examples we will not illustrate some of the intermediate steps.

EXAMPLE 4 Solve the equation $x^2 - 6x + 1 = 0$.

Solution:
$$x^2 - 6x + 1 = 0$$
$$x^2 - 6x = -1$$
$$x^2 - 6x + 9 = -1 + 9$$
$$(x - 3)^2 = 8$$
$$x - 3 = \pm\sqrt{8}$$
$$x - 3 = \pm 2\sqrt{2}$$
$$x = 3 \pm 2\sqrt{2}$$

The solutions are $3 + 2\sqrt{2}$ and $3 - 2\sqrt{2}$. ■

EXAMPLE 5   Solve the equation $3m^2 - 9m + 6 = 0$ by completing the square.

Solution: To solve an equation by completing the square, the numerical coefficient of the squared term must be 1. Since the numerical coefficient of the squared term is 3, we multiply both sides of the equation by $\frac{1}{3}$ to make the numerical coefficient of the squared term equal to 1.

$$3m^2 - 9m + 6 = 0$$

$$\frac{1}{3}(3m^2 - 9m + 6) = \frac{1}{3}(0)$$

$$m^2 - 3m + 2 = 0$$

Now proceed as in earlier examples.

$$m^2 - 3m = -2$$

$$m^2 - 3m + \frac{9}{4} = -2 + \frac{9}{4}$$

$$\left(m - \frac{3}{2}\right)^2 = -\frac{8}{4} + \frac{9}{4}$$

$$\left(m - \frac{3}{2}\right)^2 = \frac{1}{4}$$

$$m - \frac{3}{2} = \pm \sqrt{\frac{1}{4}}$$

$$m - \frac{3}{2} = \pm \frac{1}{2}$$

$$m = \frac{3}{2} \pm \frac{1}{2}$$

$$m = \frac{3}{2} + \frac{1}{2} \quad \text{or} \quad m = \frac{3}{2} - \frac{1}{2}$$

$$m = \frac{4}{2} = 2 \qquad m = \frac{2}{2} = 1$$

The solutions are 2 and 1.   ■

Exercise Set 10.2

Solve each equation by completing the square.

1. $x^2 + 2x - 3 = 0$
2. $x^2 - 6x + 8 = 0$
3. $x^2 - 4x - 5 = 0$
4. $x^2 + 8x + 12 = 0$
5. $x^2 + 3x + 2 = 0$
6. $x^2 + 4x - 32 = 0$
7. $x^2 - 8x + 15 = 0$
8. $x^2 - 9x + 14 = 0$
9. $x^2 = -6x - 9$
10. $x^2 + 5x + 4 = 0$
11. $x^2 = -5x - 6$
12. $x^2 = 2x + 15$
13. $x^2 + 9x + 18 = 0$
14. $x^2 - 9x + 18 = 0$
15. $x^2 = 15x - 56$
16. $x^2 = 3x + 28$
17. $-4x = -x^2 + 12$
18. $-x^2 - 3x + 40 = 0$
19. $x^2 + 2x - 6 = 0$
20. $x^2 - 4x + 2 = 0$
21. $6x + 6 = -x^2$
22. $x^2 - x - 3 = 0$
23. $-x^2 + 5x = -8$
24. $x^2 + 3x - 6 = 0$
25. $2x^2 + 4x - 6 = 0$
26. $2x^2 + 2x - 24 = 0$
27. $2x^2 + 18x + 4 = 0$

28. $2x^2 = 8x + 90$

31. $2x^2 + 10x - 3 = 0$

34. $2x^2 - x = 5$

37. $2x^2 - 4x = 0$

29. $3x^2 + 33x + 72 = 0$

32. $3x^2 - 8x + 4 = 0$

35. $x^2 + 4x = 0$

38. $3x^2 = 9x$

30. $4x^2 = -28x + 32$

33. $3x^2 + 6x = 6$

36. $2x^2 - 6x = 0$

39. When three times a number is added to the square of a number, the sum is 4. Find the number(s).

40. When five times a number is subtracted from two times the square of a number, the difference is 12. Find the number(s).

41. If the square of three more than a number is 9, find the number(s).

42. If the square of two less than an integer is 16, find the number(s).

43. The product of two positive numbers is 21. Find the two numbers if the larger is 4 greater than the smaller.

44. (a) What is a perfect square trinomial?

(b) Fill in the shaded area to make a perfect square trinomial and explain how you determined your answer.

$$x^2 + 8x \qquad$$

45. (a) Write a perfect square trinomial that has a term of $-12x$.

(b) Explain how you constructed your perfect square trinomial.

Cumulative Review Exercises

[6.5] **46.** Simplify $\dfrac{x^2}{x^2 - x - 6} - \dfrac{x - 2}{x - 3}$.

[7.4] **47.** Explain how you can determine if two equations represent parallel lines without graphing the equations.

[8.4] **48.** Solve the system of equations.

$$3x - 4y = 6$$
$$2x + y = 8$$

[9.5] **49.** Solve the equation $\sqrt{2x + 3} = 2x - 3$.

JUST FOR FUN

Solve by completing the sqaure.

1. $x^2 + \dfrac{3}{5}x - \dfrac{1}{2} = 0$.

2. $x^2 - \dfrac{2}{3}x - \dfrac{1}{5} = 0$.

3. $3x^2 + \dfrac{1}{2}x = 4$.

4. Using completing the square, solve the following equation for x in terms of a, b, and c: $ax^2 + bx + c = 0$.

10.3

Solving Quadratic Equations by the Quadratic Formula

▶ **1** Solve quadratic equations by the quadratic formula.

▶ **2** Determine the number of solutions to a quadratic equation using the discriminant.

▶ **1** A method that can be used to solve any quadratic equation is to use the quadratic formula. It is the most useful and versatile method of solving quadratic equations.

The standard form of a quadratic equation is $ax^2 + bx + c = 0$, where a is the numerical coefficient of the squared term, b is the numerical coefficient of the first-powered term, and c is the constant.

Quadratic Equation
in Standard Form   *Values of a, b, and c*

$x^2 - 3x + 4 = 0$ $a = 1$, $b = -3$, $c = 4$

$-2x^2 + \dfrac{1}{2}x - 2 = 0$ $a = -2$, $b = \dfrac{1}{2}$, $c = -2$

$3x^2 - 4 = 0$ $a = 3$, $b = 0$, $c = -4$

$5x^2 + 3x = 0$ $a = 5$, $b = 3$, $c = 0$

$-\dfrac{1}{2}x^2 + 5 = 0$ $a = -\dfrac{1}{2}$, $b = 0$, $c = 5$

$-12x^2 + 8x = 0$ $a = -12$, $b = 8$, $c = 0$

To Solve a Quadratic Equation by the Quadratic Formula

1. Write the equation in standard form, $ax^2 + bx + c = 0$, and determine the numerical values for a, b, and c.
2. Substitute the values for a, b, and c in the quadratic formula and then evaluate to obtain the solution.

The Quadratic Formula

$$x = \frac{-b \pm \sqrt{b^2 - 4ac}}{2a}$$

EXAMPLE 1   Solve the equation $x^2 + 2x - 8 = 0$ using the quadratic formula.

Solution: $1x^2 + 2x - 8 = 0$

            $\uparrow$ $\uparrow$ $\uparrow$
            a b c

$$x = \frac{-b \pm \sqrt{b^2 - 4ac}}{2a}$$

$$= \frac{-(2) \pm \sqrt{(2)^2 - 4(1)(-8)}}{2(1)}$$

$$= \frac{-2 \pm \sqrt{4 + 32}}{2}$$

$$= \frac{-2 \pm \sqrt{36}}{2}$$

$$= \frac{-2 \pm 6}{2}$$

$$x = \frac{-2 + 6}{2} \quad \text{or} \quad x = \frac{-2 - 6}{2}$$

$$x = \frac{4}{2} = 2 \qquad\qquad x = \frac{-8}{2} = -4$$

Check:

$x = 2$	$x = -4$
$x^2 + 2x - 8 = 0$	$x^2 + 2x - 8 = 0$
$(2)^2 + 2(2) - 8 = 0$	$(-4)^2 + 2(-4) - 8 = 0$
$4 + 4 - 8 = 0$	$16 - 8 - 8 = 0$
$0 = 0$ true	$0 = 0$ true

COMMON STUDENT ERROR

The **entire numerator** of the quadratic formula must be divided by $2a$.

Correct

$$x = \frac{-b \pm \sqrt{b - 4ac}}{2a}$$

Wrong

$$x = -b \pm \frac{\sqrt{b^2 - 4ac}}{2a}$$

$$x = \frac{-b}{2a} \pm \sqrt{b^2 - 4ac}$$

EXAMPLE 2   Solve the equation $6x^2 - x - 2 = 0$ using the quadratic formula.

Solution:

$$6x^2 - x - 2 = 0$$

$$a = 6, \qquad b = -1, \qquad c = -2$$

$$x = \frac{-b \pm \sqrt{b^2 - 4ac}}{2a}$$

$$= \frac{-(-1) \pm \sqrt{(-1)^2 - 4(6)(-2)}}{2(6)}$$

$$= \frac{1 \pm \sqrt{1 + 48}}{12}$$

$$= \frac{1 \pm \sqrt{49}}{12}$$

$$= \frac{1 \pm 7}{12}$$

$$x = \frac{1 + 7}{12} = \frac{8}{12} = \frac{2}{3} \quad \text{or} \quad x = \frac{1 - 7}{12} = \frac{-6}{12} = \frac{-1}{2}$$

Check:

$$x = \frac{2}{3} \qquad\qquad\qquad x = -\frac{1}{2}$$

$$6x^2 - x - 2 = 0 \qquad\qquad 6x^2 - x - 2 = 0$$

$$6\left(\frac{2}{3}\right)^2 - \frac{2}{3} - 2 = 0 \qquad 6\left(-\frac{1}{2}\right)^2 - \left(-\frac{1}{2}\right) - 2 = 0$$

$$\overset{2}{\cancel{6}}\left(\frac{4}{\underset{3}{\cancel{9}}}\right) - \frac{2}{3} - 2 = 0 \qquad \overset{3}{\cancel{6}}\left(\frac{1}{\underset{2}{\cancel{4}}}\right) + \frac{1}{2} - 2 = 0$$

$$\frac{8}{3} - \frac{2}{3} - \frac{6}{3} = 0 \qquad\qquad \frac{3}{2} + \frac{1}{2} - \frac{4}{2} = 0$$

$$0 = 0 \quad \text{true} \qquad\qquad\qquad 0 = 0 \quad \text{true}$$

EXAMPLE 3   Solve the equation $2x^2 + 4x - 5 = 0$ using the quadratic formula.

Solution:

$$a = 2, \qquad b = 4, \qquad c = -5$$

$$x = \frac{-b \pm \sqrt{b^2 - 4ac}}{2a}$$

$$= \frac{-4 \pm \sqrt{(4)^2 - 4(2)(-5)}}{2(2)}$$

$$= \frac{-4 \pm \sqrt{16 + 40}}{4}$$

$$= \frac{-4 \pm \sqrt{56}}{4}$$

$$= \frac{-4 \pm 2\sqrt{14}}{4}$$

Now factor out a 2 from both terms in the numerator; then divide out common factors.

$$x = \frac{\overset{1}{\cancel{2}}(-2 \pm \sqrt{14})}{\underset{2}{\cancel{4}}}$$

$$x = \frac{-2 \pm \sqrt{14}}{2}$$

Thus the solutions are

$$x = \frac{-2 + \sqrt{14}}{2} \quad \text{and} \quad x = \frac{-2 - \sqrt{14}}{2}$$

■

EXAMPLE 4   Solve the equation $x^2 = -4x + 6$ using the quadratic formula.

Solution: Write the equation in standard form.

$$x^2 + 4x - 6 = 0$$

$$a = 1, \qquad b = 4, \qquad c = -6$$

$$x = \frac{-b \pm \sqrt{b^2 - 4ac}}{2a}$$

$$= \frac{-4 \pm \sqrt{(4)^2 - 4(1)(-6)}}{2(1)}$$

$$= \frac{-4 \pm \sqrt{16 + 24}}{2}$$

$$= \frac{-4 \pm \sqrt{40}}{2}$$

$$= \frac{-4 \pm 2\sqrt{10}}{2}$$

$$= \frac{\overset{1}{\cancel{2}}(-2 \pm \sqrt{10})}{\underset{1}{\cancel{2}}}$$

$$= -2 \pm \sqrt{10}$$

$$x = -2 + \sqrt{10} \quad \text{or} \quad x = -2 - \sqrt{10}$$

■

COMMON STUDENT ERROR

Many students work the problems correctly until the last step, where they perform an error. Do not make the mistake of trying to simplify an answer that cannot be simplified any further. The following are answers that cannot be simplified, along with some common errors.

Answers That
Cannot Be Simplified   *Wrong*

$$\frac{3 + 2\sqrt{5}}{2} \qquad \frac{3 + 2\sqrt{5}}{2} = \frac{3 + \overset{1}{\cancel{2}}\sqrt{5}}{\underset{1}{\cancel{2}}} = 3 + \sqrt{5}$$

$$\frac{4 + 3\sqrt{5}}{2} \qquad \frac{\overset{2}{\cancel{4}} + 3\sqrt{5}}{\underset{1}{\cancel{2}}} = 2 + 3\sqrt{5}$$

$$\frac{3 + \sqrt{6}}{2} \qquad \frac{3 + \overset{3}{\cancel{\sqrt{6}}}}{\underset{1}{\cancel{2}}} = 3 + \sqrt{3}$$

When *both terms* in the numerator have a common factor, the common factor can sometimes be divided out, as follows:

Correct

$$\frac{2 + 4\sqrt{3}}{2} = \frac{\overset{1}{\cancel{2}}(1 + 2\sqrt{3})}{\underset{1}{\cancel{2}}} = 1 + 2\sqrt{3}$$

$$\frac{6 + 3\sqrt{3}}{6} = \frac{\overset{1}{\cancel{3}}(2 + \sqrt{3})}{\underset{2}{\cancel{6}}} = \frac{2 + \sqrt{3}}{2}$$

EXAMPLE 5   Solve the equation $x^2 = 4$ using the quadratic formula.

Solution: Write in standard form.

$$x^2 - 4 = 0$$
$$a = 1, \qquad b = 0, \qquad c = -4$$
$$x = \frac{-b \pm \sqrt{b^2 - 4ac}}{2a}$$
$$= \frac{0 \pm \sqrt{0^2 - 4(1)(-4)}}{2(1)}$$
$$= \frac{\pm \sqrt{16}}{2} = \frac{\pm 4}{2} = \pm 2$$

Thus the solutions are 2 and -2.

EXAMPLE 6   Solve the quadratic equation $2x^2 + 5x = -6$.

Solution:

$$2x^2 + 5x + 6 = 0$$

$$a = 2, \qquad b = 5, \qquad c = 6$$

$$x = \frac{-b \pm \sqrt{b^2 - 4ac}}{2a}$$

$$= \frac{-5 \pm \sqrt{(5)^2 - 4(2)(6)}}{2(2)}$$

$$= \frac{-5 \pm \sqrt{25 - 48}}{4}$$

$$= \frac{-5 \pm \sqrt{-23}}{4}$$

Since $\sqrt{-23}$ is not a real number, we can go no further. This equation has no real solution. *When given a problem of this type, your answer should be "no real solution." Do not leave the answer blank, and do not write 0 for the answer.* ■

▶**2** The expression under the square root sign in the quadratic formula is called the **discriminant.**

$$\underbrace{b^2 - 4ac}_{\text{discriminant}}$$

The discriminant can be used to determine the number of solutions to a quadratic equation.

When the discriminant is:

1. **Greater than zero,** $b^2 - 4ac > 0$, the quadratic equation has **two distinct real solutions.**
2. **Equal to zero,** $b^2 - 4ac = 0$, the quadratic equation has a **single unique solution.** This single solution is often referred to as a **double root.**
3. **Less than zero,** $b^2 - 4ac < 0$, the quadratic equation has **no real solution.**

$b^2 - 4ac$	Number of Solutions
Positive	Two distinct real solutions
0	A single unique solution
Negative	No real solution

EXAMPLE 7   (a) Find the discriminant of the equation $x^2 - 8x + 16 = 0$.
(b) Use the quadratic formula to find the solution.

Solution: (a) $a = 1, \quad b = -8, \quad c = 16$

$$b^2 - 4ac = (-8)^2 - 4(1)(16) = 64 - 64 = 0$$

Since the discriminant equals zero, there is a single unique solution.

(b) $x = \dfrac{-b \pm \sqrt{b^2 - 4ac}}{2a}$

$= \dfrac{-(-8) \pm \sqrt{0}}{2(1)}$

$= \dfrac{8 \pm 0}{2} = \dfrac{8}{2} = 4$

The only solution is 4.   ▪

EXAMPLE 8   Without actually finding the solutions, determine if the following equations have two distinct real solutions, a single unique solution, or no real solution.

(a) $2x^2 - 4x + 6 = 0$ (b) $x^2 - 5x - 8 = 0$ (c) $4x^2 - 12x = -9$

Solution: We use the discriminant of the quadratic formula to answer these equations.

(a) $b^2 - 4ac = (-4)^2 - 4(2)(6) = 16 - 48 = -32$

Since the discriminant is negative, this equation has no real solution.

(b) $b^2 - 4ac = (-5)^2 - 4(1)(-8) = 25 + 32 = 57$

Since the discriminant is positive, this equation has two distinct real solutions.

(c) First rewrite $4x^2 - 12x = -9$ as $4x^2 - 12x + 9 = 0$.

$$b^2 - 4ac = (-12)^2 - 4(4)(9) = 144 - 144 = 0$$

Since the discriminant is zero, this equation has a single unique solution.   ▪

Before we leave this section, let us look at one of many application problems that may be solved using the quadratic formula.

EXAMPLE 9   Mr. Jackson is planning to plant a grass walkway of uniform width around his rectangular swimming pool, which measures 18 feet by 24 feet. How far will the walkway extend from the pool if Mr. Jackson has only enough seed to plant 2000 square feet of grass?

Solution: Let us make a diagram of the situation (see Fig. 10.2). Let $x =$ the uniform width of the grass area. Then the total length of the larger rectangular area becomes $24 + 2x$ or $2x + 24$. The total width of the larger rectangular area becomes $18 + 2x$ or $2x + 18$.

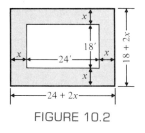

FIGURE 10.2

The grassy area can be found by subtracting the area of the pool from the larger rectangular area.

$$\text{Area of pool} = l \cdot w = (24)(18) = 432 \text{ square feet}$$
$$\text{Area of large rectangle} = l \cdot w = (2x + 24)(2x + 18)$$
$$= 4x^2 + 84x + 432 \text{ (pool plus grassy area)}$$
$$\text{Grassy area} = \text{area of large rectangle} - \text{area of pool}$$
$$= 4x^2 + 84x + 432 - (432)$$
$$= 4x^2 + 84x$$

The total grassy area must be 2000 square feet.

$$\text{Grassy area} = 4x^2 + 84x$$
$$2000 = 4x^2 + 84x$$

or
$$4x^2 + 84x - 2000 = 0$$
$$4(x^2 + 21x - 500) = 0$$

By the quadratic formula.

$$x = \frac{-b \pm \sqrt{b^2 - 4ac}}{2a}$$

$$a = 1, \qquad b = 21, \qquad c = -500$$

$$x = \frac{-21 \pm \sqrt{(21)^2 - 4(1)(-500)}}{2(1)}$$

$$= \frac{-21 \pm \sqrt{441 + 2000}}{2}$$

$$= \frac{-21 \pm \sqrt{2441}}{2}$$

$$\approx \frac{-21 \pm 49.41}{2}$$

$$x \approx \frac{-21 - 49.41}{2} \quad \text{or} \quad x \approx \frac{-21 + 49.41}{2}$$

$$x \approx \frac{-70.41}{2} \quad \text{or} \quad x \approx \frac{28.41}{2}$$

$$x \approx -35.205 \quad \text{or} \quad x \approx 14.21$$

The only possible answer is $x \approx 14.21$. Thus there will be a grass walkway about 14.2 feet wide all around the pool. ∎

Exercise Set 10.3

Determine whether each equation has two distinct real solutions, a single unique solution, or no real solution.

1. $x^2 + 3x - 5 = 0$ **2.** $2x^2 + 6x + 3 = 0$ **3.** $3x^2 - 4x + 7 = 0$ **4.** $-4x^2 + x - 8 = 0$

5. $5x^2 + 3x - 7 = 0$ **6.** $2x^2 = 16x - 32$ **7.** $4x^2 - 24x = -36$ **8.** $5x^2 - 4x = 7$

9. $x^2 - 8x + 5 = 0$ **10.** $x^2 - 5x - 9 = 0$ **11.** $-3x^2 + 5x - 8 = 0$ **12.** $x^2 + 4x - 8 = 0$

13. $x^2 + 7x - 3 = 0$ **14.** $2x^2 - 6x + 9 = 0$ **15.** $4x^2 - 9 = 0$ **16.** $6x^2 - 5x = 0$

Use the quadratic formula to solve each equation. If the equation has no real solution, so state.

17. $x^2 - 3x + 2 = 0$ **18.** $x^2 + 6x + 8 = 0$ **19.** $x^2 - 9x + 20 = 0$

20. $x^2 - 3x - 10 = 0$ **21.** $x^2 + 5x - 24 = 0$ **22.** $x^2 - 6x =$

23. $x^2 = 13x - 36$ **24.** $x^2 - 36 = 0$ **25.** $x^2 - 25 = 0$

26. $x^2 - 6x = 0$ **27.** $x^2 - 3x = 0$ **28.** $z^2 - 17z + 72 = 0$

29. $p^2 - 7p + 12 = 0$ **30.** $2x^2 - 3x + 2 = 0$ **31.** $2y^2 - 7y + 4 = 0$

32. $2x^2 - 7x = -5$ **33.** $6x^2 = -x + 1$ **34.** $4r^2 + r - 3 = 0$

35. $2x^2 - 4x - 1 = 0$ **36.** $3w^2 - 4w + 5 = 0$ **37.** $2s^2 - 4s + 3 = 0$

38. $x^2 - 7x + 3 = 0$

39. $4x^2 = x + 5$

40. $x^2 - 2x - 1 = 0$

41. $2x^2 - 7x = 9$

42. $-x^2 + 2x + 15 = 0$

43. $-2x^2 + 11x - 15 = 0$

44. $6x^2 + 5x + 9 = 0$

45. The product of two consecutive positive integers is 20. Find the two consecutive numbers.

46. The length of a rectangle is 3 feet longer than its width. Find the dimensions of the rectangle if its area is 28 square feet.

47. The length of a rectangle is 3 feet smaller than twice its width. Find the length and width of the rectangle if its area is 20 square feet.

48. Lisa wishes to plant a uniform strip of grass around her pool. If her pool measures 20 feet by 30 feet and she has only enough seed to cover 336 square feet, what will be the width of the uniform strip? See Example 9.

49. The McDonald's garden is 30 feet by 40 feet. They wish to lay a uniform border of pine bark around their garden. How large a strip should they lay if they only have enough bark to cover 296 square feet?

50. (a) What is the discriminant? (b) Explain how the discriminant can be used to determine the number of real solutions a quadratic equation has.

51. How many real solutions does a quadratic equation have if the discriminant equals

    (a) -4 **(b)** 0 **(c)** $\frac{1}{2}$?

52. Without looking at your notes, write down the quadratic formula. You must learn this formula by memory.

53. Explain in your own words why a quadratic equation will have two real solutions when the discriminant is greater than 0, one real solution when the discriminant is equal to 0, and no real solution when the discriminant is less than 0. Use the quadratic formula in explaining your answer.

Cumulative Review Exercises

[5.6, 10.2, 10.3] *Solve the following quadratic equations by (a) factoring, (b) completing the square, and (c) the quadratic formula. If the equation cannot be solved by factoring, so state.*

54. $x^2 - 13x + 42 = 0$.

55. $6x^2 + 11x - 35 = 0$.

56. $2x^2 + 3x - 4 = 0$.

57. $6x^2 = 54$.

JUST FOR FUN

1. Farmer Justina Wells wishes to form a rectangular region along a river bank by constructing fencing as illustrated in the diagram. If she only has 400 feet of fencing and wishes to enclose an area of 15,000 square feet, find the dimensions of the rectangular region.

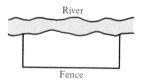

River

Fence

10.4

Graphing Quadratic Equations

▶ **1** Graph quadratic equations in two variables.

▶ **2** Find the coordinates of the vertex of a parabola.

▶ **3** Use symmetry when graphing quadratic equations.

▶ **4** Find the roots of a quadratic equation.

▶ **1** In Section 7.2 we learned how to graph linear equations in two variables. In this section we graph quadratic equations in two variables of the form

$$y = ax^2 + bx + c, \qquad a \neq 0$$

Every quadratic equation of the given form when graphed will be a **parabola.** The graph of $y = ax^2 + bx + c$ will have the general shapes indicated in Fig. 10.3.

Parabola opens upward   Parabola opens downward
(a)   (b)

FIGURE 10.3

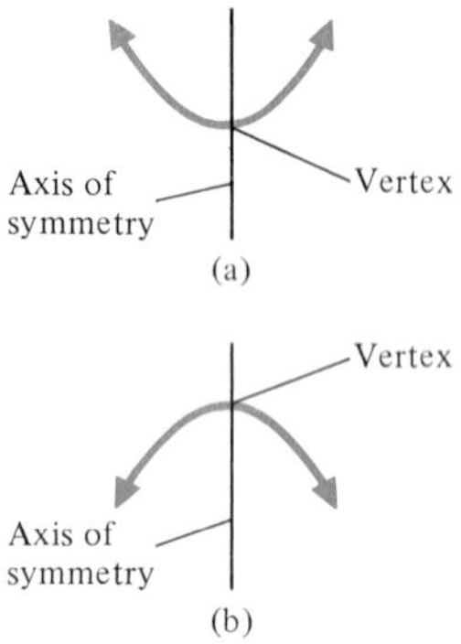

Axis of
symmetry   Vertex
(a)

Vertex

Axis of
symmetry
(b)

FIGURE 10.4

When a quadratic equation is in the form $y = ax^2 + bx + c$, the sign of the numerical coefficient of the squared term, a, will determine whether the parabola will open upward (Fig. 10.3a) or downward (Fig. 10.3b). When a is positive, the parabola will open upward, and when a is negative, the parabola will open downward. The **vertex** is the lowest point on a parabola that opens upward and the highest point on a parabola that opens downward (see Fig. 10.4).

Graphs of quadratic equations of the form $y = ax^2 + bx + c$ will have **symmetry** about a line through the vertex. This means that if we fold the paper along this imaginary line, called the **axis of symmetry,** the right and left sides of the graph will coincide.

One method that can be used to graph a quadratic equation is to plot it point by point. When determining points to plot, select values for x and determine the corresponding values for y.

EXAMPLE 1   Graph the equation $y = x^2$.

Solution: Since $a = 1$, which is greater than 0, this parabola opens upward.

	$y = x^2$		x	y
Let $x = 3$,	$y = (3)^2 = 9$		3	9
Let $x = 2$,	$y = (2)^2 = 4$		2	4
Let $x = 1$,	$y = (1)^2 = 1$		1	1
Let $x = 0$,	$y = (0)^2 = 0$		0	0
Let $x = -1$,	$y = (-1)^2 = 1$		-1	1
Let $x = -2$,	$y = (-2)^2 = 4$		-2	4
Let $x = -3$,	$y = (-3)^2 = 9$		-3	9

Connect the points with a smooth curve (Fig. 10.5). Note how the graph is symmetric about the line $x = 0$ (or the y axis).

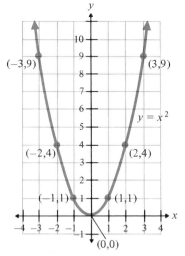

FIGURE 10.5

EXAMPLE 2 Graph the equation $y = -2x^2 + 16x - 24$.

Solution: Since $a = -2$, which is less than 0, this parabola opens downward.

$$y = -2x^2 + 16x - 24$$

		x	y
Let $x = 0$,	$y = -2(0)^2 + 16(0) - 24 = -24$	0	-24
Let $x = 1$,	$y = -2(1)^2 + 16(1) - 24 = -10$	1	-10
Let $x = 2$,	$y = -2(2)^2 + 16(2) - 24 = 0$	2	0
Let $x = 3$,	$y = -2(3)^2 + 16(3) - 24 = 6$	3	6
Let $x = 4$,	$y = -2(4)^2 + 16(4) - 24 = 8$	4	8
Let $x = 5$,	$y = -2(5)^2 + 16(5) - 24 = 6$	5	6
Let $x = 6$,	$y = -2(6)^2 + 16(6) - 24 = 0$	6	0
Let $x = 7$,	$y = -2(7)^2 + 16(7) - 24 = -10$	7	-10
Let $x = 8$,	$y = -2(8)^2 + 16(8) - 24 = -24$	8	-24

Note how the graph (Fig. 10.6) is symmetric about the line $x = 4$, which has been dashed. The vertex of this parabola is the point (4, 8). Since the y values are large, the y axis has been marked with 4-unit intervals to allow us to graph the points (0, -24) and (8, -24). The arrows on the ends of the graph indicate that the parabola continues indefinitely.

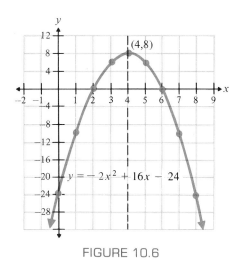

FIGURE 10.6

▶ **2** When graphing quadratic equations, how do we decide what values to use for x? When the location of the vertex is unknown, this is a difficult question to answer. When the location of the vertex is known, it becomes more obvious which values to use.

In Example 2, the axis of symmetry is $x = 4$, and the x coordinate of the vertex is also 4. When given an equation in the form $y = ax^2 + bx + c$, both the axis of symmetry and the x coordinate of the vertex can be found by using the following formula:

Axis of Symmetry and x Coordinate of Vertex
_____

$$x = \frac{-b}{2a}$$

In Example 2, $a = -2$, $b = 16$, and $c = -24$. Substituting these values in the boxed formula gives

$$x = \frac{-b}{2a} = \frac{-(16)}{2(-2)} = \frac{-16}{-4} = 4$$

Thus the graph is symmetric about the line $x = 4$, and the x coordinate of the vertex is 4.

The y coordinate of the vertex can be found by substituting the value of the x coordinate of the vertex into the original equation and solving for y.

$$
\begin{aligned}
y &= -2x^2 + 16x - 24 \\
&= -2(4)^2 + 16(4) - 24 \\
&= -2(16) + 64 - 24 \\
&= -32 + 64 - 24 \\
&= 8
\end{aligned}
$$

Thus the vertex is at the point (4, 8).

When given an equation of the form $y = ax^2 + bx + c$, the y coordinate of the vertex can also be found by the following formula.

---

y Coordinate of Vertex
_____

$$y = \frac{4ac - b^2}{4a}$$

For Example 2,

$$
\begin{aligned}
y &= \frac{4ac - b^2}{4a} \\
&= \frac{4(-2)(-24) - (16)^2}{4(-2)} \\
&= \frac{192 - 256}{-8} = \frac{-64}{-8} = 8
\end{aligned}
$$

You may use the method of your choice to find the y coordinate of the vertex.

▶ **3** One method to use in selecting points to plot when graphing parabolas is to determine the axis of symmetry and the vertex of the graph. Then select nearby values of x on any one side of the axis of symmetry. When graphing the equation, make use of your knowledge of the symmetry of the graph.

EXAMPLE 3   (a) Find the axis of symmetry of the equation $y = x^2 + 8x + 15$.
(b) Find the vertex of the graph.
(c) Graph the equation.

Solution: (a) $a = 1$, $b = 8$, $c = 15$

$$x = \frac{-b}{2a} = \frac{-8}{2(1)} = -4$$

The parabola is symmetric about the line $x = -4$. The x coordinate of the vertex is -4.

(b) Now find the y coordinate of the vertex.

$$y = x^2 + 8x + 15$$
$$y = (-4)^2 + 8(-4) + 15 = 16 - 32 + 15 = -1$$

The vertex is $(-4, -1)$.

(c) Since the axis of symmetry is $x = -4$, we will select values for x that are greater than or equal to -4. It is often helpful to plot each point as it is determined. If a point does not appear to be part of the curve of the parabola, check it.

	$y = x^2 + 8x + 15$	x	y
Let $x = -3$,	$y = (-3)^2 + 8(-3) + 15 = 0$	-3	0
Let $x = -2$,	$y = (-2)^2 + 8(-2) + 15 = 3$	-2	3
Let $x = -1$,	$y = (-1)^2 + 8(-1) + 15 = 8$	-1	8

The points are plotted in Figure 10.7a. The graph of the equation is illustrated in Figure 10.7b. Note how we make use of symmetry to complete the graph. The points $(-3, 0)$ and $(-5, 0)$ are each 1 horizontal unit from the axis of symmetry, $x = -4$. The points $(-2, 3)$ and $(-6, 3)$ are each 2 horizontal units from the axis of symmetry, and the points $(-1, 8)$ and $(-7, 8)$ are each 3 horizontal units from the axis of symmetry.

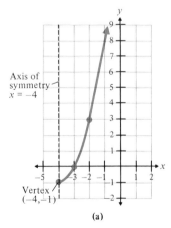

(a)

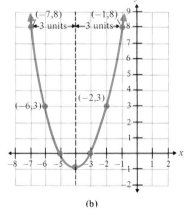

(b)

FIGURE 10.7

EXAMPLE 4   Graph the equation $y = -2x^2 + 3x - 4$.

Solution: $a = -2, \quad b = 3, \quad c = -4$

Since $a < 0$, this parabola will open downward.

$$\text{Axis of symmetry:} \quad x = \frac{-b}{2a}$$

$$x = \frac{-3}{2(-2)} = \frac{-3}{-4} = \frac{3}{4}$$

Since the x value of the vertex is a fraction, we will use the formula to find the y value of the vertex.

$$y = \frac{4ac - b^2}{4a}$$

$$y = \frac{4(-2)(-4) - 3^2}{4(-2)} = \frac{32 - 9}{-8} = \frac{23}{-8} = -\frac{23}{8}$$

The vertex of this graph is at $(\frac{3}{4}, -\frac{23}{8})$. Since the axis of symmetry is $x = \frac{3}{4}$, we will begin by selecting values of x that are greater than $\frac{3}{4}$.

$$y = -2x^2 + 3x - 4$$

			x	y
Let $x = 1$,	$y = -2(1)^2 + 3(1) - 4 = -3$		1	-3
Let $x = 2$,	$y = -2(2)^2 + 3(2) - 4 = -6$		2	-6
Let $x = 3$,	$y = -2(3)^2 + 3(3) - 4 = -13$		3	-13

The St. Louis Arch has the shape of a parabola.

When the axis of symmetry is a fractional value, be very careful when constructing the graph. You should plot as many additional points as needed. In this example, when $x = 0$, $y = -4$, and when $x = -1$, $y = -9$. Figure 10.8a shows the points found on the right side of the axes of symmetry. Figure 10.8b shows the completed graph.

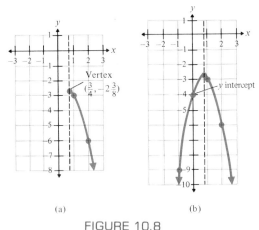

(a)   (b)

FIGURE 10.8

▶ **4**   In Example 4 the graph crossed the y axis at $y = -4$. Recall from earlier sections that to find the y intercept we let $x = 0$ and solve for y. The location of the y intercept is often helpful when graphing quadratic equations.

The value or values of x where the graph crosses the x axis (the x intercepts) are

---

Calculator Corner

Evaluating Quadratic Expressions

In this section when graphing we will often have to evaluate quadratic expressions to obtain values for y. Here we show how the polynomial $6x^2 - 3x + 4$ can be evaluated using a *scientific calculator* for the values $x = 5$ and $x = -5$.

Evaluate: $6x^2 - 3x + 4$ for $x = 5$

Evaluate: $6(5)^2 - 3(5) + 4$

$\boxed{C}$ 6 $\boxed{\times}$ 5 $\boxed{x^2}$ $\boxed{-}$ 3 $\boxed{\times}$ 5 $\boxed{+}$ 4 $\boxed{=}$ 139

Evaluate: $6x^2 - 3x + 4$ for $x = -5$

Evaluate: $6(-5)^2 - 3(-5) + 4$

$\boxed{C}$ 6 $\boxed{\times}$ 5 $\boxed{^{+}/_{-}}$ $\boxed{x^2}$ $\boxed{-}$ 3 $\boxed{\times}$ 5 $\boxed{^{+}/_{-}}$ $\boxed{+}$ 4 $\boxed{=}$ 169

Evaluate $6x^2 - 3x + 4$ for both $x = 5$ and $x = -5$ without a calculator to make sure you obtain the same answer we found with the calculator.

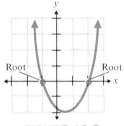

FIGURE 10.9

called the **roots** of the equation. Such points must have a y value of zero (see Fig. 10.9).

A quadratic equation of the form $y = ax^2 + bx + c$ will have either two distinct real roots (Fig. 10.10a), a double root (Fig. 10.10b), or no real roots (Fig. 10.10c). In Section 10.3 we mentioned that when the discriminant, $b^2 - 4ac$, is greater than zero, there are two distinct real solutions; when it is equal to zero, there is a single unique solution (also called a *double root*); and when it is less than zero, there is no real solution. This concept is illustrated in Figure 10.10.

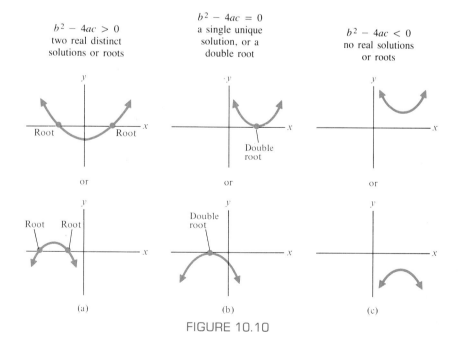

FIGURE 10.10

The roots of an equation can be found graphically or algebraically by one of the methods presented earlier in this chapter. Sometimes the roots can also be found by factoring.

EXAMPLE 5   (a) Find the roots of the equation $y = x^2 - 2x - 24$ by factoring, by completing the square, and by the quadratic formula.
(b) Graph the equation.

Solution: (a) We will find the roots using all three algebraic methods. Each root must have a y value of zero; thus, to find the roots, we set $y = 0$ and solve the equation for x.

$$0 = x^2 - 2x - 24$$

or

$$x^2 - 2x - 24 = 0$$

Method 1: Factoring.

$$x^2 - 2x - 24 = 0$$
$$(x - 6)(x + 4) = 0$$

$$x - 6 = 0 \quad \text{or} \quad x + 4 = 0$$
$$x = 6 \qquad\qquad x = -4$$

Method 2: Completing the square.

$$x^2 - 2x - 24 = 0$$
$$x^2 - 2x = 24$$
$$x^2 - 2x + 1 = 24 + 1$$
$$(x - 1)^2 = 25$$
$$x - 1 = \pm 5$$
$$x = 1 \pm 5$$
$$x = 1 + 5 = 6 \quad \text{or} \quad x = 1 - 5 = -4$$

Method 3: Quadratic formula.

$$x^2 - 2x - 24 = 0$$
$$a = 1, \qquad b = -2, \qquad c = -24$$
$$x = \frac{-b \pm \sqrt{b^2 - 4ac}}{2a}$$
$$= \frac{-(-2) \pm \sqrt{(-2)^2 - 4(1)(-24)}}{2(1)}$$
$$= \frac{2 \pm \sqrt{4 + 96}}{2}$$
$$= \frac{2 \pm \sqrt{100}}{2}$$
$$= \frac{2 \pm 10}{2}$$

$$x = \frac{2 + 10}{2} = \frac{12}{2} = 6 \quad \text{or} \quad x = \frac{2 - 10}{2} = \frac{-8}{2} = -4$$

Note that the same values were obtained by all three methods.

(b) $y = x^2 - 2x - 24$. Since $a > 0$, this parabola opens upward.

Axis of symmetry: $x = \dfrac{-b}{2a} = \dfrac{-(-2)}{2(1)} = \dfrac{2}{2} = 1$

$$y = x^2 - 2x - 24$$

x	y
1	−25
2	−24
3	−21
4	−16
5	−9
6	0

Let $x = 1$, $y = 1^2 - 2(1) - 24 = -25$

Let $x = 2$, $y = 2^2 - 2(2) - 24 = -24$

Let $x = 3$, $y = 3^2 - 2(3) - 24 = -21$

Let $x = 4$, $y = 4^2 - 2(4) - 24 = -16$

Let $x = 5$, $y = 5^2 - 2(5) - 24 = -9$

Let $x = 6$, $y = 6^2 - 2(6) - 24 = 0$

Again we make use of symmetry in completing the graph, Fig. 10.11. The roots 6 and −4 agree with the answer obtained in part (a).

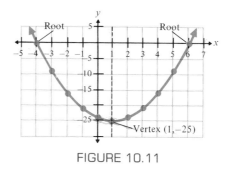

FIGURE 10.11

Exercise Set 10.4

Indicate the axis of symmetry, the coordinates of the vertex, and whether the parabola opens up or down.

1. $y = x^2 + 2x - 7$

2. $y = x^2 + 4x - 9$

3. $y = -x^2 + 5x - 6$

4. $y = 3x^2 + 6x - 9$

5. $y = -3x^2 + 5x + 8$

6. $y = x^2 + 3x - 6$

7. $y = -4x^2 - 8x - 12$

8. $y = 2x^2 + 3x + 8$

9. $y = 3x^2 - 2x + 2$

10. $y = -x^2 + x + 8$

11. $y = 4x^2 + 12x - 5$

12. $y = -2x^2 - 6x - 5$

Graph each quadratic equation, and determine the roots, if they exist.

13. $y = x^2 - 1$

14. $y = x^2 + 4$

15. $y = -x^2 + 3$

16. $y = -x^2 - 2$

17. $y = x^2 + 2x + 3$ **18.** $y = x^2 + 4x + 3$ **19.** $y = x^2 + 2x - 15$ **20.** $y = -x^2 + 10x - 21$

21. $y = -x^2 + 4x - 5$ **22.** $y = x^2 + 8x + 15$ **23.** $y = x^2 - x - 12$ **24.** $y = x^2 - 5x + 4$

25. $y = x^2 - 6x + 9$ **26.** $y = x^2 - 6x$ **27.** $y = -x^2 + 5x$ **28.** $y = x^2 - 4x + 4$

29. $y = 2x^2 - 6x + 4$ **30.** $y = x^2 - 2x + 1$ **31.** $y = -x^2 + 11x - 28$ **32.** $y = 4x^2 + 12x + 9$

33. $y = x^2 - 2x - 15$ **34.** $y = -x^2 + 5x - 4$ **35.** $y = -2x^2 + 7x - 3$ **36.** $y = 2x^2 + 3x - 2$

37. $y = -2x^2 + 3x - 2$ **38.** $y = -4x^2 - 6x + 4$ **39.** $y = 2x^2 - x - 15$ **40.** $y = 6x^2 + 10x - 4$

41. What are the coordinates of the vertex of a parabola?

42. What determines whether the graph of a quadratic equation of the form $y = ax^2 + bx + c$, $a \neq 0$, is a parabola that opens upward or downward? Explain your answer.

43. (a) What are the roots of a graph? (b) How can you find the roots of a graph algebraically?

44. How many real roots will the graph of a quadratic equation have if the discriminant has a value of

  (a) 5 **(b)** -2 **(c)** 0?

45. (a) When graphing a quadratic equation of the form $y = ax^2 + bx + c$, what is the equation of the vertical line about which the parabola will be symmetric?

Quadratic equations of the form $y = ax^2 + bx + c$ will be parabolas when graphed. If the value of a, the coefficient of the squared term in the equation, and the vertex of the parabola are as given, determine the number of x intercepts the parabola will have. Explain how you determined your answer.

46. $a = -2$, vertex at $(0, -3)$

47. $a = 5$, vertex at $(4, -3)$

48. $a = -3$, vertex at $(-4, 0)$

49. $a = -1$, vertex at $(2, -4)$

50. Will the equations below have the same x intercepts when graphed? Explain how you determined your answer.
$y = x^2 - 2x - 8$ and $y = -x^2 + 2x + 8$

51. (a) How will the graphs of the equations below compare? Explain how you determined your answer.
$y = x^2 - 2x - 8$ and $y = -x^2 + 2x + 8$

  (b) Graph $y = x^2 - 2x - 8$ and $y = -x^2 + 2x + 8$ on the same set of axes.

Cumulative Review Exercises

[6.3] **52.** Add $\dfrac{3}{x + 3} - \dfrac{x - 2}{x - 4}$.

[6.7] **53.** Solve the equation $\dfrac{1}{3}(x + 6) = 3 - \dfrac{1}{4}(x - 5)$.

[7.2] **54.** Graph $4x - 6y = 20$.

[7.6] **55.** Graph $y < -2$.

SUMMARY

GLOSSARY

Axis of symmetry *(436):* The imaginary line in a graph about which symmetry occurs.

Parabola *(435):* The graph of a quadratic equation is a parabola.

Perfect square trinomial *(422):* A trinomial that can be expressed as the square of a binomial.

Root *(440):* The value of x where a graph crosses the x axis.

Standard form of a quadratic equation *(418):* $ax^2 + bx + c = 0$, $a \neq 0$.

Vertex of a parabola *(436):* The lowest point on a parabola that opens upward and the highest point on a parabola that opens downward.

IMPORTANT FACTS

Square root property: If $x^2 = a$, then $x = \sqrt{a}$ or $x = -\sqrt{a}$ (or $x = \pm\sqrt{a}$).

Quadratic formula: $x = \dfrac{-b \pm \sqrt{b^2 - 4ac}}{2a}$

Discriminant: $b^2 - 4ac$

  If $b^2 - 4ac > 0$ the quadratic equation has two distinct real solutions.

  If $b^2 - 4ac = 0$ the quadratic equation has a single unique solution.

  If $b^2 - 4ac < 0$ the quadratic equation has no real solution.

Coordinates of the vertex of a parabola:
$\left(\dfrac{-b}{2a}, \dfrac{4ac - b^2}{4a} \right)$.

Review Exercises

[10.1] *Solve each equation.*

1. $x^2 = 25$

2. $x^2 = 8$

3. $2x^2 = 12$

4. $x^2 + 3 = 9$

5. $x^2 - 4 = 16$

6. $2x^2 - 4 = 10$

7. $3x^2 + 8 = 32$

8. $(x - 3)^2 = 12$

9. $(2x + 4)^2 = 30$

10. $(3x - 5)^2 = 50$

[10.2] *Solve each equation by completing the square.*

11. $x^2 - 10x + 16 = 0$

12. $x^2 - 8x + 15 = 0$

13. $x^2 - 14x + 13 = 0$

14. $x^2 + x - 6 = 0$

15. $x^2 - 3x - 54 = 0$

16. $x^2 = -5x + 6$

17. $x^2 + 2x - 5 = 0$

18. $x^2 - 3x - 8 = 0$

19. $2x^2 - 8x = 64$

20. $2x^2 - 4x = 30$

21. $4x^2 + 2x - 12 = 0$

22. $6x^2 - 19x + 15 = 0$

[10.3] *Determine whether each equation has two distinct real solutions, a single unique solution, or no real solution.*

23. $3x^2 - 4x - 20 = 0$

24. $-3x^2 + 4x = 9$

25. $2x^2 + 6x + 7 = 0$

26. $x^2 - x + 8 = 0$

27. $x^2 - 12x = -36$

28. $3x^2 - 4x + 5 = 0$

29. $-3x^2 - 4x + 8 = 0$

30. $x^2 - 9x + 6 = 0$

Solve each equation using the quadratic formula. If an equation has no real solution, so state.

31. $x^2 - 9x + 14 = 0$

32. $x^2 + 7x - 30 = 0$

33. $x^2 = 7x - 10$

34. $5x^2 - 7x = 6$

35. $x^2 - 18 = 7x$

36. $x^2 - x - 30 = 0$

37. $6x^2 + x - 15 = 0$

38. $2x^2 + 4x - 3 = 0$

39. $-2x^2 + 3x + 6 = 0$

40. $x^2 - 6x + 7 = 0$

41. $3x^2 - 4x + 6 = 0$

42. $3x^2 - 6x - 8 = 0$

43. $2x^2 + 3x = 0$

44. $2x^2 - 5x = 0$

[10.1–10.3] *Find the solution to each quadratic equation using the method of your choice.*

45. $x^2 - 11x + 24 = 0$

46. $x^2 - 16x + 63 = 0$

47. $x^2 = -3x + 40$

48. $x^2 + 6x = 27$

49. $x^2 - 4x - 60 = 0$

50. $x^2 - x - 42 = 0$

51. $x^2 + 11x - 12 = 0$

52. $x^2 = 25$

53. $x^2 + 6x = 0$

54. $2x^2 + 5x = 3$

55. $2x^2 = 9x - 10$

56. $6x^2 + 5x = 6$

57. $x^2 + 3x - 6 = 0$

58. $3x^2 - 11x + 10 = 0$

59. $-3x^2 - 5x + 8 = 0$

60. $-2x^2 + 6x = -9$

61. $2x^2 - 5x = 0$

62. $3x^2 + 5x = 0$

[10.4] *Indicate the axis of symmetry, the coordinates of the vertex, and whether the parabola opens upward or downward.*

63. $y = x^2 - 2x - 3$

64. $y = x^2 - 10x + 24$

65. $y = x^2 + 7x + 12$

66. $y = -x^2 - 2x + 15$

67. $y = x^2 - 3x$

68. $y = 2x^2 + 7x + 3$

69. $y = -x^2 - 8$

70. $y = -4x^2 + 8x + 5$

71. $y = -x^2 - x + 20$

72. $y = 3x^2 + 5x - 8$

Graph each quadratic equation, and determine the real roots if they exist. If they do not exist, so state.

73. $y = x^2 + 6x$

74. $y = -2x^2 + 8$

75. $y = x^2 + 2x - 8$

76. $y = x^2 - x - 2$

77. $y = x^2 + 5x + 4$

78. $y = x^2 + 4x + 3$

79. $y = -2x^2 + 3x - 2$

80. $y = 3x^2 - 4x + 1$

81. $y = 4x^2 - 8x + 6$

82. $y = -3x^2 - 14x + 5$

83. $y = -x^2 - 6x - 4$

84. $y = 2x^2 + 5x - 12$

[10.2–10.3] *Solve each problem.*

85. The product of two positive integers is 88. Find the two numbers if the larger one is 3 greater than the smaller.

86. The length of a rectangle is 5 feet less than twice its width. Find the length and width of the rectangle if its area is 63 square feet.

Practice Test

1. Solve the equation $x^2 + 1 = 21$.

2. Solve the equation $(2x - 3)^2 = 35$.

3. Solve by completing the square: $x^2 - 4x = 60$.

4. Solve by completing the square: $x^2 = -x + 12$.

5. Solve by the quadratic formula: $x^2 - 5x - 6 = 0$.

6. Solve by the quadratic formula: $2x^2 + 5 = -8x$.

7. Solve by the method of your choice: $3x^2 - 5x = 0$.

8. Solve by the method of your choice: $2x^2 + 9x = 5$.

9. Determine whether the following equation has two distinct real solutions, a single unique solution, or no real solution: $3x^2 - 4x + 2 = 0$.

10. Indicate the axis of symmetry, the coordinates of the vertex, and whether the graph opens upward or downward: $y = -x^2 + 3x + 8$.

11. Graph the following equation, and determine the roots, if they exist: $y = x^2 + 2x - 8$.

12. Graph the following equation, and determine the roots, if they exist: $y = -x^2 + 6x - 9$.

13. The length of a rectangle is 1 foot greater than three times its width. Find the length and width of the rectangle if its area is 30 square feet.

Cumulative Review Test

1. Evaluate $-x^2y + y^2 - 3xy$ when $x = -3$ and $y = 4$.

2. Solve the equation $\frac{1}{4}x + \frac{3}{5}x = \frac{1}{3}(x + 2)$.

3. Find the length of side x.

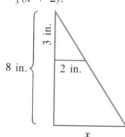

4. Solve the inequality and graph the solution on the number line: $2(x - 3) \le 6x - 5$.

5. Solve the formula $A = \dfrac{m + n + P}{3}$ for P.

6. Simplify $(6x^2y^4)^3(2x^4y^5)^2$.

7. Divide $\dfrac{x^2 + 6x + 5}{x + 2}$.

8. Factor by grouping $2x^2 - 3xy - 4xy + 6y^2$.

9. Factor $4x^2 - 14x - 8$.

10. Add $\dfrac{4}{a^2 - 16} + \dfrac{2}{(a - 4)^2}$.

11. Solve the equation $x + \dfrac{48}{x} = 14$.

12. Graph the equation $3x + 5y = 10$.

13. Solve the system of equations by the addition method.

$$3x - 4y = 12$$
$$4x - 3y = 6$$

14. Simplify $\sqrt{\dfrac{3x^2y^3}{54x}}$.

15. Add $2\sqrt{28} - 3\sqrt{7} + \sqrt{63}$.

16. Solve the equation $\sqrt{x^2 + 5} = x + 1$.

17. Solve the equation $2x^2 - 4x - 5 = 0$ using the quadratic formula.

18. If 4 pounds of fertilizer can fertilize 500 square feet of lawn, how many pounds of fertilizer are needed to fertilize 3,200 square feet of lawn?

19. The length of a rectangle is 3 feet less than three times its width. Find the width and length of the rectangle if its perimeter is 74 feet.

20. Willie jogs 3 miles per hour faster than she walks. She jogs for 2 miles and then walks for 2 miles. If the total time of her outing is 1 hour, find the rate at which she walks and jogs.

Appendices

A. Review of Decimals and Percent
B. Finding the Greatest Common Factor and Least Common Denominator
C. Geometry
D. Squares and Square Roots

Review of Decimals and Percent

> **To Add or Subtract Numbers Containing Decimal Points**
>
> 1. Align the numbers by the decimal points.
> 2. Add or subtract the numbers in the corresponding columns.
> 3. Place the decimal point in the answer directly below the decimal points in the numbers being added or subtracted.

EXAMPLE 1   Add $4.6 + 13.813 + 9.02$.

Solution:
$$\begin{array}{r} 4.600 \\ 13.813 \\ +\ 9.020 \\ \hline 27.433 \end{array}$$

EXAMPLE 2   Subtract 3.062 from 25.9.

Solution:
$$\begin{array}{r} 25.900 \\ -\ 3.062 \\ \hline 22.838 \end{array}$$

449

> **To Multiply Numbers Containing Decimal Points**
>
> 1. Multiply as if the factors were whole numbers.
> 2. Determine the total number of digits to the right of the decimal points in the factors.
> 3. Place the decimal point in the product so that the product contains the same number of digits to the right of the decimal as the total found in step 2. For example, if there are a total of three digits to the right of the decimal points in the factors, there must be three digits to the right of the decimal point in the product.

EXAMPLE 3   Multiply 2.34×1.9.

Solution:

$$
\begin{array}{r}
2.34 \longleftarrow \text{two digits to the right of the decimal point} \\
\times\ 1.9 \longleftarrow \text{one digit to the right of the decimal point} \\
\hline
2106 \\
234 \\
\hline
4.446 \longleftarrow \text{three digits to the right of the decimal point in the answer}
\end{array}
$$

EXAMPLE 4   Multiply 2.13×0.02.

Solution:

$$
\begin{array}{r}
2.13 \longleftarrow \text{two digits to the right of the decimal point} \\
\times\ 0.02 \longleftarrow \text{two digits to the right of the decimal point} \\
\hline
0.0426 \longleftarrow \text{four digits to the right of the decimal point in the answer}
\end{array}
$$

Note that it was necessary to add a zero preceding the digit 4 in the answer in order to have four digits to the right of the decimal point.

> **To Divide Numbers Containing Decimal Points**
>
> 1. Multiply both the dividend and divisor by a number that will result in the divisor being a whole number.
> 2. Divide as if working with whole numbers.
> 3. Place the decimal point in the quotient directly above the decimal point in the dividend.

To make the divisor a whole number, multiply *both* the dividend and divisor by 10 if the divisor is given in tenths, by 100 if the divisor is given in hundredths, by 1000 if the divisor is given in thousandths, and so on. Multiplying both the numerator and denominator by the same nonzero number is the same as multiplying the fraction by 1. Therefore, the value of the fraction is unchanged.

EXAMPLE 5   Divide $\dfrac{1.956}{0.12}$.

Solution: Since the divisor, 0.12, is twelve-hundredths, we multiply both the divisor and dividend by 100.

$$
\frac{1.956}{0.12} \times \frac{100}{100} = \frac{195.6}{12}.
$$

Now divide.

$$
\begin{array}{r}
16.3 \\
12. \overline{)195.6} \\
\underline{12} \\
75 \\
\underline{72} \\
36 \\
\underline{36} \\
0
\end{array}
$$

The decimal point in the answer is placed directly above the decimal point in the dividend. Thus $1.956/0.12 = 16.3$. ∎

EXAMPLE 6   Divide 0.26 by 10.4.

Solution: First, multiply both the dividend and divisor by 10.

$$
\frac{0.26}{10.4} \times \frac{10}{10} = \frac{2.6}{104.}
$$

Now divide.

$$
\begin{array}{r}
0.025 \\
104\overline{)2.600} \\
\underline{2\ 08} \\
520 \\
\underline{520} \\
0
\end{array}
$$

Note that a zero had to be placed before the digit 2 in the quotient.

$$
\frac{0.26}{10.4} = 0.025
$$ ∎

Percent

One of the main uses of decimals comes from percent problems. The word percent means "per one hundred." The symbol % means percent. One percent means "one per hundred," or

$$
1\% = \frac{1}{100} \quad \text{or} \quad 1\% = 0.01
$$

EXAMPLE 7   Convert 16% to a decimal.

Solution: Since $1\% = 0.01$
$16\% = 16(0.01) = 0.16$ ∎

EXAMPLE 8   Convert 2.3% to a decimal.

Solution: $2.3\% = 2.3(0.01) = 0.023.$ ∎

EXAMPLE 9   Convert 1.14 to a percent.

Solution: To change a decimal number to a percent, we multiply the number by 100%.

$$1.14 = 1.14 \times 100\% = 114\%$$

Often you will need to find an amount that is a certain percent of a number. For example, when you purchase an item in a state or county that has a sales tax you must often pay a percent of the item's price as the sales tax. Examples 10 and 11 show how to find a certain percent of a number.

EXAMPLE 10   Find 12% of 200.

Solution: To find a percent of a number, use multiplication. Change 12% to a decimal number, then multiply by 200.

$$(0.12)(200) = 24$$

Thus 12% of 200 is 24.

EXAMPLE 11   In Monroe County in New York State there is a 7% sales tax.
(a) Find the sales tax on a stereo system that cost $580.
(b) Find the total cost of the system, including tax.

Solution: (a) The sales tax is 7% of 580.

$$(0.07)(580) = 40.60$$

The sales tax is $40.60.
(b) The total cost is the purchase price plus the sales tax:

$$\text{Total cost} = \$580 + \$40.60 = \$620.60$$

Appendix B

Finding the Greatest Common Factor and Least Common Denominator

Prime Factorization

In Section 1.2 we mentioned that to reduce fractions to their lowest terms you can divide both the numerator and denominator by the *greatest common factor* (GCF). One method to find the GCF is to use *prime factorization*. Prime factorization is the process of writing a given number as a product of prime numbers. *Prime numbers* are natural numbers, excluding 1, that can be divided by only themselves and 1. The first ten prime numbers are 2, 3, 5, 7, 11, 13, 17, 19, 23, and 29. Can you find the next prime number? If you answered 31, you answered correctly.

To write a number as a product of primes, we can use a *tree diagram*. Begin by selecting any two numbers whose product is the given number. Then continue breaking down each of these numbers into prime numbers, as shown in Example 1.

EXAMPLE 1   Determine the prime factorization of the number 120.

Solution: We will use three different tree diagrams to illustrate the prime factorization of 120.

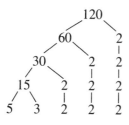   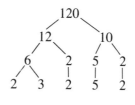   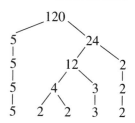

or $2 \cdot 2 \cdot 2 \cdot 3 \cdot 5$ or $2 \cdot 2 \cdot 2 \cdot 3 \cdot 5$ or $2 \cdot 2 \cdot 2 \cdot 3 \cdot 5$

Note that no matter how you start, if you do not make a mistake, you find that the prime factorization of $120 = 2 \cdot 2 \cdot 2 \cdot 3 \cdot 5$. There are many other ways 120 could be broken down, but all will lead to the prime factorization $2 \cdot 2 \cdot 2 \cdot 3 \cdot 5$. ■

Greatest Common Factor

The greatest common factor (GCF) of two natural numbers is the greatest integer that is a factor of both numbers. We use the GCF when reducing fractions to lowest terms.

To Find the Greatest Common Factor of a Given Numerator and Denominator

1. Write both the numerator and the denominator as a product of primes.
2. Determine all the prime factors that are common to both prime factorizations.
3. Multiply the prime factors found in step 2 to obtain the GCF.

EXAMPLE 2   Consider the fraction $\dfrac{108}{156}$.

(a) Find the GCF of 108 and 156.

(b) Reduce $\dfrac{108}{156}$ to lowest terms.

Solution: (a) First determine the prime factorizations of both 108 and 156.

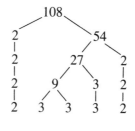   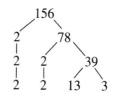

$2 \cdot 2 \cdot 3 \cdot 3 \cdot 3$ $2 \cdot 2 \cdot 3 \cdot 13$

There are two 2's and one 3 common to both prime factorizations; thus

$$\text{GCF} = 2 \cdot 2 \cdot 3 = 12$$

The greatest common factor of 108 and 156 is 12. Twelve is the greatest integer that divides both 108 and 156.

(b) To reduce $\dfrac{108}{156}$, we divide both the numerator and denominator by the GCF, 12.

$$\frac{108 \div 12}{156 \div 12} = \frac{9}{13}$$

Thus $\dfrac{108}{156}$ reduces to $\dfrac{9}{13}$. ∎

Least Common Denominator

When adding two or more fractions, you must write each fraction with a common denominator. The best denominator to use is the *least common denominator* (LCD). The LCD is the smallest number that each denominator divides. Sometimes the least common denominator is referred to as the *least common multiple* of the denominators.

To Find the Least Common Denominator of Two or More Fractions

1. Determine the prime factorization of each denominator.
2. For each prime number, determine the maximum number of times that prime number appears in any of the prime factorizations.
3. Multiply together all the prime numbers, including each prime number the maximum number of times it appears in any of the prime factorizations. The product of all these prime numbers will be the LCD.

Example 3 illustrates the procedure to determine the LCD.

EXAMPLE 3 Consider $\dfrac{7}{108} + \dfrac{5}{156}$.

(a) Determine the least common denominator.
(b) Add the fractions.

Solution: (a) We determined in Example 2 that

$$108 = 2 \cdot 2 \cdot 3 \cdot 3 \cdot 3 \quad \text{and} \quad 156 = 2 \cdot 2 \cdot 3 \cdot 13$$

We can see that the maximum number of 2's that appear in either prime factorization is two (there are two 2's in both factorizations), the maximum number of 3's is three, and the maximum number of 13's is one. Multiply as follows:

$$2 \cdot 2 \cdot 3 \cdot 3 \cdot 3 \cdot 13 = 1404$$

Thus the least common denominator is 1404. This is the smallest number that both 108 and 156 divide.

(b) To add the fractions, we need to write both fractions with a common denominator. The best common denominator to use is the LCD. Since $1404 \div 108 = 13$, we will multiply $\dfrac{7}{108}$ by $\dfrac{13}{13}$. Since $1404 \div 156 = 9$, we will multiply $\dfrac{5}{156}$ by $\dfrac{9}{9}$.

$$\frac{7}{108} \cdot \frac{13}{13} + \frac{5}{156} \cdot \frac{9}{9} = \frac{91}{1404} + \frac{45}{1404} = \frac{136}{1404} = \frac{34}{351}$$

Thus $\dfrac{7}{108} + \dfrac{5}{156} = \dfrac{34}{351}$. ∎

Appendix C

Geometry

Angles

This appendix introduces or reviews important geometric concepts. Table C1 gives the names and descriptions of various types of angles.

TABLE C.1

Angle	Sketch of Angle
An **acute angle** is an angle whose measure is between 0° and 90°.	
A **right angle** is an angle whose measure is 90°.	
An **obtuse angle** is an angle whose measure is between 90° and 180°.	
A **straight angle** is an angle whose measure is 180°.	
Two angles are **complementary angles** when the sum of their measures is 90°. Each angle is the complement of the other. Angles A and B are complementary angles.	
Two angles are **supplementary angles** when the sum of their measures is 180°. Each angle is the supplement of the other. Angles A and B are supplementary angles.	

When two lines intersect, four angles are formed as shown in Figure C1. The pair of opposite angles formed by the intersecting lines are called **vertical angles.**

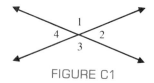

FIGURE C1

Angles 1 and 3 are vertical angles. Angles 2 and 4 are also vertical angles. *Vertical angles have equal measures.* Thus angle 1, symbolized by $\angle 1$, is equal to angle 3, symbolized by $\angle 3$. We can write $\angle 1 = \angle 3$. Similarly, $\angle 2 = \angle 4$.

Parallel and Perpendicular Lines

Parallel lines are two lines in the same plane that do not intersect (see Figure C2). **Perpendicular lines** are lines that intersect at right angles (see Figure C3).

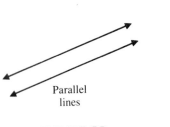

Parallel lines

FIGURE C2

Perpendicular lines

FIGURE C3

A **transversal** is a line that intersects two or more lines at different points. When a transversal line intersects two other lines, eight angles are formed, as illustrated in Figure C4. Some of these angles are given special names.

Interior angles: 3, 4, 5, 6

Exterior angles: 1, 2, 7, 8

Pairs of corresponding angles: 1 and 5; 2 and 6; 3 and 7; 4 and 8

Pairs of alternate interior angles: 3 and 6; 4 and 5

Pairs of alternate exterior angles: 1 and 8; 2 and 7

FIGURE C4

Parallel Lines Cut by a Transversal

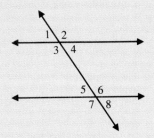

When two parallel lines are cut by a transversal:

1. Corresponding angles are equal ($\angle 1 = \angle 5$, $\angle 2 = \angle 6$, $\angle 3 = \angle 7$, $\angle 4 = \angle 8$).
2. Alternate interior angles are equal ($\angle 3 = \angle 6$, $\angle 4 = \angle 5$).
3. Alternate exterior angles are equal ($\angle 1 = \angle 8$, $\angle 2 = \angle 7$).

EXAMPLE 1   If line 1 and line 2 are parallel lines and $\angle 1 = 112°$, find the measure of angles 2 through 8.

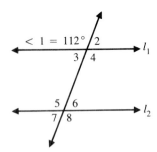

Solution: Angles 1 and 2 are supplementary. So $\angle 2$ is $180° - 112° = 68°$. Angles 1 and 4 are equal since they are vertical angles. Thus $\angle 4 = 112°$. Angles 1 and 5 are corresponding angles. Thus $\angle 5$ and its vertical angle, $\angle 8$, both measure $112°$. Angles 2, 3, 6, and 7 are all equal and measure $68°$. ■

Polygons

A **polygon** is a closed figure in a plane determined by three or more line segments. Some polygons are illustrated in Figure C5.

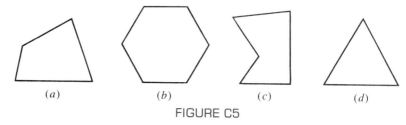

(a)   (b)   (c)   (d)

FIGURE C5

A **regular polygon** has sides that are all the same length, and interior angles that all have the same measure. Figures C5(b) and (d) are regular polygons.

Sum of the Interior Angles of a Polygon

The sum of the interior angles of a polygon can be found by the formula

$$\text{Sum} = (n - 2)180°$$

where n is the number of sides of the polygon.

EXAMPLE 2 Find the sum of the measures of the interior angles of (a) a triangle; (b) a quadrilateral (4 sides), (c) an octagon (8 sides).

Solution: (a) Since $n = 3$, we write
Sum $= (n - 1)180°$
$= (3 - 2)180° = 1(180°) = 180°$
The sum of the measures of the interior angles in a triangle is $180°$.
(b) Sum $= (n - 2)180°$
$= (4 - 2)180° = 2(180°) = 360°$
The sum of the measures of the interior angles in a quadrilateral is $360°$.
(c) Sum $= (n - 2)(180°) = (8 - 2)180° = 6(180°) = 1040°$
The sum of the measures of the interior angles in an octagon is $1040°$. ■

Triangles

Now we will briefly describe several types of triangles in Table C2.

TABLE C.2

Triangle	Sketch of Triangle
An **acute triangle** is one that has three acute angles (angles of less than $90°$).	
An **obtuse triangle** has one obtuse angle (an angle of greater than $90°$).	

A **right triangle** has one right angle (an angle equal to 90°). The longest side of a right triangle is opposite the right angle and is called the **hypotenuse.** The other two sides are called the **legs.**

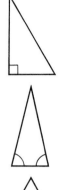

An **isosceles triangle** has two sides of equal length. The angles opposite the equal sides have the same measure.

An **equilateral triangle** has three sides of equal length. It also has three equal angles that measure 60° each.

When two sides of a *right triangle* are known, the third side can be found using the **Pythagorean theorem,** $a^2 + b^2 = c^2$, where a and b are the legs and c is the hypotenuse of the triangle.

Congruent and Similar Figures

If two triangles are **congruent**, it means that the two triangles are identical in size and shape. Two congruent triangles could be placed one on top of the other if we were able to move and rearrange them.

Two triangles are congruent if any one of the following statements is true.

1. Two angles of one triangle equal two angles of the other triangle, and the lengths of the sides between each pair of angles are equal. This method of showing that triangles are congruent is called the *angle, side, angle* method.

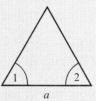 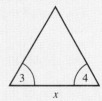

$$< 1 = < 3$$
$$< 2 = < 4$$
$$a = x$$

2. Corresponding sides of both triangles are equal. This is called the *side, side, side* method.

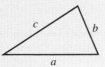

$$a = x$$
$$b = y$$
$$c = z$$

3. Two corresponding pairs of sides are equal, and the angle between them is equal. This is referred to as the *side, angle, side* method.

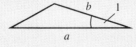 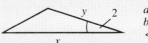

$$a = x$$
$$b = y$$
$$< 1 = < 2$$

EXAMPLE 3 Determine if the two triangles are congruent.

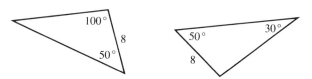

Solution: The unknown angle in the figure on the right must measure 100° since the sum of the angles of a triangle is 180°. Both triangles have the same two angles (100° and 50°), with the same length side between them, 8 units. Thus, these two triangles are congruent by the angle, side, angle method. ■

Two triangles are **similar** if three angles of one triangle equal the three angles of the other triangle and corresponding sides are in proportion. Similar figures do not have to be exactly the same size but must have the same general shape.

Two triangles are similar if any one of the following statements is true.

1. Two angles of one triangle equal two angles of the other triangle.

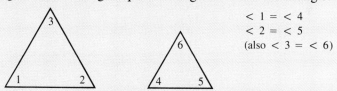

$$< 1 = < 4$$
$$< 2 = < 5$$
$$(\text{also } < 3 = < 6)$$

2. Corresponding sides of the two triangles are proportional.

$$\frac{a}{x} = \frac{b}{y} = \frac{c}{z}$$

3. Two corresponding pairs of sides are proportional, and the angle between them is equal.

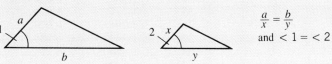

$$\frac{a}{x} = \frac{b}{y}$$
$$\text{and } < 1 = < 2$$

EXAMPLE 4 Are the triangles are *ABC* and *AB′C′* similar?

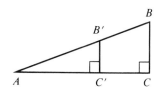

Solution: Angle *A* is common to both triangles. Since angle *C* and angle *C′* are equal (both 90°), then ∠*B* and ∠*B′* must be equal. Since the three angles of triangle *ABC* equal the three angles of triangle *AB′C′*, the two triangles are similar. ■

Appendix D

Squares and Square Roots

Number	Square	Square Root	Number	Square	Square Root
1	1	1.000	51	2,601	7.141
2	4	1.414	52	2,704	7.211
3	9	1.732	53	2,809	7.280
4	16	2.000	54	2,916	7.348
5	25	2.236	55	3,025	7.416
6	36	2.449	56	3,136	7.483
7	49	2.646	57	3,249	7.550
8	64	2.828	58	3,364	7.616
9	81	3.000	59	3,481	7.681
10	100	3.162	60	3,600	7.746
11	121	3.317	61	3,721	7.810
12	144	3.464	62	3,844	7.874
13	169	3.606	63	3,969	7.937
14	196	3.742	64	4,096	8.000
15	225	3.873	65	4,225	8.062
16	256	4.000	66	4,356	8.124
17	289	4.123	67	4,489	8.185
18	324	4.243	68	4,624	8.246
19	361	4.359	69	4,761	8.307
20	400	4.472	70	4,900	8.367
21	441	4.583	71	5,041	8.426
22	484	4.690	72	5,184	8.485
23	529	4.796	73	5,329	8.544
24	576	4.899	74	5,476	8.602
25	625	5.000	75	5,625	8.660
26	676	5.099	76	5,776	8.718
27	729	5.196	77	5,929	8.775
28	784	5.292	78	6,084	8.832
29	841	5.385	79	6,241	8.888
30	900	5.477	80	6,400	8.944
31	961	5.568	81	6,561	9.000
32	1,024	5.657	82	6,724	9.055
33	1,089	5.745	83	6,889	9.110
34	1,156	5.831	84	7,056	9.165
35	1,225	5.916	85	7,225	9.220
36	1,296	6.000	86	7,396	9.274
37	1,369	6.083	87	7,569	9.327
38	1,444	6.164	88	7,744	9.381
39	1,521	6.245	89	7,921	9.434
40	1,600	6.325	90	8,100	9.487
41	1,681	6.403	91	8,281	9.539
42	1,764	6.481	92	8,464	9.592
43	1,849	6.557	93	8,649	9.644
44	1,936	6.633	94	8,836	9.695
45	2,025	6.708	95	9,025	9.747
46	2,116	6.782	96	9,216	9.798
47	2,209	6.856	97	9,409	9.849
48	2,304	6.928	98	9,604	9.899
49	2,401	7.000	99	9,801	9.950
50	2,500	7.071	100	10,000	10.000

Answers

Exercise Set 1.2

1. $\frac{1}{3}$ **3.** $\frac{3}{4}$ **5.** $\frac{1}{2}$ **7.** $\frac{3}{7}$ **9.** $\frac{5}{8}$ **11.** Lowest terms **13.** $\frac{4}{3}$ or $1\frac{1}{3}$ **15.** $\frac{3}{10}$ **17.** $\frac{5}{28}$ **19.** $\frac{1}{12}$ **21.** $\frac{5}{3}$ or $1\frac{2}{3}$
23. $\frac{5}{16}$ **25.** 6 **27.** $\frac{5}{12}$ **29.** $\frac{13}{45}$ **31.** $\frac{19}{14}$ or $1\frac{5}{14}$ **33.** 12 **35.** 1 **37.** 2 **39.** $\frac{3}{5}$ **41.** $\frac{1}{4}$ **43.** 1
45. $\frac{3}{29}$ **47.** $\frac{37}{30}$ or $1\frac{7}{30}$ **49.** $\frac{1}{5}$ **51.** $\frac{4}{15}$ **53.** $\frac{2}{9}$ **55.** $\frac{29}{24}$ or $1\frac{5}{24}$ **57.** $\frac{1}{6}$ **59.** $\frac{47}{12}$ or $3\frac{11}{12}$ **61.** $\frac{23}{6}$ or $3\frac{5}{6}$
63. $\frac{52}{15}$ or $3\frac{7}{15}$ **65.** $\frac{81}{20}$ or $4\frac{1}{20}$ **67.** $13\frac{3}{4}$ yd **69.** $13\frac{11}{16}$ in. **71.** $8\frac{7}{16}$ ft **73.** $11\frac{7}{8}$ ft **75.** $\frac{27}{8}$ or $3\frac{3}{8}$ tsp
77. $\frac{37}{24}$ or $1\frac{13}{24}$ acre **79.** **(a)** $\frac{5}{16}$ **(b)** $\frac{20}{9}$ or $2\frac{2}{9}$ **(c)** $\frac{29}{24}$ or $1\frac{5}{24}$ **(d)** $\frac{11}{24}$

Just for Fun **1.** Rice and water 1 cup, salt $\frac{3}{8}$ tsp, butter $1\frac{1}{2}$ tsp

Exercise Set 1.3

1. $\{\ldots, -3, -2, -1, 0, 1, 2, 3, \ldots\}$ **3.** $\{1, 2, 3, 4, \ldots\}$ **5.** $\{\ldots, -3, -2, -1\}$ **7.** T **9.** T **11.** F
13. F **15.** T **17.** F **19.** T **21.** F **23.** T **25.** T **27.** F **29.** T **31.** T **33.** T **35.** F
37. F **39.** **(a)** 7, 9 **(b)** 7, 0, 9 **(c)** $-6, 7, 0, 9$ **(d)** $-6, 7, 12.4, -\frac{9}{5}, -2\frac{1}{4}, 0, 9, 0.35$ **(e)** $\sqrt{3}, \sqrt{7}$
(f) $-6, 7, 12.4, -\frac{9}{5}, -2\frac{1}{4}, \sqrt{3}, 0, 9, \sqrt{7}, 0.35$ **41.** **(a)** 5 **(b)** 5 **(c)** -300 **(d)** $5, -300$
(e) $\frac{1}{2}, 4\frac{1}{2}, \frac{5}{12}, -1.67, 5, -300, -9\frac{1}{2}$ **(f)** $\sqrt{2}, -\sqrt{2}$ **(g)** $\frac{1}{2}, \sqrt{2}, -\sqrt{2}, 4\frac{1}{2}, \frac{5}{12}, -1.67, 5, -300, -9\frac{1}{2}$
43. $-\frac{2}{3}, \frac{1}{2}, 6.3$ **45.** $-\sqrt{7}, \sqrt{3}, \sqrt{6}$ **47.** $-5, 0, 4$ **49.** $-13, -5, -1$ **51.** $1.5, 3, 6\frac{1}{4}$ **53.** $-7, 1, 5$

Cumulative Review Exercises

55. $\frac{20}{3}$ **56.** $5\frac{1}{3}$ **57.** $\frac{49}{40}$ or $1\frac{9}{40}$ **58.** $\frac{70}{27}$ or $2\frac{16}{27}$

Exercise Set 1.4

1. $2 < 3$ **3.** $-3 < 0$ **5.** $\frac{1}{2} > -\frac{2}{3}$ **7.** $0.2 < 0.4$ **9.** $\frac{2}{5} > -1$ **11.** $4 > -4$ **13.** $-2.1 < -2$
15. $\frac{5}{9} > -\frac{5}{9}$ **17.** $-\frac{3}{2} < \frac{3}{2}$ **19.** $0.49 > 0.43$ **21.** $5 > -7$ **23.** $-0.006 > -0.007$ **25.** $\frac{3}{5} < 1$
27. $-\frac{2}{3} > -3$ **29.** $8 > |-7|$ **31.** $|0| < \frac{2}{3}$ **33.** $|-3| < |-4|$ **35.** $4 < |-\frac{9}{2}|$ **37.** $|-\frac{6}{2}| > |-\frac{2}{6}|$
39. $4, -4$ **41.** $2, -2$ **43.** The distance between 0 and the number on the number line.

Cumulative Review Exercises

44. $\frac{31}{24}$ or $1\frac{7}{24}$ **45.** $\{0, 1, 2, 3, \ldots\}$ **46.** $\{1, 2, 3, 4, \ldots\}$
47. **(a)** 5 **(b)** 5, 0 **(c)** $5, -2, 0$ **(d)** $5, -2, 0, \frac{1}{3}, -\frac{5}{9}, 2.3$ **(e)** $\sqrt{3}$ **(f)** $5, -2, 0, \frac{1}{3}, \sqrt{3}, -\frac{5}{9}, 2.3$

Exercise Set 1.5

1. -18 **3.** 32 **5.** 0 **7.** $-\frac{5}{3}$ **9.** $-\frac{3}{5}$ **11.** -0.63 **13.** $-2\frac{1}{2}$ **15.** 3.1 **17.** 7 **19.** 1 **21.** -6
23. 0 **25.** 0 **27.** -10 **29.** 0 **31.** -10 **33.** 0 **35.** -6 **37.** 3 **39.** -9 **41.** 9 **43.** -27
45. -44 **47.** -26 **49.** 5 **51.** -20 **53.** -31 **55.** 91 **57.** -140 **59.** True **61.** False
63. False **65.** \$81 **67.** 454 m **69.** 2600 ft

Cumulative Review Exercises

73. 1　　**74.** $\frac{43}{16}$ or $2\frac{11}{16}$　　**75.** $|-3| > 2$　　**76.** $8 > |-7|$

Exercise Set 1.6

1. 3　**3.** -1　**5.** 0　**7.** -3　**9.** -6　**11.** 6　**13.** -6　**15.** 6　**17.** -8　**19.** 2　**21.** 2　**23.** 9
25. 0　**27.** -18　**29.** -2　**31.** 0　**33.** 0　**35.** 0　**37.** -1　**39.** -5　**41.** -41　**43.** -3
45. -180　**47.** -110　**49.** 140　**51.** 0　**53.** -82　**55.** 5　**57.** -18　**59.** -16　**61.** 10　**63.** -2
65. 13　**67.** 0　**69.** -4　**71.** 11　**73.** 81　**75.** 7　**77.** -2　**79.** -15　**81.** -2　**83.** 0　**85.** 43
87. -6　**89.** -9　**91.** 35　**93.** -21　**95.** -12　**97.** -3　**99.** 18　**101.** -9　**103.** 3500 ft
105. 65,226 ft　**107.** **(a)** yes **(b)** yes, $-3 + (-5) = -3 - 5 = -8$　**109.** **(b)** -15

Cumulative Review Exercises

110. $\{\ldots, -3, -2, -1, 0, 1, 2, 3, \ldots\}$
111. The set of rational numbers together with the set of irrational numbers form the set of real numbers.
112. $|-3| > -5$　**113.** $|-6| < |-7|$

Just for Fun　**1.** -5　**2.** -50　**3.** 50

Exercise Set 1.7

1. 12　**3.** -9　**5.** -32　**7.** -9　**9.** 12　**11.** 36　**13.** -30　**15.** -60　**17.** 0　**19.** 16　**21.** 24
23. -15　**25.** 96　**27.** 81　**29.** -10　**31.** 36　**33.** 0　**35.** -1　**37.** -120　**39.** -12　**41.** 84
43. $-\frac{3}{10}$　**45.** $\frac{14}{27}$　**47.** 4　**49.** $-\frac{15}{28}$　**51.** 3　**53.** 4　**55.** 4　**57.** -4　**59.** -18　**61.** 5　**63.** 6
65. 5　**67.** -1　**69.** -4　**71.** 9　**73.** 0　**75.** 1　**77.** 0　**79.** -4　**81.** 3　**83.** -12　**85.** -4
87. 45　**89.** $-\frac{3}{4}$　**91.** $\frac{3}{80}$　**93.** 1　**95.** $-\frac{144}{5}$　**97.** $-\frac{36}{5}$　**99.** 0　**101.** Indeterminate　**103.** 0
105. Undefined　**107.** 0　**109.** False　**111.** True　**113.** True　**115.** False　**117.** False　**119.** True
121. Two numbers with like signs: The product or quotient is positive. Two numbers with unlike signs: the product or
quotient is negative.
123. The product will be negative. There are 17 pairs of numbers ($34 \div 2 = 17$). The product of each pair is a negative
number. The product of 17 negative numbers (an odd number of negative numbers) is a negative number.

Cumulative Review Exercises

124. $\frac{25}{7}$ or $3\frac{4}{7}$　**125.** -2　**126.** -3　**127.** 3

Just for Fun　**1.** 1　**2.** -1

Exercise Set 1.8

1. 9　**3.** 8　**5.** 27　**7.** 216　**9.** -8　**11.** -1　**13.** 27　**15.** -36　**17.** 36　**19.** 16　**21.** 5
23. 16　**25.** -16　**27.** -64　**29.** 225　**31.** 80　**33.** 32　**35.** -75　**37.** 144　**39.** xyz^2　**41.** x^4z
43. a^3b^3　**45.** x^2yz^2　**47.** x^3y^2　**49.** xy^4　**51.** 5^2y^2z　**53.** xxy　**55.** $xyyy$　**57.** $xyyzzz$
59. $3 \cdot 3yz$　**61.** $2 \cdot 2 \cdot 2xxxy$　**63.** $(-2)(-2)yyyz$　**65.** 9, -9　**67.** 16, -16　**69.** 4, -4
71. 49, -49　**73.** 1, -1　**75.** $\frac{1}{4}$, $-\frac{1}{4}$　**77.** False　**79.** False　**81.** True　**83.** True　**85.** True
87. Any nonzero number will be positive when squared.
89. Positive; an even number of negative numbers are being multiplied.

Cumulative Review Exercises

90. 18　**91.** -5　**92.** $\frac{10}{3}$ or $3\frac{1}{3}$　**93.** 0

Just for Fun
1. 2^5　**2.** 3^5　**3.** x^{m+n}　**4.** $2^1 = 2$　**5.** 3^2　**6.** x^{m-n}　**7.** 2^6　**8.** 3^6　**9.** x^{mn}　**10.** 2^2x^2　**11.** 3^2x^2
12. a^2x^2

Exercise Set 1.9

1. 23 **3.** 5 **5.** 13 **7.** −4 **9.** 16 **11.** 29 **13.** −13 **15.** −2 **17.** 18 **19.** 10 **21.** 7
23. 36 **25.** −441 **27.** $\frac{1}{2}$ **29.** 10 **31.** 12 **33.** 9 **35.** 169 **37.** 112 **39.** 25 **41.** 129.81
43. 26.04 **45.** $\frac{71}{112}$ **47.** $\frac{1}{4}$ **49.** $\frac{170}{9}$ **51.** $[(6 \cdot 3) - 4] - 2$, 12 **53.** $9[[(20 \div 5) + 12] - 8]$, 72
55. $(\frac{4}{5} + \frac{3}{7}) \cdot \frac{2}{3}$, $\frac{86}{105}$ **57.** 2 **59.** 10 **61.** 3 **63.** −7 **65.** −25 **67.** 75 **69.** −20 **71.** 0 **73.** −5
75. 21 **77.** 33 **79.** −18 **81.** −3 **83.** 49 **87. (b)** −91

Cumulative Review Exercises

88. 144 **89. (a)** 25 **(b)** −25 **90.** 16 **91.** −16

Just for Fun **1.** 160 **2.** 177 **3.** −312

Exercise Set 1.10

1. Distributive property **3.** Commutative property of multiplication **5.** Distributive property
7. Associative property of multiplication **9.** Distributive property **11.** 4 + 3 **13.** $(−6 \cdot 4) \cdot 2$ **15.** $(y)(6)$
17. $1 \cdot x + 1 \cdot y$ or $x + y$ **19.** $3y + 4x$ **21.** $5(x + y)$ **23.** $3(x + 2)$ **25.** $3x + (4 + 6)$ **27.** $(x + y)3$
29. $4x + 4y + 12$ **31.** Commutative property of addition **33.** Distributive property
35. Commutative property of addition **37.** Distributive property **39.** yes **41.** no

Cumulative Review Exercises

43. $\frac{49}{15}$ or $3\frac{4}{15}$ **44.** $\frac{23}{16}$ or $1\frac{7}{16}$ **45.** 45 **46.** −25

Chapter 1 Review Exercises

1. $\frac{1}{2}$ **2.** $\frac{9}{25}$ **3.** $\frac{25}{36}$ **4.** $\frac{7}{6}$ or $1\frac{1}{6}$ **5.** $\frac{19}{72}$ **6.** $\frac{17}{15}$ or $1\frac{2}{15}$ **7.** $\{1, 2, 3, \ldots\}$ **8.** $\{0, 1, 2, 3, \ldots\}$
9. $\{\ldots, −3, −2, −1, 0, 1, 2, 3, \ldots\}$ **10.** {quotient of two integers, denominator not 0}
11. {all numbers that can be represented on the real number line} **12. (a)** 3, 426 **(b)** 3, 0, 426 **(c)** 3, −5, −12, 0, 426
(d) 3, −5, −12, 0, $\frac{1}{2}$, −0.62, 426, −$3\frac{1}{4}$ **(e)** $\sqrt{7}$ **(f)** 3, −5, −12, 0, $\frac{1}{2}$, −0.62, $\sqrt{7}$, 426, −$3\frac{1}{4}$ **13. (a)** 1 **(b)** 1
(c) −8, −9, **(d)** −8, −9, 1 **(e)** −2.3, −8, −9, $1\frac{1}{2}$, 1, −$\frac{3}{17}$ **(f)** −2.3, −8, −9, $1\frac{1}{2}$, $\sqrt{2}$, −$\sqrt{2}$, 1, −$\frac{3}{17}$ **14.** >
15. < **16.** > **17.** > **18.** < **19.** > **20.** < **21.** > **22.** < **23.** = **24.** 3 **25.** −9
26. 0 **27.** −5 **28.** −3 **29.** −6 **30.** −6 **31.** −5 **32.** 8 **33.** −2 **34.** −9 **35.** −10
36. 5 **37.** −5 **38.** 4 **39.** −12 **40.** 5 **41.** −4 **42.** −12 **43.** −7 **44.** 6 **45.** 6 **46.** −1
47. 9 **48.** −28 **49.** 27 **50.** −36 **51.** −6 **52.** −$\frac{6}{35}$ **53.** −$\frac{6}{11}$ **54.** $\frac{15}{56}$ **55.** 0 **56.** 48
57. 12 **58.** −70 **59.** −60 **60.** −24 **61.** 144 **62.** −5 **63.** −3 **64.** −4 **65.** 18 **66.** 0
67. 0 **68.** −8 **69.** 5 **70.** 9 **71.** −$\frac{3}{32}$ **72.** −$\frac{3}{4}$ **73.** $\frac{56}{27}$ **74.** −$\frac{35}{9}$ **75.** 1 **76.** 0 **77.** 0
78. Undefined **79.** Undefined **80.** Indeterminate **81.** 0 **82.** 24 **83.** −8 **84.** 1 **85.** 3 **86.** −8
87. 18 **88.** −2 **89.** −4 **90.** 10 **91.** 1 **92.** 15 **93.** −4 **94.** 16 **95.** 36 **96.** 729 **97.** 1
98. 81 **99.** 16 **100.** −27 **101.** −1 **102.** −32 **103.** $\frac{4}{49}$ **104.** $\frac{9}{25}$ **105.** $\frac{8}{125}$ **106.** x^2y
107. xy^2 **108.** x^3y^2 **109.** y^2z^2 **110.** $2^2 \cdot 3^3xy^2$ **111.** $5 \cdot 7^2x^2y$ **112.** x^2y^2z **113.** xxy **114.** $xzzz$
115. $yyyz$ **116.** $2xxxyy$ **117.** −9 **118.** −16 **119.** −27 **120.** −16 **121.** 23 **122.** −2
123. 23 **124.** 22 **125.** 26 **126.** −19 **127.** −39 **128.** −3 **129.** −4 **130.** −60 **131.** 10
132. 20 **133.** 20 **134.** 114 **135.** 9 **136.** −35 **137.** 14 **138.** 2 **139.** 26 **140.** 9 **141.** 0
142. −3 **143.** −11 **144.** −3 **145.** 21 **146.** Associative property of addition
147. Commutative property of multiplication **148.** Distributive property **149.** Commutative property of multiplication
150. Commutative property of addition **151.** Associative property of addition
152. Commutative property of addition

Chapter 1 Practice Test

1. (a) 42 **(b)** 42, 0 **(c)** −6, 42, 0, −7, −1 **(d)** −6, 42, −$3\frac{1}{2}$, 0, 6.52, $\frac{5}{9}$, −7, −1 **(e)** $\sqrt{5}$
(f) −6, 42, −$3\frac{1}{2}$, 0, 6.52, $\sqrt{5}$, $\frac{5}{9}$, −7, −1 **2.** < **3.** > **4.** −12 **5.** −11 **6.** 16 **7.** −14 **8.** 8
9. −24 **10.** $\frac{16}{63}$ **11.** −2 **12.** −69 **13.** −2 **14.** 12 **15.** 81 **16.** $\frac{27}{125}$ **17.** $2^2 5^2 y^2 z^3$
18. $2 \cdot 2 \cdot 3 \cdot 3 \cdot 3xxxyy$ **19.** 26 **20.** 10 **21.** 11 **22.** Commutative property of addition
23. Distributive property **24.** Associative property of addition **25.** Commutative property of multiplication

CHAPTER 2

Exercise Set 2.1

1. $8x$ **3.** $-x$ **5.** $x + 9$ **7.** $3x$ **9.** $4x - 7$ **11.** $2x + 3$ **13.** $5x + 5$ **15.** $5x + 3y + 3$
17. $2x - 4$ **19.** $x + 7$ **21.** $-8x + 2$ **23.** $-2x + 11$ **25.** $3x - 6$ **27.** $-5x + 3$ **29.** $6y + 6$
31. $4x - 10$ **33.** $3x - 4$ **35.** $x + \frac{5}{12}$ **37.** $48.5x + 8.3$ **39.** $x + \frac{1}{8}y$ **41.** $-4x - 8.3$ **43.** $-2x + 7$
45. $7x - 16$ **47.** $7x - 1$ **49.** $x - 8$ **51.** $21.72x - 7.11$ **53.** $-\frac{23}{20}x - 5$ **55.** $2x + 8$ **57.** $4x + 20$
59. $-2x + 8$ **61.** $-x + 2$ **63.** $x - 4$ **65.** $\frac{1}{4}x - 3$ **67.** $-1.8x + 3$ **69.** $-x + 3$ **71.** $0.8x - 0.2$
73. $x - y$ **75.** $-2x + 6y - 8$ **77.** $12 - 6x + 3y$ **79.** $x - 8y + \frac{1}{2}$ **81.** $x + 3y - 9$ **83.** $x - 4 - 2y$
85. $3x - 8$ **87.** $2x - 5$ **89.** $14x + 18$ **91.** $4x - 2y + 3$ **93.** $-x + y + 3$ **95.** $7x + 3$ **97.** $x - 9$
99. $6x - 12$ **101.** $-x - 2$ **103.** $-x + 6$ **105.** $8x - 19$ **107.** $x + 6.8$ **109.** $3x$ **111.** $x + 15$
113. $0.2x + 4y + 0.4$ **115.** $-6x + 3y - 3$ **117.** $x - 5$ **119.** $\frac{1}{6}x - \frac{10}{3}$
121. The signs of all the terms inside the parentheses are changed when the parentheses are removed.
123. (a) $2x^2$, $3x$, -5; The terms are the part of the expression that are added or subtracted.   **(b)** The factors of $2x^2$ are
$1, 2, x, 2x, x^2$ and $2x^2$. Note that $1 \cdot 2x^2 = 2x^2$, $2 \cdot x^2 = 2x^2$ and $x \cdot 2x = 2x^2$. Expressions that are multiplied are factors
of the product.

Cumulative Review Exercises

124. 7   **125.** -16 **127.** -12

Just for Fun   **1.** $18x - 25y + 3$ **2.** $9x^2 - 7x - 8$ **3.** $6x^2 + 5y^2 + 3x + 7y$ **4.** $9x - 39$

Exercise Set 2.2

1. solution   **3.** not solution   **5.** solution   **7.** not solution   **9.** solution   **11.** solution   **13.** 5   **15.** -10
17. -7 **19.** -61 **21.** 18   **23.** 43   **25.** -12 **27.** 72   **29.** -26 **31.** -58 **33.** 10   **35.** -12
37. 3   **39.** -9 **41.** 49   **43.** 3   **45.** 5   **47.** 1   **49.** 11   **51.** -26 **53.** -36 **55.** -47.5
57. 46.5   **59.** -21.58 **61.** 0   **63.** 720   **65.** Two or more equations with the same solution   **67.** Subtract 3

Cumulative Review Exercises

68. 18   **69.** -8 **70.** $2x - 13$ **71.** $10x - 32$

Just for Fun   **1.** All values satisfy the equation.   **2.** (a) and (c)

Exercise Set 2.3

1. 3   **3.** 8   **5.** -2 **7.** -12 **9.** 5   **11.** 3   **13.** $-\frac{3}{2}$ **15.** 4   **17.** 2   **19.** 49   **21.** $-\frac{1}{2}$ **23.** 6
25. $\frac{35}{19}$ **27.** 2   **29.** -1 **31.** $-\frac{3}{40}$ **33.** -75 **35.** 125   **37.** -35 **39.** 24   **41.** -5 **43.** 12
45. -16 **47.** -36 **49.** 6   **51.** -20.2 **53.** $-\frac{5}{4}$ **55.** 9   **57.** Divide by 3   **59.** Multiply by $\frac{3}{2}$, 6
61. Multiply by $\frac{7}{3}$, $\frac{28}{15}$

Cumulative Review Exercises

62. -9 **63.** 0   **64.** $-11x + 38$ **65.** -57

Exercise Set 2.4

1. 2   **3.** -6 **5.** 5   **7.** $\frac{12}{5}$ **9.** -12 **11.** 3   **13.** $\frac{15}{2}$ **15.** $-\frac{21}{32}$ **17.** $-\frac{56}{9}$ **19.** 5   **21.** $-\frac{7}{6}$
23. 3   **25.** 20   **27.** 6.8   **29.** 32   **31.** 2   **33.** 0   **35.** -1 **37.** 0   **39.** $-\frac{7}{6}$ **41.** 1   **43.** -4
45. 4   **47.** 3   **49.** 0.6   **51.** -1 **53.** 6   **55.** 3   **57.** Addition property   **59. (b)** 5

Cumulative Review Exercises

60. $\frac{49}{40}$ or $1\frac{9}{40}$ **61.** 64
62. Isolate the variable on one side of the equation.   **63.** Divide both sides of the equation by -4.

Just for Fun   **1.** $\frac{35}{6}$ **2.** $\frac{4}{5}$ **3.** -4

Exercise Set 2.5

1. 4   **3.** 1   **5.** $\frac{3}{5}$ **7.** 3   **9.** -21 **11.** 1   **13.** 4.16   **15.** 4   **17.** $-\frac{17}{7}$ **19.** No solution   **21.** $\frac{34}{5}$
23. -4 **25.** 25   **27.** All real numbers   **29.** All real numbers   **31.** 0   **33.** All real numbers   **35.** $-\frac{112}{15}$
37. 14   **39.** $-\frac{5}{3}$ **41.** 12   **43.** 16   **45.** $-\frac{10}{3}$ **47.** Same expression will appear on both sides of equation.
49. (b) 21

Cumulative Review Exercises

51. Numbers or letters multiplied together are factors; numbers or letters added or subtracted are terms.   **52.** $7x - 10$
53. $\frac{10}{7}$ **54.** -3

Just for Fun   **1.** $\frac{1}{4}$ **2.** $-\frac{17}{13}$ **3.** -4

Exercise Set 2.6

1. $5:8$ **3.** $2:1$ **5.** $25:4$ **7.** $5:3$ **9.** $1:3$ **11.** $6:1$ **13.** $13:32$ **15.** 16   **17.** 45   **19.** -100
21. 5   **23.** -30 **25.** 6   **27.** 384 min   **29.** 12.5 min   **31.** \$1004.67   **33.** 0.14 m   **35.** 208 lb
37. 2.7 lb   **39.** 495   **41.** 4.75 ft   **43.** 3.4 mi   **45.** 9.15 lb   **47.** 5.3 ℓ   **49.** 15.63 mi
51. \$1.04   **53.** 5 points   **55.** 1,481,481.5 Lire   **57.** 32 in   **59.** 11.2 ft   **61.** 5.6 in
63. Yes, her ratio is $2.12:1$

Cumulative Review Exercises

67. Commutative property of addition   **68.** Associative property of multiplication   **69.** Distributive property   **70.** $\frac{3}{4}$

Just for Fun   **1.** $\frac{1}{3}$ cup flour, $\frac{1}{6}$ tsp salt, $1\frac{1}{3}$ tbsp butter, $\frac{2}{3}$ tsp nutmeg, $\frac{2}{3}$ tsp cinnamon, 1 cup sugar   **2.** 0.625 cc

Exercise Set 2.7

1. $x > 4$ **3.** $x \geq -2$ **5.** $x > -5$ **7.** $x < 10$

9. $x \leq -4$ **11.** $x > -\frac{3}{2}$ **13.** $x \leq 1$ **15.** $x < -3$

17. $x < \frac{3}{2}$ **19.** $x > \frac{35}{9}$ **21.** $x > -\frac{8}{3}$ **23.** $x \leq -\frac{11}{3}$

25. $x \geq -6$ **27.** $x < 1$ **29.** $x < 2$ **31.** All real numbers

33. $x > \frac{3}{4}$ **35.** $x > \frac{23}{10}$ **37.** No solution   **39.** $x \geq -\frac{7}{11}$

41. All real numbers   **43.** The inequality has no solution.   **45.** When multiplying or dividing by a negative number

Cumulative Review Exercises

47. -9 **48.** -25 **49.** $\frac{14}{5}$ or $2\frac{4}{5}$ **50.** 500 kwh

Just for Fun   **1.** $x \geq -2$ **2.** $x > \frac{3}{4}$ **3.** $\neq$

Chapter 2 Review Exercises

1. $2x + 8$ **2.** $3x - 6$ **3.** $8x - 6$ **4.** $-2x - 8$ **5.** $-x - 2$ **6.** $-x + 2$ **7.** $-16 + 4x$ **8.** $18 - 6x$
9. $20x - 24$ **10.** $-6x + 15$ **11.** $36x - 36$ **12.** $-4x + 12$ **13.** $-3x - 3y$ **14.** $-6x + 4$
15. $-3 - 2y$ **16.** $-x - 2y + z$ **17.** $3x + 9y - 6z$ **18.** $-4x + 6y - 14$ **19.** $5x$ **20.** $7y + 2$
21. $-2y + 7$ **22.** $5x + 1$ **23.** $6x + 3y$ **24.** $-3x + 3y$ **25.** $6x + 8y$ **26.** $6x + 3y + 2$ **27.** $-x - 1$
28. $3x + 3y + 6$ **29.** 3   **30.** $-12x + 3$ **31.** $5x + 6$ **32.** $-2x$ **33.** $5x + 7$ **34.** $-10x + 12$
35. $5x + 3$ **36.** $4x - 4$ **37.** $22x - 42$ **38.** $3x - 3y + 6$ **39.** $-x + 5y$ **40.** $3x + 2y + 16$ **41.** 3
42. $-x - 2y + 4$ **43.** 2   **44.** -8 **45.** 11   **46.** -27 **47.** 2   **48.** $\frac{11}{2}$ **49.** -2 **50.** -3 **51.** 12

52. 1   **53.** 6   **54.** -3 **55.** $-\frac{21}{5}$ **56.** -5 **57.** -19 **58.** -1 **59.** $\frac{2}{3}$ **60.** $\frac{1}{5}$ **61.** $\frac{9}{2}$ **62.** $\frac{10}{7}$
63. -3 **64.** -1 **65.** -8 **66.** $-\frac{23}{5}$ **67.** -10 **68.** $-\frac{4}{3}$ **69.** No solution   **70.** All real numbers
71. $\frac{17}{3}$ **72.** $-\frac{20}{7}$ **73.** $\frac{22}{5}$ **74.** $3:4$ **75.** $5:12$ **76.** $25:12$ **77.** $1:1$ **78.** 3   **79.** 3   **80.** 9
81. $\frac{135}{4}$ **82.** -4 **83.** -24 **84.** 36   **85.** 90   **86.** 40 in.   **87.** 1 ft   **88.** $x \geq 2$ ⟶
89. $x < 3$ ⟶   **90.** $x \geq -\frac{12}{5}$ ⟶   **91.** No solution ⟶   **92.** All real numbers ⟶

93. $x < -3$ ⟶   **94.** $x \leq \frac{9}{5}$ ⟶   **95.** $x > \frac{8}{5}$ ⟶   **96.** $x < -\frac{5}{3}$ ⟶

97. $x \leq \frac{5}{11}$ ⟶   **98.** No solution ⟶   **99.** All real numbers ⟶   **100.** $x \leq \frac{11}{2}$ ⟶

101. 67.5 mi   **102.** $6\frac{1}{3}$ in   **103.** 9.45 ft   **104.** 1131.77 Canadian dollars   **105.** approximately $0.00036
106. 57.3°   **107.** 0.03 slugs

Chapter 2 Practice Test

1. $4x - 8$ **2.** $-x - 3y + 4$ **3.** $2x + 4$ **4.** $-x + 10$ **5.** $-6x + y - 6$ **6.** $7x - 5y + 3$ **7.** $8x - 1$
8. 4   **9.** -2 **10.** 2   **11.** -1 **12.** $-\frac{1}{7}$ **13.** No solution   **14.** All real numbers   **15.** -45
16. $x > -7$ ⟶   **17.** $x \leq 12$ ⟶   **18.** No solution ⟶   **19.** $\frac{32}{3}$ or $10\frac{2}{3}$ ft

20. 150 gal

Cumulative Review Test

1. $\frac{16}{25}$ **2.** $\frac{1}{2}$ **3.** $>$ **4.** -6 **5.** -8 **6.** 7   **7.** 3   **8.** 1   **9.** Associative Property of Addition
10. $10x + y$ **11.** $3x + 16$ **12.** 3   **13.** -40 **14.** 2   **15.** -1 **16.** 6
17. $x > 10$, ⟶   **18.** $x \geq -12$, ⟶   **19.** 158.4 lbs   **20.** $42

CHAPTER 3

Exercise Set 3.1

1. 16   **3.** 26   **5.** 126   **7.** 12.56   **9.** 10   **11.** 6   **13.** 1080   **15.** 56   **17.** 5   **19.** 60   **21.** 16

23. 6   **25.** $y = -2x + 8$, 4   **27.** $y = \dfrac{2x + 4}{6}$, 4   **29.** $y = \dfrac{-3x + 6}{2}$, 0   **31.** $y = \dfrac{4x - 20}{5}$, $-\frac{4}{5}$

33. $y = \dfrac{3x + 18}{6}$, 3   **35.** $y = \dfrac{-x + 8}{2}$, 6   **37.** $t = \dfrac{d}{r}$ **39.** $p = \dfrac{i}{rt}$ **41.** $d = \dfrac{C}{\pi}$ **43.** $b = \dfrac{2A}{h}$

45. $w = \dfrac{P - 2\ell}{2}$ **47.** $n = \dfrac{m - 3}{4}$ **49.** $b = y - mx$ **51.** $r = \dfrac{I - P}{Pt}$ **53.** $d = \dfrac{3A - m}{2}$

55. $b = d - a - c$ **57.** $y = \dfrac{-ax + c}{b}$ **59.** $h = \dfrac{V}{\pi r^2}$ **61.** 35   **63.** 10° C   **65.** 95° F   **67.** $P = 10$

69. $K = 4$ **71.** 30   **73.** $1440   **75.** $5,000   **77.** 30 in   **79.** 24 sq cm   **81.** 50.24 sq in
83. 25.12 in   **85.** 8 ft   **87.** (a) 62.1 ft   (b) 124.2 ft   **89.** 18,237.12 cu in

91. When you multiply a unit by the same unit you get a square unit.   **93.** (a) $\pi = \dfrac{C}{2r}$ or $\pi = \dfrac{C}{d}$ (b) π or about 3.14

Cumulative Review

94. 0   **95.** $3:2$ **96.** 1620 min or 27 hrs   **97.** $x \leq -17$

Just for Fun   **1.** (a) $A = d^2 - \pi\left(\dfrac{d}{2}\right)^2$ (b) 3.44 sq ft   (c) 7.74 sq ft   **2.** (a) $V = 18x^3 - 3x^2$

(b) 6027 cu cm   (c) $S = 54x^2 - 8x$ (d) 2590 sq cm

Exercise Set 3.2

1. $x + 5$ **3.** $4x$ **5.** $0.70x$ **7.** $0.10\ C$ **9.** $6x - 3$ **11.** $\frac{3}{4}x + 7$ **13.** $2(x + 8)$ **15.** $25x$ **17.** $12x$
19. $16a + b$ **21.** three more than a number **23.** three times a number, decreased by four
25. the difference of twice a number and three **27.** the difference of five and a number
29. the sum of four and six times a number **31.** three times the sum of a number and two **33.** $x, x + 12$
35. $x, x + 1$ **37.** $x, 100 - x$ **39.** $x, 4x - 5$ **41.** $x, 1.7x$ **43.** $x, 80 - x$ **45.** $x, x + 0.12x$
47. $x, x + 0.07x$ **49.** $4x$ **51.** $0.23x$ **53.** $15x$ **55.** $10x$ **57.** $300n$ **59.** $0.075x$ **61.** $10a$ **63.** $5p$
65. $x + 5x = 18$ **67.** $x + (x + 1) = 47$ **69.** $2x - 8 = 12$ **71.** $\frac{1}{5}(x + 10) = 150$ **73.** $x + (2x - 8) = 1000$
75. $x + 0.08x = 92$ **77.** $x - 0.25x = 65$ **79.** $x - 0.20x = 215$ **81.** $x + (2x - 3) = 21$ **83.** $40t = 180$
85. $15y = 215$ **87.** $25q = 150$ **89.** three more than a number is six.
91. three times a number, decreased by one, is four more than twice the number
93. four times the difference of a number and one is six.
95. six more than five times a number is the difference of six times the number and one
97. the sum of a number and the number increased by four is eight
99. the sum of twice a number and the number increased by three is five

Cumulative Review Exercises

103. 3.35 tsp **104.** $\frac{1}{6}$ cup **105.** 15 **106.** $y = \dfrac{3x - 6}{2}\left(\text{or } y = \dfrac{3}{2}x - 3\right), 6$

Just For Fun **1. (a)** $P = 100\left(\dfrac{9f}{c}\right)$ **(b)** 48% of calories are from fat **(c)** 42.4% of calories are from fat.
2. (a) $86{,}400d + 3600h + 60m + s$ **(b)** $368{,}125$ sec

Exercise Set 3.3

1. 35, 36 **3.** 37, 39 **5.** 12, 31 **7.** 27, 29, 31 **9.** 8, 20 **11.** 14 years **13.** 4 months **15.** 200 miles
17. \$750 **19.** \$29,000 **21.** \$1.19 **23.** 10 French fries **25.** Younger workers 13 hr; third 26 hr, fourth 39 hr
27. \$16.39 **29.** 5 hrs (more than 4.83) **31.** \$320 **33.** 30 terminals

Cumulative Review Exercises

36. $\frac{17}{12}$ or $1\frac{5}{12}$ **37.** Associative property of addition
38. Commutative property of multiplication **39.** Distributive property **40.** 56 lbs **41.** $b = 2M - a$

Just For Fun **1. (a)** $80 = \dfrac{74 + 88 + 76 + x}{4}$ **2.** 3 pointers: 6 **3.** n
(b) 82   2 pointers: 24 $4n$
$4n + 6$
$(4n + 6)/2 = 2n + 3$
$2n + 3 - 3 = 2n$

Exercise Set 3.4

1. 9.5 in **3.** $A = 47°, B = 133°$ **5.** 50°, 60°, 70° **7.** 4m, 4m, 2m **9.** $w = 48$ ft, $\ell = 72$ ft
11. two smaller angles, 69°; two larger angles, 111° **13.** $w = 2\frac{2}{3}$ ft, $h = 4\frac{2}{3}$ ft **15.** $h = 3$ ft, $\ell = 9$ ft
17. The area remains the same. **19.** The volume becomes eight times as great.

Cumulative Review Exercises

22. $<$ **23.** $>$ **24.** -8 **25.** $-2x - 4y + 6$

Just For Fun **1. (a)** $A = S^2 - s$ **(b)** 45 sq in **2.** $ac + ad + bc + bd$ **3. (a)** 342.56 ft **(b)** 192 ft

Chapter 3 Review Exercises

1. 12.56 **2.** 48 **3.** 20 **4.** 300 **5.** 240 **6.** 28.26 **7.** 113.04 **8.** 20 **9.** 21 **10.** -11

11. -8 **12.** 15 **13.** 4.5 **14.** $y = 2x - 12, 8$ **15.** $y = \dfrac{3x + 4}{2}, 5$ **16.** $y = \dfrac{3x - 5}{2}, -7$

17. $y = -3x - 10, -10$ **18.** $y = \dfrac{-3x - 6}{2}, 6$ **19.** $y = \dfrac{4x - 3}{3}, \dfrac{5}{3}$ **20.** $m = \dfrac{F}{a}$ **21.** $h = \dfrac{2A}{b}$

22. $t = \dfrac{i}{pr}$ **23.** $w = \dfrac{P - 2l}{2}$ **24.** $y = \dfrac{2x - 6}{3}$ **25.** $B = 2A - C$ **26.** $h = \dfrac{3V}{4\pi r^2}$ **27.** \$180   **28.** 6 in

29. 29 and 33   **30.** 127 and 128   **31.** 38 and 7   **32.** \$8,000   **33.** \$2,000   **34.** \$650   **35.** 45°, 55°, 80°
36. 30°, 40°, 150°, 140°   **37.** $w = 15.5$ ft, $l = 19.5$ ft   **38.** 103 and 105   **39.** 400 miles   **40.** \$450
41. \$12,000   **42.** 42°, 50°, 88°   **43.** 8 years   **44.** 70°, 70°, 110°, 110°

Practice Test

1. 18 ft   **2.** 145   **3.** 100.48   **4.** $R = \dfrac{P}{I}$ **5.** $y = \dfrac{3x - 6}{2}$ **6.** $a = 3A - b$

7. $c = \dfrac{D - Ra}{R}$ or $c = \dfrac{D}{R} - a$ **8.** 56 and 102   **9.** 13, 14, and 15   **10.** \$16.39   **11.** 15, 30, 30 in
12. 50°, 50°, 130°, 130°

CHAPTER 4

Exercise Set 4.1

1. x^7 **3.** y^3 **5.** 243   **7.** y^5 **9.** y^5 **11.** x^8 **13.** 25   **15.** x^4 **17.** y **19.** 1   **21.** 1   **23.** 3
25. 1   **27.** 1   **29.** x^{10} **31.** x^{25} **33.** x^3 **35.** x^{12} **37.** x^8 **39.** $1.69x^2$ **41.** x^2 **43.** $64x^6$

45. $-27x^9$ **47.** $8x^6y^3$ **49.** $73.96x^4y^{10}$ **51.** $-216x^9y^6$ **53.** $-x^{12}y^{15}z^{18}$ **55.** $\dfrac{x^2}{y^2}$ **57.** $\dfrac{x^3}{125}$ **59.** $\dfrac{y^5}{x^5}$

61. $\dfrac{216}{x^3}$ **63.** $\dfrac{27x^3}{y^3}$ **65.** $\dfrac{4x^2}{25}$ **67.** $\dfrac{64y^9}{x^3}$ **69.** $\dfrac{-27x^9}{64}$ **71.** $\dfrac{x^2}{y^3}$ **73.** $\dfrac{y^4}{x^7}$ **75.** $\dfrac{5x^2}{y^2}$ **77.** $\dfrac{1}{4x^2y}$

79. $\dfrac{7}{2x^5y^5}$ **81.** $\dfrac{-3y^4}{x^3z}$ **83.** $\dfrac{-3}{x^3y^2z}$ **85.** $\dfrac{8}{x^6}$ **87.** $64y^{12}$ **89.** $\dfrac{81x^8}{100}$ **91.** $\dfrac{x^4}{y^4}$ **93.** $\dfrac{z^{24}}{16y^{20}}$ **95.** $27x^6y^9$

97. x^4y^{12} **99.** $\dfrac{9}{16x^{12}y^4}$ **101.** $3x^7y^4$ **103.** $-18x^3y^9$ **105.** $12x^5y^8$ **107.** $10x^2y^7$ **109.** $12x^3y^4$

111. $3x^{14}y^{23}$ **113.** $54x^{17}y^{17}$ **115.** $x^{11}y^{13}$ **117.** $18x^{10}y^{28}$ **119.** cannot be simplified

121. cannot be simplified   **123.** cannot be simplified   **125.** y^2 **127.** cannot be simplified   **129.** $\dfrac{x^2}{y}$ **131.** 0

133. Negative, exponent is odd   **135.** Positive, exponent is even

Cumulative Review Exercises

137. An equation of the form $ax + b = c$ **138.** A linear equation that has only one solution
139. An equation that is true for any value of the variable.   **140.** $C = 18.84$ in, $A = 28.26$ sq in

141. $y = \dfrac{2x - 6}{5}$ or $y = \dfrac{2}{5}x - \dfrac{6}{5}$

Just For Fun   **1.** $\dfrac{9x^2}{8y^3}$ **2.** $576x^4y^{23}$

Exercise Set 4.2

1. $\dfrac{1}{x^2}$ **3.** $\dfrac{1}{5}$ **5.** x^4 **7.** x **9.** 25   **11.** $\dfrac{1}{x^6}$ **13.** $\dfrac{1}{y^{21}}$ **15.** $\dfrac{1}{x^{10}}$ **17.** 64   **19.** x^3 **21.** x^2 **23.** 9

25. $\dfrac{1}{x^3}$ **27.** y^9 **29.** $\dfrac{1}{x^4}$ **31.** 27   **33.** $\dfrac{1}{27}$ **35.** z^9 **37.** $\dfrac{1}{x^{25}}$ **39.** y^6 **41.** $\dfrac{1}{x^4}$ **43.** $\dfrac{1}{x^{19}}$ **45.** $\dfrac{1}{x^8}$

47. y^{10} **49.** 1 **51.** $\dfrac{1}{z^7}$ **53.** 1 **55.** x^4 **57.** 1 **59.** $\dfrac{1}{4}$ **61.** $\dfrac{1}{36}$ **63.** x^3 **65.** $\dfrac{1}{9}$ **67.** 125

69. $\dfrac{1}{4}$ **71.** 1 **73.** $\dfrac{5y}{x}$ **75.** $\dfrac{1}{3x^3}$ **77.** $\dfrac{5x^4}{y}$ **79.** $\dfrac{1}{9x^4y^6}$ **81.** $\dfrac{y^9}{x^{15}}$ **83.** $\dfrac{15}{x^3}$ **85.** $\dfrac{6}{x}$ **87.** $\dfrac{-27}{x^2}$

89. $\dfrac{2x}{y}$ **91.** $\dfrac{15y}{x}$ **93.** $\dfrac{x^7}{2}$ **95.** $\dfrac{1}{3y^{10}}$ **97.** $\dfrac{4}{x^2}$ **99.** $\dfrac{x^4}{2y^5}$ **101.** $8x^6y$

103. (a) Yes, $a^{-1}b^{-1} = \dfrac{1}{a} \cdot \dfrac{1}{b} = \dfrac{1}{ab}$ **(b)** No, $a^{-1} + b^{-1} = \dfrac{1}{a} + \dfrac{1}{b} \neq \dfrac{1}{a+b}$

Cumulative Review Exercises

106. -18 **107.** 18 **108.** 6.67 oz **109.** 9, 28

Just For Fun **1. (a) and (b)** $\dfrac{z^2}{9x^4y^6}$

Exercise Set 4.3

1. 4.2×10^4 **3.** 9×10^2 **5.** 5.3×10^{-2} **7.** 1.9×10^4 **9.** 1.86×10^{-6} **11.** 9.14×10^{-6}
13. 1.07×10^2 **15.** 1.53×10^{-1} **17.** 4200 **19.** 40,000,000 **21.** 0.0000213 **23.** 0.312
25. 9,000,000 **27.** 535 **29.** 35,000 **31.** 10,000 **33.** 120,000,000 **35.** 0.0153 **37.** 320
39. 0.0021 **41.** 20 **43.** 4.2×10^{12} **45.** 4.5×10^{-7} **47.** 2×10^3 **49.** 2×10^{-7} **51.** 3×10^8
53. 9.2×10^{-5}, 1.3×10^{-1}, 8.4×10^3, 6.2×10^4 **55.** 3.2×10^7 seconds **57.** 8,640,000,000 cu ft **59.** -8

Cumulative Review Exercises

61. 0 **62. (a)** $\dfrac{3}{2}$ **(b)** 0 **63.** 2 **64.** $\dfrac{-y^{12}}{64x^9}$

Just For Fun **1. (a)** About 5.87×10^{12} miles **(b)** About 500 seconds or $8\frac{1}{3}$ minutes

Exercise Set 4.4

1. Monomial **3.** Monomial **5.** Binomial **7.** Trinomial **9.** Not polynomial **11.** Binomial
13. Monomial **15.** Polynomial **17.** Trinomial **19.** Not polynomial **21.** First degree
23. $2x^2 + x - 6$, second **25.** $-x^2 - 4x - 8$, second **27.** Third **29.** Second
31. $-6x^3 + x^2 - 3x + 4$, third **33.** $5x^2 - 2x - 4$, second **35.** $-2x^3 + 3x^2 + 5x - 6$, third **37.** $6x + 1$
39. $-2x + 11$ **41.** $-x - 2$ **43.** $21x - 21$ **45.** $x^2 + 6x + 0.8$ **47.** $2x^2 + 8x + 5$ **49.** $5x^2 + x + 20$
51. $-5x^2 + x + \dfrac{17}{2}$ **53.** $5.4x^2 - 5x + 4$ **55.** $-2x^3 - 3x^2 + 4x - 3$ **57.** $3x^2 - 2xy$ **59.** $7x^2y - 3x + 2$
61. $7x - 1$ **63.** $x^2 + x + 16$ **65.** $-3x^2 + 2x - 12$ **67.** $7x^2 + 7x - 13$ **69.** $2x^3 - x^2 + 6x - 2$
71. $5x^3 - 7x^2 - 2$ **73.** $3xy + 3x + 3$ **75.** $x - 6$ **77.** $3x + 4$ **79.** -5 **81.** $-7x + 3$
83. $6x^2 + 7x - 8.5$ **85.** $8x^2 + x + 4$ **87.** $5x^2 - 2x + 7$ **89.** $2x^2 + 5x + 2$ **91.** $4x^3 - \dfrac{20}{3}x^2 - x - 4$
93. $9x^3 - x^2 - 5x - \dfrac{1}{5}$ **95.** $-x + 11$ **97.** $2x^2 - 9x + 14$ **99.** $-x^3 + 11x^2 + 9x - 7$ **101.** $3x + 17$
103. $4x + 7$ **105.** $5x^2 + 7x - 2$ **107.** $4x^2 - 5x - 6$ **109.** $4x^3 - 7x^2 + x - 2$
111. Sum of a finite number of terms of the form ax^n, where a is a real number and n is a whole number.
113. (a) The exponent on the variable is the degree of the term **(b)** It is the same as the degree of the highest degree term in the polynomial.
115. Write the polynomial with exponents on the variable decreasing from left to right.

Cumulative Review Exercises

120. $|-4| < |-6|$ **121.** False **122.** True **123.** False **124.** False **125.** $\dfrac{y^3}{8x^9}$

Just For Fun **1.** $-12x + 18$ **2.** $3x^2y - 13xy + 9xy^2 + 3x$ **3.** $8x^2 + 28x - 24$

se Set 4.5

1. $3x^3y$　　**3.** $30x^5y^7$　　**5.** $-28x^6y^{15}$　　**7.** $54x^6y^{14}$　　**9.** $3x^6y$　　**11.** $3x + 12$　　**13.** $2x^2 - 6x$　　**15.** $8x^2 - 24x$
17. $2x^3 + 6x^2 - 2x$　　**19.** $-2x^3 + 4x^2 - 10x$　　**21.** $-20x^3 + 30x^2 - 20x$　　**23.** $24x^3 + 32x^2 - 40x$
25. $0.6x^2y + 1.5x^2 - 1.8xy$　　**27.** $xy - y^2 - 3y$　　**29.** $x^2 + 7x + 12$　　**31.** $6x^2 + 3x - 30$　　**33.** $4x^2 - 16$
35. $-6x^2 - 8x + 30$　　**37.** $-2x^2 + x + 15$　　**39.** $x^2 + 7x + 12$　　**41.** $x^2 + 2x - 8$　　**43.** $6x^2 + 23x + 20$
45. $6x^2 - x - 12$　　**47.** $x^2 - 1$　　**49.** $4x^2 - 12x + 9$　　**51.** $-2x^2 + 5x + 12$　　**53.** $-4x^2 + 2x + 12$
55. $x^2 - y^2$　　**57.** $6x^2 - 5xy - 6y^2$　　**59.** $8xy - 12x - 6y^2 + 9y$　　**61.** $x^2 + 0.9x + 0.18$　　**63.** $2x^2 + 5x + 2$
65. $x^2 - 16$　　**67.** $4x^2 - 1$　　**69.** $x^2 + 2xy + y^2$　　**71.** $x^2 - 0.4x + 0.04$　　**73.** $9x^2 - 25$
75. $2x^3 + 10x^2 + 11x - 3$　　**77.** $5x^3 - x^2 + 16x + 16$　　**79.** $-14x^3 - 22x^2 + 19x - 3$
81. $18x^3 - 69x^2 + 54x - 27$　　**83.** $6x^4 + 5x^3 + 5x^2 + 10x + 4$　　**85.** $x^4 - 3x^3 + 5x^2 - 6x$
87. $3x^4 - 7x^3 - 7x^2 + 3x$　　**89.** $a^3 + b^3$　　**91.** Yes
93. No, the product of the sum and difference of the same two terms will be a binomial.

Cumulative Review Exercises

95. 13 miles　　**96.** $\dfrac{1}{16y^4}$　　**97. (a)** -216　**(b)** $\dfrac{1}{216}$　　**98.** $-6x^2 - 2x + 8$

Just For Fun　　**1.** $2x^3\sqrt{5} + 5x^2 - \dfrac{x\sqrt{5}}{2}$　　**2.** $\dfrac{1}{3}x^2 + \dfrac{11}{45}x - \dfrac{4}{15}$　　**3.** $6x^6 - 18x^5 + 3x^4 + 35x^3 - 54x^2 + 38x - 12$

Exercise Set 4.6

1. $x + 2$　　**3.** $x + 3$　　**5.** $\frac{3}{2}x + 4$　　**7.** $-3x + 2$　　**9.** $3x + 1$　　**11.** $3 + \dfrac{6}{x}$　　**13.** $-\dfrac{3}{x} + 1$　　**15.** $x^2 + 2x - 3$

17. $-2x^2 + 3x + 4$　　**19.** $x + 4 - \dfrac{3}{x}$　　**21.** $3x - 2 + \dfrac{6}{x}$　　**23.** $-x^2 - \dfrac{3}{2}x + \dfrac{2}{x}$　　**25.** $3x + 1 - \dfrac{4}{x^2}$　　**27.** $x + 3$

29. $2x + 3$　　**31.** $2x + 4$　　**33.** $x - 2$　　**35.** $x + 5 - \dfrac{3}{2x - 3}$　　**37.** $2x + 3$　　**39.** $2x - 3 + \dfrac{2}{4x + 9}$

41. $4x - 3 - \dfrac{3}{2x + 3}$　　**43.** $x^2 + 2x + 3$　　**45.** $2x^2 - x - 4 + \dfrac{2}{x - 1}$　　**47.** $2x^2 - 8x + 38 - \dfrac{156}{x + 4}$

49. $x^2 - 2x + 4$　　**51.** $x^2 - 3x + 9$　　**53.** $3x^2 + 2x + 1 + \dfrac{5}{3x - 2}$　　**55.** 7,6,4,2 from left to right

Cumulative Review Exercises

56. (a) 2　**(b)** 2, 0　**(c)** $2, -5, 0, \dfrac{2}{5}, -6.3, -\dfrac{23}{34}$　**(d)** $\sqrt{7}, \sqrt{3}$　**(e)** $2, -5, 0, \sqrt{7}, \dfrac{2}{5}, -6.3, \sqrt{3}, -\dfrac{23}{34}$
57. (a) 0　**(b)** undefined
58. Parentheses, exponents, multiplication or division left to right, addition or subtraction left to right.　　**59.** $-\dfrac{2}{3}$

Just For Fun　　**1.** $2x^2 - 3x + \dfrac{5}{2} - \dfrac{3}{2(2x + 3)}$　　**2.** $x^2 + \dfrac{2}{3}x + \dfrac{4}{9} - \dfrac{37}{9(3x - 2)}$

Exercise Set 4.7

1. 50 mph　　**3.** 13 hr　　**5.** 2.4 cm　　**7.** 250 cm³/hr　　**9.** 6 m/hr　　**11.** 1 mile　　**13.** 3.5 hr
15. Santa Fe Special, 52 mph; Amtrak, 82 mph　　**17.** Brease Along, 5 mph; Lorelei, 9 mph　　**19.** 2.24 ft/min
21. Chestnut, 4 mph; Midnight, 7 mph　　**23.** 5400 ft　　**25. (a)** 120 ft/min　**(b)** 275 ft
27. $3500 at 8%, $5400 at 11%　　**29.** $3000 at 10%, $2000 at 15%　　**31.** 23 dimes, 5 quarters　　**33.** 9 ones, 3 tens
35. 6 hours at $6, 12 hours at $6.50　　**37.** 9 lb　　**39.** $1\frac{2}{3}\ell$　　**41.** 11.1%　　**43.** 2.86 lb Family, 7.14 lb Spot Filler
45. (a) 13 Shares Getty, 52 shares United　**(b)** $434

Cumulative Review Exercises

46. (a) $\dfrac{22}{13}$ or $1\dfrac{9}{13}$　**(b)** $\dfrac{35}{8}$ or $4\dfrac{3}{8}$　　**47.** All real numbers　　**48.** $\dfrac{3}{4}$ or 0.75　　**49.** $x \le \dfrac{1}{4}$

Just For Fun   **1. (a)** $52,800 = 4416t$ **(b)** approximately 11.96 hr   **2.** 46.2 sec   **3.** 0.35 ft/sec   **4.** 6 qt

Chapter 4 Review Exercises

1. x^6 **2.** x^8 **3.** 243   **4.** 32   **5.** x^3 **6.** 1   **7.** 9   **8.** 16   **9.** $\frac{1}{x^2}$ **10.** x^5 **11.** 1   **12.** 3
13. 1   **14.** 1   **15.** $4x^2$ **16.** $27x^3$ **17.** $4x^2$ **18.** $-27x^3$ **19.** $16x^8$ **20.** $-x^{12}$ **21.** x^{12} **22.** $\frac{4x^6}{y^2}$
23. $\frac{27x^{12}}{8y^3}$ **24.** $24x^5$ **25.** $\frac{4x}{y}$ **26.** $12x^5y^2$ **27.** $9x^2$ **28.** $24x^7y^7$ **29.** $16x^8y^{11}$ **30.** $24x^{10}y^8$
31. $\frac{16x^6}{y^4}$ **32.** $\frac{x^9}{8y^9}$ **33.** $\frac{1}{x^3}$ **34.** $\frac{1}{y^7}$ **35.** $\frac{1}{25}$ **36.** x^3 **37.** x^7 **38.** 9   **39.** $\frac{1}{x^2}$ **40.** $\frac{1}{x^5}$
41. $\frac{1}{x^3}$ **42.** x^8 **43.** x^5 **44.** x^7 **45.** $\frac{1}{x^6}$ **46.** $\frac{1}{9x^8}$ **47.** $\frac{x^9}{64y^3}$ **48.** $\frac{-125}{x^6}$ **49.** $-8x^8$ **50.** $\frac{27y^3}{x^6}$
51. $\frac{x^4}{16y^6}$ **52.** $\frac{6}{x}$ **53.** $10x^2y^2$ **54.** $\frac{24y^2}{x^2}$ **55.** $\frac{6}{x^6y}$ **56.** $3y^5$ **57.** $\frac{3y}{x^3}$ **58.** $\frac{5x}{y^4}$ **59.** $\frac{4y^{10}}{x}$ **60.** $\frac{1}{2x^2y^5}$
61. 3.64×10^5 **62.** 1.64×10^6 **63.** 7.63×10^{-3} **64.** 1.76×10^{-1} **65.** 2.08×10^3 **66.** 3.14×10^{-4}
67. 0.0042   **68.** 16,500   **69.** 970,000   **70.** 0.00000438   **71.** 0.914   **72.** 536   **73.** 4,600,000   **74.** 1260
75. 19.84   **76.** 340,000   **77.** 0.09   **78.** 0.00003   **79.** 1.2×10^9 **80.** 2.4×10^1 **81.** 9.2
82. 2×10^8 **83.** 3.4×10^{-3} **84.** 5×10^8 **85.** Binomial, 1   **86.** Monomial, 0
87. $x^2 + 3x - 4$, Trinomial, second   **88.** $4x^2 - x - 3$, Trinomial, second   **89.** Binomial, second
90. Not polynomial   **91.** $-4x^2 + x$, Binomial, second   **92.** Not polynomial   **93.** $x^3 + 4x^2 - 2x - 6$, third
94. $3x + 7$ **95.** $9x + 1$ **96.** $2x - 5$ **97.** $4x^2 + 14$ **98.** $-3x^2 + 10x - 15$ **99.** $11x^2 - 2x - 3$
100. $x - 6.7$ **101.** $-2x + 2$ **102.** $9x^2 - \frac{5}{4}x + 4$ **103.** $6x^2 - 18x - 4$ **104.** $-5x^2 + 8x - 19$
105. $4x + 2$ **106.** $2x + 10$ **107.** $2x^2 - 4x$ **108.** $4.5x^3 - 13.5x^2$ **109.** $6x^3 - 12x^2 + 21x$
110. $-3x^3 + 6x^2 + x$ **111.** $24x^3 - 16x^2 + 8x$ **112.** $x^2 + 9x + 20$ **113.** $2x^2 - 2x - 12$
114. $16x^2 + 48x + 36$ **115.** $-6x^2 + 14x + 12$ **116.** $x^2 - 16$ **117.** $3x^3 + 7x^2 + 14x + 4$
118. $3x^3 + x^2 - 10x + 6$ **119.** $10x^3 - 19x^2 + 36x - 12$ **120.** $x + 2$ **121.** $x - 2$ **122.** $8x + 4$
123. $2x^2 + 3x - \frac{4}{3}$ **124.** $8x + 6 - \frac{4}{x}$ **125.** $4x - 2$ **126.** $-8x + 2$ **127.** $-2x - 4$ **128.** $\frac{5}{2}x + 5 + \frac{1}{x}$
129. $x + 4$ **130.** $2x - 3$ **131.** $5x - 2 + \frac{2}{x + 6}$ **132.** $2x^2 + 3x - 4$ **133.** $2x^2 + x - 2 + \frac{2}{2x - 1}$
134. 4 hrs   **135.** 538.46 mph   **136.** 2 hrs   **137.** \$4000 at 8%, \$8,000 at $7\frac{1}{4}$%   **138.** 1.2 ℓ of 10%, 0.8 ℓ of 5%
139. 6.5 mph   **140.** 4 hr   **141.** 60 lb of \$3.50, 20 lb of \$4.10   **142.** 29¢ stamps, 20; 19¢ stamps, 12
143. older brother, 55 mph; younger brother, 60 mph   **144.** 0.4 ℓ

Chapter 4 Practice Test

1. $6x^6$ **2.** $27x^6$ **3.** $4x^3$ **4.** $\frac{x^3}{8x^6}$ **5.** $\frac{y^4}{4x^6}$ **6.** $\frac{1}{5x^3y^6}$ **7.** Trinomial   **8.** Monomial

9. Not polynomial   **10.** $6x^3 - 2x^2 + 5x - 5$, third degree   **11.** $3x^2 - 3x + 1$ **12.** $-2x^2 + 4x$
13. $3x^2 - x + 3$ **14.** $12x^3 - 6x^2 + 15x$ **15.** $8x^2 + 2x - 21$ **16.** $-12x^2 - 2x + 30$
17. $6x^3 - 4x^2 - 28x + 24$ **18.** $4x^2 + 2x - 1$ **19.** $-x + 2 - \frac{5}{3x}$ **20.** $4x + 5$ **21.** 57.14 hr
22. 80 mph   **23.** 20 ℓ   **24.** 1.26×10^9 **25.** 2×10^{-7}

Cumulative Review Test

1. -40 **2.** 20   **3.** All real numbers   **4.** $x < -5$,   ![number line]   **5.** $w = \frac{v}{lh}$ **6.** $y = \frac{4x - 6}{3}$ or $y = \frac{4}{3}x - 2$
7. $6x^9$ **8.** $135x^8y^{13}$ **9.** $3x^2 - 2x - 5$, second degree   **10.** $-4x^2 - 6x - 11$ **11.** $8x^2 - 3x + 9$
12. $x^2 - 3x - 7$ **13.** $6x^2 - 19x + 15$ **14.** $2x^3 - 6x^2 - 12x - 40$ **15.** $3x - 2 + \frac{8}{3x}$ **16.** $2x - 3$
17. \$3.33   **18.** 4   **19.** $w = 3$ ft, $l = 10$ ft   **20.** 2 hr

CHAPTER 5

Exercise Set 5.1

1. $2^3 \cdot 5$ **3.** $2 \cdot 3^2 \cdot 5$ **5.** $2^3 \cdot 5^2$ **7.** 4   **9.** 12   **11.** 18   **13.** x **15.** $3x$ **17.** 1   **19.** xy
21. x^3y^5 **23.** 8   **25.** $3x^3y^7$ **27.** x^2 **29.** $x + 3$ **31.** $2x - 3$ **33.** $3x - 4$ **35.** $2(x + 2)$
37. $5(3x - 1)$ **39.** Cannot be factored   **41.** $4x(4x - 3)$ **43.** $2p(10 - 9p)$ **45.** $2x(3x^2 - 4)$
47. $12x^8(3x^4 - 2)$ **49.** $3y^3(8y^{12} - 3)$ **51.** $x(1 + 3y^2)$ **53.** Cannot be factored   **55.** $4xy(4yz + x^2)$
57. $2xy^2(17x + 8y^2)$ **59.** $36xy^2z(z^2 + x^2)$ **61.** $y^3z^5(14 - 9x)$ **63.** $3(x^2 + 2x + 3)$ **65.** $3(3x^2 + 6x + 1)$
67. $3x(x^2 - 2x + 4)$ **69.** Cannot be factored   **71.** $3(5p^2 - 2p + 3)$ **73.** $4x^3(6x^3 + 2x - 1)$
75. $xy(48x + 16y + 33)$ **77.** $(x + 3)(x + 2)$ **79.** $(7x - 4)(4x - 3)$ **81.** $(4x + 1)(2x + 1)$
83. $(4x + 1)(2x + 1)$ **85.** An expression written as a product of factors.

Cumulative Review Exercises

89. $-2x + 18$ **90.** 1   **91.** $\dfrac{1}{2}$ **92.** $h = \dfrac{2A}{b}$

Just For Fun   **1.** $2(x - 3)[2x^2(x - 3)^2 - 3x(x - 3) + 2]$ **2.** $2x^2(2x + 7)(3x^3 + 2x - 1)$
3. (a) $3(0) + 3(1) + 3(2) + 3(3) + 3(4)$ **(b)** $3(0 + 1 + 2 + 3 + 4)$ **(c)** $3(10) = 30$ **(d)** $3(55) = 165$

Exercise Set 5.2

1. $(x + 3)(x + 4)$ **3.** $(x + 5)(x + 2)$ **5.** $(x + 2)(x + 3)$ **7.** $(x - 5)(x + 3)$ **9.** $(2x - 3)(2x + 3)$
11. $(3x + 1)(x + 3)$ **13.** $(2x - 1)(2x - 1) = (2x - 1)^2$ **15.** $(8x + 1)(x + 4)$ **17.** $(x + 1)(3x - 2)$
19. $(x - 1)(3x - 2)$ **21.** $(3x - 4)(5x - 6)$ **23.** $(x - 3y)(x + 2y)$ **25.** $(3x + y)(2x - 3y)$
27. $(2x - 5y)(5x - 6y)$ **29.** $(x + a)(x + b)$ **31.** $(x - 2)(y + 4)$ **33.** $(a + b)(a + 2)$ **35.** $(x + 5)(y - 1)$
37. $(2 - x)(3 + 2y)$ **39.** $(a^2 + 1)(a + 2)$ **41.** $(x^2 - 3)(x + 4)$ **43.** $2(x + 4)(x - 6)$
45. $4(x + 2)(x + 2) = 4(x + 2)^2$ **47.** $x(3x - 1)(2x + 3)$ **49.** $2(x + 4y)(x - 2y)$
51. Determine if all terms have a common factor; if so factor out the GCF.
53. $x^2 + 4x - 2x - 8$, multiply the factors using the FOIL method.

Cumulative Review Exercises

55. 7.14 years   **56.** 18 days   **57.** $5x^2 - 2x - 3 + \dfrac{5}{3x}$ **58.** $x + 3$

Just For Fun   **1.** $x(3x^2 + 2)(x^2 - 5)$ **2.** $(x - y)(x^2 + y)$ **3.** $(3a - x)(6a + x^2)$

Exercise Set 5.3

1. $(x + 2)(x + 5)$ **3.** $(x + 3)(x + 2)$ **5.** $(x + 4)(x + 3)$ **7.** Cannot be factored   **9.** $(y - 15)(y - 1)$
11. $(x + 3)(x - 2)$ **13.** $(k - 5)(k + 3)$ **15.** $(b - 9)(b - 2)$ **17.** Cannot be factored   **19.** $(a + 11)(a + 1)$
21. $(x + 15)(x - 2)$ **23.** $(x + 2)^2$ **25.** $(x + 3)^2$ **27.** $(x + 5)^2$ **29.** $(w - 15)(w - 3)$
31. $(x + 24)(x - 2)$ **33.** $(x - 5)(x + 4)$ **35.** $(y - 7)(y - 2)$ **37.** $(x + 16)(x - 4)$ **39.** $(x - 2)(x - 12)$
41. $(x + 8)(x - 10)$ **43.** $(x - 5)(x - 12)$ **45.** $(x + 2)(x + 28)$ **47.** $(x - 2y)^2$ **49.** $(x + 5y)(x + 3y)$
51. $2(x - 6)(x - 1)$ **53.** $5(x + 3)(x + 1)$ **55.** $2(x - 4)(x - 3)$ **57.** $x(x - 6)(x + 3)$
59. $2x(x + 7)(x - 4)$ **61.** $x(x + 2)^2$ **63.** Multiplying factors using the FOIL method.
65. $x^2 - 11x + 24$, multiply the factors and combine like terms.
67. $2x^2 - 8xy - 10y^2$, multiply the factors and combine like terms.

Cumulative Review Exercises

69. $2x^3 + x^2 - 16x + 12$ **70.** $3x + 2 - \dfrac{2}{x - 4}$ **71.** 19.6%   **72.** $(x - 2)(3x + 5)$

Just For Fun   **1.** $(x + 0.4)(x + 0.2)$ **2.** $(x - 0.6)(x + 0.1)$ **3.** $(x + \frac{1}{5})(x + \frac{1}{5})$ **4.** $(x - \frac{1}{3})(x - \frac{1}{3})$

Exercise Set 5.4

1. $(3x + 2)(x + 1)$ **3.** $(2x + 3)(3x + 2)$ **5.** $(2x + 3)(x + 1)$ **7.** $(2x + 5)(x + 3)$ **9.** $(3x + 2)(x - 4)$
11. $(5y - 3)(y - 1)$ **13.** Cannot be factored   **15.** $(4x + 1)(x + 3)$ **17.** Cannot be factored
19. $(5y - 1)(y - 3)$ **21.** $(3x - 1)(x + 5)$ **23.** $(7x - 2)(x - 2)$ **25.** $(x - 1)(3x - 7)$ **27.** $(5z + 2)(z - 7)$
29. $(4x + 3)(2x - 1)$ **31.** $(5x - 1)(2x - 5)$ **33.** $(4x + 5)(2x - 3)$ **35.** $3(2x + 1)(x + 5)$
37. $2(3x + 5)(x - 1)$ **39.** $x(2x - 1)(3x + 4)$ **41.** $2x(2x + 3)(x - 1)$ **43.** $2x(3x + 5)(x - 1)$
45. $5(6x + 1)(2x + 1)$ **47.** $(2x + y)(x + 2y)$ **49.** $(2x - y)(x - 3y)$ **51.** $2(3x - y)(3x + 4y)$
53. Factor out the GCF if there is one.

Cumulative Review Exercises

56. $\dfrac{1}{2}$ **57.** $w = 3$ ft, $l = 8$ ft   **58.** $12xy^2(3x^3y - 1 + 2x^4y^4)$ **59.** $(x - 9)(x - 6)$

Just For Fun   **1.** $(6x - 5)(3x + 4)$ **2.** $(8x - 3)(x - 12)$

Exercise Set 5.5

1. $(x + 2)(x - 2)$ **3.** $(y + 5)(y - 5)$ **5.** $(x + 7)(x - 7)$ **7.** $(x + y)(x - y)$ **9.** $(3y + 4)(3y - 4)$
11. $4(4a + 3b)(4a - 3b)$ **13.** $(5x + 4)(5x - 4)$ **15.** $(z^2 + 9x)(z^2 - 9x)$ **17.** $9(x^2 + 3y)(x^2 - 3y)$
19. $(7m^2 - 4n)(7m^2 + 4n)$ **21.** $20(x + 3)(x - 3)$ **23.** $(x + y)(x^2 - xy + y^2)$ **25.** $(a - b)(a^2 + ab + b^2)$
27. $(x + 2)(x^2 - 2x + 4)$ **29.** $(x - 3)(x^2 + 3x + 9)$ **31.** $(a + 1)(a^2 - a + 1)$ **33.** $(2x + 3)(4x^2 - 6x + 9)$
35. $(3a - 4)(9a^2 + 12a + 16)$ **37.** $(3 - 2y)(9 + 6y + 4y^2)$ **39.** $(2x - 3y)(4x^2 + 6xy + 9y^2)$
41. $2(x - 3)(x + 2)$ **43.** $y(x + 4)(x - 4)$ **45.** $3(x + 1)^2$ **47.** $5(x + 3)(x - 1)$ **49.** $3(x + 3)(y - 2)$
51. $2(x + 6)(x - 6)$ **53.** $3y(x + 3)(x - 3)$ **55.** $3y^2(x + 1)(x^2 - x + 1)$ **57.** $2(x - 2)(x^2 + 2x + 4)$
59. $2(x + 4)(3x - 2)$ **61.** $x(3x + 2)(x - 4)$ **63.** $(x + 2)(4x - 3)$ **65.** $25(b + 2)(b - 2)$
67. $a^3b^2(a + 2b)(a - 2b)$ **69.** $3x^2(x - 3)^2$ **71.** $x(x^2 + 25)$ **73.** $(y^2 + 4)(y + 2)(y - 2)$
75. $5(2a - 3b)(a + 4b)$ **77.** $(3x + 5)(3x - 1)$ **79.** $x(x + 5)(x - 5)$ **81.** **(a)** $a^2 - b^2 = (a + b)(a - b)$
83. **(a)** $a^3 - b^3 = (a - b)(a^2 + ab + b^2)$

Cumulative Review Exercises

84. All real numbers   **85.** $y = \dfrac{2x - 6}{5}$ or $y = \dfrac{2}{5}x - \dfrac{6}{5}$ **86.** $\dfrac{8x^9}{27y^{12}}$ **87.** $\dfrac{1}{x^5}$

Just for Fun   **1.** $(x^2 + 1)(x^4 - x^2 + 1)$ **2.** $(x^2 - 3y^3)(x^4 + 3x^2y^3 + 9y^6)$
3. Cannot divide both sides of equation by $a - b$ since it equals 0.

Exercise Set 5.6

1. $0, -3$ **3.** $0, 9$ **5.** $-\dfrac{5}{2}, 3$ **7.** $4, -4$ **9.** $0, 12$ **11.** $0, -2$ **13.** $-4, 3$ **15.** $2, 10$ **17.** $3, -6$
19. $-4, 6$ **21.** $2, -21$ **23.** $-5, -6$ **25.** $4, -2$ **27.** $32, -2$ **29.** $6, -3$ **31.** $\dfrac{1}{3}, 7$ **33.** $\dfrac{2}{3}, -1$
35. $-4, 3$ **37.** $\dfrac{4}{3}, -\dfrac{1}{2}$ **39.** $2, 3$ **41.** $0, 16$ **43.** $6, -6$ **45.** $3, -3$ **47.** $8, 10$ **49.** $7, 9$ **51.** $5, 7$
53. $w = 3$ ft, $l = 12$ ft   **55.** 2 m   **57.** 3
59. **(a)** Zero-factor property may only be used when one side of the equation is equal to 0. **(b)** $-2, 3$

Cumulative Review Exercises

61. $-x^2 + 7x - 4$ **62.** $6x^3 + x^2 - 10x + 4$ **63.** $2x - 3$ **64.** $2x - 3$

Just for Fun   **1.** **(a)** 192 ft   **(b)** 8 sec   **2.** $0, 2, -5$

Chapter 5 Review Exercises

1. x^2 **2.** $3p$ **3.** 6   **4.** $4x^2y^2$ **5.** 1   **6.** $8x^2$ **7.** $2x - 5$ **8.** $x + 5$ **9.** $5(x - 4)$
10. $3(3x + 11)$ **11.** $4y(4y - 3)$ **12.** $5p^2(11p - 4)$ **13.** $6x^2y(4 + 3xy)$ **14.** $9xy(2x - 1)$
15. $2(x^2 + 2x - 4)$ **16.** $6x^4y^2(10y^2 + x^5y - 3x)$ **17.** Cannot be factored   **18.** $(x - 2)(5x + 3)$
19. $(3x - 2)(x - 1)$ **20.** $(2x + 1)(4x - 3)$ **21.** $(x + 2)(x + 3)$ **22.** $(x + 3)(x - 5)$ **23.** $(x + 7)(x - 7)$

24. $(2a - 1)(a - b)$ **25.** $(3x + 2)(y + 1)$ **26.** $(x - 2y)(x + 3)$ **27.** $(5x - 1)(x + 4)$
28. $(x + 4y)(5x - y)$ **29.** $(4x + 5y)(3x - 2y)$ **30.** $(3x - 2y)(4x + 5y)$ **31.** $(a + 1)(b - 1)$
32. $(3x + 2y)(x - 3y)$ **33.** $(4x + 3)(5x - 3)$ **34.** $(3x - 1)(2x + 3)$ **35.** $(x + 2)(x + 4)$
36. $(x - 5)(x - 3)$ **37.** $(x - 5)(x + 4)$ **38.** $(x + 5)(x - 4)$ **39.** $(x - 6)(x + 3)$ **40.** $(x - 7)(x - 2)$
41. $(x - 15)(x + 3)$ **42.** $(x + 8)(x + 3)$ **43.** $x(x + 4)(x + 1)$ **44.** $x(x - 8)(x + 5)$ **45.** $(x + 3y)(x - 5y)$
46. $4x(x + 5y)(x + 3y)$ **47.** $(2x - 1)(x + 4)$ **48.** $(3x + 1)(x + 4)$ **49.** $(4x - 5)(x - 1)$
50. $(3x + 4)(x - 3)$ **51.** $(2x - 3)(2x + 5)$ **52.** $(5x - 2)(x - 6)$ **53.** $(3x + 4)(x + 3)$
54. $(6x + 1)(x + 5)$ **55.** $(2x - 5)(x + 7)$ **56.** $(3x - 2)(2x + 5)$ **57.** $(4x + 5)(2x - 7)$ **58.** $(2x + 5)^2$
59. $x(3x - 2)^2$ **60.** $2x(3x + 1)(3x - 5)$ **61.** $(2x - 3y)(2x - 5y)$ **62.** $(8x + y)(2x - 3y)$
63. $(x + 5)(x - 5)$ **64.** $(x + 8)(x - 8)$ **65.** $4(x + 2)(x - 2)$ **66.** $9(3x + y)(3x - y)$
67. $(8x^2 + 9y^2)(8x^2 - 9y^2)$ **68.** $(4 + 5y)(4 - 5y)$ **69.** $(2x^2 + 3y^2)(2x^2 - 3y^2)$
70. $(10x^2 + 11y^2)(10x^2 - 11y^2)$ **71.** $(x - y)(x^2 + xy + y^2)$ **72.** $(x + y)(x^2 - xy + y^2)$
73. $(a + 2)(a^2 - 2a + 4)$ **74.** $(a - 1)(a^2 + a + 1)$ **75.** $(a + 3)(a^2 - 3a + 9)$ **76.** $(x - 2)(x^2 + 2x + 4)$
77. $(2x - y)(4x^2 + 2xy + y^2)$ **78.** $(3 - 2y)(9 + 6y + 4y^2)$ **79.** $8(x + 3)(x - 1)$ **80.** $2(x - 4)^2$
81. $4(x + 3)(x - 3)$ **82.** $3(y + 3)(y - 3)$ **83.** $(x - 6)(x - 4)$ **84.** $(x - 9)(x + 3)$ **85.** $(2x + 3)(2x - 5)$
86. $3(2x - 3)(x - 4)$ **87.** $8(x - 1)(x^2 + x + 1)$ **88.** $y(x - 3)(x^2 + 3x + 9)$ **89.** $y(x + 4)(x - 1)$
90. $3x(2x + 3)(x + 5)$ **91.** $(x + 3y)(x + 2y)$ **92.** $(2x - 5y)(x + 2y)$ **93.** $(2x - 5y)^2$
94. $(4y + 7z)(4y - 7z)$ **95.** $(a + 6)(b + 7)$ **96.** $y^5(4 + 5y)(4 - 5y)$ **97.** $2x(x + 4y)(x + 2y)$
98. $(2x - 3y)(3x + 7y)$ **99.** $2x(4x + 1)(4x + 3)$ **100.** $(y^2 + 1)(y + 1)(y - 1)$ **101.** $0, 4$ **102.** $-3, -4$
103. $5, -\frac{2}{3}$ **104.** $0, 3$ **105.** $0, -4$ **106.** $6, -4$ **107.** $-3, -5$ **108.** $-4, 2$ **109.** $-4, 3$
110. $-2, -5$ **111.** $2, 4$ **112.** $1, -2$ **113.** $\frac{1}{4}, -\frac{3}{2}$ **114.** $\frac{1}{2}, -8$ **115.** $2, -2$ **116.** $\frac{7}{6}, -\frac{7}{6}$
117. $10, 11$ **118.** $6, 8$ **119.** $5, 8$ **120.** $w = 7$ ft, $l = 9$ ft **121.** 5 in, 9 in

Chapter 5 Practice Test

1. $2x^2$ **2.** $3xy^2$ **3.** $4xy(x - 2)$ **4.** $3x(8xy - 2y + 3)$ **5.** $(x + 2)(x - 3)$ **6.** $(3x + 1)(x - 4)$
7. $(5x - 3y)(x - 3y)$ **8.** $(x + 4)(x + 8)$ **9.** $(x + 8)(x - 3)$ **10.** $(x - 5y)(x - 4y)$ **11.** $2(x - 5)(x - 6)$
12. $x(2x - 1)(x - 1)$ **13.** $(3x + 2y)(4x - 3y)$ **14.** $(x + 3y)(x - 3y)$ **15.** $(x + 3)(x^2 - 3x + 9)$ **16.** $2, \frac{5}{2}$
17. $-2, -3$ **18.** $-5, 1$ **19.** $4, 9$ **20.** $l = 6$ m, $w = 4$ m

CHAPTER 6

Exercise Set 6.1

1. $x \neq 0$ **3.** $x \neq 6$ **5.** $x \neq 2, x \neq -2$ **7.** $x \neq 2, x \neq -8$ **9.** $\dfrac{1}{1 + y}$ **11.** 4 **13.** $\dfrac{x^2 + 6x + 3}{2}$

15. $x + 1$ **17.** $\dfrac{x}{x - 2}$ **19.** $\dfrac{x - 3}{x - 2}$ **21.** $2(x + 1)$ **23.** -1 **25.** $-(x + 2)$ **27.** $\dfrac{-(x + 6)}{2x}$

29. $-(x + 3)$ **31.** $3x + 2$ **33.** $3x - 4$ **35.** $2x - 3$ **37.** $x + 4$ **39.** $\dfrac{x - 4}{x + 4}$ **41.** $x^2 + 2x + 4$

43. Denominator cannot be 0 for any real value of x. **45.** $x \neq 4$, denominator cannot have a value of 0. **47.** Yes
49. $x + 2$

Cumulative Review Exercises

51. $y = x - 2z$ **52.** $28°, 58°, 94°$ **53.** $\dfrac{x^4}{9y^2}$ **54.** $9x^2 - 10x - 17$

Exercise Set 6.2

1. $\dfrac{xy}{4}$ **3.** $\dfrac{80x^4}{y^6}$ **5.** $\dfrac{36x^9y^2}{25z^7}$ **7.** $\dfrac{-3x + 2}{3x + 2}$ **9.** 1 **11.** $(a + b)(a - b)$ or $a^2 - b^2$ **13.** $\dfrac{x + 3}{2(x + 2)}$

15. $x + 3$ **17.** $3x^2y$ **19.** $\dfrac{10z}{x}$ **21.** $6a^2b$ **23.** $\dfrac{3x^2}{2}$ **25.** $\dfrac{x}{x + 6}$ **27.** $\dfrac{x - 8}{x + 2}$ **29.** $\dfrac{x + 3}{x - 1}$ **31.** 1

33. $6x^3y^3$ **35.** $\dfrac{5}{3ab^2c^2}$ **37.** $\dfrac{3y^2}{a^2}$ **39.** $\dfrac{32x^7}{35my^2}$ **41.** $\dfrac{1}{2}$ **43.** $3x$ **45.** 1 **47.** 1 **49.** $\dfrac{a}{2}$

51. $\dfrac{ab(x-y)}{6}$ **53.** $\dfrac{x+3}{x-4}$ **55.** $\dfrac{a+3}{a+2}$ **57.** 1 **59.** $(3x+2)(4x-3)$ **61.** $(3x-2)(x-3)$ **63.** 1

65. $-3x(x+2)$ **67.** $\dfrac{1}{2}$ **69.** $(x-4)(x+4)$ **71.** $2x^2+x-6$, numerator must be $(x+2)(2x-3)$.

Cumulative Review Exercises

74. $20x^4y^5z^{11}$ **75.** $2x^2+x-2-\dfrac{2}{2x-1}$ **76.** $3(x-5)(x+2)$ **77.** $5, -2$

Just for Fun **1.** $\dfrac{x-3}{x+3}$ **2.** $\dfrac{x-1}{x-3}$

Exercise Set 6.3

1. $\dfrac{2x-1}{6}$ **3.** $\dfrac{x-11}{3}$ **5.** $\dfrac{x-3}{x}$ **7.** $\dfrac{x+3}{x}$ **9.** $\dfrac{x+7}{x+2}$ **11.** $-\dfrac{8}{x}$ **13.** $\dfrac{2x-11}{x-7}$ **15.** $\dfrac{2x+1}{2x^2}$

17. $\dfrac{1}{x+1}$ **19.** $\dfrac{1}{x-3}$ **21.** 0 **23.** $\dfrac{-(4x+1)}{x-7}$ **25.** $x-3$ **27.** 3 **29.** $\dfrac{x^2+9}{x+3}$ **31.** -1

33. $\dfrac{x-3}{x-1}$ **35.** $x-4$ **37.** $\dfrac{3}{4}$ **39.** $\dfrac{x-8}{x+8}$ **41.** $\dfrac{3x+2}{x-4}$ **43.** $\dfrac{6x+1}{x-8}$ **45.** The signs change

47. Should be $\dfrac{6x-2-(3x^2-4x+5)}{x^2-4x+3}$ **49.** x^2+x-9, the sum of the numerators must be $2x^2-5x-6$

51. $2x^2-7x-4$, the difference of the numerators must be $2x^2+x-3$

Cumulative Review Exercises

54. $-\dfrac{1}{5}$ **55.** $1\tfrac{1}{8}$ or 1.125 ounces **56. (a)** 75 hours **(b)** Plan 2 **57.** $\dfrac{(x-2)^2}{(2x+1)(x+3)}$

Just for Fun **1.** $\dfrac{-4x^2+8x+3}{x^2-9}$ **2.** $\dfrac{-5x-2}{x+2}$

Exercise Set 6.4

1. 3 **3.** $6x$ **5.** $10x$ **7.** x^2 **9.** $2x+3$ **11.** $x^2(x+1)$ **13.** $144x^3y$ **15.** $36x(x+5)$

17. $x(x+1)$ **19.** $180x^2y^3$ **21.** $6(x+4)(x-3)$ **23.** $(x+6)(x+5)$ **25.** $(x-8)(x+3)(x+8)$

27. $(x+3)(x-3)(x-1)$ **29.** $(x-2)(x+1)(x+3)$ **31.** $(x+2)^2$ **33.** $(3x-2)(x+6)(3x-1)$

35. $(2x+1)^2(4x+3)$

Cumulative Review Exercises

37. $\tfrac{92}{45}$ or $2\tfrac{2}{45}$ **38.** Commutative Property of addition **39.** Distributive Property **40.** Associative Property of addition **41. (b)** $-\tfrac{5}{3}$ **42. (b)** $1, -2$

Just for Fun **1.** $30x^{12}y^9$ **2.** $(x-2)(x+2)$ **3.** $(x-4)(x+3)(x-2)$

Exercise Set 6.5

1. $\dfrac{11}{2x}$ **3.** $\dfrac{3x+12}{2x^2}$ **5.** $\dfrac{2x^2-1}{x^2}$ **7.** $\dfrac{3x+5}{5x^2}$ **9.** $\dfrac{28x+15y}{20x^2y^2}$ **11.** $\dfrac{x(y+1)}{y}$ **13.** $\dfrac{9x-1}{3x}$

15. $\dfrac{5x^2+y^2}{xy}$ **17.** $\dfrac{4y-30x^2}{5x^2y}$ **19.** $\dfrac{8x-10}{x(x-2)}$ **21.** $\dfrac{11a+6}{a(a+3)}$ **23.** $\dfrac{-2x^2+4x+8}{3x(x+2)}$ **25.** $\dfrac{2}{x-2}$

27. $\dfrac{9}{x+3}$ **29.** $\dfrac{7x+1}{(x+1)(x-1)}$ **31.** $\dfrac{20x}{(x-5)(x+5)}$ **33.** $\dfrac{5x-12}{(x+3)(x-3)}$ **35.** $\dfrac{-x+6}{(x+2)(x-2)}$

37. $\dfrac{11}{(x-3)(x-4)}$ **39.** $\dfrac{2(3x-4)}{(x+4)(x-2)}$ **41.** $\dfrac{(x+3)(x-1)}{(x+2)(x+2)}$ **43.** $\dfrac{5(x+1)}{(x+4)(x-2)(x-1)}$

45. $\dfrac{4x-3}{(x+3)(x+2)(x-2)}$ **47.** $\dfrac{(2x-1)(x-4)}{(3x-1)(x+2)(2x+3)}$ **49.** $\dfrac{2x^2-6x+3}{(3x-1)(x+2)(2x+3)}$

Cumulative Review Exercises

51. 810   **52.** $x > -8$ $\xrightarrow{\hspace{0.3cm}\circ\hspace{1cm}}$ $\underset{-8}{}$ **53.** $4x - 3 - \dfrac{4}{2x+3}$ **54.** 1

Just for Fun   **1.** $\dfrac{x^2+5x-2}{(x+2)(x-2)}$ **2.** $\dfrac{x^2-7x-11}{(x+3)(x-2)}$ **3.** $\dfrac{2x-3}{2-x}$ **4.** $\dfrac{8x^3-8x^2-14x+33}{(x+2)(x-3)(2x+3)}$

Exercise Set 6.6

1. $\dfrac{8}{11}$ **3.** $\dfrac{57}{32}$ **5.** $\dfrac{25}{1224}$ **7.** $\dfrac{x^3y}{8}$ **9.** $6xz^2$ **11.** $\dfrac{xy+1}{x}$ **13.** $\dfrac{3}{x}$ **15.** $\dfrac{3y-1}{2y-1}$ **17.** $\dfrac{x-y}{y}$ **19.** $-\dfrac{a}{b}$

21. -1 **23.** $\dfrac{2(x+2)}{x^3}$ **25.** $a+b$ **27.** $\dfrac{a^2+b}{b(b+1)}$ **29.** $\dfrac{y-x}{x+y}$ **31.** $\dfrac{y^2(x+1)}{x^2(y+1)}$ **33.** (b) $\dfrac{2y-3x}{x^2y+x}$

Cumulative Review Exercises

34. All real numbers
35. An expression containing a finite number of terms of the form ax^n, for any real number a and any whole number n.
36. $48x^5y^{10}$ **37.** $\dfrac{x^2-9x+2}{(3x-1)(x+6)(x-3)}$

Just for Fun   **1.** (a) $\dfrac{2}{7}$ (b) $\dfrac{4}{13}$

Exercise Set 6.7

1. 4   **3.** 48   **5.** 30   **7.** -1 **9.** 1   **11.** 4   **13.** 3   **15.** $-\dfrac{1}{5}$ **17.** $\dfrac{1}{4}$ **19.** $\dfrac{14}{3}$ **21.** No solution
23. 2   **25.** $-\dfrac{12}{7}$ **27.** No solution   **29.** 8   **31.** -14 **33.** $-2, -3$ **35.** 4   **37.** $-\dfrac{5}{2}$ **39.** No solution
41. 5   **43.** No solution   **45.** -3 **47.** 2   **49.** (b) $\dfrac{2}{3}$
51. (a) The problem on the right is an equation, while the one on the left is not.   (b) Problem on left: write each fraction with the least common denominator $12(x-1)$; then combine numerators. Problem on right: multiply both sides of the equation by the LCD $12(x-1)$ to eliminate fractions; then solve the remaining equation.
c) Problem on left: $\dfrac{x^2-x+12}{12(x-1)}$; Problem on right: 4, -3

Cumulative Review Exercises

52. 30 rides   **53.** 50°, 130°   **54.** 75 min
55. Linear equation: $ax + b = c$, $a \neq 0$; quadratic equation $ax^2 + bx + c = 0$, $a \neq 0$

Just for Fun   **1.** 15 cm   **2.** (a) 120 ohms   (b) 900 ohms

Exercise Set 6.8

1. $80 = \frac{1}{2}(h+6)h$; $h = 10$ cm, $b = 16$ cm   **3.** $\dfrac{1}{x} + \dfrac{1}{3x} = \dfrac{4}{3}$; 1, 3   **5.** $\dfrac{1}{3} + \dfrac{1}{5} = \dfrac{1}{x}$; $\dfrac{15}{8}$ **7.** $\dfrac{1}{x} + \dfrac{1}{x+4} = \dfrac{2}{3}$; 2, 6

9. $\dfrac{9}{r-2} = \dfrac{11}{r+2}$; 20 mph   **11.** $\dfrac{1800}{4r} + \dfrac{300}{r} = 5$; 150 mph, 600 mph

13. $\dfrac{d}{2} + \dfrac{d+3}{4} = 3$; walked 3 mi; jogged 6 mi   **15.** $\dfrac{d}{8} + \dfrac{6-d}{4} = 1.2$; Jogs 2.4 mi, walks 3.6 mi

17. $\dfrac{d}{600} + \dfrac{2800-d}{500} = 5$; 3 hr at 600 mph, 2 hr at 500 mph   **19.** $\dfrac{d}{30} = \dfrac{d}{40} + 20$; 2400 meters

21. $\dfrac{t}{6} + \dfrac{t}{7} = 1$; $3\dfrac{3}{13}$ hr **23.** $\dfrac{t}{3} - \dfrac{t}{4} = 1$; 12 hr **25.** $\dfrac{3}{8} + \dfrac{3}{t} = 1$; $4\dfrac{4}{5}$ hr **27.** $\dfrac{4}{6} + \dfrac{t}{10} = 1$; $3\dfrac{1}{3}$ days

29. $\dfrac{t}{60} + \dfrac{t}{50} - \dfrac{t}{30} = 1$; 300 hr **31.** $\dfrac{t}{25} + \dfrac{t + 11}{20} = 1$; 5 hr

Cumulative Review Exercises

32. 8 **33.** $-\dfrac{3}{2}x - \dfrac{9}{2}$ **34.** 1 **35.** $\dfrac{3(x^2 - 3x - 5)}{(2x + 3)(3x - 5)(3x + 1)}$

Just for Fun **1.** All real numbers except 3 **2.** 1 or $\frac{2}{3}$ **3.** 9 buckets

Chapter 6 Review Exercises

1. $x \ne 4$ **2.** $x \ne 3, x \ne 4$ **3.** $\dfrac{1}{1 - y}$ **4.** $x^2 + 4x + 12$ **5.** $3x + 2y$ **6.** $x + 4$ **7.** $x + 2$

8. $-(2x - 1)$ **9.** $\dfrac{x - 6}{x + 2}$ **10.** $\dfrac{3x + 4}{x - 4}$ **11.** $\dfrac{8xy^2}{3}$ **12.** $6xz^2$ **13.** $\dfrac{16b^3c^2}{a^2}$ **14.** $-\frac{1}{2}$ **15.** $-2x$

16. $\dfrac{(x + y)y^2}{2x^3}$ **17.** 1 **18.** $\dfrac{x + y}{x}$ **19.** 36 **20.** $\dfrac{32z}{x^3}$ **21.** $\dfrac{3}{x - y}$ **22.** $\dfrac{1}{3(a + 3)}$ **23.** $\dfrac{a - 2}{2x(a + 2)}$

24. 1 **25.** $2x(x - 5y)$ **26.** $\dfrac{16(x - 2y)}{3(x + 2y)}$ **27.** 1 **28.** $\dfrac{x - 2}{x + 2}$ **29.** 4 **30.** $\dfrac{2(3x - 4)}{3y}$ **31.** 9

32. $\dfrac{4}{x + 10}$ **33.** $3x + 2$ **34.** $3x + 4$ **35.** 24 **36.** $15x^2$ **37.** $x(x + 1)$ **38.** $(x + 2)(x - 3)$

39. $x(x + 1)$ **40.** $(x + y)(x - y)$ **41.** $x - 7$ **42.** $(x + 7)(x - 5)(x + 2)$ **43.** $\dfrac{3}{x}$ **44.** $\dfrac{24x + y}{4xy}$

45. $\dfrac{5x^2 - 12y}{3x^2y}$ **46.** $\dfrac{7x + 12}{x + 2}$ **47.** $\dfrac{5x + 12}{x + 3}$ **48.** $\dfrac{a^2 + c^2}{ac}$ **49.** $\dfrac{7x + 12}{x(x + 3)}$ **50.** $\dfrac{-(x + 4)}{3x(x - 2)}$

51. $\dfrac{8x + 25}{(x + 3)(x + 4)}$ **52.** $\dfrac{4x + 26}{(x + 5)^2}$ **53.** $\dfrac{3(x - 1)}{(x + 3)(x - 3)}$ **54.** $\dfrac{4}{(x + 2)(x - 3)(x - 2)}$ **55.** $\dfrac{2(x - 4)}{(x - 3)(x - 5)}$

56. $\dfrac{22x + 5}{(x - 5)(x - 10)(x + 5)}$ **57.** $\dfrac{34}{9}$ **58.** $\dfrac{55}{26}$ **59.** $\dfrac{5yz}{6}$ **60.** $\dfrac{16x^3z^2}{y^3}$ **61.** $\dfrac{xy + 1}{y^3}$ **62.** $\dfrac{xy - x}{x + 1}$

63. $\dfrac{4x + 2}{x(6x - 1)}$ **64.** $\dfrac{2}{x}$ **65.** a **66.** $\dfrac{2a + 1}{2}$ **67.** $\dfrac{x + 1}{-x + 1}$ **68.** $\dfrac{x^2(3 - y)}{y(y - x)}$ **69.** 9 **70.** 1 **71.** 6
72. 6 **73.** 52 **74.** -20 **75.** No solution **76.** 18 **77.** $\frac{1}{2}$ **78.** -6 **79.** -18 **80.** $2\frac{2}{9}$ hr
81. $16\frac{4}{5}$ hr **82.** $\frac{5}{2}$, 10 **83.** Bus 80 km/hr, train 120 km/hr

Chapter 6 Practice Test

1. $\dfrac{2x^3z}{3y^3}$ **2.** $a + 3$ **3.** $\dfrac{x - 3y}{3}$ **4.** $\dfrac{4}{y(y + 5)}$ **5.** $\dfrac{7x - 2}{2y}$ **6.** $\dfrac{7x^2 - 6x - 11}{x + 3}$ **7.** $\dfrac{10x + 3}{2x^2}$

8. $\dfrac{-x + 10}{x + 2}$ **9.** $\dfrac{-1}{(x + 4)(x - 4)}$ **10.** $\dfrac{29}{10}$ **11.** $\dfrac{x^2(1 + y)}{y}$ **12.** 60 **13.** 12 **14.** $3\frac{1}{13}$ hr

Cumulative Review Test

1. 50 **2.** -8 **3.** $-\frac{5}{2}$ **4.** $\dfrac{27y^6}{x^9}$ **5.** $R = \dfrac{P - 2E}{3}$ **6.** $3x^2 - 11x + 4$ **7.** $12x^3 - 38x^2 + 39x - 15$

8. $(6a - 5)(a - 1)$ **9.** $5(2x^2 - x + 1)$ **10.** $(x - 6)(x - 4)$ **11.** $(3x + 2)(2x - 5)$ **12.** 4, $\frac{3}{2}$

13. $\dfrac{x + 3}{2x + 1}$ **14.** $\dfrac{x^2 - 8x - 12}{(x + 4)(x - 5)}$ **15.** $\dfrac{6x + 2}{(x - 5)(x + 2)(x + 3)}$ **16.** $-\frac{3}{2}$ **17.** 5 **18.** \$2000 **19.** 1.5 lb

20. 3 mi at 4 mph, 3 mi at 12 mph

CHAPTER 7

Exercise Set 7.1

1. $A(3, 1)$, $B(-3, 0)$, $C(1, -3)$, $D(-2, -3)$, $E(0, 3)$, $F(\frac{3}{2}, -1)$ **3.**   **5.** Yes
7. The x coordinate

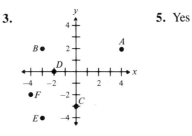

Cumulative Review Exercises

9. $-8x^2 - x + 4$ **10.** $6x^3 - 17x^2 + 22x - 15$ **11.** $(x + 3y)(x - 2)$ **12.** $\dfrac{-x - 5}{(x + 2)(x + 1)}$

Exercise Set 7.2

1. a, d, e   **3.** b, c, d   **5. (a)** 2   **(b)** 8   **(c)** $\frac{11}{2}$ **(d)** $\frac{9}{2}$ **(e)** 3   **(f)** 5

7.   **9.**   **11.**   **13.**

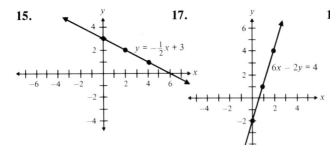

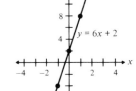

15.   **17.**   **19.**   **21.**

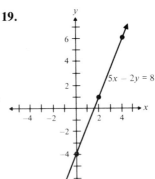
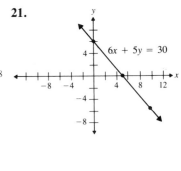

23.   **25.**   **27.**   **29.**

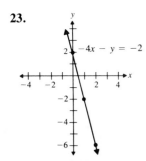
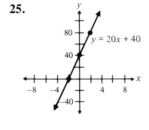
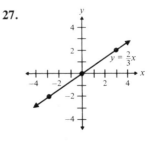
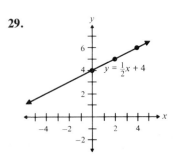

31.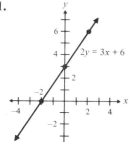

33.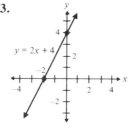

35.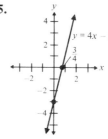

37.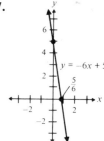

39.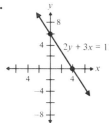

41.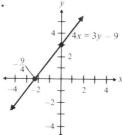

43.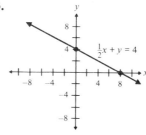

45.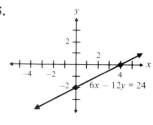

47.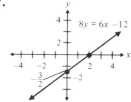

49. 51.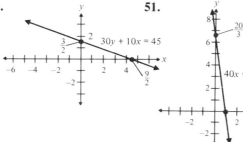

53.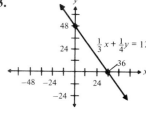

55.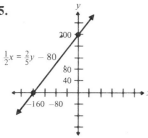

57. $x = -3$ **59.** $y = 3$ **61.** The set of points whose coordinates satisfy the equation.
63. 2, 3   **65.** A vertical line   **67.** $-6x, 3y$ **69.** $-6x, 2y$

Cumulative Review Exercises

70. 5   **71.** 4.5 mph, 7.5 mph   **72.** $\frac{1}{2}, -12$ **73.** 6, 8

Exercise Set 7.3

1. 5   **3.** $\frac{1}{2}$ **5.** 0   **7.** 0   **9.** undefined   **11.** $-\frac{5}{9}$ **13.** -2 **15.** $m = 2$ **17.** $m = -\frac{3}{2}$
19. $m = -\frac{1}{7}$ **21.** $m = \frac{7}{4}$ **23.** $m = 0$ **25.** slope is undefined
27. The ratio of the vertical change to the horizontal change between any two points on the line.
29. Lines that rise from left to right have a positive slope. Lines that fall from left to right have a negative slope.
31. No, since we cannot divide by 0. We say that the slope of a vertical line is undefined.

Cumulative Review Exercises

32. (a) An equation that contains only one variable and that variable has an exponent of 1.
(b) $2x + 3 = 5x - 6$ (answers will vary)

33. (a) An equation that contains only one variable and the greatest exponent on that variable is 2.
(b) $x^2 + 2x - 3 = 0$ (answers will vary)
34. (a) An equation that contains only one variable and one or more fractions. **(b)** $\frac{x}{3} + \frac{x}{4} = 12$ (answers will vary)
35. (a) An equation that contains two variables and the exponent on both variables is 1.
(b) $y = 3x - 2$ (answers will vary)

Just for Fun **1.** $\dfrac{225}{68}$ **2.** (3, 1), Other answers are possible. **3.** 7

4. (a) The angle (less than or equal to 90°) that the hill makes with the horizontal is measured to determine the slope.
(b) The slope of a line has no specific units. The slope of a hill is usually measured in degrees.

Exercise Set 7.4

1.
$m = 2, b = -1$
$y = 2x - 1$

3.
$m = -1, b = 5$
$y = -x + 5$

5.
$m = -4, b = 0$
$y = -4x$

7.
$m = 2, b = -3$
$-2x + y = -3$

9.
$m = -1, b = 3$
$3x + 3y = 9$

11.
$m = \frac{1}{2}, b = 4$
$-x + 2y = 8$

13.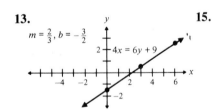
$m = \frac{2}{3}, b = -\frac{3}{2}$
$4x = 6y + 9$

15.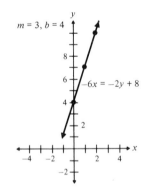
$m = 3, b = 4$
$-6x = -2y + 8$

17.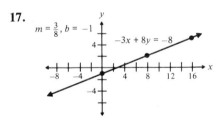
$m = \frac{3}{8}, b = -1$
$-3x + 8y = -8$

19.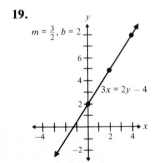
$m = \frac{3}{2}, b = 2$
$3x = 2y - 4$

21.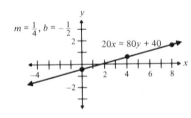
$m = \frac{1}{4}, b = -\frac{1}{2}$
$20x = 80y + 40$

23. $y = x + 2$ **25.** $y = -\frac{1}{3}x + 2$ **27.** $y = -\frac{3}{2}x + 15$ **29.** $y = -\frac{1}{2}x - 2$ **31.** Yes   **33.** No   **35.** No
37. Yes   **39.** In slope-intercept form $4x + 3y = 6$ becomes $y = -\frac{4}{3}x + 2$.

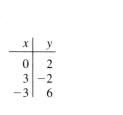
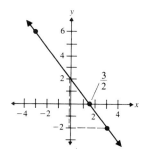

x	y
0	2
3	-2
-3	6

41. The values of y decrease as the values of x increase (graph falls as it moves to the right).
43. Compare their slopes. If the slopes are the same and their y-intercepts are different, the lines are parallel.

Cumulative Review Exercises

45. 375 rpm   **46.** 1.5 ℓ **47.** $4(x + 2y^2)(x - 2y^2)$ **48.** $\dfrac{4(2x - 1)}{(x - 2)(x + 2)}$

Exercise Set 7.5

1. $y = 5x + 4$ **3.** $y = -2x - 3$ **5.** $y = \frac{1}{2}x - \frac{9}{2}$ **7.** $y = \frac{3}{5}x - \frac{22}{5}$ **9.** $y = 3x + 10$ **11.** $y = -\frac{3}{2}x$
13. $y = \frac{1}{2}x - 2$
15. Standard form: $6x + 2y = 4$; slope-intercept form: $y = -3x + 2$, point-slope form: $y - 2 = -3(x - 0)$
17. $5x - 3y = 8$, $y = \frac{5}{3}x - \frac{8}{3}$, $y + \frac{8}{3} = \frac{5}{3}(x - 0)$

Cumulative Review Exercises

18. $3x^2 - 3x - 1 + \dfrac{6}{3x + 2}$ **19.** $\dfrac{x - 2}{2x + 1}$ **20.** 5, 6   **21.** $b = 8$ ft, $h = 9$ ft

Just for Fun   **1.** $y = \frac{3}{4}x + 2$ **2.** $y = -\frac{2}{5}x + \frac{13}{10}$
3. Factor $\frac{3}{4}$ from the x term and the constant term. Express the left side of the equation as $y - 0$.

Exercise Set 7.6

1.
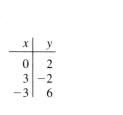

3.

5.

7.
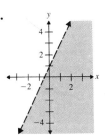

9.
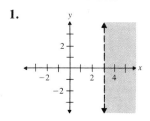

11.
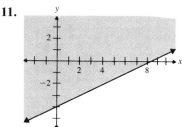

13.
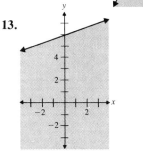

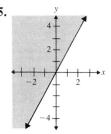
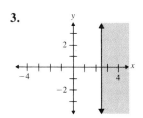
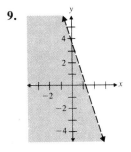

482   **Answers**

15.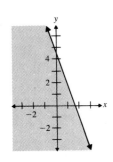

17.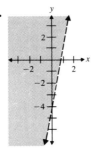

19.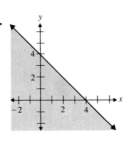

21.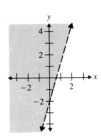

23.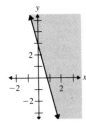

25.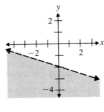

27. Points on line satisfy equation ($=$) but not inequality that is strictly greater than or less than.

Cumulative Review Exercises

28. 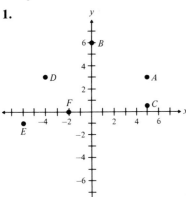   **29.** $1250   **30.** $F = \frac{9}{5}C + 32$ **31.** $m = \frac{6}{5}, b = -\frac{9}{5}$

Chapter 7 Review Exercises

1.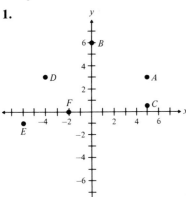

2. No
3. b, d
4. (a) -1 (b) -4
     (c) $\frac{16}{3}$ (d) $\frac{8}{3}$

5.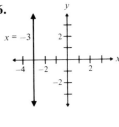

6.

7.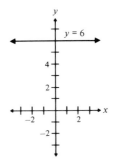

8.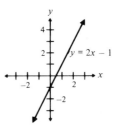

9.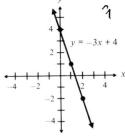

10.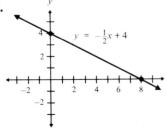

11.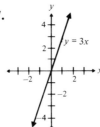

12.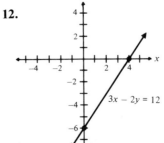

13.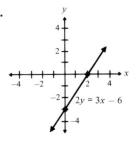

14.
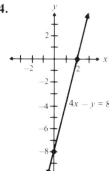

15.
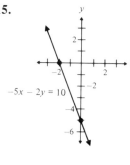

16.
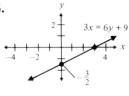

17.
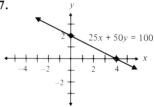

18.
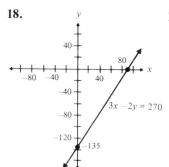

19.
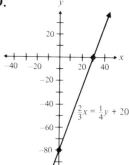

20. $-\frac{12}{5}$ **21.** $-\frac{1}{12}$ **22.** -2 **23.** 0 **24.** undefined **25.** $-\frac{5}{7}$ **26.** $\frac{9}{4}$ **27.** $\frac{1}{4}$ **28.** $m = -1$, $b = 4$
29. $m = 3$, $b = 5$ **30.** $m = -4$, $b = \frac{1}{2}$ **31.** $m = -\frac{2}{3}$, $b = \frac{8}{3}$ **32.** $m = -\frac{1}{2}$, $b = \frac{3}{2}$ **33.** $m = \frac{3}{2}$, $b = 3$
34. $m = -\frac{3}{5}$, $b = \frac{12}{5}$ **35.** $m = -\frac{9}{7}$, $b = \frac{15}{7}$ **36.** $m = \frac{1}{2}$, $b = -2$ **37.** Slope is undefined, no y intercept
38. $m = 0$, $b = -3$ **39.** $y = 2x + 2$ **40.** $y = x - \frac{5}{2}$ **41.** $y = -\frac{1}{2}x + 2$ **42.** Yes **43.** No
44. Yes **45.** Yes **46.** $y = 2x - 2$ **47.** $y = -3x + 2$ **48.** $y = -\frac{2}{3}x + 4$ **49.** $y = 2$ **50.** $x = 3$
51. $y = x - 1$ **52.** $y = -\frac{7}{2}x - 4$ **53.** $x = -4$

54.

55.
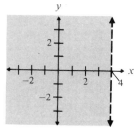

56.
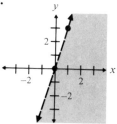

57.
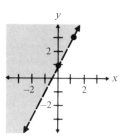

58.
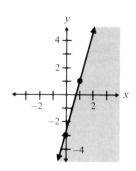

59.
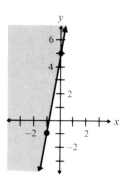

60.

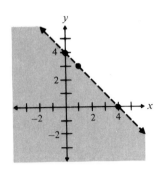

61.

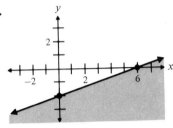

Chapter 7 Practice Test

1. a, b   **2.** $-\frac{4}{3}$ **3.** $m = \frac{4}{9}$, $b = -\frac{5}{3}$ **4.** $y = -x - 1$ **5.** $y = 3x$ **6.** $y = -\frac{3}{7}x + \frac{2}{7}$
7. Yes, the slope of both lines is $\frac{3}{2}$ and y-intercepts are different.

8.

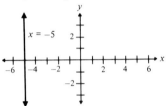

9.

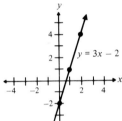

10.

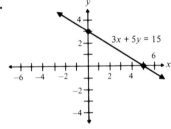

11.

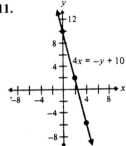

12.

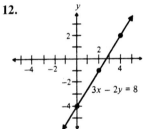

13.

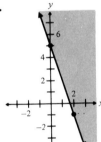

14.

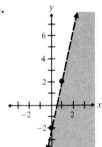

CHAPTER 8

Exercise Set 8.1

1. c   **3.** a   **5.** a   **7.** a, c   **9.** a, c   **11.** b

Cumulative Review Exercises

13. (a) 6 (b) 6, 0 (c) 6, -4, 0 (d) 6, -4, 0, $2\frac{1}{2}$, $-\frac{9}{5}$, 4.22 (e) $\sqrt{3}$, $-\sqrt{7}$ (f) 6, -4, 0, $\sqrt{3}$, $2\frac{1}{2}$, $-\frac{9}{5}$, 4.22, $-\sqrt{7}$
14. >   **15.** >   **16.** =   **17.** -27 **18.** -27 **19.** -81 **20.** 81   **21.** $\dfrac{9y^4}{x^2}$

Exercise Set 8.2

1. Consistent—one solution   **3.** Dependent—infinite number of solutions   **5.** Consistent—one solution
7. Dependent—infinite number of solutions   **9.** One solution   **11.** No solution   **13.** One solution
15. Infinite number of solutions   **17.** No solution

19.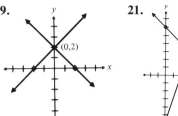
(0,2)

21.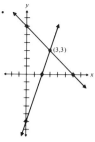
(3,3)

23.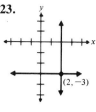
(2, −3)

25.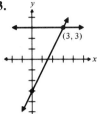
(1, 3)

27.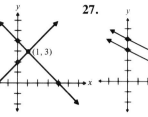
Inconsistent, no solution

29.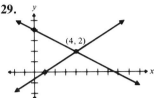
(4, 2)

31.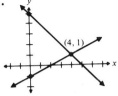
(4, 1)

33.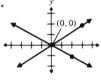
(3, 3)

35.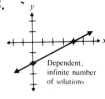
Dependent, infinite number of solutions

37.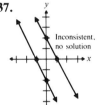
Inconsistent, no solution

39.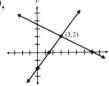
(3,2)

41.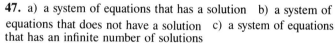
(0, 0)

43.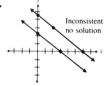
Inconsistent, no solution

45.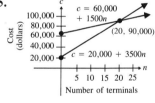
, 20 terminals
$c = 60,000 + 1500n$
(20, 90,000)
$c = 20,000 + 3500n$
Cost (dollars)
Number of terminals

47. a) a system of equations that has a solution   b) a system of equations that does not have a solution   c) a system of equations that has an infinite number of solutions
49. A single line

Cumulative Review Exercises

50. $(x + 3)(y - 4)$ **51.** $(2x + 3)(4x - 5)$ **52.** $\dfrac{-x - 6}{x + 3}$ **53.** -6

Exercise Set 8.3

1. $(2, 1)$ **3.** $(-1, -1)$ **5.** Inconsistent—no solution   **7.** $(4, -9)$ **9.** Dependent—infinite number of solutions
11. $(-1, 2)$ **13.** $(5, -3)$ **15.** $(2, 1)$ **17.** Inconsistent—no solution   **19.** $(-4, -2)$ **21.** $(6, -2)$
23. The x in the first equation because both 6 and 9 are divisible by 3.
25. You will obtain a true statement, such as $2 = 2$.

Cumulative Review Exercises

26. 6 eggs   **27.** $5\frac{1}{3}$ in   **28.** $576x^{14}y^{19}$ **29.** 10

Exercise Set 8.4

1. $(6, 2)$ **3.** $(-2, 3)$ **5.** $(4, \frac{11}{2})$ **7.** Inconsistent—no solution   **9.** $(-8, -26)$ **11.** $(4, -2)$ **13.** $(5, 4)$
15. $(\frac{3}{2}, -\frac{1}{2})$ **17.** Dependent—infinite number of solutions   **19.** $(\frac{2}{5}, \frac{8}{5})$ **21.** $(10, \frac{15}{2})$ **23.** $(\frac{31}{28}, -\frac{23}{28})$
25. $(\frac{20}{39}, -\frac{16}{39})$ **27.** $(\frac{14}{5}, -\frac{12}{5})$ **29.** You will obtain a false statement like $0 = 6$.   **31. (b)** $(4,1)$
33. There are many possible answers. Write the x and y terms with any rational coefficients, then substitute $x = 4$ and $y = -2$ to obtain the constant. Repeat process to get the second equation.

Cumulative Review Exercises

34. $\dfrac{64y^3}{x^6y^{12}}$ **35.** 22 days   **36.** $\dfrac{2x+3}{2x+1}$ **37.** $\dfrac{x^2-18x-3}{(x+1)(x-1)(x-15)}$

Just for Fun   **1.** $(8, -1)$ **2.** $\left(-\frac{105}{41}, \frac{447}{82}\right)$

Exercise Set 8.5

1. $x + y = 37$; 12, 25   **3.** $x + y = 76$; 37, 39   **5.** $x - y = 28$; 14, 42   **7.** $x + y = 25$; 8- \$2 bills, 17-\$5 bills
   $x = 2y + 1$ $y = x + 2$ $x = 3y$ $2x + 5y = 101$

9. $p + w = 540$; plane 515 mph, wind 25 mph   **11.** $T = J + 4$; Jill's boat 60 mph, Teresa's boat 64 mph
   $p - w = 490$ $3T = 3.2J$

13. $U = D + 100$; Delta 400 mph, United 500 mph   **15.** $M = P + 0.3$; 0.5 hours
   $3U + 3D = 2700$ $5M = 8P$

17. $x + y = 8000$; \$2500 at 8%, \$5500 at 10%   **19.** $x + y = 10$; 4ℓ of 25%, 6ℓ of 50%
   $0.1x + 0.08y = 750$ $0.25x + 0.50y = 0.40(10)$

21. $x + 1000y = 550$; fixed \$300, \$0.25/booklet   **23.** $x + y = 100$; 70 gal 5%, 30 gal skim
   $x + 2000y = 800$ $0.05x + 0.00y = 0.035(100)$

25. $A = 5C$; 150 shares Avon, 30 shares Coca-Cola   **27.** $x + y = 16$; 5.96 oz heavy cream; 10.04 oz half and half
   $37A + 75C = 7800$ $0.36x + 0.105y = 0.20(16)$

Cumulative Review Exercises

29. 234   **30.** $\dfrac{1}{9x^8}$

31. **(a)** Yes, third   **(b)** Yes, fourth   **(c)** No, a polynomial cannot have a negative exponent on the variable.

32. $2x - 3 + \dfrac{2}{x+4}$

Just for Fun   **1.** $9t = d$; 1.125 mi   **2.** $0.7x + 0.4y = 0.6(300)$; 200g first alloy, 100g second alloy
   $5t = d - \frac{1}{2}$ $0.3x + 0.6y = 0.4(300)$

Exercise Set 8.6

1. 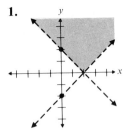   **3.** 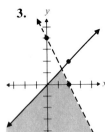   **5.** 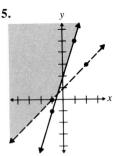   **7.** 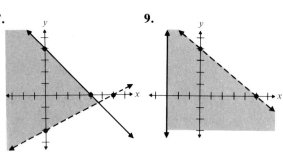   **9.**

11. 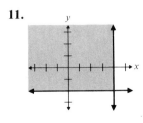   **13.** 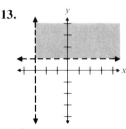   **15.**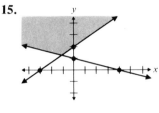

Cumulative Review Exercises

17. $3x^3 - 10x^2 + 29x - 28$ **18.** $(x - 3)(y + 1)$ **19.** $(x - 6)(x - 7)$ **20.** $(3x - 2)(2x + 1)$

Just for Fun **1.**

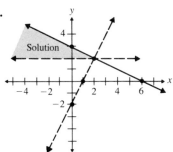

Chapter 8 Review Exercises

1. c   **2.** a   **3.** Consistent, one   **4.** Inconsistent, none   **5.** Dependent, infinite number   **6.** Consistent, one
7. No solution   **8.** One solution   **9.** Infinite number of solutions   **10.** One solution
11.   **12.**   **13.**   **14.**

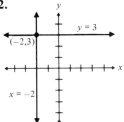   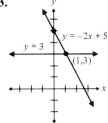   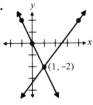   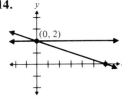

15.   **16.**   **17.**   **18.**

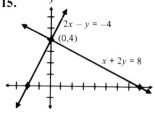   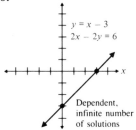   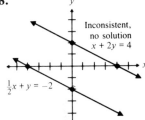

19. $(5, 2)$ **20.** $(-3, 2)$ **21.** $(2, 1)$ **22.** $(5, 4)$ **23.** $(5, 2)$ **24.** $(-18, 6)$ **25.** no solution
26. infinite number of solutions   **27.** $\left(\frac{5}{2}, -1\right)$ **28.** $\left(\frac{20}{9}, \frac{26}{9}\right)$ **29.** $(8, -2)$ **30.** $(1, -2)$ **31.** $(26, -16)$
32. $(-7, 19)$ **33.** $\left(\frac{32}{13}, \frac{8}{13}\right)$ **34.** $\left(-1, \frac{13}{3}\right)$ **35.** no solution   **36.** $\left(\frac{7}{5}, \frac{13}{5}\right)$ **37.** infinite number of solutions
38. $\left(-\frac{78}{7}, -\frac{48}{7}\right)$ **39.** $x + y = 48$; 17, 31   **40.** $x + y = 15$; 11 dimes, 4 quarters
$\qquad\qquad\qquad\quad y = 2x - 3 \qquad\qquad\qquad\quad 0.10x + 0.25y = 2.10$
41. $G + A = 40$; 15 lb of Green Turf, 25 lb of Agway   **42.** $x + y = 6$; 3 liters of each
$\qquad 0.06G + 0.45A = 20.25 \qquad\qquad\qquad\qquad 0.3x + 0.5y = 0.4(6)$
43. $p + w = 600$; 565 mph, plane; 35 mph, wind
$\qquad p - w = 530$

44.   **45.**   **46.**   **47.**

Chapter 8 Practice Test

1. b   **2.** Inconsistent—no solution   **3.** Consistent—one solution   **4.** Dependent—infinite number of solutions
5. no solution   **6.** one solution   **7.** 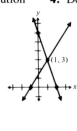   **8.**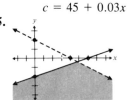

9. $\left(\frac{7}{2}, -\frac{5}{2}\right)$ **10.** $(3, 1)$ **11.** $(5, -5)$ **12.** $\left(\frac{44}{19}, \frac{48}{19}\right)$ **13.** $c = 40 + 0.08x$; 100 mi
$\qquad\qquad\qquad\qquad\qquad\qquad\qquad\qquad\qquad\qquad\qquad\quad c = 45 + 0.03x$

14. $x + y = 20$; $13\frac{1}{3}$ lb of peanuts, $6\frac{2}{3}$ lb of cashews   **15.**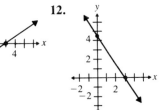
$\qquad\;\; 4x + 2.5y = 3(20)$

Cumulative Review Test

1. $\frac{8}{3}$ **2.** 2   **3.** $3, \frac{4}{3}$ **4.** $\frac{85}{4}$ **5.** $\frac{1}{2}$ **6.** $\frac{20}{3} \approx 6.67$ in   **7.** $2x^{23}y^{17}$ **8.** $(3x - 4)(2x - 1)$
9. $\dfrac{x - 33}{(x + 3)(x - 3)(x - 6)}$ **10.** $\dfrac{(x - 4)(x + 4)}{(2x - 3)(x + 3)}$ **11.**   **12.**

13.   **14.** No solution, the lines are parallel (same slope)   **15.**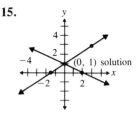

16. $\left(\frac{34}{11}, \frac{-8}{11}\right)$ **17.** 60 min   **18.** (a) \$500,000   (b) PCR   **19.** $6\frac{2}{3}\,\ell$ of 20%, $3\frac{1}{3}\,\ell$ of 35%   **20.** $\frac{300}{11} \approx 27.3$ min

CHAPTER 9

Exercise Set 9.1

1. 1   **3.** 0   **5.** -9 **7.** 11   **9.** -4 **11.** 12   **13.** 13   **15.** -1 **17.** 9   **19.** -11 **21.** $\frac{1}{2}$
23. $\frac{3}{4}$ **25.** $-\frac{2}{5}$ **27.** 2.828   **29.** 3.873   **31.** 8.944   **33.** 9.000   **35.** 9.849   **37.** 1.732   **39.** True
41. True   **43.** False   **45.** True   **47.** False   **49.** True   **51.** $7^{1/2}$ **53.** $(17)^{1/2}$ **55.** $(5x)^{1/2}$
57. $(12x^2)^{1/2}$ **59.** $(19xy^2)^{1/2}$ **61.** $(40x^3)^{1/2}$ **63.** It is nonnegative. The square root of a negative number is not a real number.   **65.** no real number when squared will be a negative number.

Cumulative Review Exercises

66. -6 **67.** -1 **68.** $\frac{30}{11}$ or $2\frac{8}{11}$ hr **69.** $\frac{4}{11}$

Exercise Set 9.2

1. 4 **3.** $2\sqrt{2}$ **5.** $4\sqrt{6}$ **7.** $4\sqrt{2}$ **9.** $4\sqrt{10}$ **11.** $4\sqrt{3}$ **13.** $6\sqrt{3}$ **15.** $2\sqrt{39}$ **17.** 16 **19.** 30
21. y^3 **23.** xy^2 **25.** $x^4y^6\sqrt{x}$ **27.** $ab^2\sqrt{c}$ **29.** $x\sqrt{3x}$ **31.** $5xy\sqrt{2y}$ **33.** $10y^2z^6\sqrt{2y}$
35. $9qb\sqrt{3bc}$ **37.** $8xz^2\sqrt{2xyz}$ **39.** $5x^2\sqrt{10yz}$ **41.** $2\sqrt{6}$ **43.** $3\sqrt{6}$ **45.** $15\sqrt{2}$ **47.** $x\sqrt{15}$
49. $2x^2\sqrt{10x}$ **51.** $6xy^2\sqrt{2x}$ **53.** $3xy^2\sqrt{6xy}$ **55.** $3xy^3\sqrt{10yz}$ **57.** $6x^3y^5$ **59.** $4x$ **61.** $13x^4y^6$
63. 4 **65.** exponent on x, 6; on y, 5 **67.** coefficient, 8; exponent on x, 12; on y, 7

Cumulative Review Exercises

69. $\dfrac{3x+1}{x-1}$ **70.** $y=-\frac{1}{2}x+\frac{3}{2}$, $m=-\frac{1}{2}$, $b=\frac{3}{2}$ **71.** **72.** $\left(\frac{2}{11}, -\frac{15}{11}\right)$

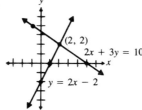

Exercise Set 9.3

1. 2 **3.** 3 **5.** 3 **7.** $\frac{1}{5}$ **9.** $\frac{3}{7}$ **11.** $\frac{1}{7}$ **13.** $2x\sqrt{5}$ **15.** $\sqrt{3xy}$ **17.** $\dfrac{\sqrt{5}}{3y}$ **19.** $\dfrac{1}{xy^2}$ **21.** $2x^2$
23. $\dfrac{4}{5xy}$ **25.** $\dfrac{y\sqrt{5}}{z}$ **27.** $\dfrac{3\sqrt{2}}{2}$ **29.** $\sqrt{2}$ **31.** $\dfrac{\sqrt{10}}{2}$ **33.** $\dfrac{\sqrt{10}}{5}$ **35.** $\dfrac{\sqrt{5}}{5}$ **37.** $\dfrac{x\sqrt{2}}{2}$ **39.** $\dfrac{x\sqrt{2}}{4}$
41. $\dfrac{x^2\sqrt{5}}{5}$ **43.** $\dfrac{x^3\sqrt{15y}}{15y}$ **45.** $\dfrac{x\sqrt{y}}{2y}$ **47.** $\dfrac{\sqrt{6}}{5x^2y^2z}$ **49.** $\dfrac{3\sqrt{5x}}{xy^2}$
51. 1. No perfect square factors in any radicand. 2. No radicand contains a fraction. 3. No square roots in any denominator.

Cumulative Review Exercises

53. $3(x+4)(x-8)$ **54.** $\dfrac{1}{x+1}$ **55.** $4, 6$ **56.**

Exercise Set 9.4

1. $2\sqrt{3}$ **3.** $-2\sqrt{7}$ **5.** $5-4\sqrt{3}$ **7.** $5\sqrt{x}$ **9.** $3\sqrt{x}$ **11.** $3+2\sqrt{y}$ **13.** $x+\sqrt{x}+4\sqrt{y}$
15. $3-2\sqrt{x}$ **17.** $2\sqrt{2}-2\sqrt{3}$ **19.** $4\sqrt{2}$ **21.** $7\sqrt{5}$ **23.** $16\sqrt{2}$ **25.** $-30\sqrt{3}+20\sqrt{5}$
27. $25\sqrt{10}-36\sqrt{5}$ **29.** $64-4\sqrt{6}$ **31.** 7 **33.** 31 **35.** $x-9$ **37.** $6-x^2$ **39.** $x-y^2$
41. $4x-9y$ **43.** $4(2-\sqrt{3})$ **45.** $\dfrac{-3(\sqrt{2}-5)}{23}$ **47.** $-2(\sqrt{2}-\sqrt{3})$ **49.** $\dfrac{-8(\sqrt{5}+2\sqrt{2})}{3}$
51. $\dfrac{2(6-\sqrt{x})}{36-x}$ **53.** $\dfrac{6(4+\sqrt{y})}{16-y}$ **55.** $\dfrac{4(\sqrt{x}+y)}{x-y^2}$ **57.** $\dfrac{x(\sqrt{x}-\sqrt{y})}{x-y}$ **59.** $\dfrac{\sqrt{5x}-x}{5-x}$ **61.** (b) $\dfrac{ab-a\sqrt{c}}{b^2-c}$

Cumulative Review Exercises

62. 18 **63.** $\frac{9}{2}, -4$ **64.** $\dfrac{-2x-3}{(x+2)(x-2)}$ **65.** 30 minutes

Exercise Set 9.5

1. 64   **3.** No solution   **5.** 4   **7.** No solution   **9.** 4   **11.** 4   **13.** No solution   **15.** 7   **17.** 7
19. 2   **21.** 9   **23.** $\frac{14}{3}$ **25.** 1   **27.** 1, 9   **29.** No solution   **31.** -2 **33.** 1, 9
35. Because there may be extraneous roots

Cumulative Review Exercises

37.

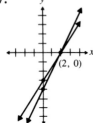

38. (2, 0)   **39.** (2, 0)   **40.** Boat 8 mph, current 4 mph

Just for Fun   **1.** 9   **2.** $\frac{9}{16}$ **3.** 9

Exercise Set 9.6

1. $\sqrt{119} \approx 10.91$ **3.** $\sqrt{164} \approx 12.81$ **5.** $\sqrt{175} \approx 13.23$ **7.** $\sqrt{41} \approx 6.40$ **9.** $\sqrt{149} \approx 12.21$
11. $\sqrt{128} \approx 11.31$ **13.** $\sqrt{17,240.89} \approx 131.30$ yd   **15.** $\sqrt{18.25} \approx 4.27$ m   **17.** $\sqrt{10} \approx 3.16$
19. $\sqrt{288} \approx 16.97$ **21.** $\sqrt{144} = 12$ in   **23.** $\sqrt{25.48} \approx 5.05$ ft   **25.** $\sqrt{225} = 15$ in
27. $\sqrt{5120} \approx 71.55$ ft/sec   **29.** $\sqrt{80,000} \approx 282.84$ ft/sec   **31.** $6.28\sqrt{1.25} \approx 7.02$ sec
33. $6.28\sqrt{0.3125} \approx 3.51$ sec   **35.** $0.2(\sqrt{1418})^3 \approx 10,679.34$ days   **37.** $\sqrt{19.5(6,370,000)} \approx 11,145.18$ m/sec

Cumulative Review Exercises

38. $x > 6$ **39.** $\dfrac{x^4}{4y^3}$ **40.** $-2x^3 + 4x^2 - 6x - 4$ **41.** $\dfrac{5}{3}x^2 - 3x + 2 - \dfrac{4}{3x} - \dfrac{1}{x^2}$

Just for Fun   **1.** 9 in by 12 in   **2.** 32.66 ft/sec   **3.** 3.25 ft   **4.** 40.93 in

Exercise Set 9.7

1. 2   **3.** -2 **5.** 2   **7.** 3   **9.** -1 **11.** 4   **13.** $3\sqrt[3]{2}$ **15.** $2\sqrt[3]{2}$ **17.** $3\sqrt[3]{3}$ **19.** $2\sqrt[4]{2}$
21. $2\sqrt[3]{5}$ **23.** x **25.** y^3 **27.** x^4 **29.** y **31.** x^5 **33.** 16   **35.** 8   **37.** 1   **39.** 27   **41.** 8
43. 625   **45.** $\frac{1}{9}$ **47.** $\frac{1}{32}$ **49.** $x^{7/3}$ **51.** $x^{4/3}$ **53.** $y^{15/4}$ **55.** $y^{21/4}$ **57.** $x^{2/3}$ **59.** x **61.** x^4
63. x^2 **65.** Both equal 4   **67.** **(b)** $x^{9/4}$

Cumulative Review Exercises

68. -42 **69.** $(3x - 4)(x - 8)$ **70.**

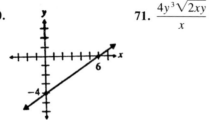

71. $\dfrac{4y^3\sqrt{2xy}}{x}$

Just for Fun   **1.** xy **2.** $3x^2y$ **3.** $\sqrt[4]{2}$ **4.** $3x\sqrt[3]{3y}$ **5.** $\dfrac{\sqrt[3]{4}}{2}$ **6.** $\dfrac{\sqrt[3]{x^2}}{x}$

Chapter 9 Review Exercises

1. 5   **2.** 6   **3.** -9 **4.** $8^{1/2}$ **5.** $(26x)^{1/2}$ **6.** $(20xy^2)^{1/2}$ **7.** $4\sqrt{2}$ **8.** $2\sqrt{11}$ **9.** $3x^2y^2\sqrt{5x}$
10. $5x^2y^3\sqrt{5}$ **11.** $x^2z\sqrt{15xyz}$ **12.** $4b^2c^2\sqrt{3ac}$ **13.** $4\sqrt{6}$ **14.** $5x$ **15.** $6x\sqrt{y}$ **16.** $5xy\sqrt{3}$

17. $10xy^3\sqrt{y}$ **18.** $2xy^2\sqrt{6xy}$ **19.** 4   **20.** $\frac{1}{5}$ **21.** $\frac{1}{2}$ **22.** $\frac{3\sqrt{5}}{5}$ **23.** $\frac{\sqrt{15x}}{6}$ **24.** $\frac{\sqrt{6x}}{6}$ **25.** $\frac{x\sqrt{2}}{2}$

26. $\frac{x^2\sqrt{2x}}{4}$ **27.** $\frac{y\sqrt{15}}{x^2}$ **28.** $\frac{x\sqrt{2y}}{y^2}$ **29.** $\frac{3\sqrt{5x}}{2x^2y}$ **30.** $\frac{z\sqrt{14x}}{7x}$ **31.** $-3(1-\sqrt{2})$ **32.** $\frac{5(3+\sqrt{6})}{3}$

33. $\frac{2\sqrt{3}-\sqrt{3x}}{4-x}$ **34.** $\frac{2(\sqrt{x}+5)}{x-25}$ **35.** $\frac{\sqrt{5x}-\sqrt{15}}{x-3}$ **36.** $4\sqrt{3}$ **37.** $-\sqrt{2}$ **38.** $-2\sqrt{x}$ **39.** 0

40. $\sqrt{2}$ **41.** $12\sqrt{10}$ **42.** $-10\sqrt{2}$ **43.** $26\sqrt{2}$ **44.** $16\sqrt{3}+20\sqrt{5}$ **45.** 81   **46.** No solution

47. 39   **48.** 8   **49.** 9   **50.** 4   **51.** 0   **52.** 7   **53.** 3   **54.** 10   **55.** $\sqrt{88}$ **56.** $\sqrt{12}$ **57.** $\sqrt{61}$

58. $\sqrt{135}\approx 11.62$ ft   **59.** $\sqrt{261}\approx 16.16$ in   **60.** 10   **61.** $\sqrt{153}\approx 12.37$ **62.** 11 ft   **63.** 2   **64.** -3

65. 2   **66.** $2\sqrt[3]{2}$ **67.** $2\sqrt[3]{3}$ **68.** $2\sqrt[4]{2}$ **69.** $2\sqrt[3]{6}$ **70.** $3\sqrt[3]{2}$ **71.** $2\sqrt[4]{6}$ **72.** x^5 **73.** x^4 **74.** y^4

75. y^5 **76.** 4   **77.** 4   **78.** $\frac{1}{9}$ **79.** 16   **80.** $\frac{1}{8}$ **81.** 125   **82.** $x^{5/3}$ **83.** $x^{10/3}$ **84.** $y^{9/4}$ **85.** $x^{5/2}$

86. $y^{11/2}$ **87.** $x^{7/4}$ **88.** x **89.** $\sqrt[3]{x^2}$ **90.** x^3 **91.** x^2 **92.** x^2 **93.** x^2 **94.** x^3 **95.** x^6

Chapter 9 Practice Test

1. $(3xy)^{1/2}$ **2.** $x+3$ **3.** $4\sqrt{6}$ **4.** $2x\sqrt{3}$ **5.** $4x^2y^2\sqrt{2y}$ **6.** $4xy\sqrt{3x}$ **7.** $5x^2y^2\sqrt{3y}$ **8.** $\frac{1}{5}$

9. $\frac{y}{4x}$ **10.** $\frac{\sqrt{2}}{2}$ **11.** $\frac{2\sqrt{5x}}{5}$ **12.** $\frac{2\sqrt{15x}}{3xy}$ **13.** $-3(2-\sqrt{5})$ **14.** $\frac{6(\sqrt{x}+3)}{x-9}$ **15.** $11\sqrt{3}$ **16.** $-3\sqrt{x}$

17. 76   **18.** 4, 8   **19.** $\sqrt{106}\approx 10.30$ **20.** $\sqrt{74}\approx 8.60$ **21.** $\frac{1}{81}$ **22.** x^5

CHAPTER 10

Exercise Set 10.1

1. 4, -4 **3.** 10, -10 **5.** 6, -6 **7.** $\sqrt{10}, -\sqrt{10}$ **9.** $2\sqrt{2}, -2\sqrt{2}$ **11.** 2, -2 **13.** $\sqrt{17}, -\sqrt{17}$

15. 3, -3 **17.** $\sqrt{15}/2, -\sqrt{15}/2$ **19.** $\sqrt{195}/5, -\sqrt{195}/5$ **21.** 1, -3 **23.** 7, -1 **25.** 2, -10

27. $1+2\sqrt{3}, 1-2\sqrt{3}$ **29.** $-6+2\sqrt{5}, -6-2\sqrt{5}$ **31.** 3, -7 **33.** 19, -1

35. $(-3+3\sqrt{2})/2, (-3-3\sqrt{2})/2$ **37.** $(-1+2\sqrt{5})/4, (-1-2\sqrt{5})/4$ **39.** $(6+3\sqrt{2})/2, (6-3\sqrt{2})/2$

41. $w=2\sqrt{10}\approx 6.32$ ft, $l=4\sqrt{10}\approx 12.65$ ft   **43.** $x^2=36$, other answers are possible

45. 9, Need an equation equivalent to $x^2=36$.

Cumulative Review Exercises

47. $2(2x+3)(x-4)$ **48.** No solution   **49.** $2000 at 6%, $8000 at 8%   **50.** $y=4x-1$

Exercise Set 10.2

1. 1, -3 **3.** 5, -1 **5.** $-2, -1$ **7.** 5, 3   **9.** -3 **11.** $-2, -3$ **13.** $-3, -6$ **15.** 7, 8

17. 6, -2 **19.** $-1+\sqrt{7}, -1-\sqrt{7}$ **21.** $-3+\sqrt{3}, -3-\sqrt{3}$ **23.** $(5+\sqrt{57})/2, (5-\sqrt{57})/2$

25. 1, -3 **27.** $(-9+\sqrt{73})/2, (-9-\sqrt{73})/2$ **29.** $-8, -3$ **31.** $(-5+\sqrt{31})/2, (-5-\sqrt{31})/2$

33. $-1+\sqrt{3}, -1-\sqrt{3}$ **35.** 0, -4 **37.** 0, 2   **39.** 1, -4 **41.** 0, -6 **43.** 3, 7

45. (a) $x^2-12x+36$

Cumulative Review Exercises

46. $\frac{4}{(x+2)(x-3)}$

47. Write the equations in slope-intercept form and compare the slopes. If the slopes are the same and the y-intercepts are different, the equations represent parallel lines.   **48.** $\left(\frac{38}{11}, \frac{12}{11}\right)$ **49.** 3

Just for Fun   **1.** $(-3+\sqrt{59})/10, (-3-\sqrt{59})/10$ **2.** $(5+\sqrt{70})/15, (5-\sqrt{70})/15$

3. $(-1+\sqrt{193})/12, (-1-\sqrt{193})/12$ **4.** $x=\dfrac{-b\pm\sqrt{b^2-4ac}}{2a}$

Exercise Set 10.3

1. Two solutions   **3.** No solution   **5.** Two solutions   **7.** Single solution   **9.** Two solutions   **11.** No solution
13. Two solutions   **15.** Two solutions   **17.** 1, 2   **19.** 4, 5   **21.** −8, 3   **23.** 4, 9   **25.** 5, −5   **27.** 0, 3
29. 3, 4   **31.** $(7 + \sqrt{17})/4, (7 - \sqrt{17})/4$ **33.** $\frac{1}{3}, -\frac{1}{2}$ **35.** $(2 + \sqrt{6})/2, (2 - \sqrt{6})/2$ **37.** No solution
39. $\frac{5}{4}, -1$ **41.** $\frac{9}{2}, -1$ **43.** $3, \frac{5}{2}$ **45.** 4, 5   **47.** 4 ft, 5 ft   **49.** 2 ft   **51. (a)** none   **(b)** one   **(c)** two

Cumulative Review Exercises

54. 6, 7   **55.** $\frac{5}{3}, -\frac{7}{2}$ **56.** Cannot be solved by factoring; $\dfrac{-3 + \sqrt{41}}{4}, \dfrac{-3 - \sqrt{41}}{4}$ **57.** 3, −3

Just for Fun   **1.** 300 ft long by 50 ft wide, or 100 ft long by 150 ft wide

Exercises Set 10.4

1. $x = -1, (-1, -8)$, up   **3.** $x = \frac{5}{2}, (\frac{5}{2}, \frac{1}{4})$, down   **5.** $x = \frac{5}{6}, (\frac{5}{6}, \frac{121}{12})$, down   **7.** $x = -1, (-1, -8)$, down
9. $x = \frac{1}{3}, (\frac{1}{3}, \frac{5}{3})$, up   **11.** $x = -\frac{3}{2}, (-\frac{3}{2}, -14)$, up

13.
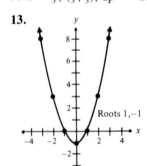
Roots 1,−1

15.
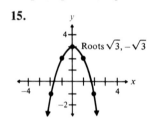
Roots $\sqrt{3}, -\sqrt{3}$

17.
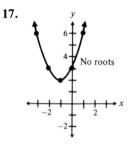
No roots

19.
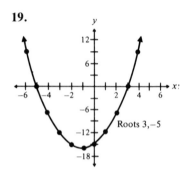
Roots 3,−5

21.
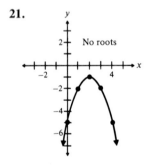
No roots

23.
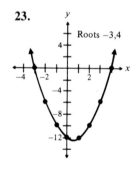
Roots −3,4

25.
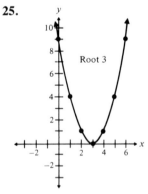
Root 3

27.
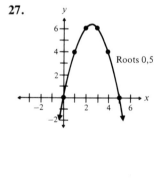
Roots 0,5

29.
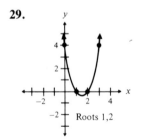
Roots 1,2

31.
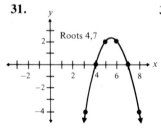
Roots 4,7

33.
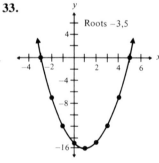
Roots −3,5

35.
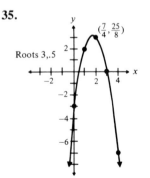
Roots 3,.5 $(\frac{7}{4}, \frac{25}{8})$

37.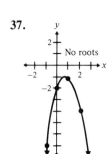

39.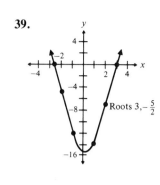

41. $\left(\dfrac{-b}{2a}, \dfrac{4ac - b^2}{4a}\right)$

43. **(a)** The values of x where the graph crosses the x axis.
(b) Let $y = 0$ and solve the resulting equation for x.

45. $x = \dfrac{-b}{2a}$ **47.** Two, the vertex is below the x axis and the parabola opens up.

49. None, the vertex is below the x axis and the parabola opens downward

51. **(a)** the graphs will be reflections of each other
(b)

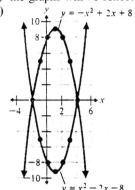

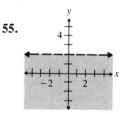

Cumulative Review Exercises

52. $\dfrac{-x^2 + 2x - 6}{(x + 3)(x - 4)}$ **53.** $\dfrac{27}{7}$ **54.**

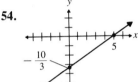

55.

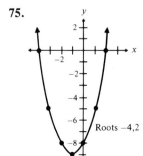

Chapter 10 Review Exercises

1. $5, -5$ **2.** $2\sqrt{2}, -2\sqrt{2}$ **3.** $\sqrt{6}, -\sqrt{6}$ **4.** $\sqrt{6}, -\sqrt{6}$ **5.** $2\sqrt{5}, -2\sqrt{5}$ **6.** $\sqrt{7}, -\sqrt{7}$
7. $2\sqrt{2}, -2\sqrt{2}$ **8.** $3 + 2\sqrt{3}, 3 - 2\sqrt{3}$ **9.** $(-4 + \sqrt{30})/2, (-4 - \sqrt{30})/2$
10. $(5 + 5\sqrt{2})/3, (5 - 5\sqrt{2})/3$ **11.** $2, 8$ **12.** $3, 5$ **13.** $1, 13$ **14.** $2, -3$ **15.** $9, -6$ **16.** $1, -6$
17. $-1 + \sqrt{6}, -1 - \sqrt{6}$ **18.** $(3 + \sqrt{41})/2, (3 - \sqrt{41})/2$ **19.** $-4, 8$ **20.** $5, -3$ **21.** $\frac{3}{2}, -2$ **22.** $\frac{5}{3}, \frac{3}{2}$
23. Two solutions **24.** No real solution **25.** No real solution **26.** No real solution **27.** One solution
28. No real solution **29.** Two solutions **30.** Two solutions **31.** $2, 7$ **32.** $-10, 3$ **33.** $2, 5$ **34.** $2, -\frac{3}{5}$
35. $-2, 9$ **36.** $6, -5$ **37.** $\frac{3}{2}, -\frac{5}{3}$ **38.** $(-2 + \sqrt{10})/2, (-2 - \sqrt{10})/2$ **39.** $(3 + \sqrt{57})/4, (3 - \sqrt{57})/4$
40. $3 + \sqrt{2}, 3 - \sqrt{2}$ **41.** No real solution **42.** $(3 + \sqrt{33})/3, (3 - \sqrt{33})/3$ **43.** $0, -\frac{3}{2}$ **44.** $0, \frac{5}{2}$
45. $3, 8$ **46.** $7, 9$ **47.** $5, -8$ **48.** $-9, 3$ **49.** $-6, 10$ **50.** $7, -6$ **51.** $1, -12$ **52.** $5, -5$
53. $0, -6$ **54.** $\frac{1}{2}, -3$ **55.** $\frac{5}{2}, 2$ **56.** $\frac{2}{3}, -\frac{3}{2}$ **57.** $(-3 + \sqrt{33})/2, (-3 - \sqrt{33})/2$ **58.** $2, \frac{5}{3}$ **59.** $1, -\frac{8}{3}$
60. $(3 + 3\sqrt{3})/2, (3 - 3\sqrt{3})/2$ **61.** $0, \frac{5}{2}$ **62.** $0, -\frac{5}{3}$ **63.** $x = 1, (1, -4)$, up **64.** $x = 5, (5, -1)$, up
65. $x = -\frac{7}{2}, (-\frac{7}{2}, -\frac{1}{4})$, up **66.** $x = -1, (-1, 16)$, down **67.** $x = \frac{3}{2}, (\frac{3}{2}, -\frac{9}{4})$, up
68. $x = -\frac{7}{4}, (-\frac{7}{4}, -\frac{25}{8})$, up **69.** $x = 0, (0, -8)$, down **70.** $x = 1, (1, 9)$, down **71.** $x = -\frac{1}{2}, (-\frac{1}{2}, \frac{81}{4})$, down
72. $x = -\frac{5}{6}, (-\frac{5}{6}, -\frac{121}{12})$, up

73.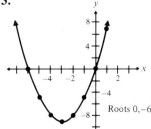

74.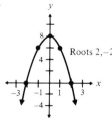

75.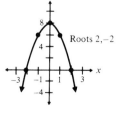

76.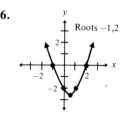

77.
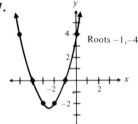
Roots −1,−4

78.
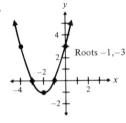
Roots −1,−3

79.
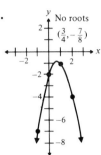
No roots
$(\frac{3}{4}, -\frac{7}{8})$

80.
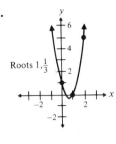
Roots $1, \frac{1}{3}$

81.
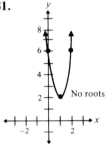
No roots

82.
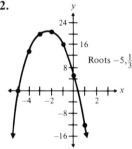
Roots $-5, \frac{1}{3}$

83.
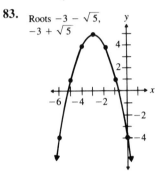
Roots $-3 - \sqrt{5}$, $-3 + \sqrt{5}$

84.
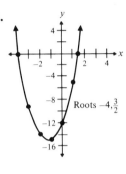
Roots $-4, \frac{3}{2}$

85. 8, 11   **86.** $w = 7$ ft, $l = 9$ ft

Chapter 10 Practice Test

1. $2\sqrt{5}, -2\sqrt{5}$ **2.** $(3 + \sqrt{35})/2, (3 - \sqrt{35})/2$ **3.** 10, −6   **4.** −4, 3   **5.** 6, −1
6. $(-4 + \sqrt{6})/2, (-4 - \sqrt{6})/2$ **7.** $0, \frac{5}{3}$ **8.** $-5, \frac{1}{2}$ **9.** No real solution   **10.** $x = \frac{3}{2}, (\frac{3}{2}, \frac{41}{4})$, down
11.

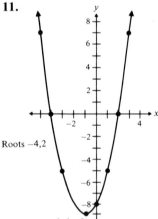
Roots −4,2
$(-1,-9)$

12.

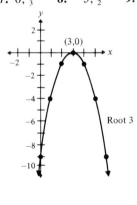
(3,0)
Root 3

13. $w = 3$ ft, $l = 10$ ft

Cumulative Review Test

1. 16   **2.** $\frac{40}{31}$ **3.** $5\frac{1}{3}$ in   **4.** $x \geq -\frac{1}{4}$ **5.** $P = 3A - m - n$ **6.** $864x^{14}y^{22}$
7. $x + 4 - \dfrac{3}{x + 2}$ **8.** $(x - 2y)(2x - 3y)$ **9.** $2(2x + 1)(x - 4)$ **10.** $\dfrac{6a - 8}{(a + 4)(a - 4)^2}$ **11.** 6, 8
12.

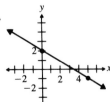

13. $(-\frac{12}{7}, -\frac{30}{7})$ **14.** $\dfrac{y\sqrt{2xy}}{6}$ **15.** $4\sqrt{7}$ **16.** 2   **17.** $\dfrac{2 + \sqrt{14}}{2}, \dfrac{2 - \sqrt{14}}{2}$
18. 25.6 lb   **19.** Width = 10 ft; length = 27 ft   **20.** walks, 3 mph; jogs, 6 mph

Index

A

Absolute value, 21,
Addition:
    associative property of, 50
    commutative property of, 50
    of fractions, 10
    of polynomials, 162
    of rational expressions, 257, 264
    of real numbers, 23
    of square roots, 391
Addition method for linear systems, 351
Addition property, 67, 71
Additive inverse, 25
Algebraic expression, 59
Applications (*see also* Word problems):
    of algebra, 111, 121, 128
    of factoring, 239
    of proportions, 93
    of quadratic equations, 239, 420, 433
    of radical equations, 402
    of rational equations, 257
    of ratios, 93
    of systems of equations, 358
Approximately equal to, 378
Area, 113
Associative property:
    of addition, 50
    of multiplication, 50
Axes, 295
Axis of symmetry, 436

B

Base, 40, 142
Binomial, 161
    square of, 171
Braces, 16
Brackets, 44

C

Calculator corners, 42, 47, 69, 115, 379, 410, 441
Cartesian coordinate system, 295
Checking:
    equations, 68, 84, 276, 398
    factoring, 203, 212
Circumference, 114
Coefficient, 59
    numerical, 59
Collinear, 299
Combining like terms, 61
Common denominator, 10, 257, 261, 264
Common factor, 7, 199
Commutative property:
    of addition, 49
    of multiplication, 50
Complementary angles, 137, Appendix C
Completing the square, 422
Complex fractions, 270
Composite number, 199
Conditional equation, 90
Conjugate, 393
Consecutive integers, 124
Consistent system of equations, 340
Constant: 60
    term, 60
Coordinates, 295
Counting number, 16
Cross multiplication, 93
Cube root, 376, 409
Cubes:
    difference of, 231
    sum of, 231

D

Decimals, 16, 378, Appendix A
    repeating, 378
    terminating, 378

Degree:
  of polynomial, 162
  of term, 162
Denominator, 7
  least common, 10, 261, Appendix B
Dependent system of equations, 341
Descartes, René, 295
Descending order, 161
Diameter, 114
Difference of two cubes, 231
Difference of two squares, 171, 230
Discriminant, 432, 441
Distance formula, 182, 404
Distributive property, 51, 62
Division:
  of fractions, 9
  of polynomials, 175
  of rational expressions, 253
  of real numbers, 33, 35
  of square roots, 386
  by zero, 37, 245
Divisor, 35
Double root, 432, 441

E

Elements of set, 16
Elimination method to solve a system of
  equations, 351
Equation of a line:
  point-slope form, 326
  slope-intercept form, 317
  standard form, 298
Equations:
  checking, 68
  conditional, 90
  consistent, 340
  containing fractions, 75, 274
  dependent, 340
  equivalent, 70
  graph of, 301
  identity (identical equations), 90
  inconsistent, 340
  linear, 67
  quadratic, 235, 417
  radical, 397
  rational, 274
  solution of, 60, 81, 86
  system of, 338
  in two variables, 298
Equilateral triangle, 137, Appendix C
Equivalent equations, 70
Evaluate 9, 84, 112
Exponent, 40, 142
  expanded power rule, 146
  fractional, 379, 411
  negative exponent rule, 150
  power rule, 146
  product rule, 142
  quotient rule, 143
  zero exponent rule, 145
Exponential form, 379

Expression, 30, 59
Extraneous roots (or solutions), 276, 398

F

Factor, 7, 65, 198
  greatest common, 7, Appendix B
Factoring, 197
  difference of two cubes, 231
  difference of two squares, 230
  general review of, 229
  grouping, by, 204, 224
  a monomial from a polynomial, 201
  sum of two cubes, 231
  trial and error, by, 214, 217
  trinomials, 209, 217
Factors, 7, 199, 200
FOIL method of multiplying binomials, 169
Formulas, 111, 112
  solving for a variable in, 116
Fractional exponents, 379, 411
Fractions, 6, 245
  addition of, 10, 257, 264
  complex, 270
  denominator, 7
  division of, 9
  equations containing, 75, 274
  lowest terms, 7, 245
  multiplication of, 8
  numerator, 7
  reducing, 7, 245
  subtraction of, 10, 257, 264

G

Geometric formulas, 113, 114, 115
Geometric problems, 113, 135
Golden rectangle, 420
Graphical solution to system of linear equations,
  340
Graphical solution to system of linear
  inqualities, 367
Graphs, 294, 301
  of linear equations, 297
  of linear inequalities, 329
  of parabolas, 435
  of quadratic equations, 435
  of system of equations, 340
Greater than, 20
Greatest common factor, 7, Appendix B, 199,
  200
Grouping, factoring by, 204, 224

H

Horizontal axis, 295
Horizontal line, slope of, 313
Hypotenuse, 402

I

Identity, 90
Imaginary numbers, 377

Inconsistent system of equations, 340
Indeterminate, 38
Index, 376
Inequalities, 20, 102, 329
    in one variable, 102
    sense of, 102
    system of linear, 367
    in two variables, 329
Integers, 16, 124
Intercepts, 302
Interest, 112
Irrational numbers, 17, 378
Isosceles triangle, 37, Appendix C

L

Least common denominator, 10, Appendix B,
    261
Least common multiple, 261
Leg of a right triangle, 402
Less than, 20
Like square roots, 391
Like terms, 60
Linear equations, 58, 67
    graphs of, 294, 297
    in one variable, 67
    systems of, 337
    in two variables, 298
Linear inequalities:
    in one variable, 102
    in two variables, 329
Lowest terms, 7, 245

M

Mixed number, 11
Mixture problem, 186
Monomial, 161
Motion (or Rate) problems, 181, 183
Multiplication:
    associative property of, 50
    commutative property of, 50
    of fractions, 8,
    of polynomials, 167
    of rational expressions, 250
    of real numbers, 33
    of square roots, 381
    by zero, 340
Multiplication property, 75
Multiplication symbols, 6

N

Natural number, 16
Negative:
    exponent, 150
    numbers, 16
    slope, 312
    square root, 377, 418
Number:
    additive inverse of number, 25
    composite number, 199

counting number, 16
integers, 16
irrational number, 17, 378
natural number, 16
negative number, 16
opposite of, 25
positive number, 16
prime number, 199
rational number, 17, 378
real numbers, 17
whole numbers, 9, 16
Number line, 16
Numerator, 7
Numerical coefficient, 59

O

Operations, 23
    order of, 44
Opposites, 25
Order of operations, 44
Ordered pair, 295
Origin, 295

P

Parabola, 435
Parallel lines, 320
Parentheses, 44
Percent, 122, 124, Appendix A
Perfect cubes, 410
Perfect square trinomials, 422
Perfect squares, 377
Perimeter, 113
Perpendicular lines, 328
Plotting points, 295
Point-slope form, 326
Polynomials, 161
    addition of, 162
    degree of, 162
    descending order of, 161
    division of, 175
    multiplication of, 167
    subtraction of, 163
Positive:
    integer, 16
    number, 16
    slope, 312
    square root, 376, 418
Power, 40, 142
    of zero, 145
Power rule for exponents, 146
Prime number, 199
Principal square root, 376
Product, 6
Product rule for exponents, 142
Product rule for radicals, 381, 411
Product of sum and difference of two
    quantities, 171
Properties of the real number system, 49
Proportion, 92, 93
Pythagorean theorem, 402

Q

Quadrants, 295
Quadratic equations, 235, 417
  graphing, 435
  solution by completing the square, 422
  solution by factoring, 235
  solution by the quadratic formula, 427
  standard form of, 236, 418
Quadratic formula, 428
  discriminant of, 432
Quadrilateral, 113
Quotient rule for exponents, 143
Quotient rule for radicals, 386

R

Radical equation, 397
Radical expression, 376
Radical sign, 376
Radicals, 375
  evaluating, 377
  product rule for, 381, 411
  quotient rule for, 387
  simplifying, 382, 386
Radicand, 376
Radius, 114
Rate problems, 181, 183
Ratio, 92
Rational equations, 274
Rational expressions, 244
  adding, 257, 264
  dividing, 253
  multiplying, 250
  subtracting, 257, 264
  where defined, 245
Rational numbers, 17, 378
Rationalizing the denominator:
  binomial denominator, 393
  monomial denominator, 388
Real number, 1, 17
Real number line, 16
Real number system, 15
Real numbers, properties of, 49
Reciprocal, 75
Rectangular coordinate system, 295
Reducing:
  fractions, 7
  rational expressions, 247
Right triangle, 402
Roots, 375, 441
  cube, 376, 409
  square, 376
  written with fractional exponents, 379, 409

S

Scientific notation, 157
Secondary fractions, 270
Sense of inequality, 102

Set, 16
Similar figures, 97
Simple interest, 112
Simplifying:
  algebraic expressions, 65, 84
  a complex fraction, 271
  square roots, 382, 386
Simultaneous linear equations, 338
Slope-intercept form, 317
Slope of line, 310
Solution:
  to an equation, 68
  extraneous, 276
  to an inequality, 102
  to a system of equations, 338
  to a system of inequalities, 367
Solving equations, 58, 84
  containing fractions, 75, 274
  containing radicals, 397
  linear, 58, 81, 86
  quadratic, by completing the square, 422
  quadratic, by factoring, 235
  quadratic, by the quadratic formula, 428
Solving for a variable, 116
Square of a binomial, 171
Square root property, 418
Square roots, 376
  addition of, 391
  division of, 386
  index of, 376
  like, 391
  multiplication of, 381
  negative, 377, 418
  positive, 376, 418
  principal, 376
  simplifying, 382, 386
  subtraction of, 391
  table, Appendix D
  unlike, 392
Standard form:
  of linear equation, 298
  of quadratic equation, 236, 418
Study skills, 2
Substitution method to solve a system of
  equations, 346
Subtraction:
  definition of, 28
  of fractions, 10
  of polynomials, 163
  of rational expressions, 257, 264
  of real numbers, 28
  of square roots, 391
Sum of two cubes, 231
Supplementary angles, 137, Appendix C
System of linear equations, 337
  applications of, 358
  solution by addition (or elimination)
    method, 351
  solution by graphing, 340
  solution by substitution, 346
System of linear inequalities, 367

T

Terms, 59
  combining like, 61
  degree of, 162
  like, 60
  of ratio, 92
  unlike, 60
Trial and error, factoring by, 214, 217
Trinomial, 161
  factoring, 209, 217
  perfect square, 422

U

Undefined, 37
Unit, 199
Unlike:
  square roots, 392
  terms, 60

V

Variable, 6, 59
Vertex, 436, 438
Vertical axis, 295
Vertical line, slope of, 314

W

Whole numbers, 9, 16
Word problems (*see* also applications):
  changing to equations, 121–126

financial problems, 30, 95, 97, 112,
  131–133, 343, 362
with fractions, 281–287
geometric problems, 94, 97, 113–116,
  135–137, 239, 281, 403, 420, 433
mixture problems, 186–189, 362–365
number problems, 130, 131, 239, 358, 359
percent problems, 122, 132
scientific problems, 405, 406
work problems, 284–287
Work problems, 284–287

X

X axis, 295
X intercept, 302

Y

Y axis, 295
Y intercept, 302

Z

Zero, 16
  division by, 37
  as an exponent, 145
Zero exponent rule, 145
Zero-factor property, 236

PHOTO CREDITS

Chapter 1	Chapter Opener Exercise Set 1.4, Problem 67 Exercise Set 1.5, Problem 104	Sylvie Chappaz/Photo Researchers Allen R. Angel Alex Von Koschembar/Photo Researchers
Chapter 2	Chapter Opener Exercise Set 2.6, Problem 34 Exercise Set 2.6, Problem 49	Superstock Allen R. Angel David R. Frazier Photolibrary
Chapter 3	Chapter Opener Exercise Set 3.1, Problem 87	Nancy Sartullo/The Stock Market Allen R. Angel
Chapter 4	Chapter Opener Exercise Set 4.3, Problem 54	Ron Dahl/Superstock K. Iwasaki/The Stock Market
Chapter 5	Chapter Opener Section 5.6, Just for Fun (1)	Superstock Ted Eckhart/Photo Researchers
Chapter 6	Chapter Opener Exercise Set 6.3, Problem 56	Allen R. Angel DRS Productions/The Stock Market
Chapter 7	Chapter Opener Exercise Set 7.4, Problem 45	Four by Five/Superstock Four by Five/Superstock
Chapter 8	Chapter Opener Exercise Set 8.5, Problem 10 Section 8.5, Just for Fun (1)	Peter Christopher/Masterfile Philippe Blondel/Allsport/Vandystadt Allen R. Angel
Chapter 9	Chapter Opener Exercise Set 9.5, Problem 40 Exercise Set 9.6, Problem 13	R. Berenholtz/The Stock Market Superstock Superstock
Chapter 10	Chapter Opener Section 10.4, Example 4	Hans Namuth/Photo Researchers GALA/Superstock